GANGLIOSIDES AND NEURONAL PLASTICITY

GANGLIOSIDES AND NEURONAL PLASTICITY

Edited by

Guido Tettamanti
Dipartimento di Chimica
e Biochimica Medica
Facoltà di Medicina e Chirurgia
Università di Milano
Milano, Italy

Robert W. Ledeen
Dept of Neurology and Biochemistry
Albert Einstein College of Medicine
Bronx, New York, USA

Konrad Sandhoff
Institut für Organische
Chemie und Biochemie
Universität Bonn
Bonn, FRG

Yoshitaka Nagai
Dept of Biochemistry
University of Tokyo, Tokyo
and Dept of Neurobiology
Brain Research Institute
Niigata University, Niigata
Japan

Gino Toffano
Fidia Neurobiological
Research Laboratories
Abano Terme, Italy

FIDIA RESEARCH SERIES
Volume 6

LIVIANA PRESS
Padova

SPRINGER-VERLAG BERLIN HEIDELBERG GMBH

FIDIA RESEARCH SERIES

An open-end series of publications on international biomedical research, with special emphasis on the neurosciences, published by LIVIANA Press, Padova, Italy, in cooperation with FIDIA Research Labs, Abano Terme, Italy.

The series will be devoted to advances in basic and clinical research in the neurosciences and other fields.

The aim of the series is the rapid and worldwide dissemination of up-to-date, interdisciplinary data as presented at selected international scientific meetings and study groups.

Each volume is published under the editorial responsibility of scientists chosen by organizing committees of the meetings on the basis of their active involvement in the research of the field concerned.

Originally published by Springer-Verlag Berlin Heidelberg New York Tokyo in 1986
Softcover reprint of the hardcover 1st edition 1986

ISBN 978-1-4757-5311-0 ISBN 978-1-4757-5309-7 (eBook)
DOI 10.1007/978-1-4757-5309-7

CONTENTS

Introduction IX

I. BIOCHEMICAL AND TECHNOLOGICAL ASPECTS

H. Egge, J. Peter-Katalinić, *A review of mass spectrometric techniques in the structural analysis of native and derivatized gangliosides.* . . 1

S. Sonnino, D. Acquotti, G. Kirschner, G. Fronza, G. Tettamanti, *Advances in ganglioside chemistry* 17

K. Bock, *The solution conformation of gangliosides inferred from HSEA calculations and high field NMR spectroscopy.* 47

T. Taki, Y. Hirabayashi, H. Ishikawa, S. Ando, K. Kon, K. Tanaka, M. Matsumoto, *A novel disialoganglioside (GD1α) with N-acetylneuraminyl (α2→6)-N-acetylgalactosamine linkage in rat ascites hepatoma cells* 57

I. Kracun, H. Rösner, C. Cosovic, *Topographical distribution of the gangliosides in the developing and adult human brain* 67

D.M. Marcus, *The use of antibodies to identify glycosphingolipids and to localize them in tissues*. 77

H. Rösner, C.J. Willibald, S. Henke-Fahle, *Expression of ganglioside-antigens during neuronal differentiation studied by use of monoclonal antibodies* 83

II. FUNCTIONAL ASPECTS

R.K. Yu, J.R. Goldenring, J.Y.H. Kim, R.J. DeLorenzo, *Gangliosides as differential modulators of membrane-bound protein-kinase systems* 95

S. Ando, Y. Tanaka, K. Kon, *Membrane aging of the brain synaptosomes with special reference to gangliosides* 105

G. Dawson, L.W. Hancock, A.L. Horwitz, R. Wollman, N. Cashman, J. Antel, *Possible association of degenerative motor neuron disease (ALS) with abnormal ganglioside metabolism: abnormal gangliosides in ALS spinal cord and ALS-like symptoms in a unique partial HexB deficiency syndrome* 113

H. Rahmann, W. Probst, *Ultrastructural localization of calcium at synapses and modulatory interactions with gangliosides* 125

A. Wieraszko, W. Seifert, *Involvement of gangliosides in the synaptic transmission in the hippocampus and striatum of the rat brain.* . 137

H. Baba, N. Miyatani, S. Sato, K. Nakamura, T. Yuasa, T. Miyatake, *Antibody to glycolipid in a patient with IgM paraproteinemia and polyradiculoneuropathy* 153

III. METABOLIC ASPECTS

D.A. Aquino, M.A. Bisby, R.W. Ledeen, *Axonal transport of gangliosides and neutral glycolipids in the peripheral nervous system. Identification of ganglioside types in motoneurons of the PNS* 161

U. Hinrichs, S. Sonderfeld, G. Schwarzmann, E. Conzelmann, K. Sandhoff, *Concepts of ganglioside metabolism* 171

R. Ghidoni, M. Trinchera, B. Venerando, A. Fiorilli, G. Tettamanti, *Metabolism of exogenous GM1 and related glycolipids in the rat* 183

IV. GANGLIOSIDES AND NEURONAL PLASTICITY IN VITRO

S. Hakomori, E. Bremer, Y. Okada, *Ganglioside-mediated modulation of growth factor receptor function* 201

S. Varon, S.D. Skaper, R. Katoh-Semba, *Neuritic responses to GM1 ganglioside in several in vitro systems* 215

Y. Nagai, S. Tsuji, J. Nakajima, T. Sasaki, *Functional analysis of ganglioside-respondable human neuroblastoma cell lines* 231

R. Dal Toso, D. Presti, D. Benvegnù, G. Tettamanti, G. Toffano, A. Leon, *Primary neural cell cultures and GM1 monosialoganglioside: a model for comprehension of the mechanisms underlying GM1 effects in CNS repair process in vivo* 245

E. Yavin, *Modulation of neurite outgrowth and ganglioside associated changes in cerebral neurons and PC12 pheochromocytoma cells* 257

S.D. Skaper, R. Katoh-Semba, L. Facci, S. Varon, *Ganglioside effects on astroglial cells in vitro* 271

F.J. Roisen, S.G. Matta, G. Yorke, M.M. Rapport, *The role of gangliosides in neurotrophic interaction in vitro* 283

M. Durand, B. Guérold, D. Lombard-Golly, H. Dreyfus, *Evidence for the effects of gangliosides on the development of neurons in primary cultures* 295

P.E. Spoerri, *Facilitated-establishment of contacts and synapses in neuronal cultures: ganglioside-mediated neurite sprouting and outgrowth* 309

R.E. Baker, *Effect of added gangliosides on the development of fetal mouse spinal cord-dorsal root ganglion explants cultured in a chemically defined medium* 327

P. Doherty, J.G. Dickson, T.P. Flanigan, F.S. Walsh, *Molecular specificity of ganglioside action on neurite regeneration in cell cultures of sensory neurons* 335

V. GANGLIOSIDES AND NEURONAL PLASTICITY IN VIVO

K. Fuxe, L.F. Agnati, F. Benfenati, I. Zini, G. Gavioli, G. Toffano, *New evidence for the morphofunctional recovery of striatal function by ganglioside GM1 treatment following a partial hemitransection of rats. Studies on dopamine neurons and protein phosphorylation* 347

G.L. Dunbar, W.M. Butler, B. Fass, D.G. Stein, *Behavioral and neurochemical alterations induced by exogenous gangliosides in brain damaged animals: problems and perspectives*. 365

I. Zini, M. Zoli, L.F. Agnati, K. Fuxe, E. Merlo Pich, P. Davalli, A. Corti, G. Gavioli, G. Toffano, *Studies on the involvement of polyamines for the trophic actions of the ganglioside GM1 in mechanically and 6-hydroxydopamine lesioned rats. Evidence for a permissive role of putrescine* 381

J.H. Greenberg, M. Reivich, R. Urbanics, K. Tanaka, E. Dora, G. Toffano, *The effect of GM1 on cerebral metabolism, microcirculation and histology in focal ischemia* 397

S.E. Karpiak, Y.S. Li, P. Aceto, S.P. Mahadik, *Acute effects of gangliosides on CNS injury* 407

G. Vantini, B. Figliomeni, R. Zanoni, A. Gorio, G. Jonsson, M. Fusco, *Effect of GM1 on the alterations induced by selective neurotoxins in the developing CNS*. 415

B. Oderfeld-Nowak, M. Skup, M. Gradkowska, L. Kiedrowski, *Early biochemical effects of GM1 ganglioside treatment in lesioned brain: dependence on degree of fiber degeneration* 427

J. Cahn, M.G. Borzeix, G. Toffano, *Effect of GM1 ganglioside and of its inner ester derivative in a model of transient cerebral ischemia in the rat* 435

M. Raiteri, P. Versace, M. Marchi, *Early recovery of striatal dopamine uptake in rats with unilateral nigro-striatal lesion following treatment with monosialoganglioside*. 445

P. Pinelli, C. Pasetti, L. Mazzini, F. Pisano, A. Villani, *Motorneuron sprouting and spinal plasticity in Amyotrophic Lateral Sclerosis: the "window of opportunity" for a ganglioside treatment* 453

S. Bassi, M.G. Albizzati, M. Sbacchi, L. Frattola, M. Massarotti, *Subacute phase of stroke treated with ganglioside GM1* 461

M. Massarotti, *Ganglioside therapy of peripheral neuropathies: a review of clinical literature*. 465

Subject Index 481

INTRODUCTION

This volume contains the Proceedings of the Meeting "Neuronal plasticity and gangliosides" which was held at Mantova, Italy, on May 29-31, 1985, as a satellite to the Tenth Meeting of the International Society for Neurochemistry, (Riva del Garda, Italy, May 19-24, 1985). The Symposium took place in the "Teatro Bibiena", one of the oldest and most beautiful theatres in Italy, known for having been the place for some of the very first performances of W.A. Mozart.

The number of registered participants in the Meeting was about 180 including 40 accompanying persons. The persons invited as Speakers or Chairmen of Colloquia and Workshops were 48. Eighteen countries were represented with the largest contingent coming from Italy, followed by Federal Republic of Germany, The United States, France and Great Britain. The Meeting was based on Colloquia, Workshops, Round Tables and Poster Sessions, with a total of 42 Lectures and 32 Poster presentations.

The Meeting provided the opportunity for a free and fruitful interchange and fusion between the 'culture' of conventional ganglioside biochemists and that of neurobiologists and experimental pathologists. The updating on the structural, methodological, metabolic and functional aspects of gangliosides was integrated by the presentations of elegant models for the study of neuritogenesis and synaptogenesis in vitro and in vivo, and of the repair processes of both peripheral and central nervous tissue. Despite the abundance of open questions on the biochemical mechanisms underlying the various aspects of neuronal plasticity, the direct implication of gangliosides in these phenomena emerged in a definite way. A discreet but active presence of representatives of major pharmaceutical industries proved the emerging interest in the therapeutic potential of gangliosides.

The scientific sessions had a crowded audience and were of high standard. Discussions, which benefited by the abundance of the allotted time, were very stimulating and lively. The scientific interest of the participants was accompanied by relaxed and joyful entertainments and was fed by the impressive beauty of such a city of art as Mantova.

At the Meeting the atmosphere was very friendly, the dialogue between participants very intense, and satisfaction evident. Therefore it could be judged that each one would have come away with new perspectives for his own work.

The Editors wish to express their thanks to the chairmen, speakers, and discussants who gave the most important contribution to the goals reached by the Meeting.

Finally, we acknowledge and thank Fidia Research Laboratories for their generous support to the Meeting and superb assistance in the publication of this volume.

Guido Tettamanti
Robert W. Ledeen
Konrad Sandhoff
Yoshitaka Nagai
Gino Toffano

NOMENCLATURE AND SCHEMATIC STRUCTURE OF THE GANGLIOSIDES AND RELATED NEUTRAL GLYCOSPHINGOLIPIDS CITED IN THIS BOOK
(Tables 1 and 2)

Table 1. *Designation of some components of gangliosides and allied glycosphingolipids*

Trivial name	Schauer's designation (+)	IUPAC-IUB designation (*)
Glucose		Glc
Galactose		Gal
N-acetylgalactosamine		GalNAc
N-acetylglucosamine		GlcNAc
Mannose		Man
Rhamnose		Rha
Fucose		Fuc
N-acetylneuraminic acid	Neu5Ac	NeuAc
N-glycolylneuraminic acid	Neu5Gl	NeuGl
Ceramide		Cer

(+) Schauer R., (1982) Sialic acids. Chemistry, metabolism and function. Cell Biology Monographs, Springer Verlag, Berlin, Vol. 10, pp. 5-8
(*) IUPAC-IUB commission on Biochemical nomenclature (1977) The nomenclature of lipids. Lipids, 12: 455-468

Table 2. *Designation and schematic structure of gangliosides and allied glycosphingolipids*

Designation			Schematic structure (°)
Trivial	*Svennerholm* (§)	*IUPAC-IUB* (*)	
Glucosylceramide		GlcCer	Glcβ1$\rightarrow$1'Cer
	GM4	I^{3}NeuAc-GalCer	NeuAcα2$\rightarrow$3Galβ1$\rightarrow$1'Cer
Lactosylceramide	AGM3	LacCer	Galβ1$\rightarrow$4Glcβ1$\rightarrow$1'Cer
	GM3	II^{3}NeuAc-LacCer	NeuAcα2$\rightarrow$3Galβ1$\rightarrow$4Glcβ1$\rightarrow$1'Cer
	GD3	II^3(NeuAc)$_2$-LacCer	NeuAcα2$\rightarrow$8NeuAcα2$\rightarrow$3Galβ1$\rightarrow$4Glcβ1$\rightarrow$1'Cer
Trihexosylceramide	AGM2	GgOse$_3$Cer	GalNAcβ1$\rightarrow$4Galβ1$\rightarrow$4Glcβ1$\rightarrow$1'Cer
	GM2	II^{3}NeuAc-GgOse$_3$Cer	GalNAcβ1$\rightarrow$4(NeuAcα2$\rightarrow$3)Galβ1$\rightarrow$4Glcβ1$\rightarrow$ 1'Cer
Tetrahexosylceramide	AGM1	GgOse$_4$Cer	Galβ1$\rightarrow$3GalNAcβ1$\rightarrow$4Galβ1$\rightarrow$4Glcβ1$\rightarrow$1'Cer
	GM1, GM1a	IV^{3}NeuAc-GgOse$_4$Cer	Galβ1$\rightarrow$3GalNAcβ1$\rightarrow$4(NeuAcα2$\rightarrow$3)Galβ1$\rightarrow$4Glc-β1$\rightarrow$1'Cer
	GM1b	II^{3}NeuAc-GgOse$_4$Cer	NeuAcα2$\rightarrow$3Galβ1$\rightarrow$3GalNAcβ1$\rightarrow$4Galβ1$\rightarrow$4Glcβ-1$\rightarrow$1'Cer
	Fuc-GM1	IV^{2}Fuc,II^{3}NeuAc-GgOse$_4$Cer	Fucα1$\rightarrow$2Galβ1$\rightarrow$3GalNAcβ1$\rightarrow$4(NeuAcα2$\rightarrow$3)Gal-β1$\rightarrow$4Glcβ1$\rightarrow$1'Cer
	GD1a	IV^{3}NeuAc,II^{3}NeuAc-GgOse$_4$Cer	NeuAcα3$\rightarrow$2Galβ1$\rightarrow$3GalNAcβ1$\rightarrow$4(NeuAcα2$\rightarrow$3) Galβ1$\rightarrow$4Glcβ1$\rightarrow$1'Cer
	GD1b	II^3(NeuAc)$_2$-GgOse$_4$Cer	Galβ1$\rightarrow$3GalNAcβ1$\rightarrow$4(NeuAcα2$\rightarrow$8NeuAcα2$\rightarrow$3) Galβ1$\rightarrow$4Glcβ1$\rightarrow$1'Cer
	Fuc-GD1b	IV^{2}Fuc,II^3(NeuAc)$_2$-GgOse$_4$Cer	Fucα1$\rightarrow$2Galβ1$\rightarrow$3GalNAcβ1$\rightarrow$4(NeuAcα2$\rightarrow$8Neu-Acα2$\rightarrow$3)Galβ1$\rightarrow$4Glcβ1$\rightarrow$1'Cer
	GT1a	IV^3(NeuAc)$_2$,II^{3}NeuAc-GgOse$_4$Cer	NeuAcα2$\rightarrow$8NeuAcα2$\rightarrow$3Galβ1$\rightarrow$3GalNAcβ1$\rightarrow$4(NeuAcα2$\rightarrow$3)Galβ1$\rightarrow$4Glcβ1$\rightarrow$1'Cer
	GT1b	IV^{3}NeuAc,II^3(NeuAc)$_2$-GgOse$_4$-Cer	NeuAcα2$\rightarrow$3Galβ1$\rightarrow$3GalNAcβ1$\rightarrow$4(NeuAcα2$\rightarrow$8 NeuAcα2$\rightarrow$3)Galβ1$\rightarrow$4Glcβ1$\rightarrow$1'Cer

	GT1c	II^3(NeuAc)$_3$-GgOse$_4$Cer	Galβ1→3GalNAcβ1→4(NeuAcα2→8NeuAcα2→8NeuAcα2→3)Galβ1→4Glcβ1→1'Cer
	GQ1b	IV^3(NeuAc)$_2$,II^3(NeuAc)$_2$-GgOse$_4$Cer	NeuAcα2→8NeuAcα2→3Galβ1→3GalNAcβ1→4(NeuAcα2→8NeuAcα2→3)Galβ1→4Glcβ1→1'Cer
	GQ1c	IV^{3}NeuAc,II^3(NeuAc)$_3$-GgOse$_4$-Cer	NeuAcα2→3Galβ1→3GalNAcβ1→4(NeuAcα2→8NeuAcα2→8NeuAcα2→3)Galβ1→4Glcβ1→1'Cer
	GP1c	IV^3(NeuAc)$_2$,II^3(NeuAc)$_3$-GgOse$_4$Cer	NeuAcα2→8NeuAcα2→3Galβ1→3GalNAcβ1→4(NeuAcα2→8NeuAcα2→8NeuAcα2→3)Galβ1→4Glcβ1→1'Cer
Globotetraosylceramide		GbOse$_4$Cer	GalNAcβ1→3Galα1→4Galβ1→4Glcβ1→1'Cer
Globopentaosylceramide		GgOse$_5$Cer	Galβ1→3GalNAcβ1→3Galα1→4Galβ1→4Glcβ1→1'Cer
Forsmann antigen		IV^{3}GalNAc-GbOse$_4$Cer	GalNAcα1→3GalNAcβ1→3Galα1→4Galβ1→4Glcβ1→1'Cer
Paragloboside		nLcOse$_4$Cer	Galβ1→4GlcNAcβ1→3Galβ1→4Glcβ1→1'Cer
Sialosylparagloboside		{ IV^{3}NeuAc-nLcOse$_4$Cer	NeuAcα2→3Galβ1→4GlcNAcβ1→3Galβ1→4Glcβ1→1'Cer
		IV^{6}NeuAc-nLcOse$_4$Cer	NeuAcα2→6Galβ1→4GlcNAcβ1→3Galβ1→4Glcβ1→1'Cer
Sialosyllactosaminyl-paragloboside		{ VI^{3}NeuAc-nLcOse$_6$Cer	NeuAcα2→3Galβ1→4GlcNAcβ1→3Galβ1→4GlcNAcβ1→3Galβ1→4Glcβ1→1'Cer
		VI^{6}NeuAc-nLcOse$_6$Cer	NeuAcα2→6Galβ1→4GlcNAcβ1→3Galβ1→4GlcNAcβ1→3Galβ1→4Glcβ1→1'Cer

(*) IUPAC-IUB commission on Biochemical nomenclature (1977) The nomenclature of lipids. Lipids, 12: 455-468
(§) Svennerholm L., (1980) Ganglioside designation. Adv. Exptl. Med. Biol. 125: 11
(°) Terms within brackets represent branching points in the molecule

Gangliosides and neuronal plasticity
G. Tettamanti, R.W. Ledeen, K. Sandhoff,
Y. Nagai, G. Toffano (eds.)
Fidia Research Series, vol. 6
Liviana Press, Padova, © 1986

A REVIEW OF MASS SPECTROMETRIC TECHNIQUES IN THE STRUCTURAL ANALYSIS OF NATIVE AND DERIVATIZED GANGLIOSIDES

H. Egge and J. Peter-Katalinić

Institut für Physiologische Chemie der Universität Bonn, Nussallee 11,
5300 Bonn, Federal Republic of Germany

INTRODUCTION

Mass spectrometry is especially qualified as an analytical tool because of its very high sensitivity and because of the abundant information provided on structural features of the analyte even if it is present only in minute quantities. Today, its application to the analysis of trace amounts of material is practically indispensable in biological and medical sciences (Burlingame, 1985).

In this respect multifunctional and structurally complex molecules like gangliosides pose a special challenge for study. Dawson and Sweeley (1971) were the first to use electron impact mass spectrometry (EI-MS) at 70 eV for the analysis of GM3, GD3 (Neu 5Gc), GM2, GM1, GD1a and GD1b as pertrimethylsilyl derivatives. They were able to show that the number of monosaccharides present as well as the number of those hexose units that are unsubstituted at C-3 could be calculated from characteristic sugar fragment ions relative to those derived from the ceramide residue. N-acetylhexosamine residues either in terminal or internal position could be identified with the aid of specific fragment ions. Information could also be obtained on the type and number of sialic acids present (Neu5Ac or Neu5Gc) as well as on the long chain base and fatty acid constituents of the ceramide residue. It has to be especially mentioned that this remarkable set of data could be obtained from fragment ions appearing in the mass range up to m/z 1000, far below the molecular weight of the analytes. During the following years a large number of pertrimethylsilylated or peracetylated derivatives of glycosphingolipids were analysed by EI-MS as reviewed by Egge (1978).

Abbreviations: MS, mass spectrometry; EI, electron impact; FD, field description; CI, chemical ionization; SI, secondary ion; FAB, fast atom bombardment; DCI, direct chemical ionization; TMS, trimethylsilyl; Neu5Ac, or NeuAc, N-acetylNeuramicin acid; Neu5Gc, or NeuGc, N-glycolylneuramicin acid; a.m.u., arbitrary mass units.

In spite of the easy preparation, the large increase in molecular weight (72 a.m.u. per O-TMS group and 42 a.m.u. per O-acetyl group) and the thermal instability limited the use of these methods of derivatization to molecules carrying at most five or six carbohydrate residues. A major contribution to the derivatization of heat labile compounds came from the group of K.-A. Karlsson (1974). Improved methods of permethylation introduced by S.-I. Hakomori (1964) opened new pathways for obtaining heat stabile derivatives. Although being more laborious, the mass increase per OH-group is only 14 a.m.u. and the tendency to form rearrangement ions under electron impact is reduced considerably as compared to O-TMS or O-acetyl derivatives. The chemical stability of the ether bond offered additional advantages for the chemical modification of the polar N-acyl- and carboxylester groups as outlined in Scheme 1.

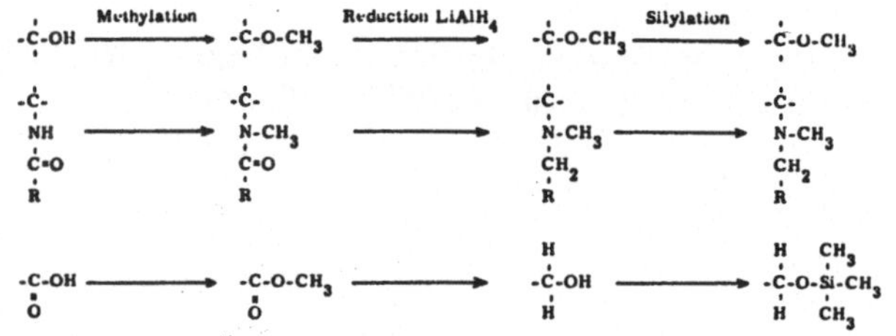

Scheme 1. Modification of Functional Groups

This scheme of derivatization, introduced by K.-A. Karlsson et al. (1974, 1978), proved to be highly effective in the structural elucidation of gangliosides carrying up to two sialic acid residues and molecular weights in the range of 2500 a.m.u. (Fig. 1) (Fredman et al., 1981).

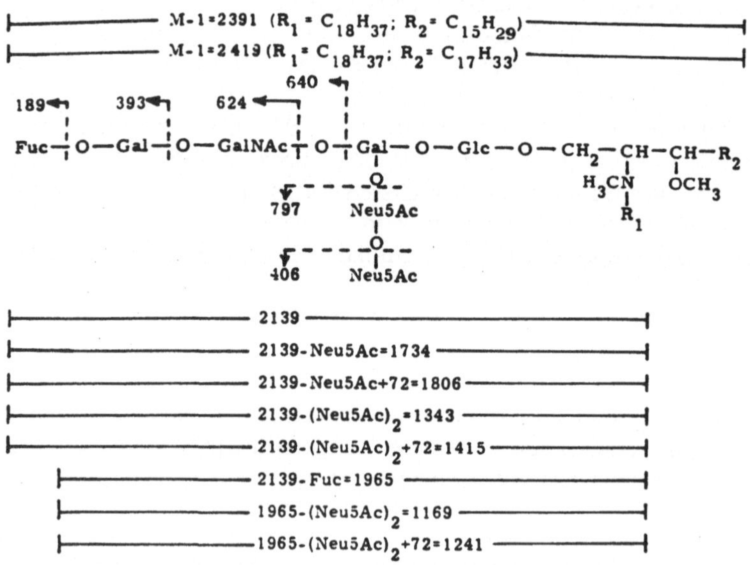

Figure 1. Scheme of fragmentation observed with permethylated-reduced-trimethylsilylated ganglioside Fuc-GD1b under electron impact at 40 eV electron energy (Fredman et al., 1981).

Again, however, the heat lability also characteristic of these derivatives leads to complicated spectra of fragment ions that are often very difficult to interpret. This is especially true for the "higher" gangliosides carrying more than two sialic acid residues. It is now well recognized that the limitations of EI-MS cannot be overcome by other methods of derivatization but rather by the application of more gentle methods of ionization such as field desorption mass spectrometry (FD-MS), chemical ionization mass spectrometry (CI-MS), secondary ion mass spectrometry (SI-MS) and fast atom bombardment mass spectrometry (FAB-MS) to native or suitably derivatized gangliosides.

A number of workers in the field have shown that with the aid of these soft ionization techniques a large number of structural features of gangliosides can be determined, such as

— molecular weight,
— type of long chain bases and fatty acids,
— distribution of the ceramide residues,
— number and sequence of hexose and hexosamine residues,
— type and number of sialic acid residues,
— occurrence of O-acetyl groups or lactone ring formation,
— presence of disialo or trisialo groups and their sites of attachment to the carbohydrate backbone.

Here, the potentials and limitations of different MS techniques in establishing structure-spectra relationships, using either native or suitably derivatized gangliosides, will be reviewed.

MASS SPECTROMETRY

Chemical Ionization Mass Spectrometry (CI-MS)

Chemical ionization mass spectrometry uses gas-phase ion-molecule reactions between reagent gas and sample molecules as a "soft ionization" method (for a review see Lin and Smith, 1984). A reagent gas, methane, ethane, iso-butane or ammonia is first ionized by electron impact. The ions thus formed like CH_5^+ (methane), $C_2H_5^+$ (ethane), $C_4H_9^+$ (iso-butane) and NH_4^+ (ammonia) can react chemically with the analyte in fast acid/base-type reactions with subsequent fragmentation. The extent of fragmentation can be controlled by the appropriate choice of reagent gas and/or ion source temperature. The ammonia CI spectra of permethylated gangliosides were studied by Ariga et al. (1982, 1984). For GM3 carrying either Neu5Ac or Neu5Gc residues pseudomolecular ions M + 1 and M + 1—32 could be obtained together with fragment ions characteristic for the ceramide residue, the type of sialic acid and carbohydrate sequence as shown schematically in Figure 2. For permethylated GM1, GD1a and GD1b, however, rather complex ammonia CI spectra were obtained showing extensive secondary fragmentation which obscured essential structural features of the analytes. In particular, no primary fragment ions could be obtained that unequivocally indicated the site of attachment of the sialic acid residues. Due to either low volatility or thermal instability, no molecular or pseudomolecular ions could be observed either. Fragment ions that could be directly correlated with special structural features are shown in Figure 3.

After reduction of the permethylated ganglioside GM1 with $LiAlH_4$, followed by silylation of the primary OH-group produced at the sialic acid residue, the pattern of fragmentation is changed drastically (Ariga et al., 1982). Here pseudomolecular ions M + H and M + 73 could be observed together with ions derived from the carbohydrate residue. Hardly any fragment ions carrying the ceramide part could be observed. Again, intense secondary fragmentation of the structurally important pentasaccharide ion [Glc-Gal-(Neu5Ac)-GalNAc-Gal]$^+$ and the tetrasaccharide ion [Neu5Ac-Gal-

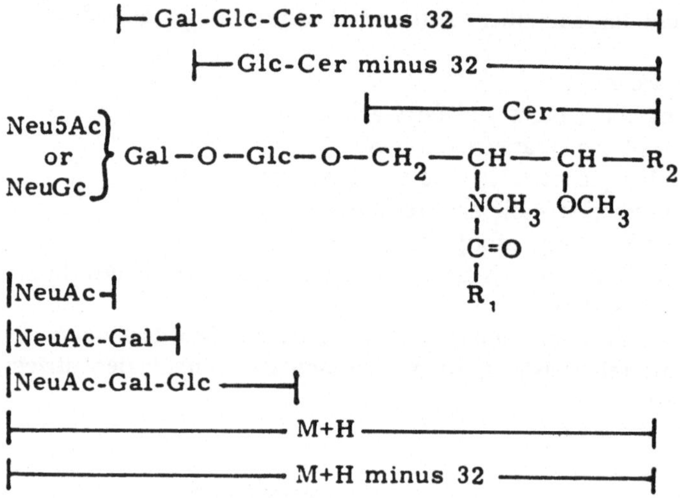

Figure 2. Schematic pattern of fragmentation observed in ammonia CI-MS of permethylated GM3.

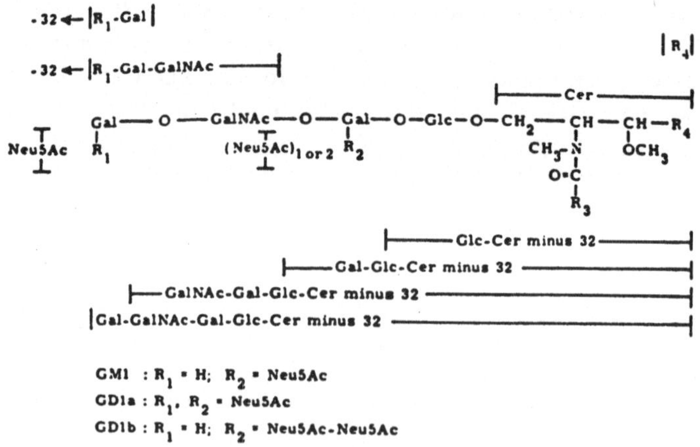

Figure 3. Schematic pattern of fragmentation of permethylated gangliosides GM1, GD1a and GD1b in ammonia CI-MS (Ariga et al., 1982; 1984).

GalNAc-Gal]$^+$, yielding tetra- and trisaccharide ions (Glc-Gal-GalNAc-Gal)$^+$ and [Gal-GalNAc-Gal]$^+$ without the neuraminic acid residue, prohibit a clear definition of the site of attachment of the sialic acid to the carbohydrate chain. No CI spectra of higher gangliosides carrying three, four or more sialic acids were reported so far.

In order to overcome the drawbacks of high molecular weight and unfavourable fragmentation, oligosaccharides were liberated by ozonolysis from 5 mg each of individual gangliosides, reduced and permethylated (Tanaka et al., 1984). The CI spectra of the permethylated oligosaccharides derived from GM3, GM2, GM1, GD1a and GD1b exhibited pseudomolecular ions M + H or $M + NH_4$ and series of fragment ions that could be assigned unequivocally. In particular, sialic acid-containing di-, tri- and tetrasaccharide ions derived from the reduced and the nonreducing end of the molecule could be observed that allowed the determination of those sugar residues to which the sialic acid is attached. The general fragmentation pattern observed by Tanaka et al. (1984) is summarized in Figure 4.

Field Desorption Mass Spectrometry (FD-MS)

FD-MS developed by the group of Beckey (Beckey and Schulten, 1975; Schulten, 1977) has recently been used with considerable success for the molecular weight determination of carbohydrates and glycolipids (Linscheid et al., 1981; Egge et al., 1983; Dell et al., 1983a, b). It is especially attractive because it also allows the analysis of underivatized samples on the µg level. The technical difficulties associated with this technique, however, have restricted the use of this method to a few highly specialized laboratories.

The group of Handa (Kushi and Handa, 1982; Handa et al., 1983; Kushi et al.,

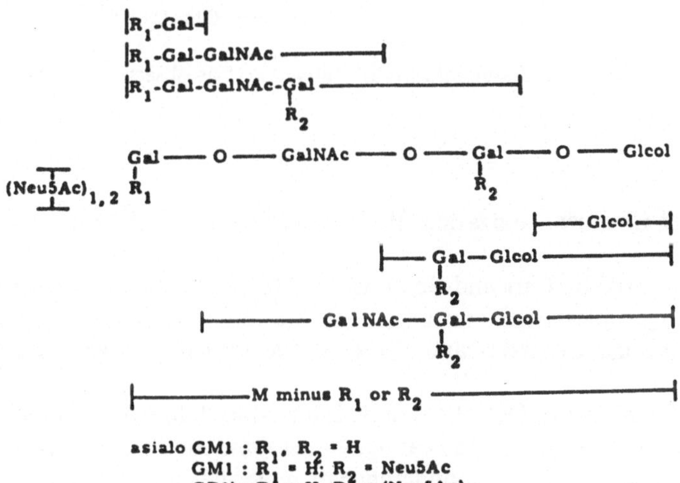

Figure 4. Schematic presentation of structurally important fragment ions observed under ammonia CI-MS of permethylated oligosaccharides released by ozonolysis from gangliosides.

1983; Handa and Kushi, 1984; Handa and Nakamura, 1984) has made extensive studies on the applicability of FD-MS and secondary ion SI-MS to the analysis of native and permethylated glycosphingolipids and the gangliosides GM3, GM2, GM1 and GD1a.

The FD-MS spectra of native gangliosides are characterized by a pronounced tendency to form ions after cleavage of the glycosidic bond of the sialic acid residues. Pseudomolecular ions M + Na—H_2O are normally present. Some of the characteristic features are exemplified in Figure 5 for GM1.

As can be deduced from the scheme in Fig. 5, extensive secondary fragmentation of the gangliotetraosyl residue takes place after the elimination of the sialic acid residue. Therefore, no information as to the site of linkage of the Neu5Ac residue can be extracted from the spectrum. This failure to produce linkage specific sequence ions is, however, not an intrinsic trait of FD-MS but rather the consequence of recording positive ion spectra (see Fig. 10).

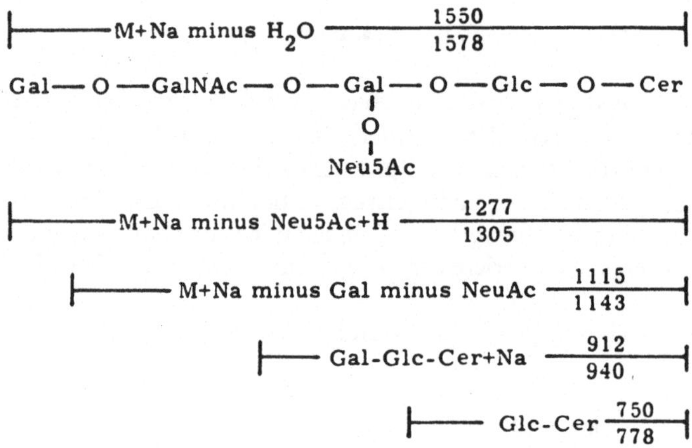

Figure 5. Schematic pattern of fragmentation of native GM1 as observed by FD-MS (Kushi et al., 1983).

Ammonia Direct Chemical Ionization Mass Spectrometry (DCI-MS)

DCI-MS was used by Carr and Reinhold (1984) for the analysis of permethylated and reduced gangliosides GM3 and GM1. Spectra structure relationship was firmly established by the use of $^{14}NH_3$ and $^{15}NH_3$ as the reagent gas and reduction with $LiAlH_4$.

Principle of analysis: In DCI the sample is introduced directly into the ion plasma of the reagent gas with programmed heating. Under these conditions a bell-shaped total ion profile is produced with a preponderance of molecular weight related ions during the early part of the heating profile, whereas during the later part of the heating program pyrolysis products predominate (Reinhold and Carr, 1982; 1983). The DCI spectra reported by Carr and Reinhold (1984) were chosen from scans showing maximum molecular weight related ions.

The DCI spectra of permethylated GM1 showed abundant cationized molecular species and a large number of fragment ions. Their interpretation was, however, due to the presence of both protonated and ammonia-cationized species, not quite as straight-forward as the spectra obtained with permethylated and reduced species. In Figure 6 the major fragment ions are shown for the permethylated/reduced species that allow an unequivocal determination of the sequence of carbohydrate residues, the site of linkage of the Neu5Ac residue and the type of long chain base and fatty acid residues.

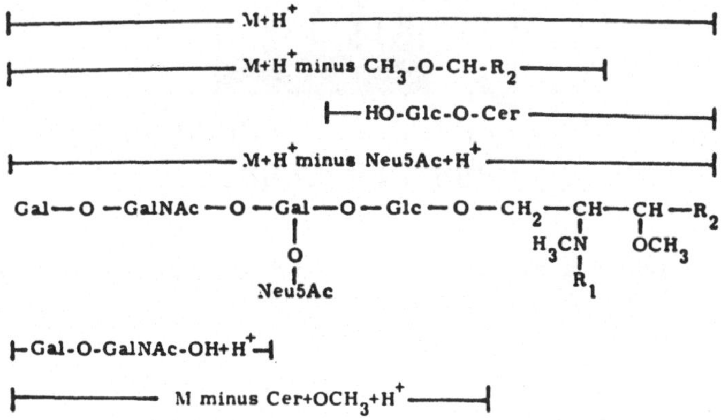

Figure 6. Pattern of fragmentation observed in the ammonia direct chemical ionization of permethylated and reduced ganglioside GM1.

Fast Atom Bombardment Mass Spectrometry (FAB-MS)
(Barber et al., 1981; Williams et al., 1981)

During the last few years, several reports described the use of SI-MS and FAB-MS for the structural characterization of native or derivatized glycolipids and gangliosides (Arita et al., 1983a, b; 1984; Dell et al., 1983a, b; Dennis et al., 1985; Egge et al., 1982; Egge et al., 1984 a, b; Egge et al., 1985 a, b; Egge and Peter-Katalinić, 1985; Fukuda et al., 1985; Handa and Kushi, 1984; Hanfland et al., 1984; Kushi and Handa, 1982; Kushi et al., 1983; Neuenhöfer et al., 1985). The results obtained by several laboratories show that by appropriate choice of the liquid matrix in which the sample is dissolved and the application of negative ion abstraction, native gangliosides carrying four or more sialic acid residues can be structurally characterized without prior derivatization. Molecular weight related ions up to m/z 3000 could be observed (Fig. 7) together with significant sequence ions that gave information on carbohydrate sequence, type and number of sialic acids, their sites of attachment and the type of ceramide residue. As matrices glycerol (Handa and Kushi, 1982), thioglycerol and triethanolamine/tetramethylurea (Arita et al., 1983; Egge et al., 1983) were used. In comparative measurements under otherwise identical conditions we found molecular ion intensities 5 to 10 times higher in thioglycerol as compared to triethanolamine/tetramethylurea and better by a factor of about 20 than in glycerol.

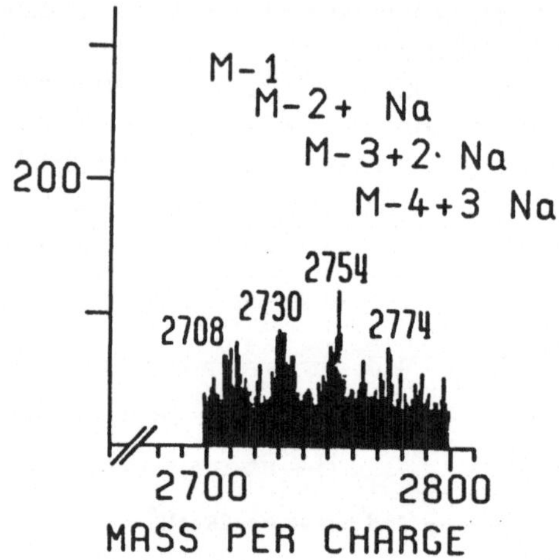

Figure 7. Partial negative ion fast atom bombardment mass spectrum showing the region of the molecular-weight-related ions of a pentasialo ganglioside.

In addition, specific information can be extracted from the spectra concerning the presence of alkali or acid-labile structural features like lactone ring formation or O-acyl groups, features which are normally lost during derivatization. The sensitivity of the method also allows the analysis of mixtures of gangliosides, thus providing a powerful tool for the assessment of the sample purity.

FAB-MS of permethylated gangliosides using positive ion abstraction, on the other hand, can furnish additional and valuable data on sample homogeneity. Since in negative ion FAB-MS of native gangliosides the pseudomolecular ions M-1 coincide with the ions M-Neu5Ac of the higher homologues that carry one, two or three more sialic acid residues, it may be difficult in the absence of other specific sequence ions to evaluate the purity of the analyte. Here positive ion FAB-MS shows great advantages. Addition of sodium acetate to the target leads to the formation of intense $M+Na^+$ ions that can be clearly discriminated from the fragment ions that do not normally carry Na^+ ions, thus allowing a quantitative evaluation of ganglioside mixtures from the relative intensity of the molecular-weight-related $M+Na^+$ ions down to a ‰ level.

a) Specifically, negative ion FAB-MS furnishes information on: the molecular weight through intense $M\text{-}1^-$ ions. Difficulties may be encountered with gangliosides carrying more than two sialic acid residues. Since only singly charged molecules are observed, the sialic acid may carry cations H^+, Na^+, K^+ etc., thus giving rise to a very complex pattern of molecular-weight-related ions. This difficulty can be overcome by adding acid to the matrix like oxalic, citric or hydrochloric acid. Under these conditions abstraction of H_2O due to the lactone ring formation may be observed in gangliosides carrying a Neu5Acα2→8 Neu5Ac group (Egge et al., 1985a) as indicated in Figure 8.

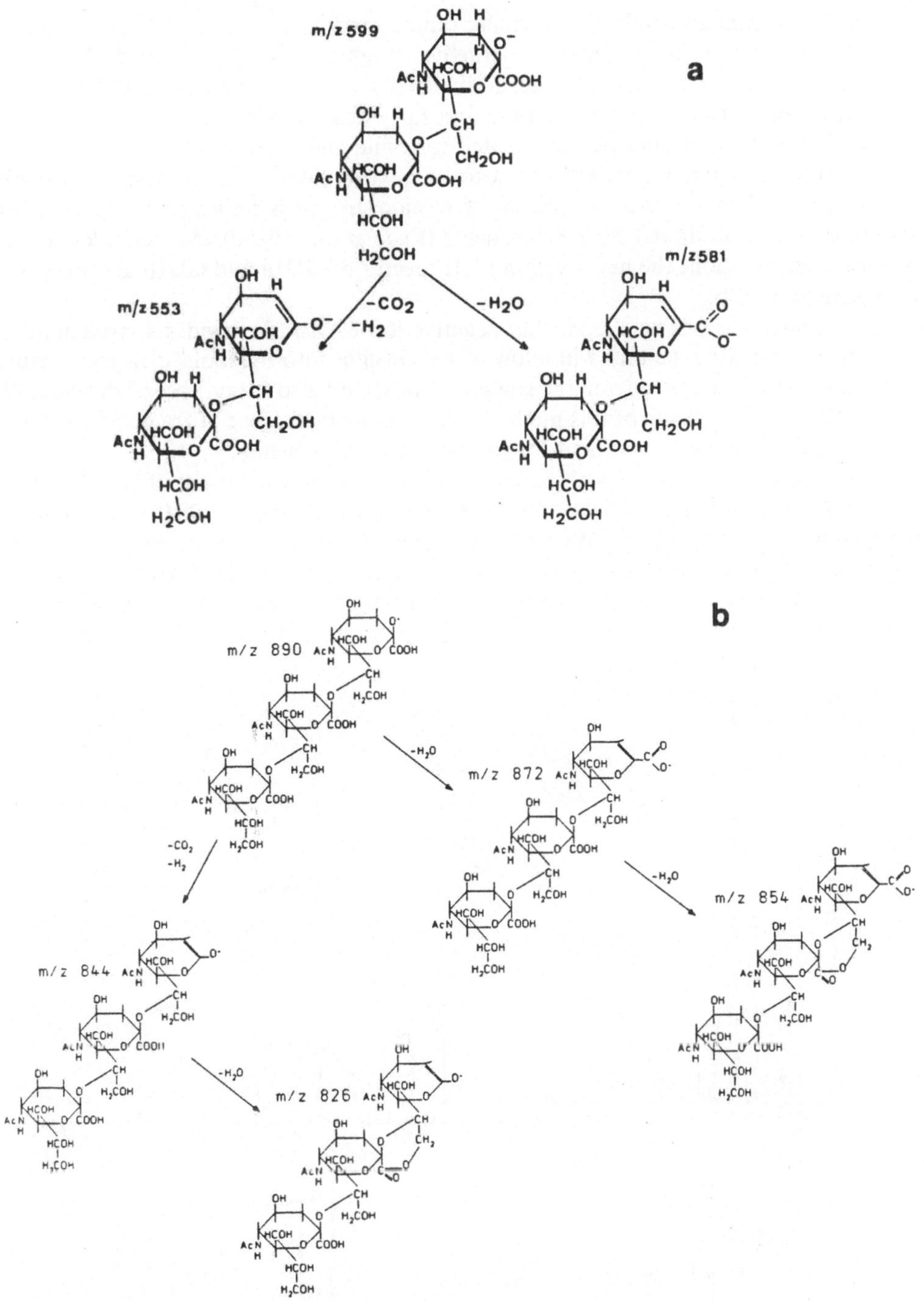

Figure 8. Fragmentation pathways of disialo- (a) and trisialo- (b) groups as observed in negative ion FAB-MS.

b) The ceramide residue is normally represented by ions carrying the oxygen at C-1 of the sphingosine. Normally no further fragmentation is observed. Therefore other methods like EI, CI or DCI of derivatized samples have to be used in order to get further insight into long chain base and fatty acid composition.

c) From the molecular-weight-related ions and the ceramide residue the overall carbohydrate composition in terms of hexoses, hexosamines and sialic acids can exactly be calculated. The presence of disialo- or trisialo groups is indicated by specific ion groups as shown in linked scan experiments (Egge et al., 1985a). As examples for sequence determination, the negative ion FAB spectra of GD1a and GD1b are presented in Figure 9a and b.

In summary it can be stated, that negative ion FAB-MS provides a large number of sequence specific ions that will allow a close insight into the molecular architecture of the gangliosides derived from the gangliotetraose and also from those of the lacto-N-series. The only drawback of this method is the complete absence of fragment ions that indicate individual fatty acid and long chain base constituents.

d) FAB-MS and SI-MS with positive ion abstraction. The group of Handa (Handa and Kushi, 1984; Kushi et al., 1983; Handa and Nakamura, 1984) has compared the results of FD-MS and SI-MS with a number of native and reduced gangliosides. As already stated above, the positive ion spectra using either FAB-, SI- or FD-ionization are characterized by a preferred formation of ions that are produced by elimination of sialic acid residues followed by fragmentation of the carbohydrate

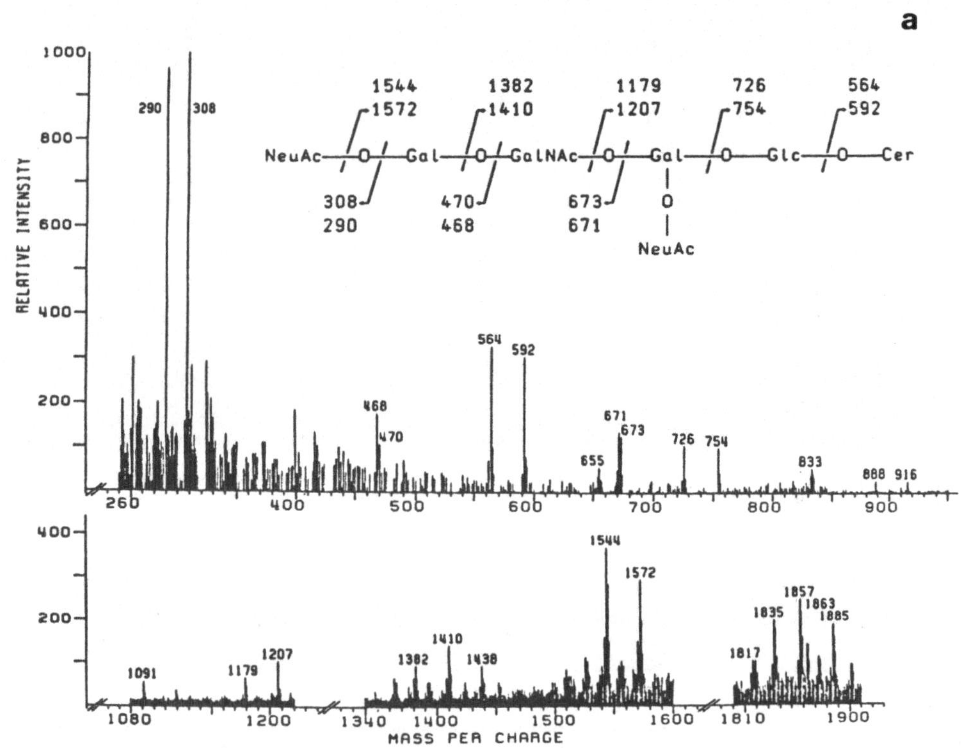

Figure 9a. Continued next page.

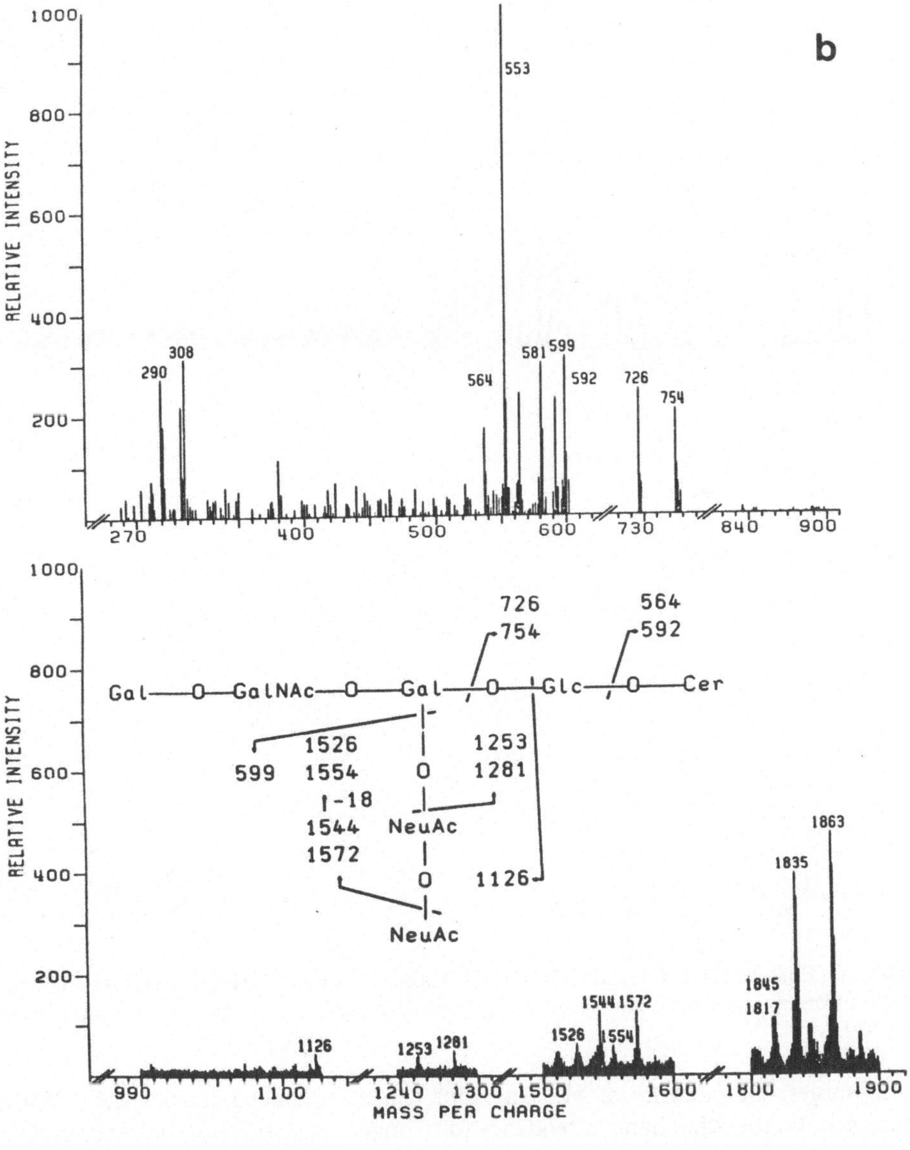

Figure 9. Negative ion FAB-MS spectrum of native GD1a (a) and GD1b (b) in a matrix of thioglycerol. Background subtraction was applied up to m/z 950.

backbone. Thus, no clearcut sequences can be determined. As an example the positive and negative ion FAB spectra of GM1 are shown in Figure 10a and b.

Positive ion FAB-MS is however very well suited for the analysis of permethylated gangliosides (Egge et al., 1985a). The spectra are characterized by a few but very prominent ions that allow the determination of specific structural features and, as already pointed out above, the homogeneity of the analyte. After addition of sodium acetate

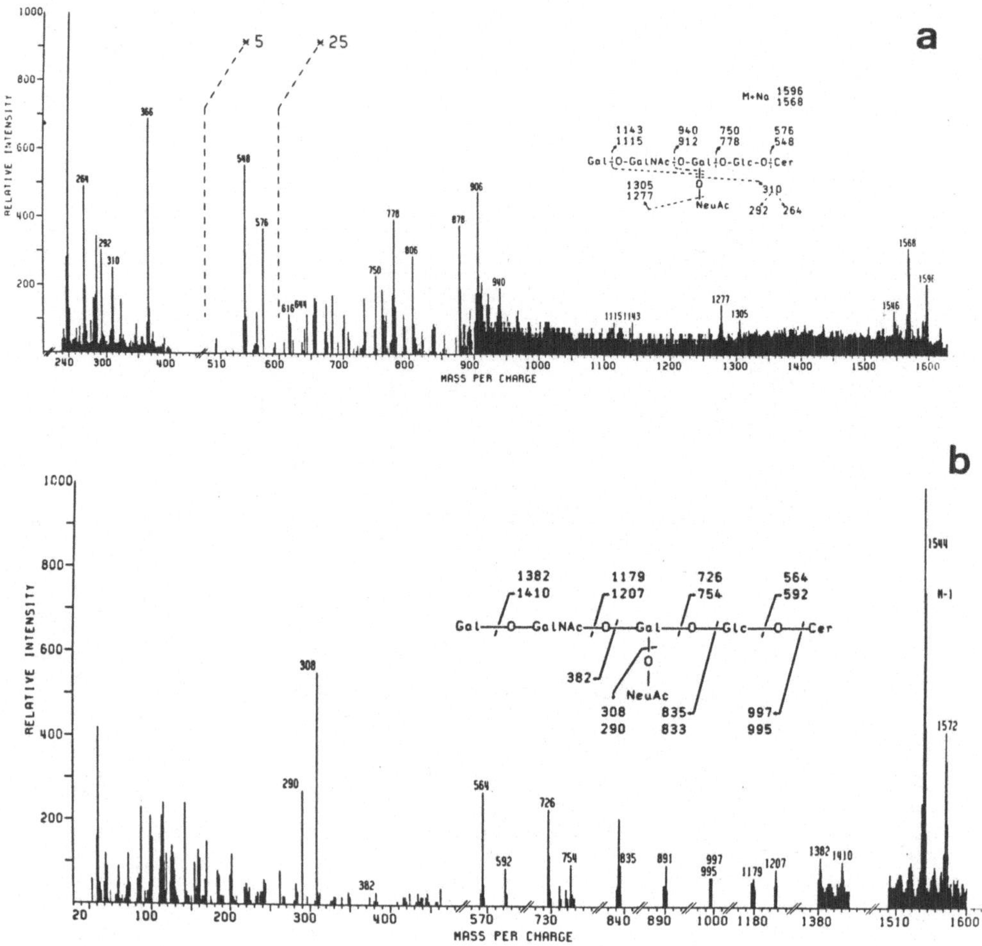

Figure 10. Positive (a) and negative (b) ion FAB spectra of underivatized ganglioside GM1. 5μg samples were analysed in a matrix of thioglycerol. Background subtraction was applied up to m/z 900.

to the target, the intensity of $M+Na^+$ ions can be enhanced considerably. Without added salt, on the other hand, a tendency to stronger fragmentation is observed. Under these conditions of measurement, fragment ions are formed by elimination of the fatty acid moiety, thus allowing also the analysis of the ceramide constituents (Fukuda et al., 1985). The major fragmentations A-E observed in the positive ion FAB spectra of permethylated gangliosides from GM1 to GP are summarized in Scheme 2.
The ions A_1 and A_2 indicate the presence of mono-, di- or trisialo groups. From ion B, generally of high intensity, the type of substitution of the terminal Gal can be deduced. The occurrence of m/z 228 points to a 1→3 link at the GalNAc residue. Fragment ions D, sometimes of lower intensity, represent the whole carbohydrate residue and from the ions C and E the ceramide composition can be deduced. As an example, the positive ion FAB-MS of permethylated GQ1b is presented in the Figure 11.

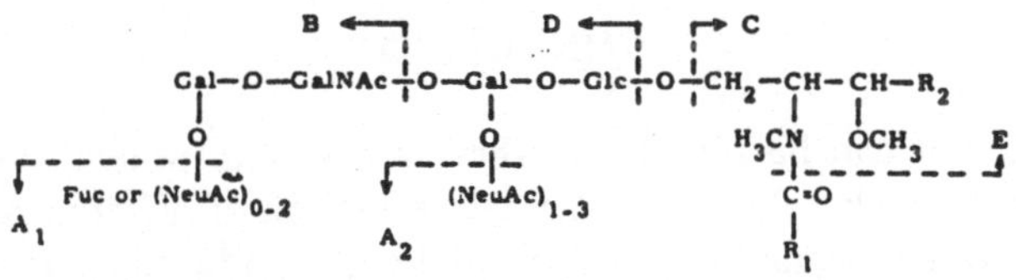

Scheme 2. Major Fragment Ions Observed in Positive Ion FAB-MS of Permethylated Gangliosides

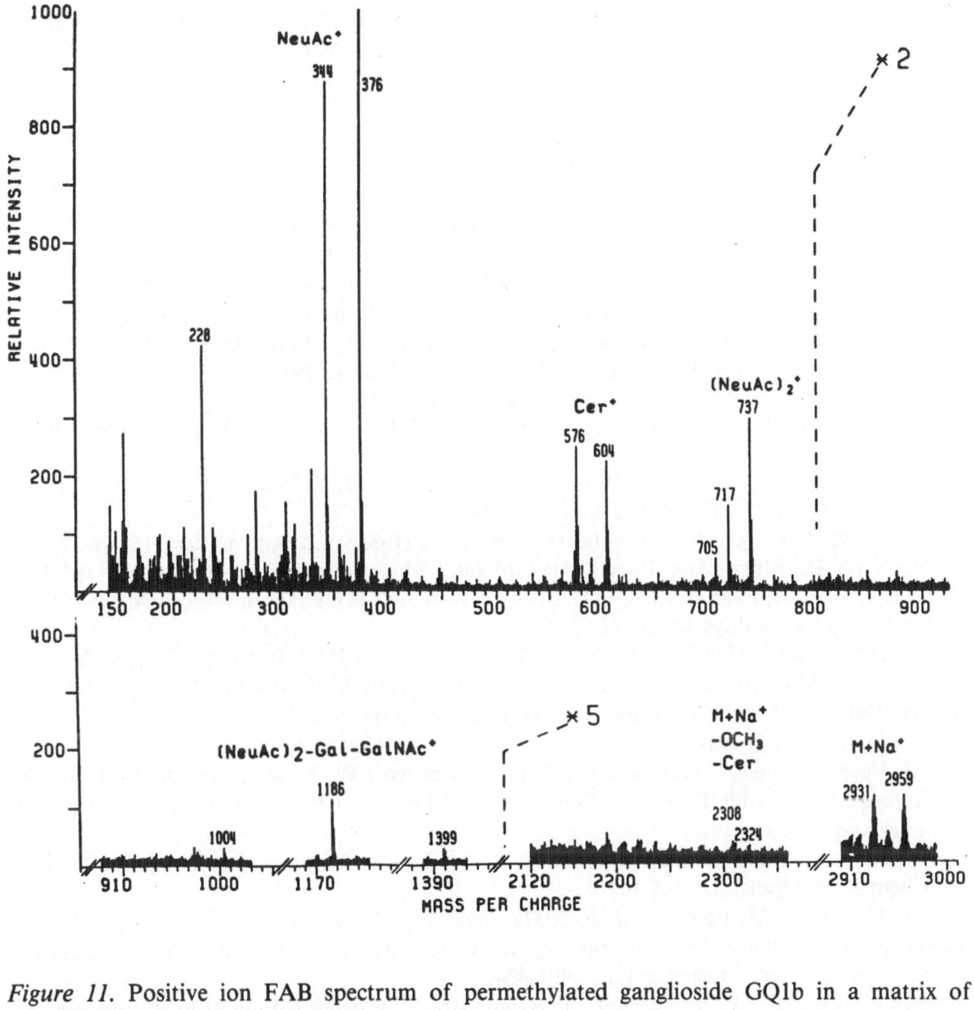

Figure 11. Positive ion FAB spectrum of permethylated ganglioside GQ1b in a matrix of thioglycerol, (Egge et al., 1985a).

CONCLUSION

In conclusion it can be stated that there are several potent mass spectrometric echniques available today that allow the determination of different structural aspects of native or derivatized gangliosides. From the data produced by several laboratories it can safely be concluded that FAB-MS, using either positive or negative ion abstraction, can be considered as the most suitable method of analysis with regard to information on purity of the sample, spectra-structure relationship, ease of handling (e.g. of labile derivatives) and sensitivity of ion detection in the high mass range.

ACKNOWLEDGMENTS

This work was supported by the Deutsche Forschungsgemeinschaft. The technical assistance of B. Barnhusen and M. Pflüger is gratefully acknowledged.

REFERENCES

Ariga T, Yu RK, Suzuki M, Ando S, Myatake T (1982) J Lipid Res 23: 437-442.
Ariga T, Yu RK, Myatake T (1984) J Lipid Res 25: 1096-1101.
Arita M, Iwamori M, Higuchi T, Nagai Y (1983a) J Biochem (Tokyo) 93: 319-322.
Arita M, Iwamori M, Higuchi T, Nagai Y (1983b) J Biochem (Tokyo) 94: 249-256.
Arita M, Iwamori M, Higuchi T, Nagai Y (1984) J Biochem (Tokyo) 95: 971-981.
Barber M. Bordoli RS, Sedgwick RD, Tyler AN (1981) Nature 293: 270-275.
Beckey HD, Schulten H-R (1975) Angew Chem 87: 425-460.
Burlingame AL (1985) Mass spectrometry in Health and Life Sciences, Elsevier, Amsterdam, XXI-XXIV.
Carr SA and Reinhold VN (1984) Biomed Mass Spectrom 11: 633-642.
Dawson G and Sweely CC (1971) J Lipid Res 12: 56-64.
Dell A, Morris HR, Egge H, v Nicolai H, Strecker G (1983a) Carbohydr Res 115: 41-52.
Dell A, Oates JE, Morris HR, Egge H (1983b) Int J Mass Spectrom & Ion Phys 46: 415-418.
Dennis RD, Geyer R, Egge H, Menges H, Stirm S, Wiegandt H (1985) Eur J Biochem 196: 51-58.
Egge H (1978) Chem Phys Lipids 21: 349-360.
Egge H, Dabrowski J, Hanfland P, Dell A, Dabrowski U (1982) in: Taketomi T, Nagai Y (eds): New Vistas in Glycolipid Research Plenum Press, New Yprk and London, pp. 33-40.
Egge H, Dell A, v Nicolai H (1983) Arch Biochem Biophys 224: 235-253.
Egge H, Dabrowski J, Hanfland P (1984a) Pure & Appl Chem 56: 807-819.
Egge H, Peter-Katalinić J, Hanfland P (1984b) in: Ledeen RW, Yu RK, Rapport MM, Suzuki K (eds): Ganglioside Structure, Function and Biomedical Potential. Plenum Press, New York, pp. 55-63.
Egge H, Peter-Katalinić J, Reuter G, Schauer R, Ghidoni R, Sonnino S, Tettamanti G (1985a) Chem Phys Lipids, 37; 127-141.
Egge H, Kordowicz M, Peter-Katalinić J, Hanfland P (1985b) J Biol Chem 260: 4927-4935.
Egge H, Peter-Katalinić J (1985c) in: Burlingame AL (ed): Mass Spectrometry in Health and Life Sciences Elsevier, Amsterdam, pp. 401-424.
Fredman P, Mansson J-E, Svennerholm L, Samuelsson B, Pascher I, Pimlott W, Karlsson K-A, Klinghardt GW (1981) Eur J Biochem 116: 553-564.

Fukuda MN, Dell A, Oates JE, Wu P, Klock JC, Fukuda M (1985) J Biol Chem 260: 1067-1082.
Hakomori S-I (1964) J Biochem (Tokyo) 55: 205-208.
Handa S, Kushi Y, Kambara H, Shizukuishi K (1983) J Biochem (Tokyo) 93: 315-381.
Handa S, Nakamura K (1984) J Biochem (Tokyo) 95: 1323-1329.
Handa S, Kushi Y (1984) in: Ledeen RW, Yu RK, Rapport MM, Suzuki K (eds): Ganglioside Structure, Function and Biomedical Potential. Plenum Press, New York, pp. 65-73.
Hanfland P, Kardowicz M, Niermann H, Egge H, Dabrowski U, Peter-Katalinić J, Dabrowski J (1984) Eur J Biochem 145: 531-542.
Karlsson K-A, Pascher I, Pimlott W, Samuelsson BE (1974) Biomed Mass Spectrom 1: 49-56.
Karlsson K-A (1978) Progr. Chem Fats other Lipids 16: 207-230.
Kushi Y, Handa S, Kambara H, Shizukuishi K (1983) J Biochem (Tokyo) 94: 1814-1815.
Kushi Y and Handa S (1982) J Biochem (Tokyo) 91: 923-931.
Lin YY and Smith LL (1984) Mass Spectrom Rev 3: 319-355.
Linscheid M, D'Angona J, Burlingame A-L, Dell A, Ballou CE (1981) Proc Natl Acad Sci USA 78: 1471-1475.
Neuenhofer S, Schwarzmann G, Egge H, Sandhoff K (1985) Biochemistry 24: 525-532.
Reinhold VN and Carr SA (1982) Anal Chem 45: 490-503.
Reinhold VN and Carr SA (1983) Mass Spectrom Rev 2: 153-221.
Schulten H-R (1977) in: Glick D (ed): Methods of Biochemical Analysis. Vol. 24, pp. 313-448.
Tanaka Y, Yu R, Ando S, Ariga T, Itoh T (1984) Carbohydr Res 126: 1-14.
Williams DH, Bradley C, Bojesen G, Santikarn S, Taylor LEC (1981) J Am Chem Soc 103: 5700.

Gangliosides and neuronal plasticity
G. Tettamanti, R.W. Ledeen, K. Sandhoff,
Y. Nagai, G. Toffano (eds.)
Fidia Research Series, vol. 6
Liviana Press, Padova, © 1986

ADVANCES IN GANGLIOSIDE CHEMISTRY

Sandro Sonnino, Domenico Acquotti, Günther Kirschner[1], Giovanni Fronza[2], Guido Tettamanti

Study Center for the Functional Biochemistry of Brain Lipids,
Department of Biological Chemistry, The Medical School, University of Milan,
Via Saldini 50, 20133 Milano, Italy;
[1] FIDIA Research Laboratories, Department of Chemistry, Abano Terme, Italy;
[2]CNR Study Center for Natural Organic Substances, Department of Chemistry,
Polytechnic School of Milan, Milano, Italy

INTRODUCTION

Gangliosides, sialic acid containing glycosphingolipids, are normal components of the plasma membrane of vertebrate cells and are particularly abundant in the nervous system. They are constituted of a hydrophilic portion, the negatively charged oligosaccharide, and a hydrophobic portion, the ceramide (N-acylated long chain base), linked together by a glycosidic linkage (Ledeen and Yu, 1978; Wiegandt, 1982). The hydrophilic portion is oriented towards the extracellular environment and is assumed to be involved in recognition phenomena mediated by specific interactions between the oligosaccharide moiety and the external ligand (Brady and Fishman, 1979; Holmgren et al., 1980). The hydrophobic portion is inserted into the lipid layer of the membrane and might participate in the process of signal transduction through the membrane

---

Abbreviations: Gangliosides are coded according to Svennerholm, (1970) and to the suggestions of the IUPAC-IUB Commission for the nomenclature of lipids (1977). Ganglioside derivatives follow the same suggestions. Neu: neuraminic acid; l.c.b.: long chain base. Deacetylated GM1, deAc-GM1: $II^3NeuGgOse_4Cer$, Galβ1 → 3GalNacβ1 → 4(Neuα2 → 3)Galβ1 → 4Glcβ1 → 1'Cer; deacetylated, deacylated GM1, deAc-deAcyl-GM1, $II^3NeuGgOse_4$ l.c.b.: Galβ1 → 3GalNacβ1 → 4(Neuα2 → 3)Galβ1 → 4Glcβ1 → 1'l.c.b. The molecular species of GM1 containing pyrene-decanoic acid, is coded GM1 (pyrene), and those containing 5-doxyl-stearic acid or 16-doxyl-stearic acid, GM1 (5-doxyl) and GM1 (16-doxyl), respectively, or generally GM1 (doxyl). The GM1 ganglioside that contains an acetyl group instead of the long acyl chain is coded GM1 (acetyl). A shorthand formula is suggested for the ganglioside molecular species having a homogeneous ceramide composition (Sonnino et al., 1985) on the basis of the following rationale: ganglioside designation (Svennerholm, 1970) is followed in parenthesis by l.c.b. configuration (e, erythro; t, threo), length, and degree of insaturation and, separated by a space-bar, fatty acid length and degree of unsaturation: for example GM1 (eC18:1/C18:0) is the species of GM1 containing erythro C18 sphingosine and stearic acid; DDQ: dicyanodichlorobenzoquinone; DMSO-d_6: deuterated dimethylsulfoxide; ESR: electron spin resonance; NMR: nuclear magnetic resonance; EPC: egg phosphatidylcholine; SUV: small unilamellar vesicles; c.m.c.: critical micellar concentration. GLC-MS, gas liquid chromatography-mass spectrometry; FAB-MS, fast atom bombardment-mass spectrometry.

(Brady and Fishman, 1979).

Gangliosides are a crowded family of compounds differing in both the oligosaccharide and ceramide compositions; species are known that have almost identical ceramide composition with different oligosaccharide moieties, and species with the same oligosaccharide but varying ceramide composition. All these molecular species are mixed together in different proportions in organs and tissues and so they occur in crude extracts obtained from the same materials. Of course the study of the functional implications of gangliosides implies verification of the specific role played by individual gangliosides in well defined membrane microdomains. When pursuing these objectives it is necessary to use pure and well characterized ganglioside species homogeneous in both their oligosaccharide and ceramide portions, and to possess ganglioside derivatives carrying proper probes or labels (fluorescent, paramagnetic, radioactive) as tools for biochemical and biophysical investigations.

For preparative purposes gangliosides are generally extracted from brains that can be obtained fresh and in large amounts at the slaughterhouse (bovine, porcine) (Tettamanti et al., 1973; Sonnino et al., 1978). The separation of gangliosides into individual entities begins with techniques (thin layer and column chromatography, including high performance column chromatography) that resolve gangliosides on the basis of their oligosaccharide moieties (Ledeen and Yu, 1978; Ghidoni et al., 1980; Gazzotti et al., 1985). The compounds thus obtained are heterogeneous in the ceramide moiety. The molecular species homogeneous in the ceramide moiety, with special reference to the long chain base which displays the greatest differences, can be separated from each other by reversed phase high performance liquid chromatography (Sonnino et al., 1984).

The procedures used to prepare homogeneous gangliosides have been described (Gazzotti et al., 1984a; 1984b; Sonnino et al., 1985) and will not be dealt with in the present report. It should be emphasized that application of these procedures enables to prepare relatively large amounts of the ganglioside species that are more abundant in the starting material (for instance ganglioside GM1, GD1a, GD1b, and GT1b, as the species which carry unsaturated long chain bases). Minor gangliosides, with regards to the oligosaccharide and ceramide portions, constitute a real problem for the high cost and huge labour required for their preparation. As well, the production of ganglioside derivatives faces the preliminary problem of the availability of sufficient amounts of the starting natural compound.

The technology of ganglioside chemical manipulation and derivatization underwent recently a tremendous improvement, which opened new trends for ganglioside production. Aim of this report is to summarize the principal aspects of this development with special reference to the preparation of: (a) gangliosides that occur in nature in very minor amounts; (b) gangliosides that carry chemical modifications in the oligosaccharide or ceramide portions of their molecule; (c) gangliosides that contain special "probes" (fluorescent, paramagnetic); (d) gangliosides that carry [^{3}H] or [^{14}C] isotopes in various parts of their molecule. The experimental conditions for the various chemical reactions and for purification of the obtained products are described in detail and the final yield of each compound is given. The yield of the different reactions and procedures for purification was calculated after thin layer chromatographic separation and densitometric quantification of the reaction products (Chigorno et al., 1982).

HYDROGENATION OF GANGLIOSIDES

The ceramide portion of gangliosides is constituted of a long chain base (l.c.b.) carboamidically linked to a fatty acid. In most gangliosides the l.c.b.'s (Fig. 1) are 18 or 20 carbon atom amino alcohols, 90% of which containing a double bond at position 4 (2-amino-1,3-dihydroxy-octadec-4t-ene, C18 - sphingosine, and 2-amino-1,3-dihydroxy-eicos-4t-ene, C20-sphingosine). The fatty acids are represented for more than 98% by a saturated species (Sonnino et al., 1984). Therefore isolation of the molecular species of gangliosides that contain saturated l.c.b. is extremely difficult. On the other hand, the presence of a C4-C5 trans double bond in l.c.b. promotes condensation of ceramides in a closely packed way owing to formation of intermolecular hydrogen bonds. This, from one side, suggests the usefulness of studies aimed at understanding the importance of the presence or absence of the double bond and, from the other, imposes the availability of both the unsaturated and saturated species of gangliosides for comparative investigations. The preparation of large amounts of saturated ganglioside species is based on chemical hydrogenation of the l.c.b. double bond starting from the most abundant unsaturated species (Sonnino et al., 1984) (Fig. 2). The procedure for hydrogenation of gangliosides at the level of the l.c.b. double bond, already applied to gangliosides GM2, GM1, Fuc-GM1, GD1a, GD1b, Fuc-GD1b, GT1b and GQ1b, is the following. Ganglioside is dissolved in ethanol-water, 7:3, by vol, (1 mg/ml) and submitted to catalytic hydrogenation in the presence of platinum dioxide (0.1-0.5 mg/mg of ganglioside), at room temperature and for 36 h under continuous stirring. The catalyst is then removed by filtration (Acro LC 25, 0.2 μm, Gelman, U.S.A.) and the solution evaporated to dryness, suspended in redistilled water, dialyzed against redistilled water and lyophilized. No further purification is necessary and the yield of the reaction is very close to 100%.

Verification of the occurred hydrogenation can be accomplished by GLC-MS. Figure 3 shows the major GLC-MS fragmentation pattern of the l.c.b.'s, released from gangliosides before and after hydrogenation, and analyzed as the N-acetyl-TMS derivatives. Fragments A, B, C and E provided by the l.c.b. with the same number of carbon atoms, (C18 or C20) differ from one ganglioside species to the other by only two units unequivocally indicating the addition of two hydrogens per molecule. NMR spectroscopy performed on GM1 ganglioside demonstrates that the only difference between the two species, following treatment, resides in the olefinic protons H-4 and H-5 that resonate at 5.32-5.43 and 5.53-5.58 ppm, respectively. In the saturated species the

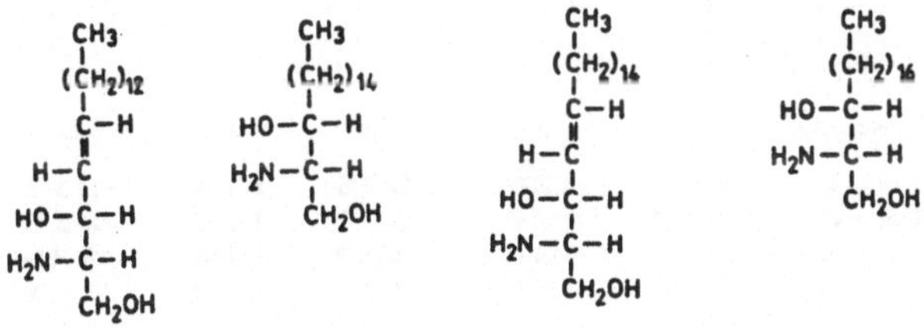

Figure 1. Chemical structure of the major long chain bases present in gangliosides from central nervous system.

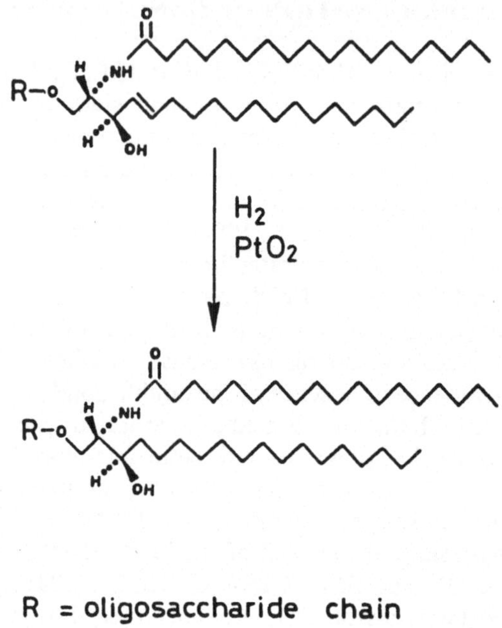

Figure 2. A general scheme for preparing gangliosides with saturated long chain base.

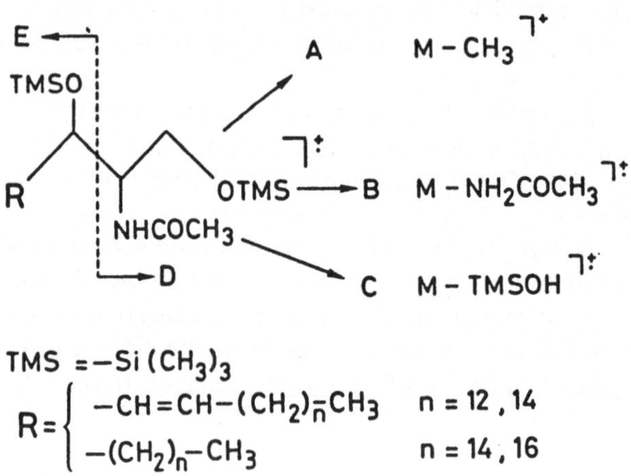

l.c.b.	A	B	C	D	E
18.1	470	426	395	174	311
18:0	472	428	397	174	313
20:1	498	454	423	174	339
20:0	500	456	425	174	341

Figure 3. Main MS fragmentation pattern of long chain bases, released from gangliosides by methanolysis and analyzed as N-acetyl, trimethylsilyl ether. Each fragmentation refers to "erythro" and/or "threo" species.

olefinic signals disappear (Fig. 4, Table 1). This confirms GLC-MS data and provides evidence for the selectivity of process.* NMR analysis provided additional and interesting evidence, particularly concerning the α-carbonyl methylene protons. These protons are diastereotopic, the nearest chiral centers being l.c.b. C-2 and C-3 carbons; therefore some difference of their chemical shift would be expected. On the contrary they are isochronous for all the unsaturated molecular species, but not for the saturated ones ($\Delta\delta$ = 0.06 ppm, J(8a,8b) = 13.5 Hz; see also Table 1). This behaviour may suggest that removal of the double bond is followed by either changes in the shielding long-range effects or by reorganization of the lateral interactions between the two carbon tails of ceramide.

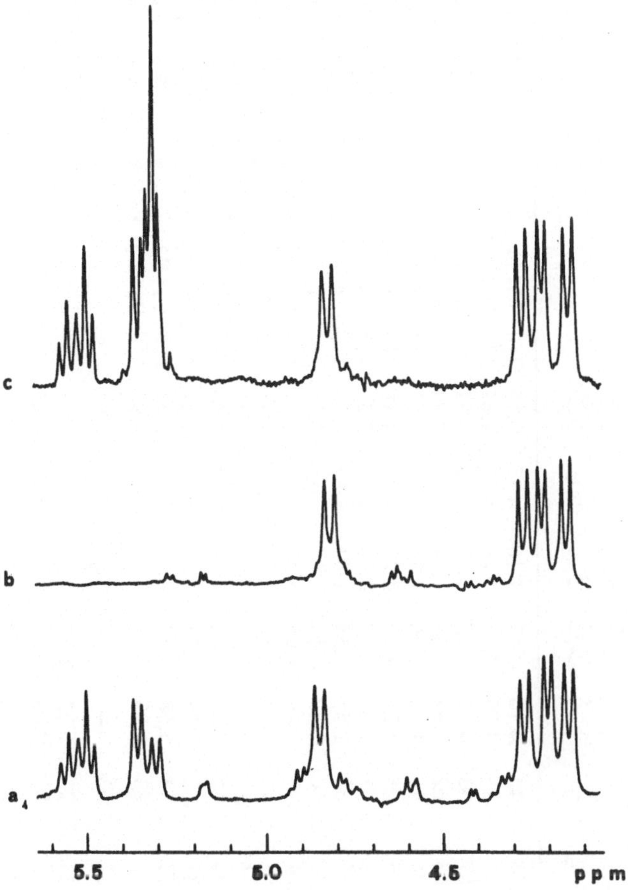

Figure 4. 300 MHz proton NMR spectra, 4-5.6 ppm, of GM1(e18:1/18:0) (a), GM1(e18:0/18:0) (b), GM1(e18:1/18:1) (c), in DMSO-d_6 at 25°C.

* Hydrogenation did also occur at the level of the double bonds carried by the fatty acid moiety. Owing to the paucity of these bonds (an average of 2% unsaturated fatty acid out of the total fatty acid content) this event is practically negligible.

Table 1. *Proton NMR Data*[a] *for GM1 and Several Derivatives*[b]

	CHEMICAL SHIFTS								COUPLING CONSTANTS						
	GM1 natural	deAc-GM1	deAc-de-Acyl-GM1	GM1 (acetyl)	GM1 (e18:0/18:0)	GM1[f] (e18:1/18:1)	GM1 inner ester		GM1 natural	deAc-GM1	deAc-de-Acyl-GM1	GM1 (acetyl)	GM1 (e18:0/18:0)	GM1[f] (e18:1/18:1)	GM1 inner ester
H-1(I)	4.15	4.15	4.16	4.15	4.16	4.15	4.15	J(1,2)(I)	7.7	7.7	7.8	7.9	7.9	8.0	7.7
H-1(II)	4.26	4.25	4.25	4.28	4.28	4.28	4.62	J(1,2)(II)	7.8	7.9	7.8	8.0	8.0	7.8	7.5
H-1(III)	4.89	4.83	4.81	4.82	4.83	4.83	4.81	J(1,2)(III)	8.7	8.8	8.9	8.9	8.5	8.7	8.3
N-H(III)	7.63	7.62	7.64	7.66	7.67	7.66	7.34	J(2,NH)(III)	9.2	8.8	9.5	9.5	9.0	9.0	9.2
$COCH_3$(III)	1.76	1.76	1.76	1.76	1.76	1.76	1.87	J(1,2)(IV)	7.0	7.1	7.2	7.2	7.2	7.0	7.5
H-1(IV)	4.19	4.25	4.25	4.22	4.23	4.22	4.16	J(3e,3a)(A)	11.5	12.0	12.0	12.0	12.0	12.0	13.7
H-3e(A)	2.53	2.47	2.42	2.55	2.54	2.55	2.37	J(3e,4)(A)	5.0	5.0	4.7	4.8	4.8	4.9	5.5
H-3a(A)	1.63	1.58	1.55	1.64	1.64	1.64	1.59	J(3a,4)(A)	11.5	12.0	11.5	12.0	12.0	12.0	10.0
H-5(A)	3.36	2.50	2.49	d	d	d	3.75	J(4,5)(A)	10.0	10.0	9.5	d	d	d	10.2
N-H(A)	8.04	—	—	8.05	8.03	8.07	8.06	J(5,6)(A)	10.0	10.0	9.5	d	d	d	10.6
$COCH_3$(A)	1.87	—	—	1.88	1.88	1.88	1.88	J(5,NH)(A)	8.3	—	—	9.5	8.5	8.0	8.5
H-2(R)	3.76[c]	d	2.75	d	d	d	d	J(2,NH)(R)	9.2	8.8	—	8.8	9.0	8.9	8.9
N-H(R)	7.49	7.51	—	7.60	7.59	7.52	7.50	J(3,4)(R)	7.0	7.0	6.8	7.0	d	7.0	7.0
H-4(R)	5.34	5.32	5.43	5.37	d	5.34	5.34	J(4,5)(R)	15.6	15.0	15.6	15.2	d	15.0	15.2
H-5(R)	5.53	5.53	5.58	5.54	d	5.53	5.53	J(5,6)(R)	6.5	6.5	6.5	6.5	d	6.5	6.5
H-6(R)	1.92	1.93	1.99	1.95	d	1.97	1.93	J(8,9)(R)	7.5	7.5	—	—	7.0	7.5	7.5
H-8(R)	2.02	2.02	—	—	2.03[e] 2.09	2.02	2.01								
H-10(R)	1.23	1.23	1.23	1.23	1.23	1.23	1.23								
H-14(R)	0.85	0.85	0.85	0.85	0.85	0.85	0.85								
$COCH_3$(R)	—	—	—	1.79	—	—	—								

a) Chemical shifts in ppm from internal TMS; coupling constants in Hz. Solvent: DMSO-d_6 to determine data on amide protons, and DMSO-d_6 + 2% D_2O for the other ones. b) The ganglioside numbering is that proposed by Koerner et al., 1983: ceramide: R, glucose: I, internal galactose: II, N-acetylgalactosamine: III, external galactose: IV, N-acetylneuraminic acid: A. c) From Koerner et al., 1983. d) Not detected. e) δ (H-8a) = 2.03, δ (H-8b) = 2.09; J (8a,8b) = 13.5 Hz. f) Olefinic protons of oleic acid (proton numbering refers to the fatty acid): δ(H-9) = δ(H-10) = 5.32; J(8,9) = J(10,11) = 4.5 Hz.

PREPARATION OF GANGLIOSIDES WITH THE L.C.B. IN THE THREO-CONFIGURATION

The l.c.b.'s, show 4 different chemical configurations because of the substituent disposition at carbon 2 and 3 (Fig. 5). Natural gangliosides were demonstrated to contain only the l.c.b. with the configuration 2S,3R = 3D(+)erythro (Sonnino et al., 1984) and, as consequence, they are coded "erythro"-gangliosides. The erythro configuration seems to allow the formation of an hydrogen bond between the hydroxyl group at position 3 of l.c.b. (donor) and the amide carbonyl group (acceptor) (Pascher, 1976) of the fatty acid residue. The availability of gangliosides with the "threo" configuration of the l.c.b. would greatly help understanding the importance of hydrogen bond formation.

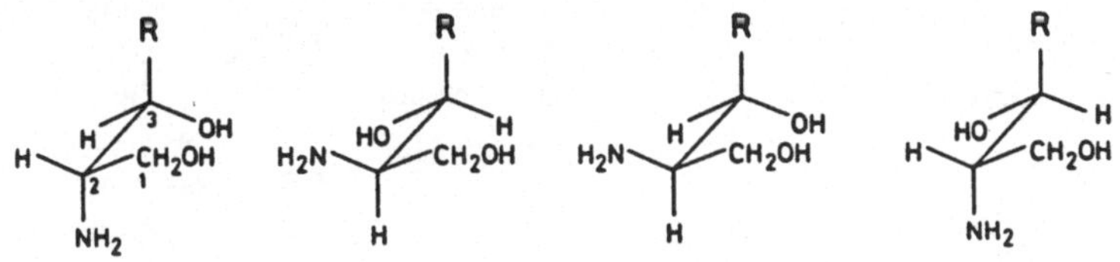

2S,3R	2R,3S	2R,3R	2S,3S
3D(+)erythro	3L(-)erythro	3D(+)threo	3L(-)threo

$$R = \begin{cases} -CH{=}CH-(CH_2)_n-CH_3 & n = 12, 14 \\ -(CH_2)_n-CH_3 & n = 14, 16 \end{cases}$$

Figure 5. Absolute configuration of natural and unnatural long chain bases.

Gangliosides containing the threo (2S,3S = 3L(—)threo) form of l.c.b. are prepared by the dicyano-dichlorobenzoquinone (DDQ/NaBH4 method; Fig. 6). The method, already applied to gangliosides GM2, GM1, Fuc-GM1 and GD1a, is the following (Ghidoni et al., 1981). A ganglioside solution in chloroform-methanol 2:1, by vol (2 mg/ml) is mixed with an equal volume of Triton X-100 solution in the same solvent (60 mg/ml). The solvent is evaporated to dryness under vacuum at 37°C. The residue is carefully dissolved in a sodium-dehydrated toluene solution of DDQ (36 mg/ml) in order to get a DDQ/ganglioside ratio of 18/1 (W/W). The mixture is allowed to react at 37°C for 40 h under continuous stirring in a screw-capped glass container, and evaporated to dryness at 37°C. The dark brown residue is suspended in acetone (0.5 ml/mg of ganglioside), briefly sonicated in an ultrasonic water bath, centrifuged at 3000 rpm for 10 min, the supernatant, containing Triton X-100 and DDQ, being discarded. This treatment is repeated four times and at the end a white precipitated is obtained containing the 3-keto-derivative of ganglioside. The oxidized ganglioside is further purified on a silica gel 100 column (3ml sedimented gel/mg starting ganglioside), equilibrated and eluted with the mixture chloroform-methanolwater, at the volume ratio of 60:35:3 for GM2, and 60:35:5 for GM1, Fuc-GM1 and GD1a. The oxidation reaction occurs only on the ganglioside species containing unsaturated l.c.b.'s The unreacted saturated species, present in the starting material, are removed during the purification of 3-keto ganglioside (Ghidoni et al., 1981).

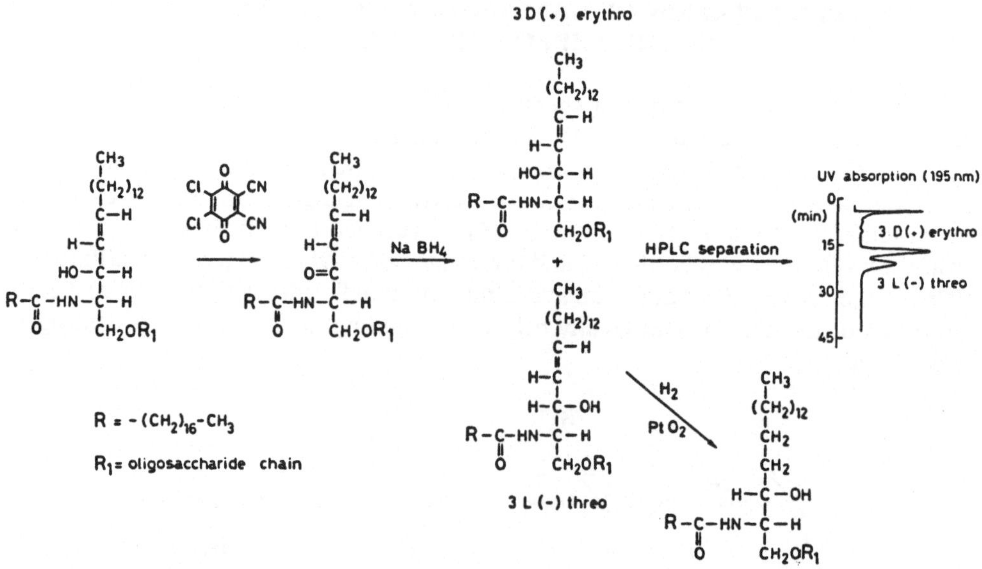

Figure 6. A general scheme for preparing gangliosides that contain the unnatural long chain base with the configuration 3L(-)threo, starting from natural compounds.

The purified 3-keto-ganglioside is dissolved in propan-1-ol water, 8:2, by vol, (0.25 mg/ml) and treated for 1 h at room temperature with solid $NaBH_4$ (3 mg/mg of ganglioside) with continuous stirring, in order to reduce the compound to the previous alcohol. At the end of the reaction the excess of $NaBH_4$ is destroyed by adding a few drops of 0.1 N acetic acid, the pH being adjusted to about 6. The reaction mixture is then evaporated to dryness at 37°C under vacuum, the residue dissolved in a small amount of water, dialyżed overnight against redistilled water and lyophilized. The residue, which contains a mixture of threo and erythro derivatives, is dissolved in redistilled water and submitted to reversed phase HPLC (Sonnino et al., 1984; Gazzotti et al., 1984a.; 1984b). This procedure enables to separate the ganglioside species containing l.c.b.'s with threo configuration (C18 and C20) from those having the erythro configuration (C18 and C20). The threo — ganglioside species that contain C18 or C20 saturated l.c.b.'s can be prepared from the corresponding unsaturated species by catalytic hydrogenation (see above).

The oxidation reaction that leads to formation of the 3-keto-derivative, proceeds with a 70% yield. A further loss of 3-keto ganglioside occurs during the purification on silica gel column chromatography. The reduction reaction using $NaBH_4$ is not stereospecific and provides a mixture of threo and erythro-gangliosides in which the two isomers are present in a molar ratio of about 7:13. Therefore the final yield of threo-gangliosides (the species containing C18 l.c.b. and the species containing C20 l.c.b.) is about 22%.

Verification of the occurrence of oxidation at carbon C-3 of long chain base can be obtained by I.R. analysis. As shown in Figure 7, which relates to ganglioside GM1,

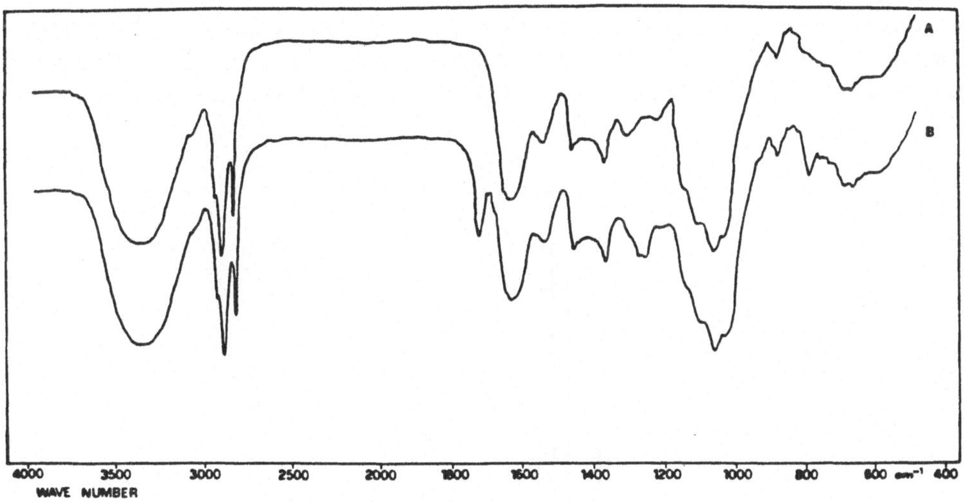

Figure 7. Infrared spectra of KBr solid suspension of GM1(A) and 3-keto-GM1(B).

the infra red spectra of the 3-keto derivative is characterized by a sharp peak at 1700 cm^{-1}, due to the absorption of the allylic ketone, and a broad band at 1250 cm^{-1} typical of carbonyl groups. The spectra of GM1 before treatment and of the corresponding 3-keto derivative after reduction were exactly the same, and in both of them the above peaks at 1700 and 1250 cm^{-1} were missing.

The erythro-threo configuration occurring in the ganglioside species obtained by reduction of the 3-keto-derivative can be assessed by GLC-MS. To this purpose l.c.b.'s must be released from gangliosides by methanolysis. The acidic conditions under which methanolysis is performed, are capable per se to produce the hybrid of resonance carbocation (-CH = CH-CH- ⟷ -CH-CH = CH-)$^+$ and, as a consequence, a partial inversion of configuration. Of course this does not occur when saturated l.c.b.'s are submitted to methanolysis, since carbocation is not formed. Therefore in order to establish the existence of a threo configuration in the starting ganglioside it is necessary to submit the sample first to hydrogenation (see above) and then to methanolysis and GLS-MS analysis. Figure 8 shows the GLC profile of the N-acetyl-TMS l.c.b. released from the GM1 molecular species that contain a C18:1 l.c.b. in the threo configuration, before and after hydrogenation. Before hydrogenation two peaks are recorded, corresponding to the threo and erythro l.b.c., respectively, indicating that some artifactual inversion took place. Both peaks show the same MS fragmentation pattern indicating that they are isomers. After hydrogenation only the threo l.c.b. appears certifying for the threo configuration in the starting ganglioside.

SYNTHESIS OF GM1 GANGLIOSIDE INNER ESTER

Gangliosides can be present in nature as inner esters, or lactones, (Gross et al., 1980), generally in low amounts. In these compounds the negative charges are partially or totally removed, owing to esterification of sialic acid(s) with hydroxyl group(s) of the same ganglioside molecule. Recently a lactonic form of ganglioside GD1b in which carboxyl group of the external sialic acid esterifies one hydroxyl group of the internal

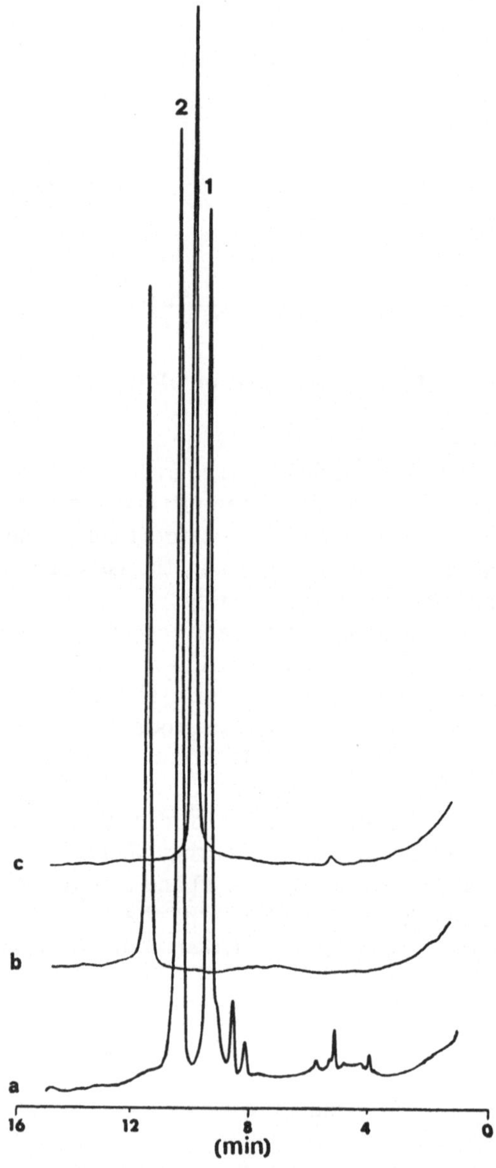

Figure 8. Gas liquid chromatographic separation of N-acetyl, trimethylsilyl ether derivatives of long chain bases obtained from GM1 containing threo C18:1 long chain base (a) (peak 1 corresponds to the threo derivative and peak 2 to the erythro one), from GM1 containing erythro C18:0 long chain base (b), and from GM1 containing threo C18:1 long chain base after hydrogenation (c). GLC was carried out isotermically at 240°C on 25 m SE-54 capillary column.

sialic acid has been isolated from adult human brain (Riboni et al., 1986). Formation and breakdown of ganglioside inner esters (or lactones) might be a functionally important process and availability of ganglioside lactones would greatly help verifying this point. Since preparation of ganglioside lactones from natural sources is not presently proposable, a synthetic process for ganglioside lactonization has been developed. This was applied to ganglioside GM1 and led to formation of a GM1 derivative, in which sialic acid esterifies the hydroxyl group in position 2 of the inner galactose residue (Fig. 9).

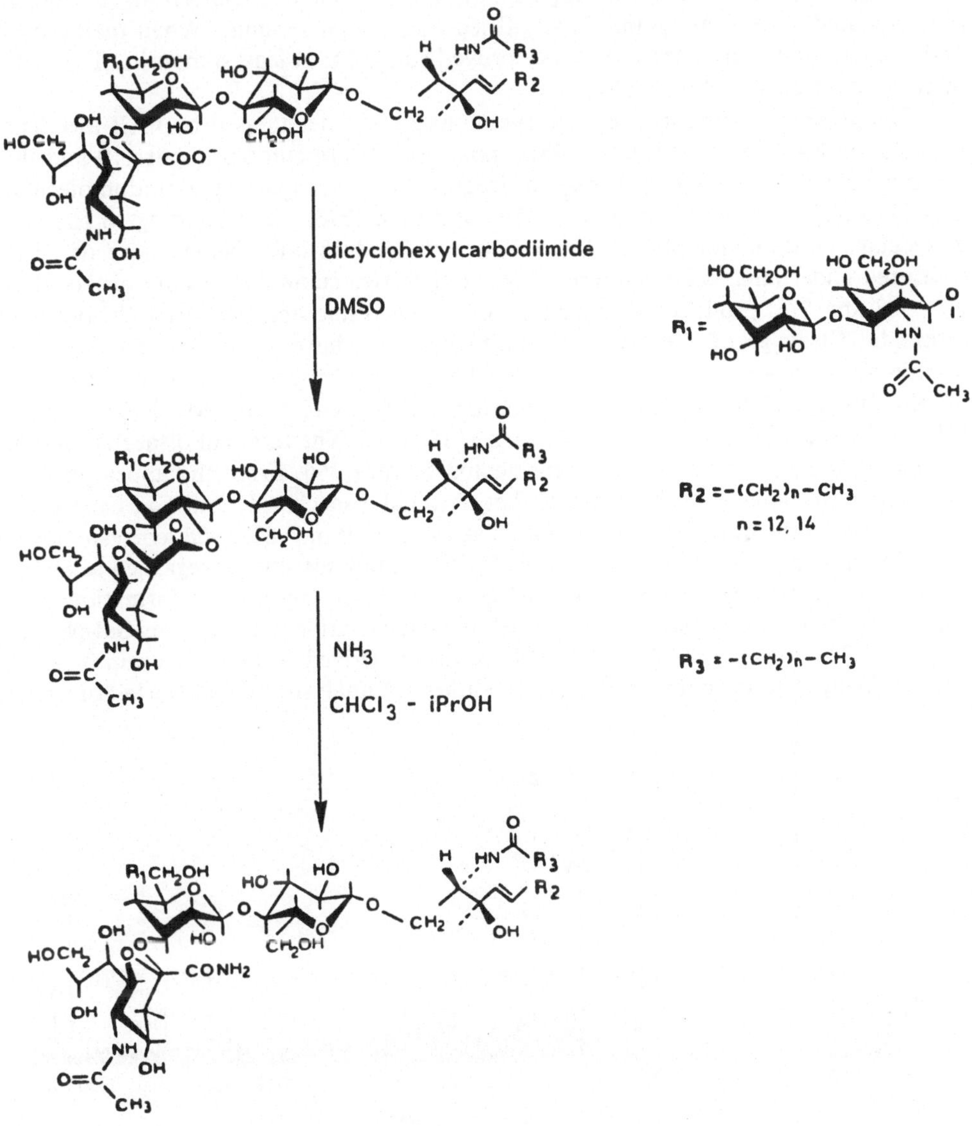

Figure 9. A general scheme for preparing GM1 inner ester and GM1 amide.

GM1, prepared as the sodium salt (Corti et al., 1981), is dissolved in dehydrated dimethylsulfoxide (100 mg/ml), and converted into the free acid form by passing through a column filled with anhydrous styrene-type resin AG 50 (Bio-Rad, Richmond, USA), H^+ form, (1 mg/4 mg of ganglioside) and eluted with dimethylsulfoxide. The eluted mixture is directly treated with dicyclohexylcarbodiimide (14.2 mg/100 mg of ganglioside) at room temperature for 1 h, under continuous stirring. The dicyclohexylurea, formed during the reaction, is removed by filtration and the solution is treated with acetone (5 mgl/100 mg of ganglioside) in order to precipitate the GM1 inner ester. This is dissolved in chloroform-propan-2-ol, 1:1, by vol, (0.5 ml/100 mg of GM1), reprecipitated with acetone and finally dried under high vacuum. When maintained under these conditions GM1 inner ester proved stable for at least 6 months. The yield of the full process is about 95%.

The presence of an inner ester in the molecule can be assessed by FAB-MS. The positive ion FAB-MS spectrum of GM1 inner ester is presented in Figure 10 and the scheme indicating the major pathways of fragmentation in Figure 11. Pseudomolecular ions $[M+Na]^+$ are observed at m/z 1550 and m/z 1578, corresponding to the two molecular species containing C18 and C20 l.c.b. and 18 units below those of GM1 measured under identical conditions. Accordingly, two ceramide ions are observed at m/z 548 and 576, produced by cleavage of the glycosidic bond between glucose and ceramide. Cleavage of the glycosidic bond between the glucose and the galactose residue furnishes ions at m/z 710 and 738. The next higher sequence ions are produced by elimination of the terminal disaccharide Gal-GalNAc from $[M+Na]^+$ at m/z 1185, 1213 and of the terminal Gal at m/z 1388, 1416. The terminal disaccharide ion Gal-GalNAc on the other hand is represented by m/z 366. From the appearance of these latter five ions it can be deduced that terminal disaccharide unit Gal-GalNAc is not involved in the inner ester formation. The ion pair at m/z 878, 906, corresponding formally to [Gal-Glc-Cer plus Na^+ minus $H_2O]^+$, could possibly be regarded as an indication for the involvement of the inner Gal residue in the lactone ring formation. The ion at m/z 656 could formally be attributed to a trisaccharide ion [GalNAc-Gal-NeuAc lactone$]^+$. The ions at m/z 310, 292, 264 are surely derived from the neuraminic acid residue, likely indicating that under the conditions of analysis, part of the lactone ring

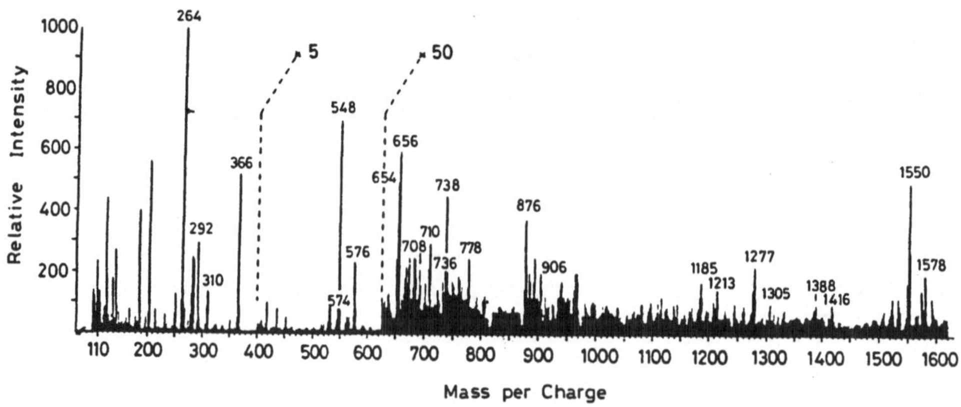

Figure 10. Postive ion FAB-MS of GM1 ganglioside inner ester.

is cleaved. The results of negative ion FAB-MS are in agreement with the above findings. In particular, sequence ions derived from both the nonreducing terminal and the ceramide residue support the existence of a lactone ring between NeuAc and the inner galactose as indicated in the scheme of fragmentation (Fig. 11).

The identification of the hydroxyl group involved in the ester linkage, has been achieved by ^{1}H-NMR analysis of the underivatized product (Fig. 12). Comparison of the chemical shifts of GM1 and GM1 inner ester shows that the proton at position 2 of internal galactose is dramatically deshielded (+1.41 ppm) in the latter compound.

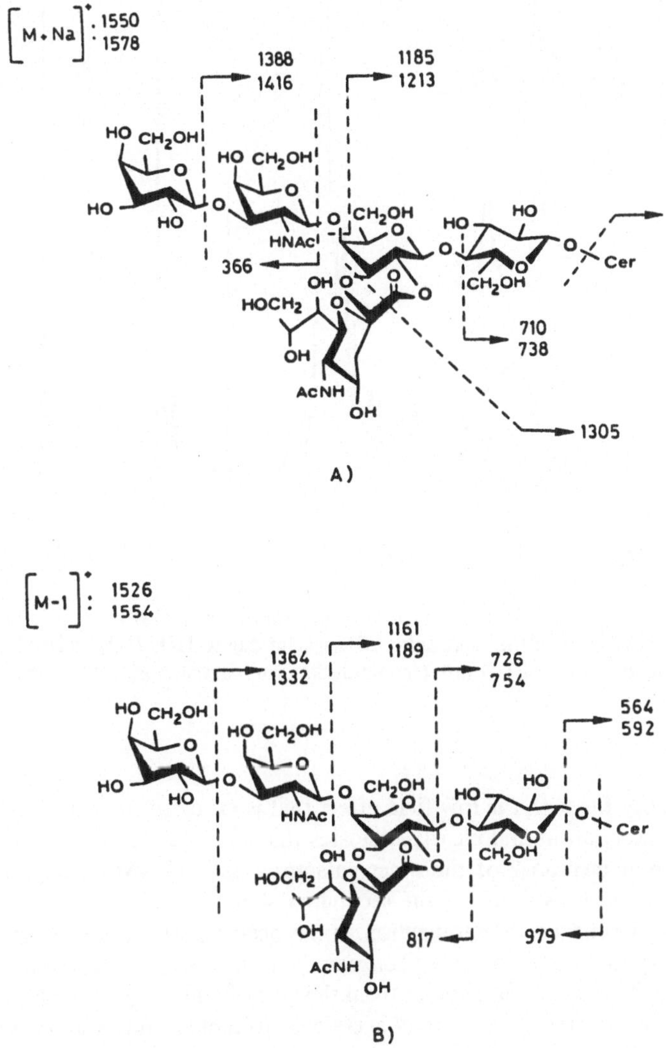

Figure 11. Positive (a) and negative (b) FAB-MS fragmentation pattern of GM1 ganglioside inner ester.

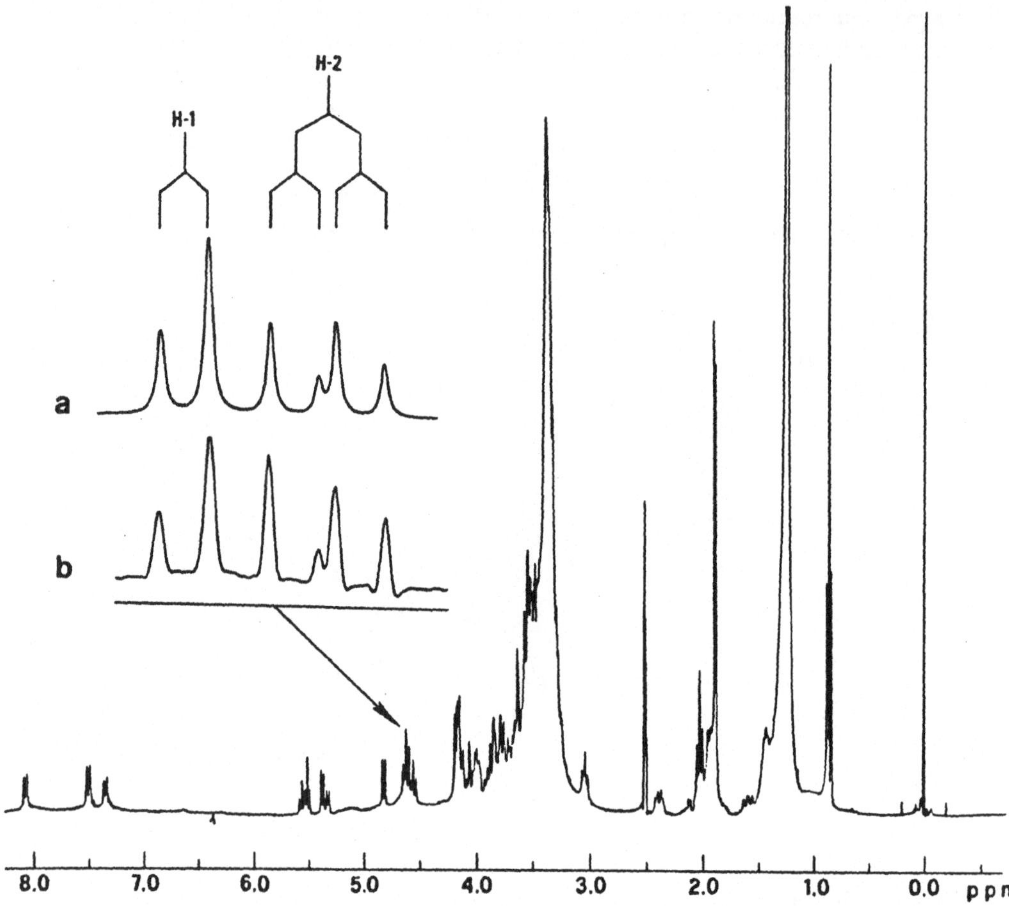

Figure 12. 300 MHz proton NMR spectrum of GM1 inner ester in DMSO-d_6 solution at 23°C. The insert shows: AB part of the three spin ABX system due to H-1, H-2, and H-3 protons of the internal galactose unit (b); computer simulation for comparison (a). Tetramethylsilane (TMS) reference.

This finding fairly fits with the low field effect that is expected to occur on secondary alcohols upon acetylation and clearly indicates that the ester formation involves the hydroxyl group at positon 2 of the inner galactose unit. ^{1}H-NMR analysis provides additional information concerning the secondary structure changes occurring in the molecule as a consequence of lactonization. It has been reported (Koerner et al., 1983) that a through space interaction exists between the NeuAc and GalNAc units of GM1. Evidence for such an interaction comes from the lack of additivity of the glycosidation shifts experienced by resonances of N-acetylgalactosamine and N-acetylneuraminic acid residues. In the case of GM1 inner ester the behaviour of the protons of GalNac, which is not involved in ester linkage formation, is different from that of the same protons in GM1. Particularly relevant are the upfield shift by 0.42 and 0.29 ppm of proton at position 2 and of the amide proton respectively, and the downfield shift by 0.14 ppm of acetoamide proton, which indicate conformational differences between GM1 and GM1 inner ester.

Table 2 reports the presently available chemical shifts of the protons of sialic acid residue for GM1 and GM1 lactone and for GM3 and GM3 lactone (Yu et al., 1985) together with the chemical shift differences ($\Delta\delta$) which represent the long range glycosidation shifts at protons of sialic acid by adding the disaccharide unit Gal-GalNAc to the Gal residue of GM3 and GM3 lactone. Comparison of the $\Delta\delta$ values shows that the substitution effects are completely different in ganglioside lactones and gangliosides. This is not surprising since lactonization gives rise to a highly rigid system with three fused rings which hinders any conformational flexibility of the NeuAc residue.

Table 2. *Comparison of the chemical shifts ($\Delta\delta$)[a] for protons of N-acetylneuraminic acid residues of gangliosides GM1 and GM3 and their lactones*

	GM1	GM3[b]	$\Delta\delta$	GM1 lactone	GM3[c] lactone	$\Delta\delta$
H-3e	2.53	2.75	−0.22	2.73	2.36	+0.01
H-3a	1.63	1.36	+0.27	1.59	1.51	+0.08
H-4	3.74[b]	3.55	+0.19	4.11	4.12	−0.01
H-5	3.35	3.46	−0.09	3.75	3.54	+0.21

a) differences between the chemical shifts of GM1 and GM3 and of GM1 lactone and GM3 lactone
b) from Koerner et al., 1983
c) from Yu et al., 1985

PREPARATION OF GM1-AMIDE

The amide derivative of GM1 is obtained by ammoniolysis of GM1 inner ester (Fig. 9). This reaction is practically quantitative, no purification being required at the end of the reaction. GM1 inner ester is dissolved in carefully dehydrated chloroform-propan-2-ol, 1:1, by vol, (1 mg/ml), and ammonia vapours, exhaustively dessiccated through soda lime, are bubbled at room temperature for one hour under continuous stirring. The mixture is then dried, the residue dissolved in water and lyophilized against redistilled water. Assessment of sialic acid amidation is accomplished by GLC-MS (Figs. 13 and 14), after release of sialic acid from GM1-amide and proper derivatization. The fragments A, C, D, E and F all containing the C1 group, differ from the corresponding ones obtained from GM1 by 15 units; conversely, the fragments B and G, which do not contain the C1 group, display the same values for both GM1 and GM1-amide. 15 mass units correspond to the difference between [-CO-O-CH_3] (59 mass units), which is present in the sialic acid released from GM1 and [-CO-NH_2] (44 mass units), present in sialic acid derived from GM1-amide.

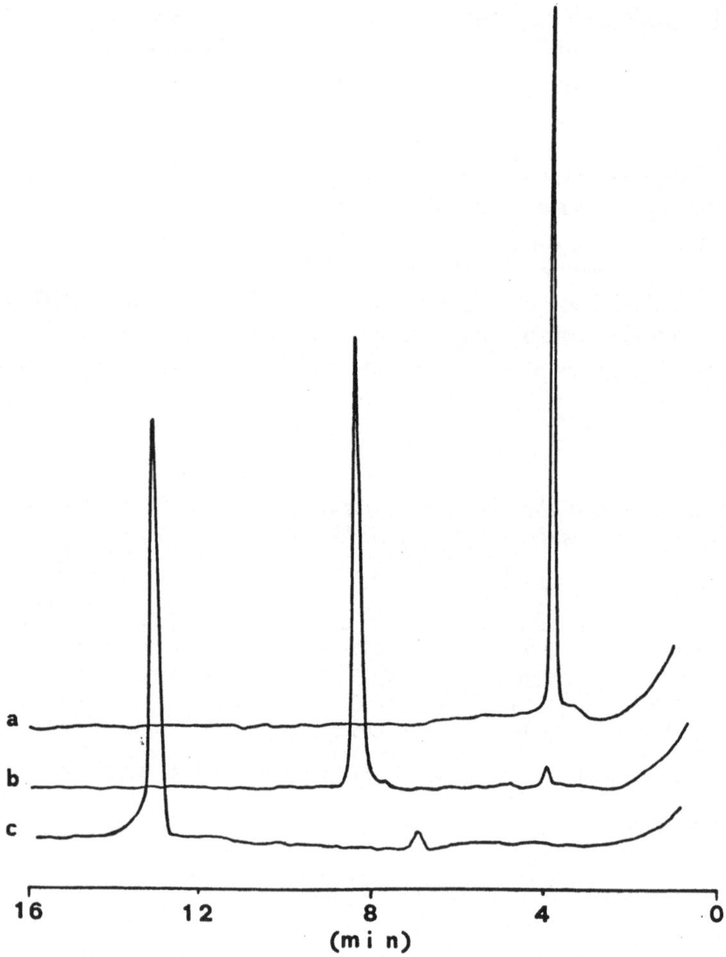

Figure 13. Gas liquid chromatographic separation of the trimethylsilyl derivatives of sialic acid residues released by mild acid methanolysis from deAc-GM1, (a), GM1 (b), and GM1 amide (c). Separation has been performed on a 2 m OV-17 glass column, at 230°C.

PREPARATION OF GM1 GANGLIOSIDE DEACETYLATED AT THE LEVEL OF SIALIC ACID, (deAc-GM1) AND OF GM1 DEACETYLATED AT THE LEVEL OF SIALIC ACID AND DEAÇYLATED AT LEVEL OF THE CERAMIDE MOIETY (deAc-deAcyl-GM1)

DeAc-GM1 and deAc-deAcyl-GM1 can be used as model compounds or as intermediates for more complex processes. They are prepared by alkaline hydrolysis of GM1 ganglioside in the presence of tetramethylammonium hydroxide (Fig. 15) (Sonnino et al., 1985).

GM1 ganglioside is dissolved in butan-1-ol and mixed at 100°C with 10M aqueous

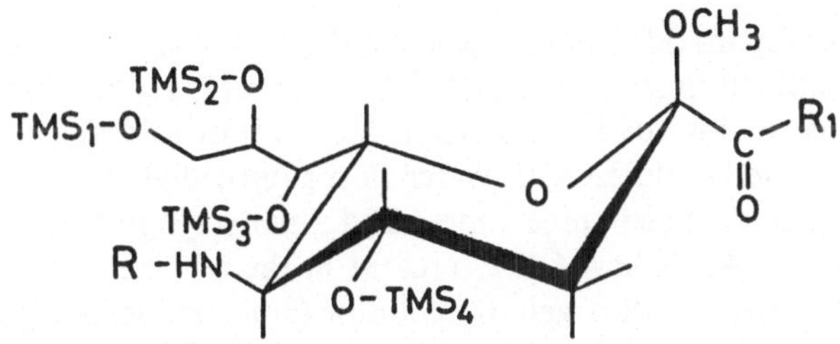

$R = -H, -COCH_3$ $R_1 = -OCH_3, NH_2$

$TMS = -Si(CH_3)_3$

	m/e	NeuSAc	NeuSAc-NH2	Neu
	M^+	625	510	583
A	$M-[-CH_3]^{\bullet}$	610	595	568
B	$M-[-\dot{C}OR_1]$	566	566	524
C	$M-[-\dot{C}HOTMS_2 - CH_2OTMS_1]$	420	405	378
D	$M-[-\dot{C}H_2OTMS_1,\ TMS_3OH,\ TMS_4OH]$	342	327	299
E	$C-[CH_3OH,\ TMS_4OH]$	298	283	256
F	$M-[-\dot{C}HOTMS_3 - CHOTMS_2 - CH_2OTMS_1,\ -NHR]$	259	244	259
G	$CH{=}\dot{N}H{-}R - CH\ OTMS_4$	173	173	131

Figure 14. Main MS fragmentation pattern of trimethylsilyl derivatives of sialic acid released from gangliosides by mild acid methanolysis.

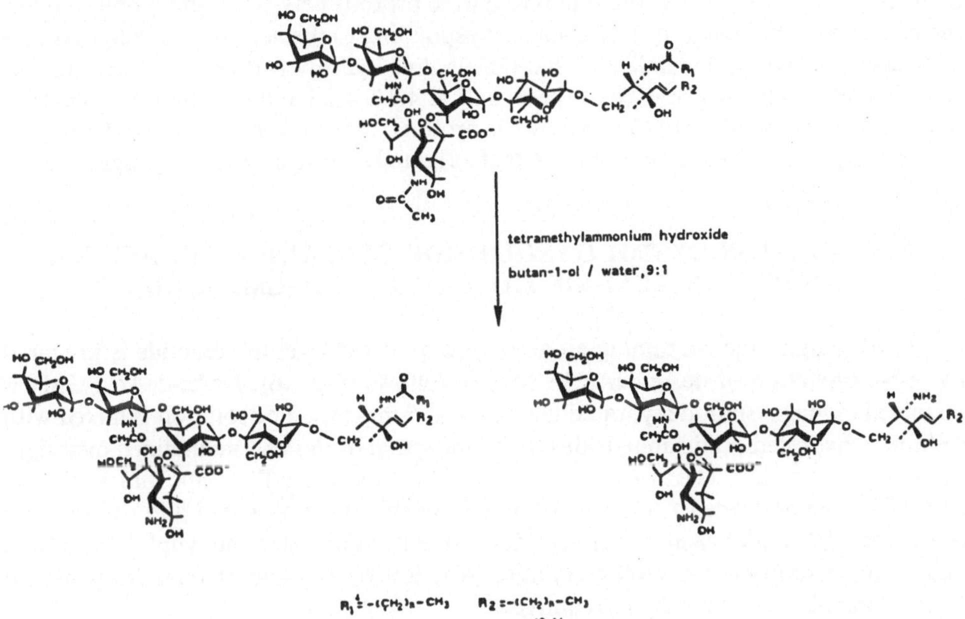

Figure 15. A general scheme representing alkaline hydrolysis of GM1 gangliosides using tetramethylammonium hydroxide.

tetramethylammoniumhydroxide in order to have the following ganglioside final concentrations: ganglioside 10 mg/ml; base, 1M. The mixture, after refluxing at 100°C and under continuous stirring for 13 h, is evaporated almost to dryness under vacuum at 50°C. The wet residue is diluted with water (5 mg/ml), dialyzed at 4°C for 1 day against redistilled water changed three times, and finally lyophilized. The two products, deAc-GM1 and deAc-deAcyl-GM1, present in the residue, are separated from each other and purified on a silica gel 100 column (3ml sedimented gel/mg starting ganglioside) previously equilibrated and eluted with methanol:butan-1-ol:water, 2:2:1, by vol. Under these conditions the yield is about 42% for both deAc-GM1 and deAc-deAcyl-GM1.

The occurrence of deacetylation at the level of the sialic acid residue can be assessed by GLC-MS analysis. As shown in Figure 14 the MS spectra of sialic acid from deAc-GM1 (or deAc-deAcyl-GM1) displays fragments A, B, C, D, E and G, that possess masses 42 units lower than the corresponding fragments of the sialic acid obtained from GM1. This reflects the difference between [-$NHCOCH_3$] (58 mass units) of N-acetylneuraminic acid and [-NH_2] (16 mass units) of neuraminic acid. Confirmation and further details are provided by ^{1}H-NMR analysis. The acetamide methyl protons and the amide proton of sialic acid, that in GM1 occur at 1.87, and 8.04 ppm, respectively, completely disappear in deAc-and deAc-deAcyl-GM1. Moreover the lower intensity of the signal at 0.85 ppm corresponding to alkyl methyl protons in deAc-deAcyl-GM1, as compared to that in GM1, indicates that only one long chain is present in the former molecule. In addition deacylation induces a strong upfield shift (0.86-1.01 ppm) of the CH groups to the nitrogen atom, which rises up from the strongly overlapped region of the core oligosaccharide protons between 3 and 4 ppm. Finally the presence of the signals at 1.76 ppm, corresponding to the acetamide methyl protons of N-acetylgalactosamine indicates that the alkaline hydrolysis does not affect this saccharide unit, while the presence of the signals at 4.16, 4.25 and 4.81 ppm (with double intensity), corresponding to the β-anomeric protons, indicates that the neutral oligosaccharide sequence remains the same as that originally present in GM1 ganglioside.

PREPARATION OF GM1 GANGLIOSIDE CONTAINING AN ACETYL GROUP INSTEAD OF THE FATTY ACID (GM1 acetyl)

GM1 ganglioside containing an acetyl group at the level of ceramide is prepared by re-N-acetylation of deAc-deAcyl-GM1, as follows (Fig. 16). DeAc-deAcyl-GM1 is dissolved in magnesium-dehydrated methanol (0.5 mg/ml). The solution is mixed with sodium-dehydrated acetic anhydride (10μL/ mg ganglioside) and maintained overnight at room temperature. The mixture is then evaporated to a small volume under vacuum at 37°C in the presence of toluene. 10 ml of toluene are then added and the solution is dried at 37°C under vacuum. The residue is dissolved in water and lyophilized. Using anhydrous conditions the re-N-acetylation reaction occurs almost quantitatively, no further purification steps being required.

The ^{1}H-NMR spectrum of this compound (Table 3) shows the presence, besides the peaks corresponding to the acetyl group of sialic acid and of N-acetylgalactosamine (that resonate at 1.88 and 1.76 ppm respectively), a new peak of the same intensity

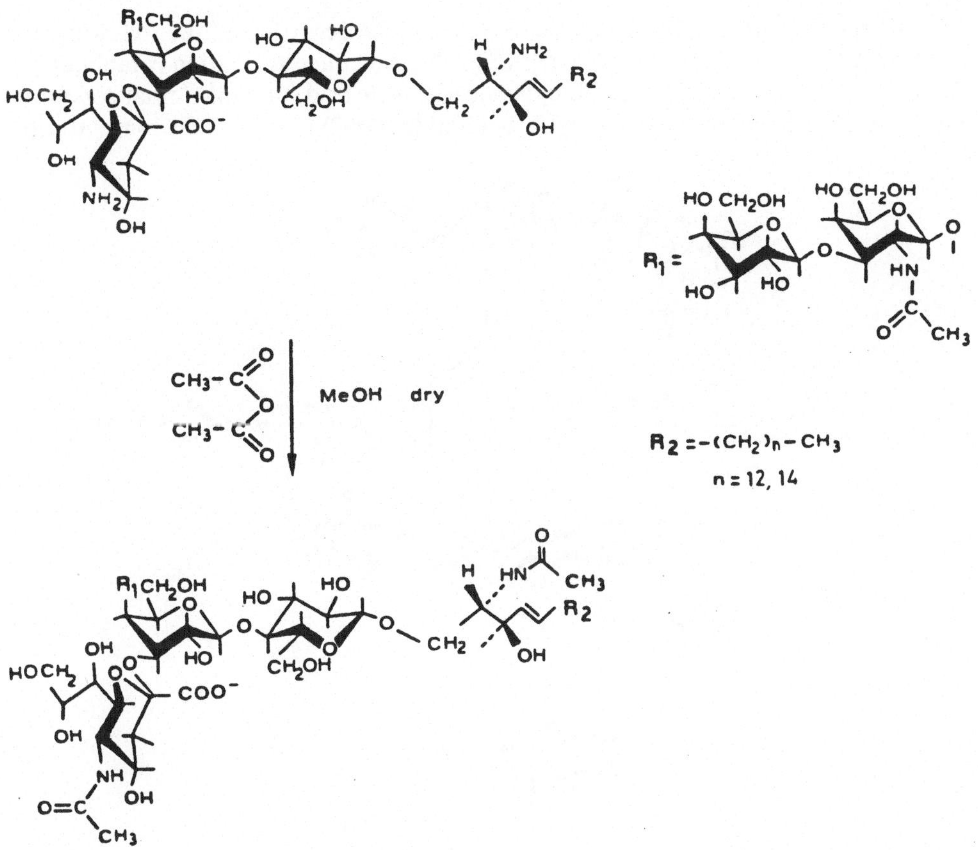

Figure 16. A general scheme for preparing GM1 ganglioside that contains an acetyl group, GM1(acetyl), instead of the fatty acid chain.

resonating at 1.78 ppm. This peak corresponds to the acetyl group inserted on the long chain base. On the other hand the intensity resonance at 0.85 ppm, due to the alkyl region, equals that of the acetamide methyl protons, indicating the presence of a single alkyl chain.

PREPARATION OF GM1 GANGLIOSIDE WITH A GIVEN AND HOMOGENEOUS FATTY ACID MOIETY

The fatty acid moiety of the gangliosides present in the nervous system is mainly represented by stearic acid. Instead the fatty acid composition of gangliosides of extraneural origin is more heterogeneous, with acyl chains differing in hydrocarbon length and presence or absence of double bonds and hydroxyl groups. The preparation of GM1 species with a defined acyl chain is performed according to the following procedure starting from deAc-deAcyl-GM1 (Fig. 17). An aqueous solution containing one fatty acid (myristic, stearic, arachidic, lignoceric, oleic acid) (4.5 mM, final concentration) and dimethylaminopropyl-ethylcarbodiimide (DEC) (5 mM, final concentration)

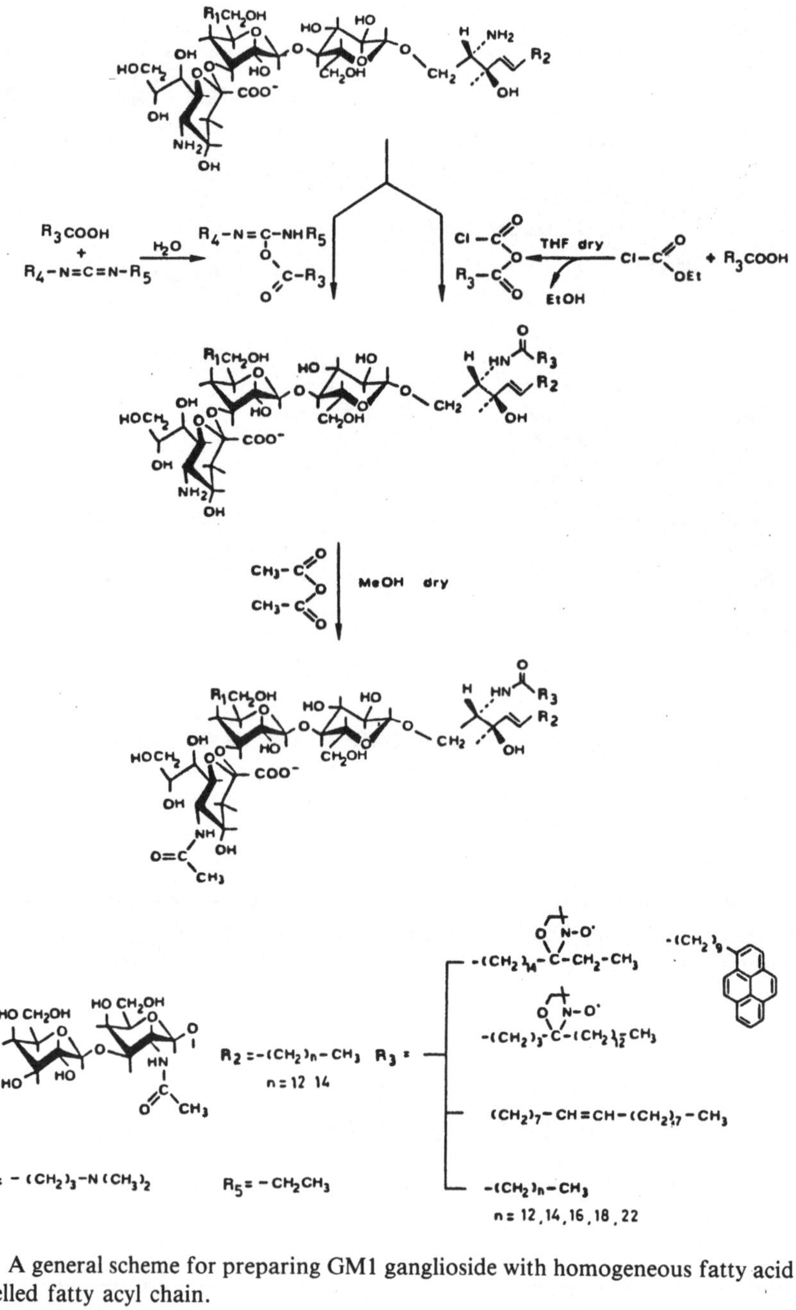

Figure 17. A general scheme for preparing GM1 ganglioside with homogeneous fatty acid moiety, or labelled fatty acyl chain.

is maintained at 50°C for 2h under continuous stirring. DeAc-deAcyl-GM1 (0.5 mg/ml, with a molar ratio ganglioside/fatty acid of 1/10) is added to the solution and the mixture allowed to react for 5 additional hours under the same conditions. At the end the mixture is chilled in an ice bath and 6N NaOH is added till pH 12, in order to remove possibly formed O-Acyl ester groups. After standing at room temperature for one night, the alkaline mixture is dialyzed at 4°C for 2 days against distilled water and lyophilized. The synthesized compound contained in the residue (deAc-GM1) is isolated and purified by chromatography on a silica gel column (3 ml sedimented gel/mg starting deAc-deAcyl-GM1), previously equilibrated and eluted with the solvent system chloroform-methanol-water, 60:35:5, by vol. Purified deAc-GM1 is then submitted to re-N-acetylation as reported above. The yield of the final compound is about 55%, regardless of the nature of the starting fatty acid.

The process of re-N-acylation and re-N-acetylation can be followed by ^{1}H-NMR analysis. When a saturated fatty acid is used, the obtained spectra fairly overlap with those of natural GM1; when an unsaturated fatty acid as oleic acid is used, a new resonance peak appears at 5.32 ppm (Table 1, Fig. 4) due to the two olefinic protons.

PREPARATION OF GM1 GANGLIOSIDE WITH A FLUORESCENT OR PARAMAGNETIC ACYL CHAIN

A fluorescent or paramagnetic fatty chain is inserted on the l.c.b. amino group of deAc-deAcyl-GM1, according to the following procedure (Fig. 17) (Acquotti et al., 1986). 1-pyrene decanoic acid, or 5-doxyl-stearic acid, or 16-doxyl-stearic acid is dissolved in dried tetrahydrofuran (50 μmol/ml). The solution is cooled in an ice-salt bath till reaching —20°C, mixed with an equimolar amount of triethylamine and ethylchloroformate, and allowed to stand for 5 min at —20°C under continuous stirring. Then an equimolar amount of deAc-deAcyl-GM1, dissolved in tetrahydrofuran-water 20:1, by vol, (20 umoles/ml) is added and the reaction mixture maintained for 10 min at room temperature under continuous stirring, and finally evaporated to dryness under vacuum at 37°C. The residue, dissolved in redistilled water, is dialyzed at 4°C for 2 days against redistilled water and lyophilized. The synthesized fluorescent or paramagnetic compound, contained in the residue, is isolated and purified by silica gel 100 column chromatography (3 ml sedimented gel/mg starting deAc-deAcyl-GM1) and submitted to re-N-acetylation (see above). The final yield for fluorescent, GM1(pyrene), or paramagnetic, GM1(5-doxyl) and GM1 (16-doxyl), GM1 is about 27%.

The fluorescence emission spectra of GM1 (pyrene) (10^{-5} M), dissolved in 50 mM aqueous KCl is shown in Figure 18. The spectrum is characterized by the presence of two emission peaks at 398 and 380 nm, and a major wide peak at 480 nm. This indicates that GM1 (pyrene) is a fluorescent compound that maintains the characteristic pyrene feature to display a different emission spectrum depending on the self-interactions occurring in solution. The two emission peaks at 398 and 380 nm are typical of non-interacting pyrene molecules, and the preponderant wide peak at 480 nm indicates the occurrence of interactions within the hydrophobic tails carrying the pyrene group.

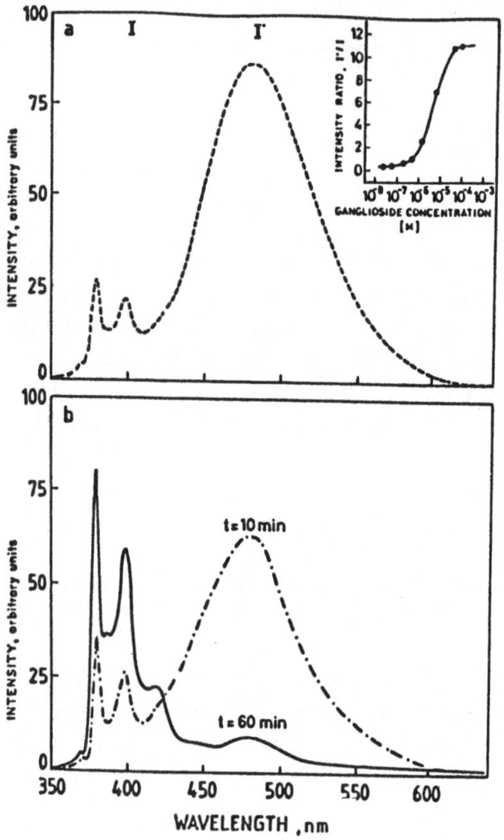

Figure 18. Fluorescence emission spectrum of GM1(pyrene) in aqueous solution; the insert shows the intensity ratio I_{480}/I_{398} versus ganglioside concentration (a); and kinetic of incorporation of GM1(pyrene) into preformed vesicles (b).

A low value of the excimer/monomer (I_{480}/I_{398}) ratio is an expression of monomeric state, and a high value of micellar state. As shown in the insert of Figure 18 the excimer/monomer ratio of GM1 (pyrene), when measured versus increasing concentrations in water, undergoes a sharp transition from low to high values at a concentration very close to 5×10^{-5} M. This is likely the concentration at which GM1 (pyrene) molecules aggregate to form micelles, in other words the c.m.c. of GM1 (pyrene). The critical micellar concentration of pyrene is much greater than that of GM1 (about 10^{-9} M) (Formisano et al., 1979; Ulrich-Bott and Wiegandt, 1984), likely reflecting the modified chemical structure of GM1 molecule due to insertion of the pyrene group.

Since ganglioside GM1 has the ability to insert into preformed monolamellar vesicles of phospholipids, this behaviour has been checked for GM1 (pyrene) using vesicles of egg phosphatidylcholine, and measuring the excimer/monomer ratio of a micellar dispersion of GM1 (pyrene) before and after incubation with vesicles. After incubation (see Fig. 18) the intensity of the excimer peak dramatically decreases with concomitant increase in the intensity of the monomer peaks reflecting a decrease in number of the ganglioside molecules aggregated in micelles and an increase of the ganglioside molecules less interacting with each other. This suggests that GM1 (pyrene) is transferred from micelles to vesicles into which it is incorporated resulting in a molecular dispersion in the phospholipid matrix. All these studies demonstrate that GM1 (pyrene) behaves similarly to natural GM1 in both the aggregation properties and the capability to be transferred from micelles to vesicular dispersion of phospholipids.

The ESR spectra (Fig. 19) of a 10^{-4} M solution of GM1 (doxyl) in 50 mM KCl consist of single, spin-spin broadened, lines indicating the occurrence of strong spin-spin interactions. This is consistent with the hypothesis that in aqueous solution and at 10^{-4} M concentration GM1 (doxyl) is present in micellar dispersion. Therefore the ESR measurements prove that GM1 (doxyl) molecules contain a paramagnetic probe retaining its characteristics and maintain the ganglioside attitude to micellize in aqueous solution.

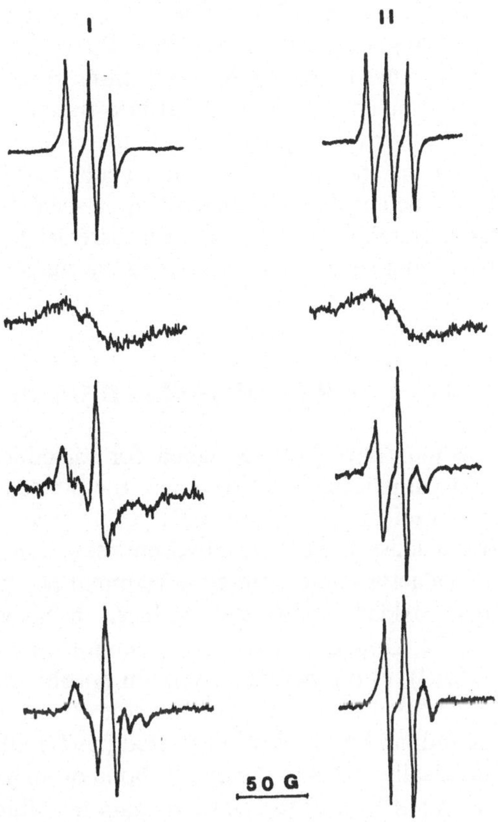

Figure 19. EPR spectra of GM1(5-doxyl), I, and GM1(16-doxyl), II, in different media; GM1 (doxyl) in CH_2Cl_2/CH_3OH, 1:1, by vol a); GM1(doxyl) in water solution b); GM1:GM1(doxyl), 100:1, by mol., in water solution c); GM1(doxyl) incorporated into preformed vesicles d).

Using aqueous solutions containing GM1 and GM1 (doxyl) the ESR spectra are similar to those of GM1 (doxyl) for mixtures where the two components are in a ratio not exceeding 10 to 1, while for ratios of the order 100 to 1 the spectra indicate a strong decrease of spin-spin interactions indicating that the paramagnetic probe is uniformly dispersed within the lipid environment of the micelle.

The features of the spectra are clearly different for the two spin-label positional isomers GM1(5-doxyl) and GM1(16-doxyl) and indicate that, as expected (Marsh, 1981), the amplitude of angular motion increases dramatically on proceeding down the lipid carbon chain toward the terminal methyl group (Fig. 19). The ability of the GM1 (doxyl) to insert into preformed unilamellar vesicles of phosphatidylcholine has been assessed by monitoring the ESR spectra at different times of incubation at 37°C of aqueous solutions of GM1 (doxyl) with EPC-SUV dispersions. The shape of the spectra gradually turned from those of pure GM1 (doxyl) in KCl solution to those obtained by dilution of GM1 (doxyl) with GM1, or by insertion of 5- and 16-doxyl stearic acid into model membranes (Bertoli et al., 1981). This strongly suggests the insertion of the spin-labeled gangliosides into the bilayer membrane. A progressive decrease of the outer hyperfine splitting (2 $A_{\parallel}$) was observed with time, reaching a minimum after 1 h of incubation at 37°C (Fig. 19). After this period the spectra remained unchanged for at least 8 h, indicating that no spin-spin interactions occurred and all the labeled ganglioside micelles could be considered inserted into the outer layer.

Therefore, also in the case of the paramagnetic species of GM1, they maintain the paramagnetic features of the starting labeled fatty acids, and resemble natural GM1 in their behaviour. It can be concluded that both fluorescent and paramagnetic GM1 species appear to be quite suitable for studying the transfer of gangliosides within membranes and the contribution given by gangliosides to the surface and inner features of the membrane.

PREPARATION OF RADIOLABELLED GANGLIOSIDES

The following methods have been designed for introducing radioactivity into gangliosides: a) the biosynthetic method, in which the labeled compound is biosynthesized starting from a radioactive precursor (Kolodny et al., 1970; Tallman et al., 1977); b) the galactose oxidase-sodium borohydride reduction method, in which the galactose (or N-acetyl- galactosamine) moiety in terminal position is first oxidized at carbon 6 by galactose oxidase, and then reduced back with tritiated sodium borohydride (Suzuki, 1972; Orlando, et al., 1979; Ghidoni et al., 1977; Novak et al., 1977; Leskawa et al., 1984); c) the periodate-sodium borohydride reduction method, in which sialic acid is oxidized at the level of the side chain, and then submitted to reduction with tritiated sodium borohydride (Veh et al., 1977); d) the reduction method in which tritium is catalytically added to the double bond of long chain bases (Schraven et al., 1977; Schwarzmann, 1978); e) the DDQ oxidation-sodium borohydride reduction method, in which the unsaturated long chain bases are specifically oxidized by DDQ at carbon-3, and then reduced with tritiated sodium borohydride (Ghidoni et al., 1981); f) the de-N-acetylation, re-N-acetylation method, in which the acetyl group of sialic acid is removed and substituted with a [^{3}H] or [^{14}C] labelled one (Chigorno et al., 1985); g) the de-N-acylation, re-N-acylation method in which the original fatty acid is substituted with [^{14}C] or [^{3}H] labelled fatty acid (Acquotti et al., 1986). Compounds of relatively low specific radioactivity and with a relevant cost are obtained by the biosynthetic method. The periodate method is valuable in terms of the obtained high specific radioactivity and easy applicability; however it cannot avoid marked structural modifications of ganglioside sialic acid residue(s) which may be important from the functional point of view. The reduction method produces completely saturated gangliosides that display physico-chemical properties quite different from those of the natural compounds (Cantù et al., 1985). Methods b), e), f) and g) appear to have a number of advantages; they provide isotopic radio labelling and allow to introduce radioactivity in the oligosaccharide or in the ceramide portion of the molecule. Furthermore double radio labelling of the molecule, and [^{3}H] or [^{14}C] labelling is possible. Depending on the specific experimental purpose, the most adeguate radio labelling approach can be adopted.

(a) *Preparation of gangliosides tritium labelled at the level of the terminal galactose and N-acetylgalactosamine residues.* The procedure for tritium labelling of ganglosides at the terminal galactose or N-acetylgalactosamine residues by the galactose-oxidase-^{3}H- $NaBH_4$ method, used in our laboratory, is briefly described. Twenty mg of ganglioside GM2, GM1 or GD1b, dissolved in 1 ml of chloroform-methanol, 2 : 1, by vol, are mixed with 10 ml of the same solvent containing 180 mg of Triton-X 100. The mixture is dried under vacuum and the residue dissolved in 8 ml of 5 mM EDTA-25 mM sodium phosphate buffer pH 7.0, containing 450 Units of galactose-oxidase. The mixture is incubated at 37° under constant stirring for 12 h, then 450 additional Units of galactose-oxidase are added, and incubation prolonged for 6 further hours. The enzymatic reaction is stopped by addition of 4 vol of tetrahydrofuran and vigorous shaking. The mixture is dried and chromatographed on a 110 cm × 2 cm silica gel 100 column, previously equilibrated with chloroform-methanol-water, 60 : 35 : 5, by volume and eluted with the same solvent. The obtained oxidized ganglioside, pure over 98%, is dissolved in 5 ml of tetrahydrofuran-water-0.1 M aqueous NaOH, 4 : 2 : 1, by vol and treated at room temperature with 100 mCi of [^{3}H]$NaBH_4$. After 5 h, 10 mg of cold $NaBH_4$ are added and the mixture allowed to react for further 30 min. The mixture is concentrated to almost dryness, diluted with distilled water and dialyzed. The radioactive ganglioside is finally purified by chromatography on a silica gel 100 column (40 cm × 1 cm), equilibrated and eluted with chloroform-methanol-water, 60 : 35 : 5, by vol. The percentage of oxidized product in the mixture is 30%, 80% and 90% for ganglioside GM2, GD1b and GM1, respectively. After reduction with tritiated sodium borohydride some highly apolar radioactive compounds (10-20% of total radioactivity) appear to be present in the mixture. Their removal is achieved by the final column chromatography purification step. The purity of the final preparation of each ganglioside is over 98%. The specific radioactivity of the labelled gangliosides is constantly in a range of 1.3-1.5 Ci/mmol.

(b) *Preparation of gangliosides tritium labelled at the 3-position of l.c.b.* Gangliosides can be tritium labelled at the 3-position of l.c.b. by the DDQ /[^{3}H]$NaBH_4$ method.

The procedure that has been applied to gangliosides GM3, GM2, GM1, Fuc-GM1, and GD1a, follows the indications already reported in this paper for the preparation of "threo"-gangliosides, with the only difference that tritiated $NaBH_4$ is employed. After column chromatographic purification, the 3-keto-derivative is reduced with $[^3H]NaBH_4$ under neutral conditions. The mixture of labelled erythro- and threo ganglioside species is submitted to HPLC fractionation (Fig. 20), in order to separate the erythro- from the threo- species of gangliosides. The corresponding saturated molecular species are prepared by catalytic hydrogenation of the unsaturated ones as described above.

All the final labelled compounds proved to be over 99% pure, and to possess a specific radioactivity ranging from 0.85 to 1.3 Ci/mmol. The variability is higher and the values are lower than those obtained with the galactose oxidase/$[^3H]NaBH_4$ methods, using $[^3H]NaBH_4$ with the same specific radioactivity. This is likely due to the starting neutral conditions used during $[^3H]NaBH_4$ treatment that cause partial degradation of $NaBH_4$. On the other hand neutral conditions during this treatment are required for preventing degradation of the β-unsaturated ketones that are highly unstable under alkaline conditions.

(c) *Preparation of GM1 ganglioside radiolabelled at the level of sialic acid.* GM1 ganglioside, ^3H- or ^{14}C- labelled at the level of acetyl group of sialic acid is prepared using radiolabelled acetic anhydride by re-N-acetylation of deAc-GM1 according to the re-N-acetylation method reported above with some modifications. Using $[1\text{-}^{14}C]$-acetic anhydride (Amersham, U.K.) at 111 mCi/mmol in toluene the preparation procedure is as follows. Seven mg of deAcGM1 are dissolved in 1 ml of 5% triethylamine in magnesium-dehydrated methanol and 11.1 μl 5% $[1\text{-}^{14}C]$-acetic anhydride in toluene are added. After stirring for 2 h at room temperature, 20 μl of dehydrated acetic anhydride are added, and the mixture stirred for 2 additional hours. The solution is then diluted with 100 ml of redistilled water and dialyzed for two days against redistilled water at 4°C. The dialyzed solution is frozen, lyophilized and the residue dissolved in 1 ml of propan-1-ol-water, 7:3, by vol. Stored at 4°C the $[^{14}C]$-labelled GM1 is stable for at least 4 months.

Paying care to perform the procedure under rigorous anhydrous conditions the re-N-acetylation reaction occurs almost quantitatively and the specific radioactivity of the final compound is 53 mCi/mmol.

(d) *Preparation of GM1 ganglioside radiolabelled at the level of acyl chain moiety.* The de-N-acylation re-N-acylation procedure described above leads to produce radiolabelled gangliosides when radioactive fatty acids are employed. GM1 ganglioside, $[^3H]$- or $[^{14}C]$-labelled at the level of the fatty acyl chain is prepared by re-N-acylation of de-Ac-deAcyl-GM1, followed by re-N-acetylation, using the same chlorocarbonate fatty acyl mixed anhydride method reported above for the preparation of fluorescent or paramagnetic GM1. Using, for istance, $[^{14}C]$ stearic acid, with a specific radioactivity of 56 mCi/mol, labelled GM1 is prepared with a purity over 99%, and with a specific radioactivity identical to that of the starting labelled fatty acid.

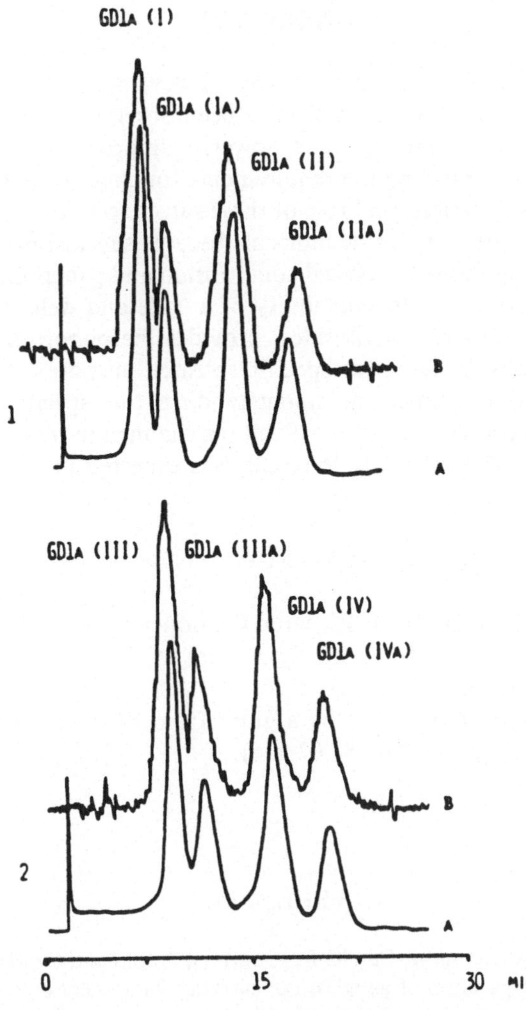

Figure 20. HPLC of DDQ-NaB^3H_4 tritiated GD1a ganglioside, before (1) and after (2) hydrogenation of the double bond at the 4-position of the long chain bases. Two mg of GD1a carrying 1.7 mCi, dissolved in 0.1 ml of water, were used. The chromatographies were performed on a Spherisorb S5 ODS2 column (250 x 10 mm) using acetonitrile-phosphate buffer, 60:40 by vol as the solvent system. The elution profile was monitored by (A) absorbance recording at 195 nm, and (B) radioactivity measuring with a Berthold LB503 HS radiospectrometer equipped with a 0.12 ml solid scintillator cell. Each peak corresponds to a GD1a molecular species that contains: I, erythro C18:1-; IA, threo C18:1-; II, erythro C20:1-; IIA, threo C20:1-; III erytro C18:0; IIIA, threo C18:0-; IV, erythro C20:0-; IVA, threo C20:0 long chain base.

CONCLUSION

The chemical technology of gangliosides is developing at a rapid rate. New ganglioside derivatives, such as lactones, have been observed to occur in nature. The application of HPLC procedures gave a powerful improvement to preparation of highly pure gangliosides, fulfilling the requirements for experimental approaches aimed at understanding the physiological role of these substances. The technology for high yield manipulations of the ganglioside molecule became successful and led to produce gangliosides with a programmed chemical composition and ganglioside derivatives carrying special probes. As well, the availability of a wide and well established strategy for radiochemical labelling of gangliosides, provides the opportunity to prepare the labelled compounds most suitable for specific research purposes. All this gives great impulse to investigations aimed at understanding the specific role played by gangliosides in the functional behaviour of cell plasma membranes. On these premises a very flourishing period in ganglioside research is expected in the near future.

ACKNOWLEDGMENTS

The authors wish to thank Prof. Riccardo Ghidoni and Prof. Giovanni Galli from the University of Milan and Prof. Heinz Egge from the University of Bonn, for their help and friendly discussion.

This work was supported in part by a grant from the Consiglio Nazionale delle Ricerche (CNR), Roma, (grant No. 83.02.855).

REFERENCES

Acquotti D, Sonnino S, Masserini M, Casella L, Fronza G, Tettamanti G (1986) A new chemical procedure for the preparation of gangliosides carrying fluorescent or paramagnetic probes on the lipid moiety. Chem Phys Lipids 40: 71-86.

Brady O and Fishman PH (1979) Biotransducers of membrane - mediated information. Adv Enzymol 50: 303-323.

Bertoli E, Masserini M, Sonnino S, Ghidoni R, Cestaro B, Tettamanti G (1981) Electron paramagnetic resonance studies on the fluidity and surface dynamics of egg phosphatidylcholine vesicles containing gangliosides. Biochim Biophys Acta 467: 196-202.

Cantù L, Corti M, Sonnino S, Ghidoni R, Gazzotti G, Tettamanti G (1985) Micellar properties of hydrogenated and natural GM1, in Glycoconjugates, proceedings of the VIII International Symposium, Houston, Texas, U S A, September 8-13, Davidson EA, Williams JC, Ferrante NM eds Praeger, p. 471.

Chigorno V, Sonnino S, Ghidoni R, Tettamanti G (1982) Densitometric quantification of brain gangliosides, separated by two dimensional thin layer chromatography. Neurochem Int 5, 397-403.

Chigorno V, Pitto M, Cardace G, Acquotti D, Kirschner G, Sonnino S, Ghidoni R, Tettamanti G (1985) Association of gangliosides to fibroblasts in culture: a study performed with GM1 [^{14}C]-labelled at the sialic acid acetyl group. Glycoconjugate J 2, 279-291.

Corti M, Degiorgio V, Sonnino S, Ghidoni R, Masserini M, Tettamanti G (1981) GM1 ganglioside - Triton X-100 mixed micelles: changes of micellar properties studied by laser-light scattering and enzymatic methods. Chem Phys Lipids 28, 197-214.

Formisano S, Johnson ML, Lee G, Aloj SM, Edelhoc H (1979) Critical micelle concentrations of gangliosides. Biochemistry 18, 1119-1124.

Gazzotti G, Sonnino S, Ghidoni R, Kirschner G, Tettamanti G (1984a) Analytical and preparative high-performance liquid chromatography of gangliosides. J Neurosc Res 12, 179-192.

Gazzotti G, Sonnino S, Ghidoni R, Orlando P, Tettamanti G (1984b) Preparation of the tritiated molecular forms of gangliosides, separated with homogeneous long chain base composition. Glycoconjugate J, 1, 111-121.

Gazzotti G, Sonnino S, Ghidoni R (1985) Normal-phase high performance liquid chromatographic separation of non-derivatized ganglioside mixtures. J Chromatogr 348, 371-378.

Ghidoni R, Tettamanti G, Zambotti V (1977) Labelling of natural substrates for the radiochemical assay of enzymes involved in the lipid storage diseases: a general procedure for tritiation of gangliosides. Biochem Exp Biol 13, 61-69.

Ghidoni R, Sonnino S, Tettamanti G, Baumann N, Reuter G, Schauer R (1980) Isolation and characterization of a trisialoganglioside from mouse brain, containing 9-0-acetyl-N-acetylneuraminic acid. J Biol Chem 255, 6990-6995.

Ghidoni R, Sonnino S, Masserini M, Orlando P, Tettamanti G (1981) Specific tritium labeling of gangliosides at the 3-position of sphingosines. J Lipid Res 22, 1286-1295.

Gross SK, Williams MA, McCluer RM (1980) Alkali labile, sodium borohydride-reducible ganglioside sialic acid residues in brain. J Neurochem 34, 1351-1361.

Holmgren JH, Elwing H, Fredman P, Strannegard O, Svennerholm L (1980) Gangliosides as receptors for bacterial toxins and Sendai virus. Adv Exp Med Biol 125, 453-470.

IUPAC-IUB Commission on Biochemical Nomenclature (1977) The nomenclature of lipids. Lipids 12, 455-468.

Koerner TAW, Prestegard JH, Demou PC, Yu RK (1983) High resolution proton NMR studies of gangliosides. 1. Use of homonuclear two-dimensional spin-echo J.-correlated spectroscopy for determination of residue composition and anomeric configuration. Biochemistry 22, 2676-2687.

Ledeen RW and Yu RK (1978) in Research Methods in Neurochemistry, Marks N and Roonight R eds, Plenum Publishing Corp N Y, 371-410.

Leskawa KC, Dasgupta S, Chien JL, Hogan EL (1984) A simplified procedure for the preparation of tritiated GM1 ganglioside and other glycosphingolipids. Anal Biochem 140, 172-177.

Marsch D (1981) in Electron spin resonance: spin lebels. E Grell ed Membrane spectroscopy, Springer-Verlag, Berlin, 51-142.

Novak A, Lowden JA, Gravel YL, Wolfe LS (1979) Preparation of radiolabeled GM2 and GA2 gangliosides. J Lipid Res 678-681.

Orlando P, Cocciante G, Ippolito G, Massari P, Roberti S, Tettamanti G (1979) The fate of tritium labeled GM1 ganglioside injected in mice. Pharmacol Res Comm 11, 759-773.

Pascher J (1976) Molecular arrangements in sphingolipids. Conformation and hydrogen bonding of ceramide and their implication on membrane stability and permeability. Biochim Biophys Acta 455, 433-451.

Riboni L, Sonnino S, Acquotti D, Malesci A, Ghidoni R, Egge H, Mingrino S, Tettamanti G (1986) Natural occurrence of ganglioside lactones: isolation and characterization of GD1b inner ester from adult human brain. J Biol Chem 261: 8514-8519.

Schraven J, Cap C, Nowoczek G, Sandhoff K (1977) A radiometric assay for sialidase activity on ganglioside GD1a. Anal Biochem 78, 333-339.

Schwarzmann (1978) A simple and novel method for tritium labeling of gangliosides and other glycosphingolipids, Biochim Biophys Acta 529, 106-114.
Sonnino S, Ghidoni R, Galli G, Tettamanti G (1978) On the structure of a new fucose containing gangliosides from pig cerebellum. J Neurochem 31, 947-956.
Sonnino S, Ghidoni R, Gazzotti G, Kirschner G, Galli G, Tettamanti G (1984) High performance liquid chromatography preparation of the molecular species of GM1 and GD1a gangliosides with homogeneous long chain base composition. J Lipid Res 25, 620-629.
Sonnino S, Kirschner G, Ghidoni R, Acquotti D, Tettamanti G (1985) Preparation of GM1 ganglioside molecular species having homogeneous fatty acid and long chain base moieties. J Lipid Res 26, 248-257.
Suzuki Y and Suzuki K (1972) Specific radioactive labeling of terminal N-acetylgalactosamine of glycosphingolipids by the galactose oxidase-sodium borohydride method. J Lipid Res 13, 687-690.
Svennerholm L (1970) in Handbook of neurochemistry. Vol. III, A Lajtha, ed, Plenum Publishing Corp, New York, 425-452.
Tallman JF, Fishman PH, Henneberry RC (1977) Determination of sialidase activities in HeLa cells using gangliosides specifically labelled in N-acetylneuraminic acid. Arch Biochem Biophys 182, 556-562.
Tettamanti G, Bonali F, Marchesini S and Zambotti V (1973) A new procedure for the extraction, purification and fractionation of brain gangliosides. Biochim Biophys Actra 296, 160-170.
Ulrich-Bott B and Wiegandt H (1984) Micellar properties of glycosphingolipids in aqueous media. J Lipid Res 25, 1233-1245.
Veh RW, Corfield AP, Sander M, Schaver R (1977) Neuraminic acid-specific modification and tritium labelling of gangliosides. Biochim Biophys Acta 486, 145-160.
Wiegandt H (1982) The gangliosides. Adv Neurochem 4, 149-223.
Yu RK, Koerner TAW, Ando S, Yome HC, Prestegard JM (1985) High-resolution proton NMR studies of gangliosides. III. Elucidation of the structure of ganglioside GM3 lactone. J Biochem 98, 1367-1373.

Gangliosides and neuronal plasticity
G. Tettamanti, R.W. Ledeen, K. Sandhoff,
Y. Nagai, G. Toffano (eds.)
Fidia Research Series, vol. 6
Liviana Press, Padova, © 1986

THE SOLUTION CONFORMATION OF GANGLIOSIDES INFERRED FROM HSEA CALCULATIONS AND HIGH FIELD NMR SPECTROSCOPY

K. Bock

Department of Organic Chemistry, The Technical University of Denmark,
DK-2800 Lyngby, Denmark

INTRODUCTION

Our understanding of the significance of carbohydrates in biological systems has increased during the last twenty years (Sharon, 1975; Sharon and Lis, 1981). Analytical techniques have improved tremendously during that period and Lindberg (1981) has reviewed how structural analysis of complex oligo- and polysaccharides can be carried out on milligram amounts of compound using modern spectroscopic and chemical tools. Through synthetic work, as demonstrated by Lemieux (1978), Paulsen (1982), Ogawa et al. (1984) and others, it is possible to prepare complex oligosaccharides in relatively large amounts. With these compounds it is possible to gain further insight into the understanding of the interaction between carbohydrates and proteins (such as enzymes, antibodies or lectins) (Lemieux, 1982; 1984). However, in order to obtain a better picture of these interactions it is necessary to have information about the preferred conformation of the oligosaccharides in solution. The present paper primarily discusses how a conformational analysis can be carried out using modern instrumentation and computers. Examples of the interpretation of the results in relation to the recognition of oligosaccharides will be discussed in the last part of the paper.

CONFORMATIONAL ANALYSIS OF OLIGOSACCHARIDES

Conformational analysis of oligosaccharides can be based on data from the following experiments:

1. X-Ray (or neutron) diffraction studies.

Abbreviations: n.m.r., nuclear magnetic resonance; HSEA, hard sphere exo- anomeric effect; EA, exo-anomeric effect; CPK, Corey-Pauling-Kaltun; NOE, nuclear Overhauser effect; PCILO, Perturbation Configuration Interaction using Localized Orbital.

2. Chiroptical methods.
3. Nuclear magnetic resonance spectroscopic data.

The different diffraction studies require the carbohydrate molecule to be crystalline, which is a limitation because many oligosaccharides do not crystallize very easily. Furthermore, lattice forces may cause deviation from the conformation which is predominating in solution. Chiroptical methods, on the other hand, yield information about the molecules in solution and have been used extensively in the study of polysaccharides (Rees, 1981). The results are, however, most easily interpreted when the compounds are simple repeating oligo- or polysaccharides (e.g. amylose).

A detailed conformational analysis of complex carbohydrate structures in solution is at present only possible using nuclear magnetic resonance (n.m.r.) data. High resolution n.m.r spectrometers operating at 500 MHz have made it possible to obtain experimental n.m.r. parameters which contain detailed conformational information about the oligo- and polysaccharides in aqueous solutions (Bock and Lemieux, 1979; Lemieux et al., 1980; Thøgersen et al., 1982; Bock et al., 1982a; 1984; Bock, 1983; Hayes et al., 1982; Paulsen et al., 1985). This point will be discussed in further detail below. However, in order to provide a complete interpretation of the n.m.r. data and in order to be able to draw relevant conclusions from the results, it is necessary to support these experimentally acquired data with a model which allows a simple theoretical evaluation of the preferred conformation of the oligosaccharides in solution.

Several theoretical approaches have been used in the study of carbohydrate conformations and the most frequently applied are the following: 1. ab initio calculations, 2. force fields calculations, 3. hard spheres calculations.

Ab initio calculations have not yet been used to calculate the preferred conformation of an oligosaccharide even though one ab initio calculation has been carried out on a fixed conformation of maltose (Melberg et al., 1979). The reason is that these calculations are very expensive and in general not practical on large molecules. Most ab initio calculations have therefore been performed on model compounds, which simulate the atoms involved in the glycosidic linkages (Jeffrey et al., 1978; Jeffrey and Yates, 1981; Wolfe et al., 1979). The molecular orbital method in the PCILO approximation has also been used in the study of the conformation of oligosaccharide linkages (Giacomini et al., 1970; Tvaroska and Kozar, 1980; Yadav and Luger, 1983).

The simpler force fields calculations, which include both bond and angle deformations, torsional terms, non-bonded interactions, coulombic terms, and hydrogen bonding, have been carried out on several mono- and some disaccharides (Marchessault et al., 1980; Bluhm et al., 1982; Melberg and Rasmussen, 1980; Jeffrey and Taylor, 1980). These calculations are, however, still rather expensive and time consuming to carry out on molecules larger than disaccharides. It has therefore been important to develop a simple and inexpensive method, which gives reliable results even on larger oligosaccharides, despite the fact that simple functions are used to describe the interaction between the component monosaccharides. Hard-sphere calculations which only take into account the non-bonded interaction between the monosaccharide units of an oligosaccharide have proven to give results which are in excellent accord with evidence from n.m.r. spectra, provided the importance of the *exo*-anomeric effect is recognized (Bock and Lemieux, 1979; Lemieux et al., 1980; Thøgersen et al., 1982). This method, the hard sphere *exo*-anomeric effect (HSEA) calculations, which has its major force in simplicity and low cost, will be described in further detail below.

Hard Spheres Calculations

The problem in a conformational analysis of a disaccharide is relatively simple if the monosaccharide units can be considered as rigid bodies, i.e. that they exist in their regular unstrained chair conformations as indicated in Figure 1 for maltose. Sheldrick and Akrigg (1980) have published an investigation of 161 X-ray structures of com-

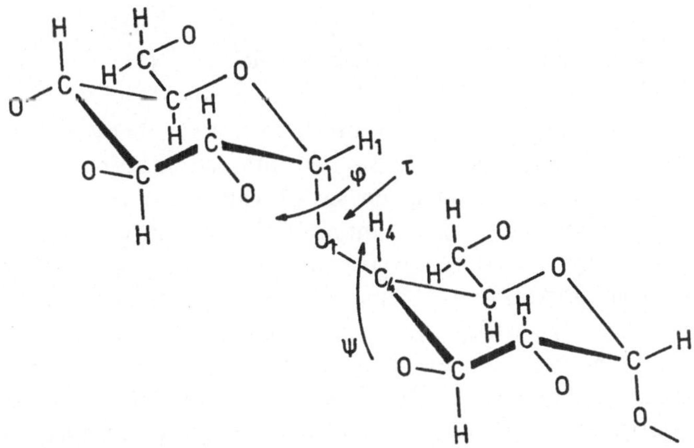

Figure 1. Maltose with the torsion angles φ, ψ and τ which defines the conformation of the disaccharide.

pounds containing pyranose rings and discussed the validity of a rigid-body assumption. Their results indicated that it is possible to establish coordinates for an average pyranose ring. Based on these results and keeping in mind that the molecules investigated are dissolved in aqueous solutions, it appears to be a reasonable assumption (Arnott and Scott, 1972). However, in order to obtaine as accurate results as possible, coordinates for the individual monosaccharide units are taken from good neutron- or X-ray diffraction experiments. Because the preferred conformation of oligosaccharides is determined to a large extent by the atoms, particularly the protons located around the glycosidic linkage, only proton coordinates from neutron diffraction studies can be used. If X-ray diffraction data are available, only the experimentally determined coordinates from the heavy atoms are used and the proton coordinates are generated using a computer program. This program positions the protons at a 1.10 A distance from the carbon atom along a vector defined by the remaining carbon-carbon and carbon-oxygen bond vectors. Hydrogen atoms of hydroxyl groups are not included in the calculations.

The three degrees of freedom which remain to be investigated are thus the rotations around the C_1-O_1 bond (φ_H rotation for the H_1-C_1-O_1-C_4 fragment), defined positive according to IUPAC recommendations (1971), and rotation around the O_1-C_4 (ψ_H rotation for the C_1-O_1-C_4-H_4 fragment) and the size of the glycosidic bond angle (τ). Experimental data from X-ray or neutron diffraction studies suggest the τ-angle in most oligosaccharides can be considered constant ($\simeq$ 117°) (Lemieux et al., 1980), but the value can of course be included as a variable in the calculations.

Hard-sphere calculations primarily take into account the non-bonded interactions between the atoms. The literature gives different potential functions for the interaction between the atoms, but in the present work the values published by Kitaygorodski (1961; 1978) have been chosen. The data have been used to estimate the non-bonded interaction between two atoms following the equation:

$$V_{pot} = 3.5\ (-0.04/z^6 + 8.5 \cdot 10^3\ e(^{-13z})\ \text{kcal/mole} \qquad (1)$$

in which $z = r_{ij}/r_O$. The r_O value is the equilibrium distance between the atoms and r_{ij} is equal to the actual distance between the two interacting atoms.

The interaction energies for a given set of values for φ_H, ψ_H and τ are then summed for all interactions between two monosaccharide units and the sum represents the pure non-bonded interaction energy for that conformation. φ_H, ψ_H is then varied through 360° in given intervals (e.g. 5°) and this leads to a series of interaction energies as a function of φ_H and ψ_H which can be inspected in different ways.

However, experimental (Lemieux et al., 1980; Thøgersen et al., 1982; Lemieux et al., 1979) and theoretical (Jeffrey et al., 1978; Wolfe et al., 1979) evidence has proven that the *exo*-anomeric effect makes an important contribution to the preferred conformation of the glycosidic linkage. The consequence of the *exo*-anomeric effect is that the aglyconic carbon prefers an orientation in which the φ_H-angle is ~60° in β-D-glycosides and —60° in α-D-glycosides, respectively, as shown in Figure 2.

Thøgersen et al. (1982) have discussed the quantitative aspects of the *exo*-anomeric effect and based on *ab-initio* calculations of dimethoxymethane, Jeffrey et al. (1978) proposed a torsional potential describing the *exo*-anomeric effect for α- and β-glycosides respectively:

$$V_{exo\text{-}\alpha} = 1.58\ (1\text{-}\cos\varphi) - 0.74\ (1\text{-}\cos 2\varphi) - 0.70\ (1\text{-}\cos 3\varphi) + 1.72\ \text{kcal/mole} \quad (2)$$
$$V_{exo\text{-}\beta} = 2.61\ (1\text{-}\cos\varphi) - 1.21\ (1\text{-}\cos 2\varphi) - 1.18\ (1\text{-}\cos 3\varphi) + 2.86\ \text{kcal/mole} \quad (3)$$

This energy contribution is calculated for each φ-value and added to the non-bonded interaction energies according to equation (1) mentioned above, giving the total energy encountered in these calculations. These are called hardsphere, *exo*-anomeric (HSEA) calculations.

The minimum energy conformation can now be inspected with respect to short proton-proton or proton-oxygen distances or hard interactions which can be related to the experimentally determined n.m.r. parameters, which thus can support the calculated results. The coordinates determined for this minimum energy conformation can furthermore be used in standard molecular plot programs. These programs allow one to plot the molecule either as ball and stick models or as CPK models as shown in Figure 3, which can be inspected from different angles to identify interactions between the individual monosaccharides (Khare et al., 1985).

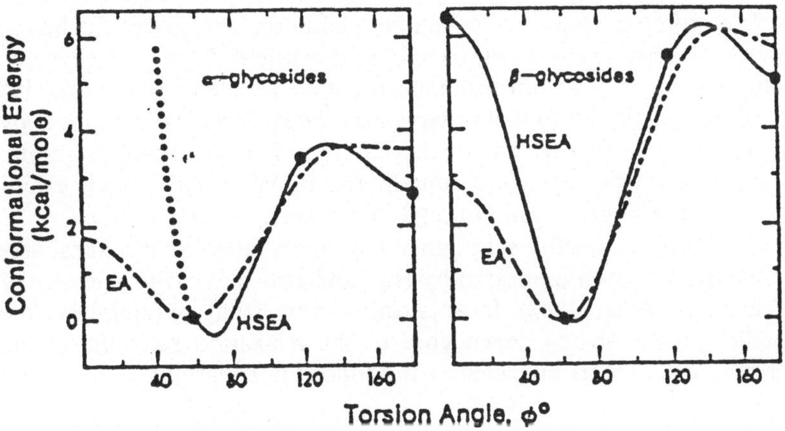

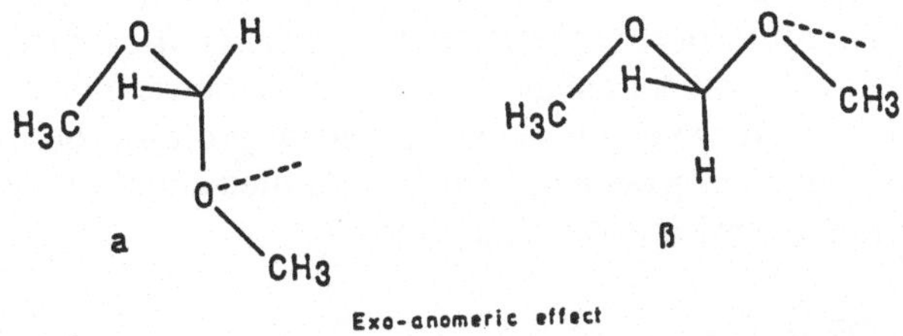

Figure 2. Potential energy functions to describe the *exo*-anomeric effect (EA) for α- and β-glycosides in HSEA calculations. The solid points are those calculated by Jeffrey et al. (1978).

NUCLEAR MAGNETIC RESONANCE SPECTROSCOPY

^{1}H and ^{13}C-n.m.r. spectroscopy is the most direct method by which information about the preferred conformation of oligosaccharides in solution is obtained.

^{1}H-n.m.r. Parameters

The ^{1}H-n.m.r. parameters which yield conformational information are as follows: 1. chemical shifts, 2. coupling constants, 3. spin lattice relaxation rates, 4. nuclear Overhauser enhancements.

After a complete assignment of the ^{1}H-n.m.r. spectrum has been done as discussed by Bock and Thøgersen (1983), it is relevant to discuss how these data can be used in a conformational analysis of an oligosaccharide.

The proton spin-spin coupling constants can be used to confirm that the chair conformations of the individual units are similar to those found in the parent monosaccharides. The proton chemical shifts can be used in the analysis of the interglycosidic conformation because protons will be shifted downfield if they are close in space to oxygen atoms (< 2.70 Å) from neighbouring units (Lemieux and Bock, 1983). This downfield shifting is similar to that observed for the H-3 and H-5 protons in pyranoses, when the anomeric configuration is changed from β- to α-. Similarly, upfield shifts may be observed if the molecules contain functional groups which exhibit strong anisotropy, like C=O groups in N-acetyl derivatives.

The application of proton spin-lattice relaxation rates for structural assignments of carbohydrates has been pioneered by Hall and coworkers (Berry et al., 1979; Hall et al., 1979; Evelyn et al., 1982). Its application in the study of interglycosidic conformations relies on the strong dependence of the relaxation rates on intramolecular proton-proton distances as indicated in the following equation:

$$\frac{1}{T_1} = R_1 = \frac{3}{2} \cdot \left(\frac{h}{2\pi}\right)^2 \cdot \gamma_H^4 \cdot \tau_c / \sum_{i \neq j} r_{ij}^{-6} = c / r_{ij}^{-6} \quad (4)$$

Generally, relaxation contributions from another monosaccharide unit can be observed if the proton-proton distances are below 3.00 Å. The problem with the application of proton-relaxation rates in a conformational analysis is that it is difficult to determine the individual relaxation contributions and thus obtain a quantitative measure for the proton-proton interactions in neighbouring pyranose rings. This problem can be solved by measurement of nuclear Overhauser enhancements (Noggle and Schirmer, 1971). In this experiment the individual proton-proton relaxation contributions can be determined between the proton saturated and the protons receiving relaxation contribution as indicated in equation (5):

$$\mathrm{NOE}\ (d)^s = r_{ds}^{-6}/c \cdot 2 \cdot \sum_{d \neq j} r_{dj}^{-6} \quad (5)$$

Performed in the difference mode, this experiment gives very reliable results and it is possible to measure enhancements as small as 2% with confidence. This approach is the most powerful tool in the conformational analysis of oligosaccharides. Furthermore, if qualitative data only can be accepted, the experiment can be performed as a 2-dimensional experiment (Jeener et al., 1979).

^{13}C-n.m.r. Parameters

Even though the carbon atoms are not located at the surface of the molecules the ^{13}C-n.m.r. data contain some conformational information. The parameters which are of importance in a conformational analysis are as follows: 1. chemical shifts, 2. long-range coupling constants, 3. spin-lattice relaxation rates.

Due to the angular dependence of the three bond C-O-C-H long range coupling constants (Lemieux et al., 1972; Hamer et al., 1978) it is possible to obtain information about the interglycosidic torsion angles from high resolution proton coupled spectra of oligosaccharides or from spectra of isotopically labelled molecules. These values have in several cases been used successfully in the conformational analysis of oligosaccharides (Lemieux et al., 1980; Bock et al., 1982; Hayes et al., 1982; Gagnaire et al., 1977). The only limitation to the use of these long-range coupling constants is that the line width of the resonance is to some extent dependent on the molecular size, which implies that it is generally not possible to use this approach for larger oligosaccharides.

Finally, relaxation rates can be used in the conformational analysis of oligosaccharides, but due to the dominating dipoledipole relaxation mechanism from the protons directly bonded to the carbon atoms, these values are mainly used in the study of the molecular motion of the individual monosaccharide units in an oligo- or polysaccharide as discussed by several authors (Allerhand and Doddrell, 1971; Czarniecki and Thornton, 1977; Berry et al., 1977; Neszmelyi et al., 1977).

APPLICATIONS

The application of the HSEA calculations combined with n.m.r. analysis in the investigation of the preferred conformation of oligosaccharides has been successful in several cases. Studies of simple disaccharides such as sucrose (Bock and Lemieux, 1982), trehalose (Bock et al., 1983), maltose (Lemieux and Bock, 1983) gentiobiose (Bock and Vignon, 1982), and a disaccharide portion of the Forsmann antigen (Bock et al., 1983) have been published.

Furthermore, the analysis of more complex oligosaccharides related to the blood group determinants (Lemieux et al., 1980; Thøgersen et al., 1982), bacterial O-antigens (Bock et al., 1982; Bock et al., 1984) and complex glycoproteins (Bock et al., 1982; Paulsen et al., 1985; Brisson and Carver, 1983) have been very convincing.

A study of the oligosaccharide components related to gangliosides has also been performed (Sabesan et al., 1984). The results from this analysis are presented in Table 1 for the GM1 structure. The minimum energy conformation of GM1 is furthermore shown in Figure 3. Veluraja and Rao (1983) have also published a force field calculation of GM1 and related structures. The results are also shown in Table 1 and it is clearly seen that the agreement is rather good. The only discrepancy is found for the β-D-galNac link to the 4-position of the internal β-D-gal unit where Veluraja and Rao's results are more than 20° different from the one published by Sabesan et al. (1984). Experimental data obtained from the ^{1}H-n.m.r. data (Sabesan et al., 1984; Koerner et al., 1983) are, however, in better agreement with the conformation suggested by Sabesan et al. An investigation of the preferred conformation of GM1 shows that the sialic acid is introduced into a region of asialo-GM1 which is very hydrophobic; however, inspection of molecular models reveals that G1 still presents two rather different surfaces, one of which is rather hydrophobic and the other much more hydrophilic.

Table 1. *Calculated Minimum Energy Conformation of GM1*[a]

	β-D-gal 1→3 β-D galNAc 1→4	[α-NeuNAc 2→3]	β-D-gal 1→4	β-D-glc 1-Cer
HSEA[b]	55/10	55/10	—165/—15	55/0
Force Field[c]	50/11	26/14	—158/—24	60/0

a: ϕ/ψ = Values as defined in text.
b: Data from Sabesan et al (1984).
c: Data from Veluraja and Rao (1983).

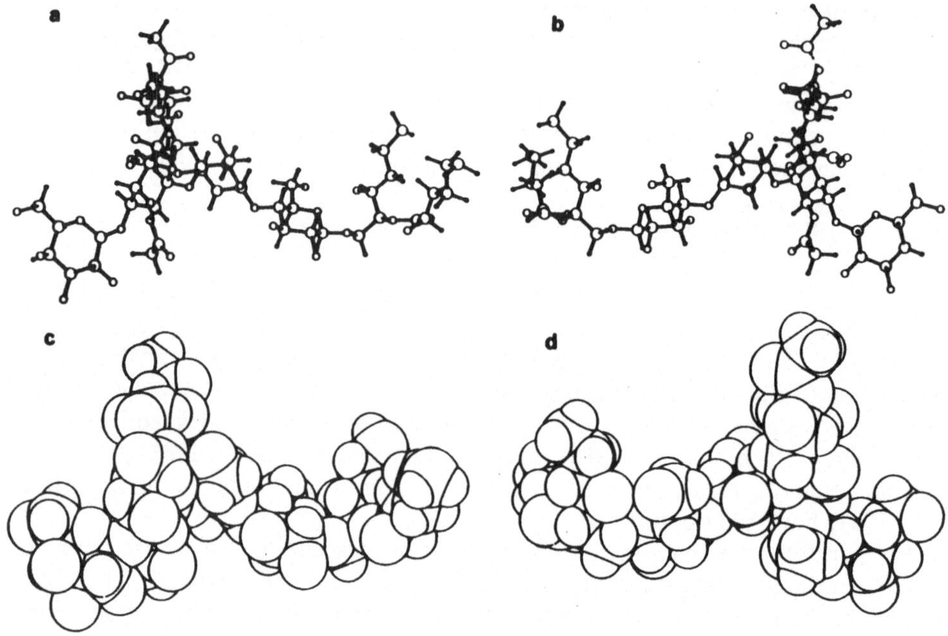

Figure 3. GM1 in its minimum energy conformation (Sabesan et al., 1984) shown from two different faces as a "Stick and Ball model" in A and B and as CPK models in C and D.

CONCLUSIONS

In summary it can be concluded that a simple method (the HSEA calculations combined with n.m.r.-analysis) has been developed which gives a three dimensional picture of the oligosaccharides involved in biological recognition processes. These can then be used in the interpretation of the results from *in vivo* or *in vitro* biological experiments where the oligosaccharides are involved.

ACKNOWLEDGMENTS

The author wishes to express his thanks to Professor R.U. Lemieux, Edmonton, Canada for a very stimulating and inspiring collaboration during the development of the HSEA calculations.

REFERENCES

Allerhand A and Doddrell D (1971) J Am Chem Soc 93: 2777-2779.
Arnott S and Scott WE (1972) J Chem Soc Perkins, II: 324-335.
Berry JM, Hall LD, Wong KF (1977) Carbohydr Res 56: C16-C20.
Berry JM, Hall LD, Welder DG, Wong KF (1979) Am Chem Soc Symposium Series 87: 30-49.

Bluhm TL, Deslandes Y, Marchessault RH, Perez S, Rinaudo M (1982) Carbohydr Res 100: 117-130.
Bock K and Lemieux RU (1979) Jap J Antibiotics 32: S163-S177.
Bock K, Bundle D, Josephson S (1982a) J Chem Soc Perkin II, 59-70.
Bock K, Arnarp J, Lönngren J (1982b) Eur J Biochem 129: 171-178.
Bock K and Vignon M (1982) Nouveau Journal de Chimie 6: 301.
Bock K (1983) Pure and Appl Chem 55: 605-622.
Bock K and Lemieux RU (1982) Carbohydr Res 100: 63-74.
Bock K and Thøgersen H (1983) Annual Reports on NMR Spectroscopy 13: 1-57.
Bock K, Defaye J, Driguez H, Bar-Guilloux E (1983) Eur J Biochem 131: 595-600.
Bock K, Meldal M, Bundle DR, Iversen T, Garegg PJ, Norberg T, Lindberg AA, Svenson SB (1984) Carbohydr Res 130: 23-34.
Brisson J-R and Carver JP (1983) Can J Biochem Cell Biol 61: 1067-1078.
Czarniecki MF and Thornton ER (1977) J Am Chem Soc 99: 8279-82.
Evelyn L, Hall LD, Stevens JD (1982) Carbohydr Res 100: 55-62.
Gagnaire DY, Nardin R., Taravel FR, Vignon MR (1977) Nouveau Journal de Chimie 1: 423-430.
Giacomini M, Pullman B, Maigret B (1970) Theoret Chim Acta 19: 347-364.
Hamer GK, Balza F, Cyr N, Perlin AS (1978) Can J Chem 56: 3109-3110.
Hayes ML, Serianni AS, Barker R (1982) Carbohydr Res 100: 87-101.
Hall LD, Wong KF, Hull WE, JD Stevens (1979) J Chem Commun 953-955.
IUPAC-IUB Commission on Biochemical Nomenclature (1971) Arch Biochem Biophys 145: 405-621.
Jeener J, Meier BH, Bachmann P, Ernst RR (1979) J Chem Phys 71: 4546-4553.
Jeffrey GA, Pople JA, Binkley JS, Vishveshwara S (1978) J Am Chem Soc 100: 373-379.
Jeffrey GA and Taylor R (1980) J Comp Chem 1: 99-109. Jeffrey GA and Yates JH (1981) Carbohydr Res 96: 205-213.
Khare DP, Hindsgaul O, Lemieux RU (1985) Carbohydr Res 136: 285-308.
Kitaygorodsky AI (1961) Tetrahedron 14: 230-236.
Kitaygorodsky AI (1978) Chem Soc Rev 7: 133-162.
Koerner Jr, TAW, Prestegard JH, Demou PC, Yu RK (1983) Biochemistry, 22: 2676-2687.
Lemieux RU, Nagabhushan TL, Paul B (1972) Can J Chem 50: 773-776.
Lemieux RU (1978) Chem Soc Rev 7: 423-452.
Lemieux RU, Koto S, Voisin D (1979) Am Chem Soc Symposium Series 87: 17-29.
Lemieux RU, Bock K, Delbaere LT, Koto S, VS Rao (1980) Can J Chem 58: 631-653.
Lemieux RU (1982) in: Laidler KJ (ed): IUPAC Frontiers of Chemistry. Pergamon Press, New York, pp. 3-24.
Lemieux RU and Bock K (1983) Archiv Biochem & Biophys 221: 125-134.
Lemieux RU (1984) VII International Symp. on Medicinal Chemistry, Aug 27-31, Uppsala, Sweden, Swedish Pharmaceutical Society.
Lindberg B (1981) Chem Soc Rev 10: 409-434.
Marchessault RH, Bleha T, Deslandes Y, Revol J-F (1980) Can J Chem 58: 2415-2422.
Melberg S, Rasmussen K, Scordamaglia R, Tosi C (1979) Carbohydr Res 76: 23-37.
Melberg S and Rasmussen K, (1980) Carbohydr Res 78: 215-224.
Neszmelyi A, Tori K, Lukacz G (1977) J Chem Soc Chem Commun, 613-614.
Noggle JH and Schirmer RE (1971) The Nuclear Overhauser Effect, Academic Press, New York, pp. 45.
Ogawa T, Yamamoto H, Nukuda T, Kitajima T, Sugimoto M (1984) Pure & Appl Chem 56: 779-795.
Paulsen H (1982) Angew Chem Int Ed 21: 155-172.

Paulsen H, Peters T, Sinnwell V, Lebuhn R, Meyer B (1985) Liebigs Ann Chem 489-509.
Rees DA (1981) Pure and Appl Chem 53: 1-14.
Sabesan S, Bock K, Lemieux RU (1984) Can J Chem 62: 1034-1045.
Sharon N (1975) Complex Carbohydrates : Their Chemistry, Biosynthesis and Function, Addison Westley, Reading Mass.
Sharon N and Lis H (1981) Chem Eng News 13: 21-44.
Sheldrick B and Akrigg D (1980) Acta Cryst B36: 1615-21.
Thøgersen H, Lemieux RU, Bock K, Meyer B (1982) Can J Chem 60: 44-57.
Tvaroska I and Kozār T (1980) Am Chem Soc 102: 6929-36.
Veluraja K and Rao VSR (1983) Carbohydr Polymers 3: 175-192.
Wolfe S, Whangbo M-H, Mitchell DJ (1979) Carbohydr Res 69: 1-26.
Yadav JS and Luger P (1983) Carbohydr Res 119: 57-73.

Gangliosides and neuronal plasticity
G. Tettamanti, R.W. Ledeen, K. Sandhoff,
Y. Nagai, G. Toffano (eds.)
Fidia Research Series, vol. 6
Liviana Press, Padova, © 1986

A NOVEL DISIALOGANGLIOSIDE (GD1α) WITH N-ACETYLNEURAMINYL (α2→6)-N-ACETYLGALACTOSAMINE LINKAGE IN RAT ASCITES HEPATOMA CELLS

Takao Taki, Yoshio Hirabayashi, Hidemi Ishikawa, Susumu Ando[1], Kazuo Kon[1], Koichi Tanaka[1], Makoto Matsumoto

Department of Biochemistry, Shizuoka College of Pharmacy, Shizuoka 422, Japan,
[1]Department of Biochemistry, Tokyo Metropolitan Institute of Gerontology,
Itabashi-ku, Tokyo 173, Japan

INTRODUCTION

Gangliosides in plasma membranes have been assumed to be involved in cellular recognition sites such as receptors of bacterial toxins (van Heyningen et al., 1971; Cuatrecasas, 1973), viruses (Haywood, 1974; Holmgren et al., 1980), hormones (Mullin et al., 1976) and chemical mediators (Woolley and Gommi, 1965). On the other hand, some special gangliosides have been proposed to be regulators of cell growth or differentiation (Bremer et al., 1984; Tsuji et al., 1983). Furthermore, alteration of gangliosides associated with oncogenic transformation are well known phenomena (Hakomori, 1984). Since sialic acid of gangliosides plays a key role in these cellular reactions, number of sialic acids, manner of linkage and location of sialyl residues in the back bone structure of gangliosides are thought to be critical for the function of the cells. Sialic acids in ganglioside are attached to the internal or terminal galactose of ganglio-N-tetraose. In the biosynthesis of ganglio series gangliosides, sialic acid is transferred to the galactose moiety of lactosyl-ceramide, then N-acetylgalactosamine and galactose are transferred in stepwise fashion to form GMla. However, in rat ascites hepatoma cells, a biosynthetic pathway of ganglioside via asialogangliosides has been demonstrated by our structural and metabolic studies (Hirabayashi et al., 1978; Taki et al., 1979a). In the tumor cells, N-acetylgalactosamine is preferentially transferred to lactosylceramide and followed by a transfer of galactose to form gangliotetraosylcera-

---

Abbreviations: NMR, nuclear magnetic resonance; NDV, Newcastle disease virus; with regards to ganglioside nomenclature GD1 means a disialoganglioside having the ganglio-N-tetraose structure and the suffix "α" is tentatively alloted to a series of gangliosides which have one sialic acid residue at the C-6 position of the penultimate N-acetylgalactosamine. Accordingly, a monosialoganglioside having a sialic acid attached to the C-6 position of N-acetylgalactosamine in the gangliotetraose should be called GM1α.

mide. Sialic acid is then transferred to the terminal galactose resulting in the formation of GMlb. From the series of study on glycolipids in the tumor cells, two major gangliosides, mono- and disialo-, were isolated. The monosialoganglioside was determined to be GMlb (NeuAc(α2→3)Gal(β1→3)GalNAc(β1→4)Gal(βl→4)Glc(βl→l)Ceramide)(Matsumoto et al., 1981). This paper concerns the structure analysis of the disialoganglioside, named GDlα, following gas chromatographic-mass spectrometric analysis of permethylated sugars, enzymatic degradation and also proton NMR spectra. The ganglioside GDlα was determined to have a unique NeuAcα 2→6 GalNAc linkage.

MATERIALS AND METHODS

Isolation of GD1α

Rat ascites hepatoma AH 7974F cells were obtained as described previously (Hirabayashi et al., 1978). The lyophilized hepatoma cells (38 g) were homogenized with 600 ml of chloroform/methanol (2:1). After filtration, the insoluble residue was reextracted three more times with 600 ml each of chloroform/methanol (1:1). The combined filtrates were evaporated in a rotary evaporator. The total lipids were dissolved in 100 ml of methanolic 0.5 N KOH, and the mixture was incubated at 37°C for 3h. After exhaustive dialysis against distilled water, the sac contents were lyophilized. The base-treated materials was peracetylated, and the glycolipid fraction was obtained by Florisil column chromatography according to the method of Saito and Hakomori (1971). Total glycosphingolipid thus obtained (250 mg) was applied to a DEAE-Cellulofine column and neutral glycolipids were eluted first with 600 ml of chloroform/methanol/water (30:60:8), and finally 300 ml of methanol. Gangliosides were then eluted with 300 ml of 0.1 M ammonium acetate in methanol. The ganglioside fraction was further purified by Iatrobeads column chromatography. After washing the column with 100 ml of chloroform, gangliosides were eluted with a linear gradient formed by 600 ml of chloroform/methanol/water, 70:20:0.2 and 780 ml of chloroform/methanol/water, 20:80:5. Fractions of 6 ml were collected and the purity of ganglioside in each fraction was checked by thin layer chromatography. The fraction containing a disialoganglioside was designated as GDlα and subjected to structure analysis. The yield of the ganglioside was about 1.6 mg from 38 g of dried hepatoma cells.

Analytical Procedures

Neutral sugars and amino sugar in the ganglioside were analyzed by gas chromatography after methanolysis, N-acetylation and trimethylsilylation being performed as described by Bhatti et al. (1970). The long chain bases were determined by gas liquid chromatography according to the method of Gaver and Sweeley (1966). The fatty acid methyl esters were also analyzed by gas chromatography. The detailed analytical conditions for gas chromatographic analysis of monosaccharides, sphingosine bases and fatty acids have been described previously (Matsumoto et al.,

1981). The sphingosine base was determined colorimetrically (Lauter and Trams, 1961).

Methylation Experiments

Gangliosides (about 200 μg) were permethylated with methyliodide and sodium hydride as described by Imanari and Tamura (1967). Aliquots of each methylated sample were methanolyzed with methanolic 3% HCl (w/v) at 100°C for 3h. The methanolysates were analyzed for methylated sialic acid by gas chromatography-mass spectrometry. The remaining permethylated samples were subjected to acetolysis to prepare alditol acetates according to Stoffel and Hanfland (1973). The methylated alditol acetates were analyzed by gas chromatography with a 1.5 % OV-225 column (180° C) for neutral sugar derivatives and with a 1 % OV-17 column (215°C) for amino sugar derivatives.

Sequential Enzymatic Hydrolysis

The carbohydrate sequence and anomeric configurations of the ganglioside were determined by degradation study with various exoglycosidases as used in the previous study (Matsumoto, et al. 1981). Neuraminidase from *Cl. perfringens* and Newcastle disease virus (NDV), β-galactosidase and β-hexosaminidase from jack bean were used.

Nuclear Magnetic Resonance

400-MHz spectra of the intact glycolipid (1.6 mg) were obtained on a JEOL FX-400 spectrometer in the Fourier transformation mode. The operation was performed at 90°C in dimethyl-d_6-sulfoxide (Me_2SO-d_6) solution and then in a Me_2SO-d_6 solution containing D_2O at the same temperature.

RESULTS AND DISCUSSION

Figure 1 shows an elution profile of the hepatoma gangliosides from the Iatrobeads column chromatography. The fractions indicated with a bar in the figure were collected and purity checked by thin layer chromatography with two different solvent systems, chloroform/methanol/water (60 : 35: 8) and chloroform/methanol/2.5 N ammonium hydroxide (60 : 35 : 8). The isolated ganglioside showed one spot between GDla and GDlb in both solvent systems.

Chemical Composition of GDlα

The ganglioside contained glucose, galactose, N-acetylgalactosamine, N-acetylneuraminic acid and sphingosine in a molar ratio of 1 : 2.1 : 0.8 : 1.9 : 0.9. The ganglioside GDlα was found to be hydrolyzed to asialo-GMl by the treatment of *Cl. perfringens* neuraminidase. The enzyme treatment suggested that both sialosyl residues

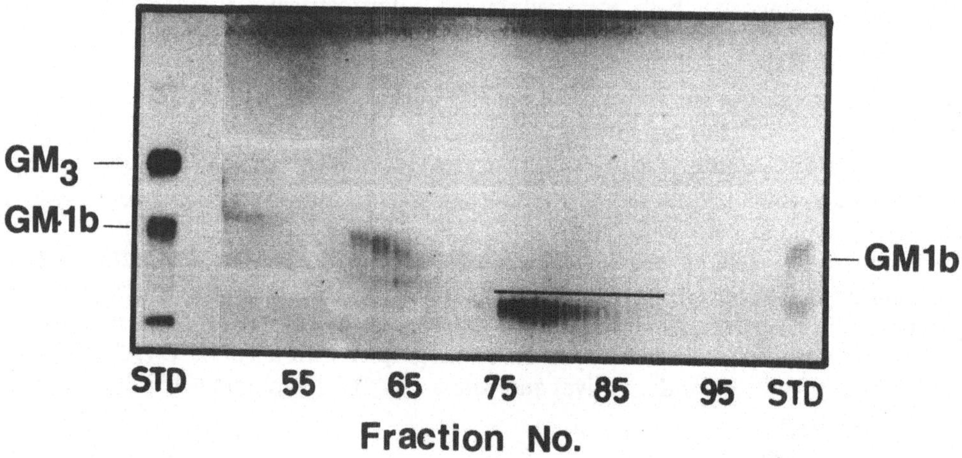

Figure 1. An elution profile of AH 7974F gangliosides from Iatrobeads column chromatography. Aliquots of each fraction was monitored by thin layer chromatography. Gangliosides were visualized with resorcinol/HCl reagent. Solvent system: chloroform/methanol/water, 65:35:8, by volume.

were attached to the terminal galactose-N-acetylgalactosamine portion of gangliotetraosylceramide.

Methylation Experiments on GDIα

Permethylated GDIα was derivatized to partially methylated alditol acetates by acetolysis, followed by reduction and acetylation. Alditol acetates thus obtained were analyzed by gas chromatography-mass spectrometry to determine the linkages of sugar components. The data are summarized in Table 1. As compared with the results on GDla and asialo-GDIα, GDIα gave a unique derivative, 4-0-methyl-1, 3, 5, 6-0-tetra-acetyl-2-methyl-acetamide-2-deoxygalactitol, which indicated the presence of a branch structure at the internal N-acetylgalactosamine. Analysis of alditol acetates of asialo-GDIα showed the appearance of 4, 6-0-dymetyl-1, 3, 5-triacetyl-2-methylacetamide-2-deoxygalactitol. This is strong evidence that one of the two sialic acid residues is attached at the C-6 position of N-acetylgalactosamine. Comparison of the alditol acetate derivatives from GDIα and asialo-GDIα revealed that another sialic acid was attached at the C-3 position of the non-reducing end galactose.

Sialic acid residues were liberated by methanolysis from the permethylated GDIα and appeared to be a single species which was proved to be 4, 7, 8, 9-0-tetramethyl-5-N-methyl, acetyl-neuraminyl methylester, methylketoside by gas chromatography-mass spectrometry. This indicates that the two sialic acid residues are separately located in the back-bone structure instead of forming a disialosyl linkage.

Table 1. *Partially O-Methylated Hexitol and Hexosaminitol Acetates Identified in the Hydrolysates of Permethylated GDlα, Asialo-GDlα and GDla*

Hexitol and hexosaminitol acetates	GDlα	asialo-GDlα	GDla
2,3,6-Tri-0-methyl-1,4,5-Tri-0-acetyl glucitol	+	+	+
2,3,4,6-Tetra-0-methyl-1,5-di-0-acetylgalactitol	-	+	-
2,3,6-Tri-0-methyl-1,4,5-Tri-0-acetylgalactitol	+	+	-
2,4,6-Tri-0-methyl-1,3,5-Tri-0-acetylgalactitol	+	-	+
2,6-Di-0-methyl-1,3,4,5-Tetra-0-acetylgalactitol	-	-	+
4,6-Di-0-methyl-1,3,5-Tri 0-acetylgalactosaminitol	-	+	+
4-Mono-0-methyl-1,3,5,6-Tetra-0-acetylgalactosaminitol	+	-	-

Hydrolysis of GDlα with NDV Neuraminidase and Exoglycosidases

In order to confirm NeuAc(α2→6)GalNAc linkage, NDV neuraminidase was employed, since this enzyme was reported to hydrolyze specifically NeuAc(α2→3)Gal and NeuAc(α2→8)NeuAc linkages, and to have little effect on NeuAc(α2→6) Gal and NeuAc (α2→6)GalNAc linkages in glycoconjugates (Paulson et al., 1982). By the treatment of GDle with NDV neuraminidase, a monosialoganglioside which moved just below GMlb was produced (Fig. 2). Based on the methylation study and the specificity of NDV neuraminidase, the chemical structure of the monosialoganglioside is presumed to be

NeuAcα2 ↘
$\quad\quad\quad\quad$ 6
$\quad\quad\quad\quad$ GalNAcβ1→4 Galβ1→4 Glcβ1→lceramide
$\quad\quad\quad\quad$ 3
$\quad\quad\quad$ ↗
Gal 1β

This ganglioside is designated as GMlα in this paper.

Exoglycosidase treatments were applied to determine the anomeric configurations of each sugar components and their arrangement. The asialo derivative obtained by hydrolysis with *Cl. perfringens* neuraminidase was hydrolyzed by jack bean β-galactosidase to form a glycolipid with the same mobility as that of gangliotriosylcera-

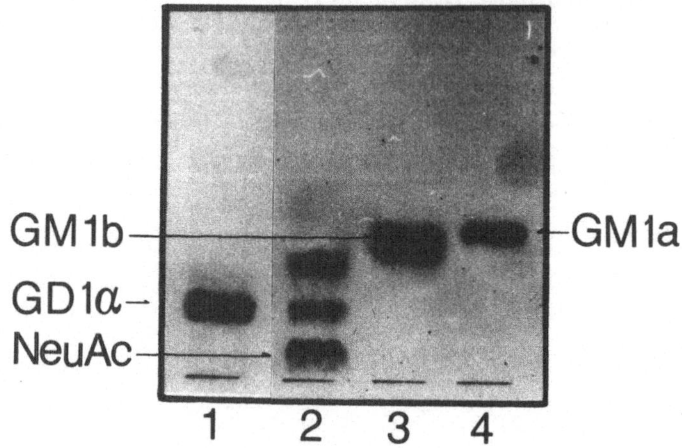

Figure 2. NDV neuraminidase treatment on GDlα. Lane 1, GDle; lane 2, hydrolysis products with NDV neuraminidase and GDlα after 60 min incubation; lane 3, GMlb; lane 4, GMla. Gangliosides and neuraminic acid were visualized with resorcinol/HCl reagent. Solvent system was the same as that in Figure 1.

mide. The glycolipid was then converted into lactosylceramide and monoglycosylceramide by jack bean β-hexosaminidase and β-galactosidase, respectively. These results established the anomeric configurations and sequential arrangement of sugars in the ganglioside as gangliotetraosylceramide.

Based on these results described above, the structure of GDlα is proposed to be:

NeuAcα2 ↘
6 GalNAc β1→4 Gal β1→4 Glcβ1→1ceramide
3
↗
NeuAcα2→3 Gal β1

Proton NMR of GDlα

Proton NMR spectrum of GDlα taken in anhydrous Me_2SO-d_6 is shown in Figure 3. Peaks at 7.716 ppm and 7.670 ppm represent the amide proton of each N-acetylneuraminic acid. Two double doublet signals at 2.688 ppm (coupling constant 11.6) and 2.709 ppm (coupling constant 11.6) were assigned as equatorial H-3 proton of N-acetylneuraminic acid attached to the C-6 position of N-acetylgalactosamine and N-acetylneuraminic acid attached to the C-3 position of galactose according to the data reported by Vliegenthart et al. (1983).

The anomeric signals were assigned from a spectrum measured in a D_2O-

containing solvent. Chemical shifts of the anomeric protons were deduced in the following structure:

NeuAcα2 ↘
6
GalNAc β1→4 Gal β1→4 Glcβ1→1ceramide
3
↗
NeuAcα2→3 Gal β1

4.263 ppm	4.514 ppm	4.213 ppm	4.192 ppm

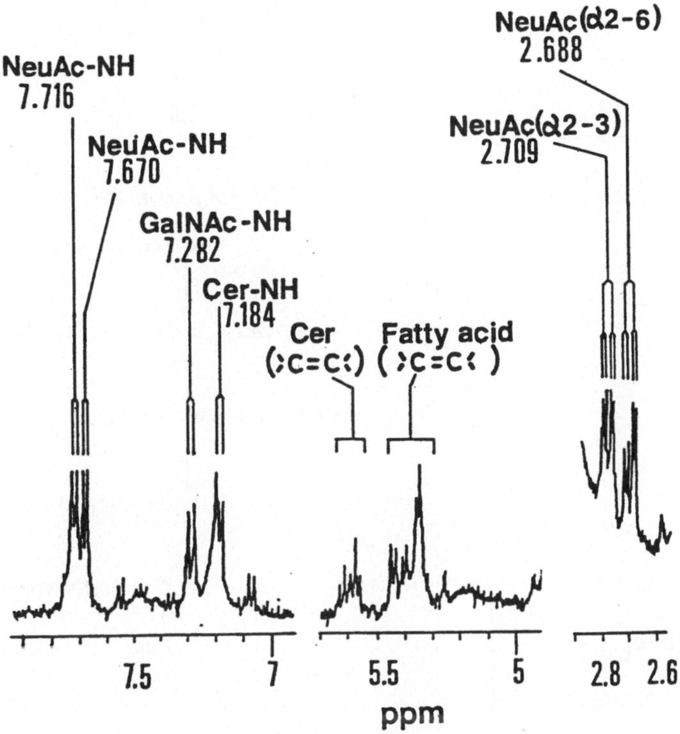

Figure 3. Proton NMR spectrum of GDlα obtained at 400 MHz at 90°C in anhydrous Me_2SO-d_6.

Fatty Acid and Sphingosine Compositions of GDlα

Fatty acid and sphingosine compositions of GDlα are listed in Table 2 and compared with those of GMlb. The major fatty acids of GDlα are $C_{20:0}$, $C_{24:0}$ and $C_{24:1}$.

Table 2. Fatty Acid and Sphingosine Base Compositions of GDIα *and GMIb*

Fatty acid	GDIα	GMIb
$C_{16:0}$	1.3 %	6.9 %
$C_{18:0}$	10.5	12.4
$C_{20:0}$	26.6	16.0
$C_{21:0}$	trace	trace
$C_{22:0}$	15.9	15.8
$C_{23:0}$	1.8	trace
$C_{24:0}$	20.5	19.7
$C_{24:1}$	21.6	26.4
others	1.9	2.8
Sphingosine base	**GDIα**	**GMIb**
$d_{18:0}$	51.0 %	57.6 %
$d_{18:1}$	39.0	27.5
$d_{20:0}$	4.8	8.9
$d_{20:1}$	3.2	6.0
others	2.0	—

Scheme 1. *Biosynthetic Pathway of Glycolipid in Rat Ascites Hepatoma AH 7974F*

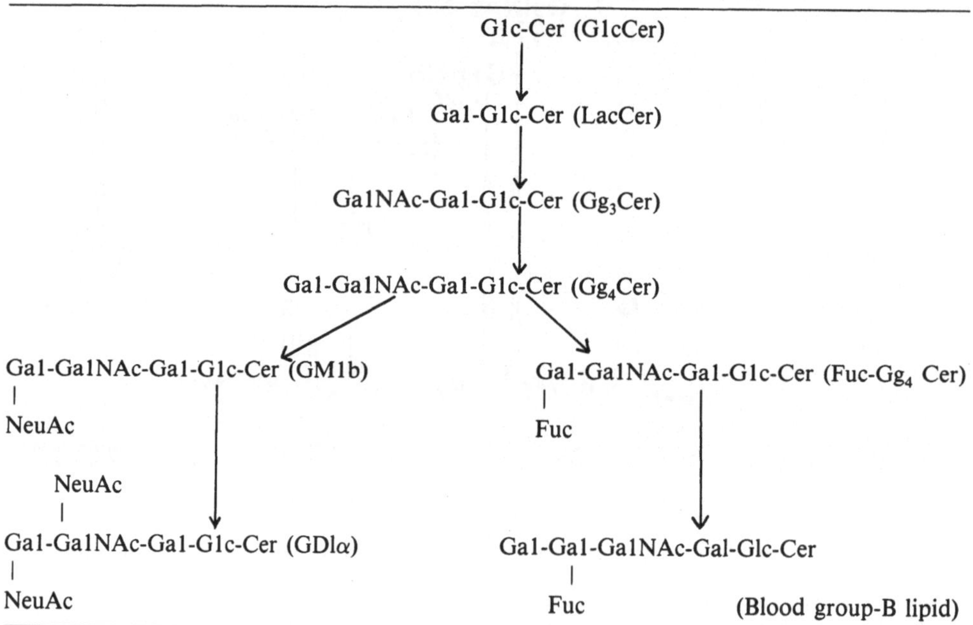

And the main long chain bases are $d_{18:0}$(51 %) and $d_{18:1}$(39 %). These compositions are similar to those of GMIb. This similarity suggested that these two compounds, GMIb and GDIα were metabolically closely related. Recently, we have found a sialyltransferase which catalyzed the transfer of sialic acid from CMP-N-acetylneuraminic acid to GMlb in AH 7974F cells. And we have already demonstrated all the enzyme activities shown in Scheme 1 (Taki et al., 1979a, b). Therefore, GDlα is thought to be synthesized by the pathway shown in Scheme 1.

The structure NeuAcα2→6GalNAc is known to be distributed among glycoproteins having O-linked oligosaccharides in mucin type glycoproteins (Kornfeld and Kornfeld, 1980). This structure, however, has rarely been found in ganglioside species. Recently, Ohashi (1981) isolated from frog brain a ganglioside similar to ours and partially characterized the chemical structure. On the other hand, the appearance of a lacto-series of ganglioside with NeuAcα2→6Gal structure was observed in human cancer tissue as a result of oncofetal expression (Hakomori et al., 1983). GDlα found in rat ascites hepatoma cells may also represent the oncofetal expression of gangliosides.

REFERENCES

Bhatti T, Chambers RE, Clamp JR (1970) The gas chromatographic properties of biologically important N-acetylglucosamine derivatives, monosaccharides, disaccharides, trisaccharides, tetra-saccharides and pentasaccharides. Biochem Biophys Acta 222: 339-347.

Bremer EG, Hakomori S, Bowen-Pope DF, Raines E, Ross R (1984) Ganglioside-mediated Modulation of cell growth, growth factor, binding, and receptor phosphorylation. J Biol Chem 259: 6818-6825.

Cuatrecasas P (1973) Gangliosides and membrane receptors for cholera toxin. Biochemistry 12: 3558-3566.

Gaver RC and Sweeley CC (1966) Chemistry and metabolism of sphingolipids. 3-Oxo derivatives of N-acetylsphingosine and N-acetyl dihydrosphingosine. J Am Chem Soc 88: 3643-3646.

Hakomori S, Patterson CM, Nudelman E, Sekiguchi K (1983) A monoclonal antibody directed to N-acetylneuraminosylα2-6-galactosyl residue in gangliosides and glycoproteins. J Biol Chem 258: 11819-11822.

Hakomori S (1984) Tumor-associated carbohydrate antigens. Ann Rev Immunol 2: 103-126.

Haywood AM (1974) Characteristics of Sendai virus receptors in a model membrane. J Mol Biol 83: 427-436.

Hirabayashi Y, Taki T, Matsumoto M, Kojima K (1978) Comparative study on glycolipid composition between two cell types of rat ascites hepatoma cells. Biochim Biophys Acta 529: 96-105.

Holmgren J, Svennerholm L, Elwing H, Fredman P, Strannegård Ö (1980) Sendai virus receptor: Proposed recognition structure based on binding to plastic adsorbed gangliosides. Proc Natl Acad Sci USA 77: 1947-1950.

Imanari T, Tamura Z (1967) Gas chromotography of glucuronides. Chem Pharm Bull 15: 1677-1681.

Kornfeld R, Kornfeld S (1980) Structure of glycoproteins and their oligosaccharide units. In: Lennarz W J (ed): Biochemistry of glycoproteins and proteoglycans. Plenum Publishing Corp, New York, pp. 1-34.

Lauter CJ, Trams EG (1961) A spectrophotometric determination of sphinogosine. J Lipid Res 3: 136-138.

Matsumoto M, Taki T, Samuelsson B, Pascher I, Hirabayashi Y, Li S- C, Li Y- T (1981) Further characterization of the structure of GMlb ganglioside from rat ascites hepatoma. J Biol Chem 256: 9737-9741.

Mullin BR, Fishman PH, Lee G, Aloj SM, Ledley FD, Winand RJ, Kohn LD, Brady RO (1976) Thyrotropin-ganglioside interactions and their relationship to the structure and function of thyrotropin receptors. Proc Natl Acad Sci USA 73: 842-846.

Ohashi M (1981) A new type of ganglioside. Carbohydrate structures of the major gangliosides from frog brain. In: Yamakawa T, Osawa T, Handa S (eds): Glycoconjugates, Proc VIth Int Symp Glycoconjugates. Japan Sci Soc Press, Tokyo, pp. 33-34.

Paulson JC, Weistein J, Dorland L, van Halbeek H, Vliegenthart JFG (1982) Newcastle disease virus contains a linkage-specific glycoprotein sialidase. J Biol Chem 257: 12734-12738.

Saito T, Hakomori S (1971) Quantitative isolation of total glycosphingolipids from animal cells. J Lipid Res 12: 257-259.

Stoffel W, Hanfland P (1973) Analysis of amino sugar-containing glycosphingolipids by combined gas-liquid chromatography and mass spectrometry. Hoppe-Seyler's Z Physiol Chem 354: 21-31.

Taki T, Hirabayashi Y, Ishiwata Y, Matsumoto M, Kojima K (1979a) Biosynthesis of different gangliosides in two types of rat ascites hepatoma cells with different degrees of cell adhesiveness. Biochim Biophys Acta 572: 113-120.

Taki T, Hirabayashi Y, Matsumoto M, Kojima K (1979b) Enzymic synthesis of a new type of fucose containing glycolipid with fucosyltransferase of rat ascites hepatoma cell AH 7974F. Biochim Biophys Acta 572: 105-112.

Tsuji S, Arita M, Nagai Y (1983) GQlb, A bioactive ganglioside that exhibits novel nerve growth factor (NGF)-like activities in the two neuroblastoma cell lines. J Biochem 94: 303-306.

van Heyningen WE, Carpentar CCJ, Pierce NF, Greenough WB (1971) Deactivation of cholera toxin by ganglioside. J Infec Dis 124: 415-418.

Vliegenthart JFG, Dorland L, van Halbeek H (1983) High-resolution, ^{1}H-Nuclear magnetic resonance spectroscopy as a tool in the structural analysis of carbohydrates related to glycoproteins. Ad Carbohyd Chem Biochem 41: 209-374.

Woolley DW, Gommi BW (1965) Serotonin receptors. VII. Activities of various gangliosides as the receptors. Proc Natl Acad Sci, USA 53: 959-963.

Gangliosides and neuronal plasticity
G. Tettamanti, R.W. Ledeen, K. Sandhoff,
Y. Nagai, G. Toffano (eds.)
Fidia Research Series, vol. 6
Liviana Press, Padova, © 1986

TOPOGRAPHICAL DISTRIBUTION OF THE GANGLIOSIDES IN THE DEVELOPING AND ADULT HUMAN BRAIN

Ivica Kracun,[1,2] Harald Rösner,[3] Cedomir Cosovic[1]

[1]Department of Chemistry and Biochemistry, Medical Faculty, Zagreb, Yugoslavia;
[2]Clinical Hospital Center, Rebro, Neurobiochemical Laboratory, Yugoslavia and
[3]Institute of Zoology, University Hohenheim, Stuttgart,
Federal Republic of Germany

INTRODUCTION

In contrast to the experimental knowledge about gangliosides, there are relatively few studies describing gangliosides in the developing (Vanier et al., 1971; Yusuf et al., 1977; Martinez and Ballabriga, 1978; Kracun et al., 1983) and adult (Suzuki, 1965; Riboni et al., 1985) human brain. The developmental studies revealed that the prenatal period of "growth spurt", occurring between 15 and 30 weeks of gestation (Sidman and Rakic, 1973), is paralleled by an increase in the content of gangliosides and a change in their composition. In animal studies similar developmental changes of gangliosides have been confirmed as consecutive phenomena of neuron genesis, neuronal differentiation, synaptogenesis and myelination (Dreyfus et al., 1980; Rösner, 1977; 1980; 1982).

In the adult human brain, regional differences in ganglioside pattern were first detected by Suzuki (1965). In all the other studies of human brain, however, mostly gross-brain regions or even whole brains have been examined. From such data a connection of the gangliosides with the cytoarchitecture of the well-defined compartments was impossible. Therefore, we tried to establish a structurally related distribution of gangliosides in the adult human brain and to compare this with the premature defined anlage of selected areas such as the occipital and frontal cortex and hippocampus.

Abbreviations: A, amygdala; As, area subcallosa; Cc, cerebellar cortex; CC, corpus callosum; CI, capsula interna; CIn, colliculi inferiores; CM, corpora mammillaria; CS, colliculi superiores; E, epiphysis (pineal gland); F, frontal cortex; FW, frontal white matter; gC, gyrus cinguli; Gd, gyrus dentatus; GP, globus pallidus; h, hippocampus; Hy, hypothalamus; I, insular cortex; LC, locus coeruleus; Nac, nucleus accumbens; NBM, nucleus basalis Meynert; NC, nucleus caudatus; NeuAc, N-acetylneuraminic acid; Nr, nucleus ruber; Oc, occipital cortex; Of, orbitofrontal cortex; oh, optic chiasma; P, putamen; Pf, prefrontal cortex; Pr, parietal cortex; Sn, substantia nigra; Sr, striatal (visual) cortex; T, temporal cortex; TLC, thin-layer chromatography; Th, thalamus; Thp, pulvinar thalami; To, paleocortex (tuberculum olfactorium).

MATERIALS AND METHODS

Four adult human brains, corresponding to the age between 40-50 years were derived from Forensic Medicine. Human fetal brains at 16, 22, and 30 weeks of gestation were obtained from legal abortions. The fetal age was calculated on the basis of crown rump length (mm, Pontineau-Olivie Index). One brain of a four months old child was included. Dissections were performed within 12 hours after death on a cold plate under stereomicroscopic control (Kracun et al., 1984).

From homogenized samples the gangliosides were extracted according to Svennerholm and Fredman (1980) and purified for thin-layer chromatography (TLC) as described previously (Rösner, 1982). Total lipid-bound sialic acid (NeuAc).was determined with the HCl/resorcinol reagent (Svennerholm, 1957) and expressed as μg NeuAc per mg protein. Total proteins were determined by the Lowry method (Lowry et al., 1951). TLC of the gangliosides was performed according to Rösner (1980). The distribution of individual gangliosides (% of total ganglioside-NeuAc) was determined by densitometric scanning. From these values the a/b ratios (% GT1a + % GD1a + % GM1/ % GQ1b + % GT1b + % GD1b) were calculated, which give an indication of a possible preponderance either of a- or b-pathway of ganglioside biosynthesis (Yu and Ando, 1980).

RESULTS

Developmental Expression of Gangliosides in Layers of the Frontal and Occipital Neocortex and Hippocampus

All developmental data are given in Table 1 and in Figure 1 a-e. In all three brain areas the most striking developmental accretion of gangliosides was established between 16 and 22 weeks of gestation. After 30 weeks to adult the content tended to drop in the hippocampus. In the occipital cortex the concentration remained rather constant. In the frontal cortex two maxima of ganglioside concentration were revealed, at 22 weeks of gestation and at 4 months after birth. Looking, however, at the developing laminas of both neocortical areas, different developmental profiles of ganglioside concentration were revealed for the sublate layer and the cortical plate. The latter showed a maximum at 22 weeks of gestation. In the sublate layer high values were found at 22 and 30 weeks. Thereafter, the gangliosides decreased remarkably (to 2 μg NeuAc/mg protein) in the sublate layer (now white matter) of the adult frontal cortex.

In spite of the differences in total ganglioside content the relative developmental profiles of individual gangliosides were very similar in the sublate layer and cortical plate of both the frontal and occipital cortex up to 4 months after birth (Fig. 1 a-e). Thereafter, up to adult, the areas developed differently. The most striking change occurred in all four areas between 16 and 22 weeks of gestation, characterized by a decrease in the percentage of GQ1b, GT1b, GD1b, and an increase of GD1a and GM1. Highest %-values for GD1a were obtained at 22 weeks (about 40% in both occipital and 50% in the frontal laminas). In the occipital cortex (both layers) the b-pathway gangliosides decreased up to 30 weeks. Thereafter a striking increase especially of

Table 1. *Total Lipid-Bound Sialic Acid Distribution and a/b Ratio of the Major Ganglioside Fractions in the Cortical Layers (Frontal, Occipital) and Hippocampus during Human Brain Development*

REGION (layer)	AGE 16 w	22 w	30 w	4 m	A
FRONTAL CORTEX	7.8 ±2.7 a/b:1.08	16.8±1.5 2.1	7.6 ±1.3 2.4	14.9 ±0.9 2.05	11.8 ±2.5 1.35
Cortical plate	7.3 ±0.7	20.0±0.3	8.86		
Subplate layer	9.1 ±1.4	16.6±2.7	16.7 ±2.6	14.8[a]	2.0 ±0.1
OCCIPITAL CORTEX	5.2 ±1.6 a/b:0.6	11.4 1.8	10.7 ±2.7 1.18	12.2 ±2.8 1.63	10.9 ±0.9 0.64
Cortical plate	4.9 ±0.7	16.0±3.8	8.9 ±1.9		
Subplate layer	8.3	15.2±1.2	16.4 ±3.0	10.68[a]	
HIPPOCAMPUS	7.9 a/b:1.2	11.4±2.6 1.7	12.02 1.6	9.69 1.6	9.1 ±0.9 3.0

Total lipid bound sialic acid is expressed as μg/mg protein: a/b ratio of the major ganglioside fractions (%GT1a + %GD1a + %GM1/%GQ1b + %GT1b + %GD1b) is given as specified in "Materials and Methods";

± mean standard error from 2-8 samples of one brain/age

w - weeks of gestation, m - months of postnatal life

[a] underlying white matter of the frontal/occipital cortex

GD1b occurred leading to a preponderance of this ganglioside (more than 30% of total ganglioside-NeuAc) in the adult occipital cortical anlage. Also in the frontal cortical plate (both layers) GD1b increased over the same period, but remained below 20% of total ganglioside-NeuAc in the adult. In the frontal cortical plate GD1a, although decreasing relative to the other gangliosides after 22 weeks of gestation, remained the predominant ganglioside in the adult. Unlike all other gangliosides there was a more or less continuous increase of the percentage of GM1 throughout whole development. In the frontal sublate layer of the adult (now white matter) GM1 became the main fraction (about 30% of total ganglioside-NeuAc).

Looking at the hippocampus pattern changes were not as pronounced as in the neocortex, but in principle very similar. This area, however, was characterized by an increasing predominance of GD1a and GM1 over GD1b and GT1b during postnatal development from 4 months to adult.

The described pattern changes are more simply demonstrated also by the a/b ratios (Tab. 1) for the total frontal and occipital cortex as well as the hippocampus. The a/b ratios increased transiently in the neocortex (maximum occipital at 22, frontal at 30 weeks of gest). In the hippocampus, however, a striking increase of this ratio was observed in the later postnatal period, indicating a maturation to a striking predominance of a-gangliosides in this structure.

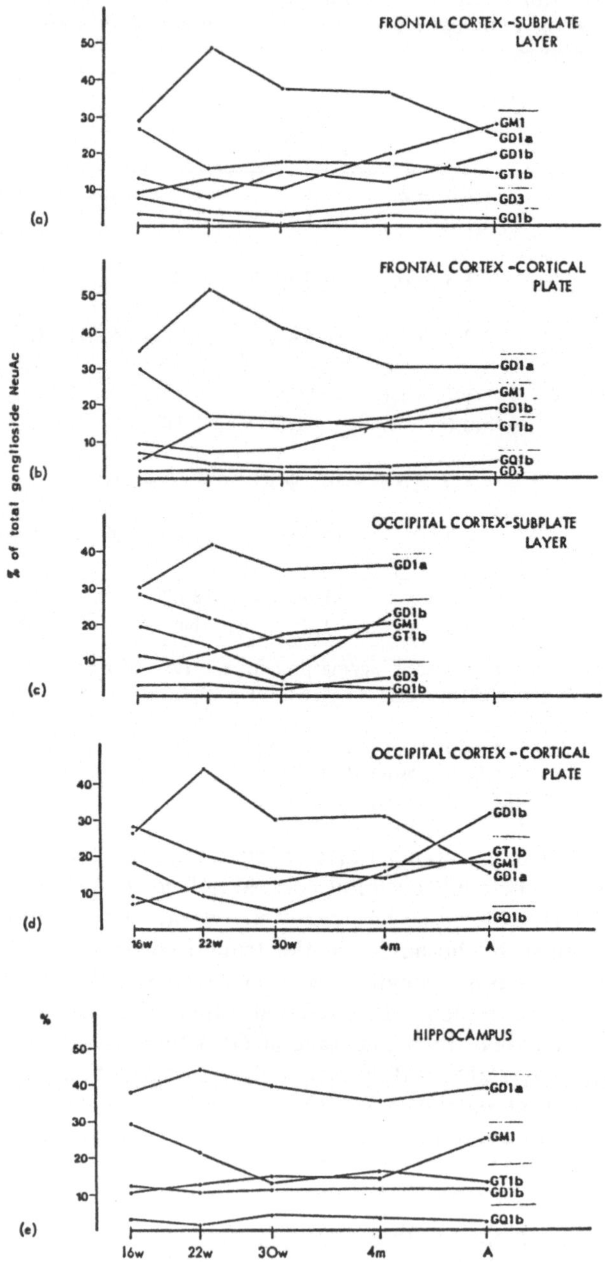

Figure 1 a-e. Developmental profiles of the individual gangliosides (% of total ganglioside-NeuAc) within neocortical layers of the fontal and occipital pole (a-d) in comparison with hippocampus (e).
w-weeks of gestation; m-months of postnatal life; A-adult human brain.
For each stage one brain was examined except adult stage (N=4).

Topographical Distribution of Gangliosides in the Adult Human Brain

All data are given in Table 2 and Figure 2a, b. In the adult human brain the highest content of gangliosides (up to 13.7 μg NeuAc/mg protein) was measured in the hemispheric cortex. The ganglioside pattern showed a shift from frontal (a/b ratio 1.1. - 1.3) to occipital (a/b ratio 0.64) mainly due to a higher percentage of GD1b within occipital brain regions, particularly the visual cortex (a/b ratio 0.55). The archicortical

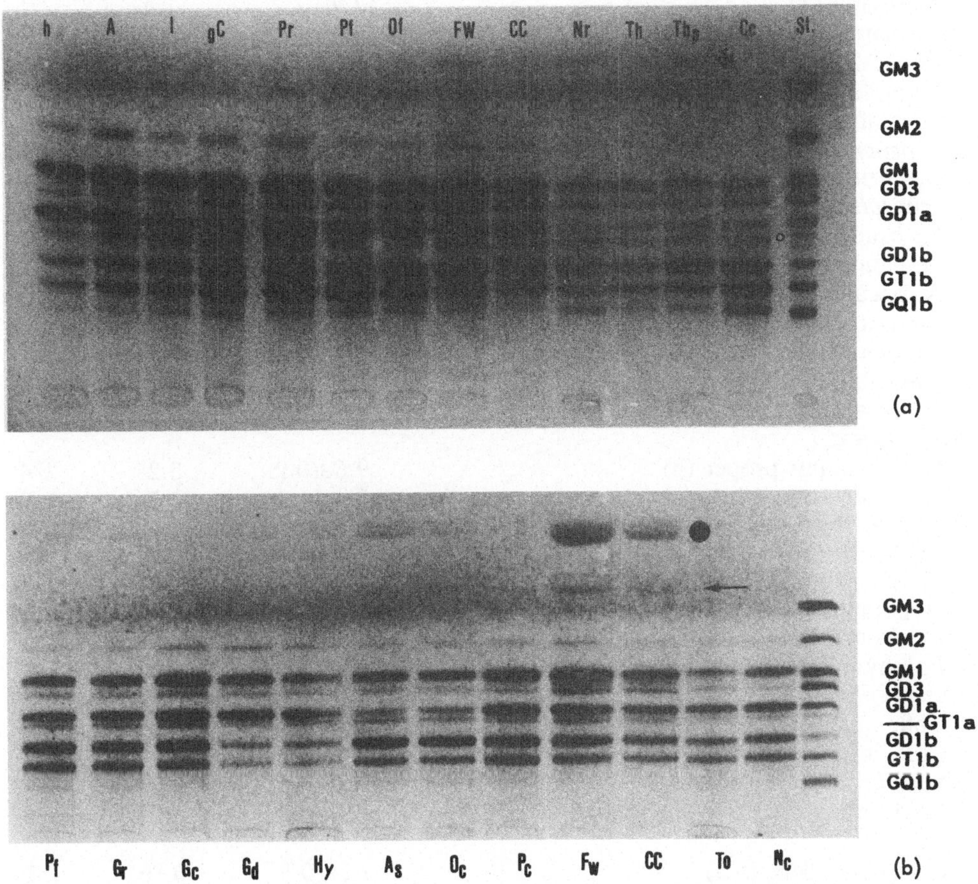

Figure 2. Thin-layer chromatography of the gangliosides extracted from the distinct regions of the adult human brain. On Fig. 2a, white round marks the spot migrating similary to GT1a. On the Fig. 2b, black point marks the asialo-lipids. Black arrow - fraction migrating ahead of GM3 corresponding to GM4. Gr - cortex of the frontal gyrus rectus; for other abbreviations see table 2. TLC plates were developed with the double-solvent system (Rösner, 1980); first run: chloroform-methanol-12 mM $MgCl_2$-15 N NH_4OH 60:36:7:3.8, by volume; second run: chloroform-methanol-12 mM $MgCl_2$ 58:40:9, by volume.

Table 2. *Total Lipid-bound Sialic Acid Distribution in the Adult Human Brain (means* ± SD)

Region	μg NeuAc/mg protein	a/b ratio[a]	n/m[b]
Neocortex			
Prefrontal (Pf)	11.3 ± 1.4	1.12	4/16
Frontal (F)	11.8 ± 2.5	1.35	2/3
Orbitofrontal (Of)	10.9 ± 1.0	1.28	3/6
Cingular (gC)	8.6 ± 1.3	1.27	2/4
Temporal (T)	13.7 ± 1.2	1.47	3/6
Insular (I)	10.3 ± 1.1	1.24	2/3
Parietal (Pr)	11.2 ± 1.1	0.95	3/8
Occipital (Oc)	10.9 ± 0.9	0.64	2/5
Striatal (Sr)	9.9 ± 0.9	0.55	2/3
Cerebellar cortex (Cc)	9.8 ± 1.6	0.36	2/8
Paleocortex (To)	6.8 ± 2.4	0.95	1/2
Telencephalon (nuclei)			
N. caudatus caput (NC)	11.4 ± 1.5	0.93	3/10
N. caudatus corpus (NC)	7.5 ± 1.2	0.99	3/4
N. accumbens (Nac)	10.6 ± 1.2	1.13	2/3
Putamen (P)	7.5 ± 1.0	1.26	2/7
Globus pallidus (GP)	7.2 ± 1.4	1.67	4/7
Amygdala (A)	11.9 ± 1.4	2.4	4/5
N. basalis Meynert (NBM)	8.9 ± 0.3	1.6	2/2
Archicortex			
Hippocampus proper (h)	9.6 ± 0.9	3.2	2/4
Gyrus dentatus (Gd)	8.4 ± 1.1	2.7	1/2
Diencephalon			
Thalamus (Th)	7.4 ± 0.7	0.7	2/5
Thalamus pulvinar (Thp)	4.1 ± 0.5	0.57	2/2
Hypothalamus (Hy)	8.8 ± 1.4	0.86	4/4
C. mammillaria (CM)	5.5 ± 0.05	0.88	2/4
Epiphysis (E)	3.9 ± 1.0	0.92	3/3
Mesencephalon			
Coll. superiores (CS)	6.5 ± 0.9	0.48	2/2
Coll. inferiores (CIn)	8.4 ± 1.0	1.22	2/2
Substantia nigra (Sn)	7.4 ± 0.5	1.28	1/3
Nucleus ruber (Nr)	3.9 ± 0.2	0.56	2/6
Pons			
Locus coeruleus (LC)	5.1 ± 0.5	0.86	1/2
Fibers			
Frontal white matter (FW)	2.0 ± 0.1	1.36	2/4
Optic chiasm (oh)	2.1 ± 0.1	0.68	2/2
Capsula interna (CI)	2.9 ± 0.8	1.45	2/4
Corpus callosum (CC)	2.4 ± 0.8	2.1	2/6

[a] a/b ratio of the major ganglioside fractions (%) (%GT1a + %GD1a + %GM1/%GQ1b + %GT1b + + %GD1b) (see "Material and Methods").
[b] n, number of brains, m, number of samples

regions (hippocampus, gyrus dentatus) showed an unusually high proportion of GD1a and GM1 (a/b ratio 3.2 and 2.7, respectively). Subcortical telencephalic nuclei (corpus striatum) had a similar ganglioside pattern as the hemispheral cortex, with relatively higher proportions of GD1a and GM1 relative to GT1b, GQ1b and GD1b. In the globus pallidus an even higher percentage of GM1 than in the hemispheral cortex as well as detectable amounts of GM4 were found. The amygdala (archistriatum) was an exception within telencephalic nuclei with very high values for GD1a and GM1 (like hippocampus) and an a/b ratio of 3.0. Diencephalic regions in general had lower amounts of total gangliosides (3.9 - 8.8 μg NeuAc/mg protein). The ganglioside patterns were characterized by a predominance of the b-pathway (a/b ratio 0.6 - 0.7). Also GM4 could be clearly recognized (Fig. 2a, b arrow).

Within the mesencephalon, the ganglioside content was similar to that of diencephalic nuclei. However, the ganglioside patterns were different. The nucleus ruber contained more GD1b and GT1b (a/b ratio 0.56), while in the adjacent substance nigra (a/b ratio 1.28) a reverse ganglioside pattern with high proportions of GD1a and GM1 was found. A similar difference was observed between the superior (a/b ratio 0.48) and the inferior (a/b ratio 1.22) colliculum of the mesencephalic tectum. Recognizable differences between white matter tracts on the basis of the ganglioside patterns were also observed comparing corpus callosum (a/b ratio 2.1), frontal white matter (a/b ratio 1.36), optic tract (a/b ratio 0.68), and internal capsule (a/b ratio 1.45).

DISCUSSION

Gangliosides of the Adult Human Brain

In this chapter we attempt to briefly describe some regional differences of gangliosides in relation to the adult human brain cytoarchitecture. Ganglioside content showed the highest values within the hemispheral neocortex and neostriatum (N. caudatus), which well correlates with the highest density of the neuronal membranes (processes, synapses) generally believed to contain the bulk of cellular gangliosides (Ledeen, 1978). Toward the brain stem gangliosides content drops parallelly to the decrease in the proportion of neuropil in favour of myelinated tracts. The higher degree of myelination in these samples correlates well with the presence of GM4, which is enriched in myelin (Ledeen et al., 1980; Rösner, 1982).

Interesting regional differences in ganglioside pattern were observed in the hemispheral cortex, showing a high percentage of GD1b in the visual cortex in comparison with other cortical fields. A high proportion of GD1b was also detected in the optic tract which terminates in the IVth layer of the visual cortex (Gennari spot), which is clearly visible to the naked eye (Filimonoff, 1947). We therefore suppose that the high level of GD1b in the visual cortex is due to terminals of optic fibers. At present it is not possible to interpretate the different distributions of gangliosides in the other cortical fields. However it seems that they correlate with ontogenetic-phylogenetic cortical divisions (Economo and Koskinas, 1925) discussed previously (Kracun et al., 1984).

Within the archicortex (hippocampus) and archistriatum (amygdala) exceptionally high percentage values of GD1a and GM1 have been revealed. A comparable preponderance of a-gangliosides has been also reported for the rat hippocampus (Irwin and Irwin, 1982) as well as for retinal ganglion cells in the chicken (Rösner, 1982). These observations suggest a possible functional relation of a-pathway gangliosides to glutaminergic synaptic transmission, as was inferred recently — especially for GM1 — in slices from rat hippocampus by electrophysiological recordings (Wieraszko and Seifert, 1984).

Comparing the telencephalic with diencephalic nuclei, some higher precentage values of GD1b characterize structures of the diencephalon, possibly because of many passing fibres containing GD1b (Seyfried et al., 1984). A good example is the nucleus ruber containing numerous passing fibres of the cerebellorubral and thalamorubral pathway and showing a high level of GD1b and appreciable amounts of GM4.

Gangliosides of the Developing Human Brain

During development, in the maturating layers of the neocortex (frontal, occipital) an increase in ganglioside content occurs between 16 and 22 weeks of gestation. Simultaneously the proportion of GD1a increases and that of GT1b decreases. This change in the ganglioside composition corresponds to the period of most intensive human cortical synaptogenesis (Kostovic and Molliver, 1974) occurring within the cortical plate-subplate layer interface and the marginal zone. This developmental result well supports our preliminary findings about ganglioside distribution within the neocortical layers of the 28 week old fetus (Kracun et al., 1983) showing an increase of GD1a and a decrease of GT1b from the proliferating (ventricular zone) to differentiating layers. The presented data show that increase of GD1a is a common feature of the cortical plate and sublate layer in both occipital and frontal cortex between 16 and 22 weeks of gestation. After 30 weeks of gestation the frontal and occipital cortex begin to develop differently with respect to ganglioside composition. However, this difference becomes clear not before 4 months after birth. In both areas a postnatal increase in GD1b (more pronounced in the occipital) and an increase in GM1 (more pronounced in the frontal) lead to the final area-specific pattern of the adult.

In the hippocampus, unlike the neocortex, there is no perinatal increase in GD1b, and GD1a remains at a high level (around 40% of total ganglioside-NeuAc) from 16 weeks of gestation to adult. This may be indicative of an earlier differentiation of the hippocampus (prenatal myelination and establishment of interneurons) as pointed out by Jacobson (1978).

One general conclusion from these developmental data is that during early differentiation (up to 22 weeks of gestation) there are similar changes in gangliosides overall in the human brain, characterized by an increase in total concentration and a shift in the pattern in favour of a-pathway gangliosides. These correspond to the well-documented developmental changes of brain gangliosides in mammals and birds.

Probably more important, from a functional point of view, seems the finding that in human brain, characterized by a "prolongated" functional maturation, a second period of ganglioside changes starts perinatally and becomes most pronounced after 4 months from birth. It is in this late period, in which the different brain regions "select" distinct ganglioside species, that a high degree of regional pattern differences result in the adult. Therefore, gangliosides in the human brain should be further studied at the level of distinct regions by correlating neuroanatomical, immunohistochemical and biochemical methods with greater attention on minor ganglioside fractions including alkali labile species (Riboni et al., 1984).

REFERENCES

Dreyfus H, Louis JC, Harth S, Mandel P (1980) Gangliosides in cultured neurons. Neurosci 5: 1647-1655.

Economo C and Koskinas GN (1925) Die Cytoarchitektonik der Hirnrinde des erwachsenen Menschen, Springer, Berlin.

Filimonoff IN (1947) A rationale subdivision of the cerebral cortex. Arch Neurol Psychiat 58: 296-311.

Irwin LN and Irwin CC (1982) Developmental changes and regional variation in the gangliosides composition of the rat hippocampus. Dev Brain Res 4: 481-485.

Jacobson M (1978) Developmental neurobiology, Plenum Press, New York.

Kostovic I and Molliver MW (1974) A new interpretation of the laminar development of cerebral cortex: synaptogenesis in different layers of the neopallium in the human fetus. Anat Rec 178, 395.

Kracun I, Rösner H, Kostovic I, Rahmann H (1983) Areal and laminar distribution of gangliosides in the fetal human neopallium at 28 weeks of gestation. Roux's Arch Dev Biol 192: 108-112.

Kracun I, Rösner H, Cosovic C, Stavljenic A (1984) Topographical atlas of the gangliosides of the adult human brain. J Neurochem 43: 979-989.

Ledeen RW (1978) Ganglioside structures and distribution: are they located at nerve endings? J Supramolec Struct 8: 1-17.

Ledeen RW, Cochran FB, Yu RK, Samuels FG, Haley JE (1980) Gangliosides of the CNS myelin membrane. Adv Exp Biol Med 125: 167-176.

Lowry OH, Rosebrough NJ, Farr AL, Randall RJ (1951) Protein measurement with Folin phenol reagent. J Biol Chem 193: 265-275.

Martinez M, Ballabriga A (1978) A Chemical study on the development of the human forebrain and cerebellum during the growth spurt period. Gangliosides and plasmalogens. Brain Res 159: 351-362.

Riboni L, Malesci A, Gaini SM, Sonnino S, Ghidoni R, Tettamanti G (1984) Ganglioside pattern of normal human brain, from samples obtained at surgery. A study especially referred to alkali labile species. J Biochem 96: 1943-1946.

Rösner H (1977) Gangliosides, sialoglycoproteins and acetylcholinesterase of the developing mouse brain. Roux's Arch Dev Biol 183: 325-335.

Rösner H (1980) A new thin-layer chromatographic approach for separation of multisialogangliosides. Analyt Biochem 109: 437-442.

Rösner H (1982) Gangliosides changes in the chicken optic lobes as biochemical indicators of brain development and maturation. Brain Res 236: 46-61.

Seyfried TN, Bernard DJ, Yu RK (1984) Cellular distribution of gangliosides in the developing mouse cerebellum: analysis using the staggerer mutant. J Neurochem 43: 1152-1162.

Suzuki K (1965) The pattern of mammalian brain gangliosides. III. Regional and developmental differences. J Neurochem 12: 969-979.

Svennerholm L (1957) Quantitative estimation of sialic acids. Biochim biophys Acta 24: 604-611.

Svennerholm L and Fredman P (1980) A procedure for quantitative isolation of brain gangliosides. Biochim Biophys Acta 617: 97-109.

Vanier MT, Holm M, Ohmann R, Svennerholm L (1971) Developmental profiles of gangliosides in human and rat brain. J Neurochem 18: 581-592.

Sidman RL and Rakic P (1973) Neuronal migration, with special references to developing human brain: A review. Brain Res 62: 1-35.

Wieraszko A and Seifert W (1984) Evidence for a functional role of gangliosides in synaptic transmission: studies on rat brain striatal slices. Neurosci Lett 52: 123-128.

Yu RK and Ando S (1980) Structures of some new complex gangliosides of fish brain. In: Svennerholm L, Dreyfus H, Urban P (eds): Structure and Function of Gangliosides, Plenum Press, New York, pp. 33-45.

Yusuf HKM, Merat A, Dickerson JWT (1977) Effect of development on the gangliosides of human brain. J Neurochem 28: 1299-1304.

Gangliosides and neuronal plasticity
G. Tettamanti, R.W. Ledeen, K. Sandhoff,
Y. Nagai, G. Toffano (eds.)
Fidia Research Series, vol. 6
Liviana Press, Padova, © 1986

THE USE OF ANTIBODIES TO IDENTIFY GLYCOSPHINGOLIPIDS AND TO LOCALIZE THEM IN TISSUES

Donald M. Marcus

Departments of Medicine, Microbiology and Immunology, Baylor College of Medicine, One Baylor Plaza, Houston, Texas 77030, USA

INTRODUCTION

Antibodies to glycosphingolipids (GSLs) are used by many investigators to identify carbohydrate structures of GSLs and for localization of these compounds in tissue sections. Proper interpretation of these immunological data requires knowledge of certain principles of carbohydrate immunochemistry and these will be reviewed in the next section. Some recent immunocytochemical studies will be presented in the final section.

CARBOHYDRATE IMMUNOLOGY

Antibodies to carbohydrate antigenic determinants "recognize" sugar sequences, the configuration of hydroxyl groups, anomeric linkages and conformation. It is not generally appreciated that there are two fundamentally different types of anti-carbohydrate binding sites, cavity and groove sites, and that quite different kinds of information are obtained by use of these antibodies. Antibodies with cavity sites bind to the terminal nonreducing portion of carbohydrate chains and most of their binding energy derives from binding to the terminal and subterminal sugar residues (Cisar et al., 1975; Kabat, 1976; Sharon et al, 1981; Berzofsky and Berkower, 1984). An example of this type of specificity is provided by antibodies directed against the human ABH blood group antigens (Kabat, 1976; Watkins, 1980). The H antigen (Table 1A) is the biosynthetic precursor of the A and B antigens. Addition of an α-galactosyl residue to the H structure creates the B determinant, which is essentially nonreactive with anti-H antibodies.

A sharp contrast is provided by antibodies directed against certain lipo-

Abbreviations: GSL, glycosphingolipid; gal, D-galactose; GlcNAc, N-acetyl-D-glucosamine; GalNAc, N-acetyl-D-galactosamine, Man, D-mannose; Rha, L-rhammose.

polysaccharide antigens of Salmonella, which are composed of repeating oligosaccharide units (Table 1B). Antibodies that react with the group E determinant

Table 1. *Structures of Carbohydrate Antigenic Determinants*

A. Blood group determinat

```
H                                      Gal (β1-4) GlcNAc——————
                                        |  (α1-2)
                                       Fuc

B                          Gal (α1-3) Gal (β1-4) GlcNAc——————
                                        |  (α1-2)
                                       Fuc

A                       GalNAc (α1-3) Gal (β1-4) GlcNAc——————
                                        |  (α1-2)
                                       Fuc
```

B. Salmonella group E lipopolysaccharide determinants

```
             Man (α1-4) Rha
Gal (α1-6) Man (α1-4) Rha
Gal (β1-6) Man (α1-4) Rha
```

designated 3 are strongly inhibited by the disaccharide Man-L-Rham (Uchida et al, 1963). These antibodies are inhibited equally well by trisaccharides that contain a terminal non-reducing α- or β-galactosyl residue linket to mannose (Fig. 1). Kabat (1976) has suggested that the combining site of these antibodies is in the shape of a shallow groove that is open at each end. Other antibodies that can bind to internal determinants include anti-dextrans (Cisar et al., 1975; Sharon et al., 1981), antibodies against globotetraosylceramide (Schwarting et al., 1979; Karol, et al., 1981) that also bind to globopentaosylceramide (Forssman), and antibodies to GM1 (Kundu et al., 1980). In the latter study antibodies elicited by immunization with GM1 and purified by elution from a GM1 affinity column reacted with GM2, among other glycolipids. The importance of this distinction is that if one wishes to use antibodies to identify sugar sequences in glycoconjugates, it is important to determine whether the antibodies used react with terminal nonreducing residues or internal determinants, because quite different information will be obtained from use of the different kinds of antibodies.

Two other points should also be kept in mind. Anti-carbohydrate antibodies will react with the appropriate sugar sequences of GSLs or glycoproteins, and other techniques need to be used to identify the carrier portion of the molecule. For example, a monoclonal antibody that identified a ganglioside in extracts of gastrointestinal tumors (Magnani, et al., 1982), was found to bind to a glycoprotein in serum (Magnani, et al., 1983). Second, the full range of crossreactivity of antibodies, including monoclonal antibodies, may not be appreciated initially either because of a lack of potentially crossreactive compounds for examination, or because of the insensitivity of certain assay techniques. For example, a monoclonal antibody (A2B5) that was thought to recognize only GQ1b ganglioside (Eisenbarth, et al., 1979) was subsequently found to bind to a number of other gangliosides (Kundu et al., 1983; Kasai and Yu, 1983; Fredman et al., 1984). Several monoclonal anti-GD3 antibodies also bind to other disialogangliosides.

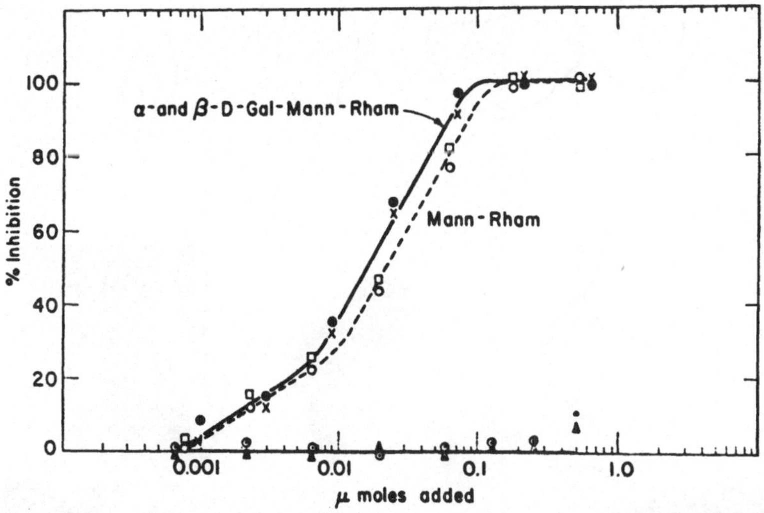

Figure 1. Inhibition of antibodies directed against Salmonella E system 3 antigen by oligosaccharides. The structures of the disaccharides and trisaccharides are presented in Table 1B. (Reprinted from Uchida et al., 1963, copyright 1963 by The American Chemical Society; reprinted by permission of the copyright owner).

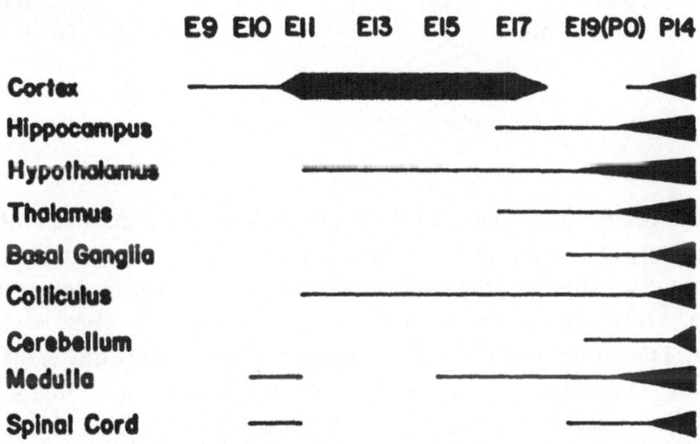

Figure 2. Summary diagram of temporal expression of the 7A antigen in the developing nervous system of the mouse. The width of the lines corresponds to intensity of tissue reactivity demonstrated by immunocytochemistry. Reprinted from Yamamoto et al., (1985) with permission of the copyright owner, The National Academy of Sciences (USA).

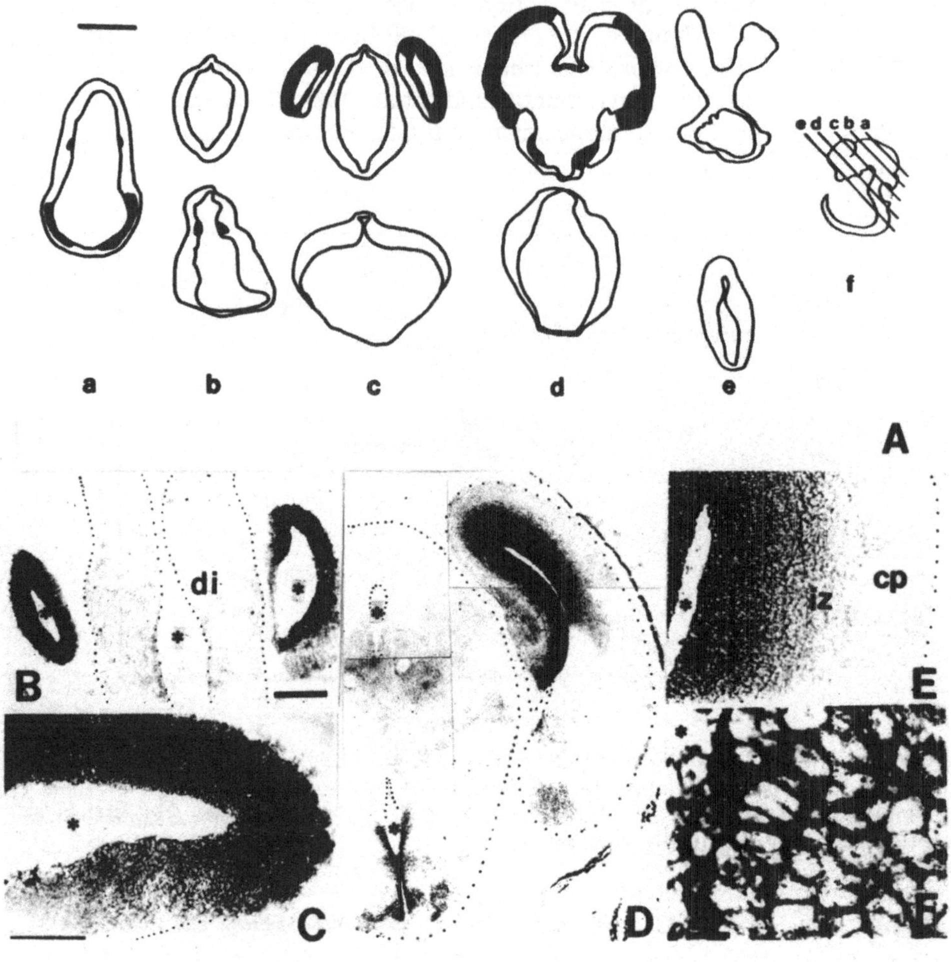

Figure 3. Immunohistochemical localization of 7A antigen in the mouse embryonic brain. (A) Camera lucida drawings of successive levels (a-e, shown in f) of a stage E11 mouse brain. Filled areas are immunoreactive. (Bar = 1mm.) (B) A section through the cortex and the diencephalon (di) of E13 mouse. Only the cortex is strongly immunoreactive, (Bar = 500μm.) (C) High magnification of B. The entire cortical wall is positive. (Bar = 200 μm.) (D) Coronal section of E17 mouse forebrain. The strongest reactivity is seen in the cortex and on a small hypothalamic region. (Bar = 500μm). (E) Higher magnification of cortex in D. Only the ventricular (vz) and subventricular zones (svz) are strongly positive. ix, Intermediate zone; cp, cortical plate. (Bar-200μm.) (F) High magnification of positive regions in D. The asterisk marks the ventricular cavity. (Bar = 10μm.) Configuration of the tissue is outlined by dotted lines. Reprinted from Yamamoto et al., (1985) with permission of the copyright owner, the National Academy of Sciences (USA).

THE USE OF ANTIBODIES TO LOCALIZE GSLs IN TISSUES

Antibodies provide a unique means of identifying minor glycolipids in tissues, and of identifying the precise sites of their cellular and subcellular localization. This is exemplified by the recent identification of neutral glycolipids in higly restricted regions of the central nervous system (Dodd et al., 1984; Jessell and Dodd, 1985; Yamamoto et al., 1985). In the last study, monoclonal antibody 7A, which reacts with the X-determinant (Gal(β1-4)[Fuc(α1-3)GlcNAc), was used to demonstrate the presence of a GSL bearing this determinant in the embryonic brain of rodents. Immunocytochemical staining revealed the presence of this antibody in the ventricular and subventricular zones of the cerebral cortex between 11 and 17 days of embryonic development (Figs. 2 and 3). Glycolipids bearing this determinant were demonstrated by binding of the 7A antibody to neutral GSLs of rat brain. In the study of Dodd et al. (1984) a monoclonal antibody against the X-determinant and antibodies against globoseries GSLs identified a subset of primary sensory neurons in dorsal root ganglia. These studies demonstrate the enormous potential applications of immunological techniques for analysis of development and differentiation.

REFERENCES

Berzofsky JA and Berkower IJ (1984) in: Paul WE (ed): Fundamental Immunology. Raven Press, New York, pp. 621-626.

Cisar J, Kabat EA, Dorner MM, Liao J (1975) Binding properties of immunoglobulin combining sites specific for terminal or nonterminal antigenic determinants in dextran. J Exp Med 142: 435-459.

Dodd J, Solter D, Jessell TM (1984) Monoclonal antibodies against carbohydrate differentation antigens identify subsets of primary sensory neurones. Nature 311: 469-472.

Eisenbarth GS, Walsh FS, Nirenberg M (1979) Monoclonal antibody to a plasma membrane antigen of neurons. Proc Natl Acad Sci (USA) 76: 4913-4917.

Fredman P, Magnani JL, Nirenberg M, Ginsburg V (1984) Monoclonal antibody A2B5 reacts with many gangliosides in neuronal tissue. Arch Biochem Biophys 233: 661-666.

Jessell TM and Dodd J (1985) Structure and expression of differentation antigens on functional subclasses of primary sensory neurons. Phil Trans Royal Soc Lond B 308: 271-281.

Kabat EA (1976) Structural Concepts in Immunology and Immunochemistry, 2nd edition. Holt, Rinehart and Winston, New York, Chapters 6 and 7.

Karol RA, Kundu SK, Marcus DM (1981) Immunochemical relationship between Forssman and globoside glycolipid antigens. Immunol Commun 10: 237-250.

Kasai N and Yu RK (1983) The monoclonal antibody A2B5 is specific to ganglioside GQ1c. brain Res 277: 155-158.

Kundu SK, Marcus DM, Veh RW (1980) Preparation and properties of antibodies to GD3 and GM1 gangliosides. J Neurochem 34: 184-188.

Kundu SK, Pleatman MA, Redwine WA, Boyd AE, Marcus DM (1983) Binding of monoclonal antibody A2B5 to gangliosides. Biochem Biophys Res Commun 116: 836-842.

Magnani JL, Nilsson B, Brockhaus M, Zopf D, Steplewski Z, Koprowski H, Ginsburg V (1982) A monoclonal antibody-defined antigen associated with gastrointestinal cancer is a ganglioside containing sialylated lacto-N-flucopentaose III. J Biol Chem 257: 14365-14369.

Magnani JL, Steplewski Z, Zoprowski H, Ginsburg V (1983) Identification of the gastrointestinal and pancreatic cancer-associated antigen detected by monoclonal antibody 19-9 in the sera of patients as a mucin. Cancer Res 43: 5489-5492.

Schwarting GA, Kundu SK, Marcus DM (1979) Reaction of antibodies that cause paroxysmal cold hemoglobinuria (PCH) with globoside and Forssman glycosphingolipidis. Blood 53: 186-192.

Sharon J, Kabat EA, Morrison SL (1981) Studies on mouse hybridomas secreting IgM or IgA antibodies to α(1-3)-linked dextran. Mol Immunol 18: 831-846.

Uchida T, Robbins PW, Luria SE (1963) Analysis of the serologic determinant groups of the Salmonella E-group O-antigens. Biochemistry 2: 663-668.

Watkins WM (1980) Biochemistry and genetics of the ABO, lewis, and P blood group systems. Adv Hum Genet 10: 1-136.

Yamamoto M, Boyer AM, Schwarting GA (1985) Fucose-containing glycolipids are stage- and region-specific antigens in developing embryonic brain of rodents. Proc Natl Acad Sci (USA) 82: 3045-3049.

Gangliosides and neuronal plasticity
G. Tettamanti, R.W. Ledeen, K. Sandhoff,
Y. Nagai, G. Toffano (eds.)
Fidia Research Series, vol. 6
Liviana Press, Padova, © 1986

EXPRESSION OF GANGLIOSIDE-ANTIGENS DURING NEURONAL DIFFERENTIATION STUDIED BY USE OF MONOCLONAL ANTIBODIES

H. Rösner, C.J. Willibald, S. Henke-Fahle[1]

Institute of Zoology, University of Stuttgart-Hohenheim, 7000 Stuttgart 70, and
[1]Max-Planck-Institute for Developmental Biology, 7400 Tübingen,
Federal Republic of Germany

INTRODUCTION

Progress in development and maturation of the brain has been shown to be paralleled by changes of gangliosides (Rösner, 1982). One generalization of these studies was that predominance of GD3 may characterize mitotically active precursor cells (Dreyfus et al., 1975; Rösner, 1980; 1982; Rösner et al., 1985; Goldmann et al., 1984) and accretion of more complex gangliosides indicates the transition to postmitotic young neurons (Rösner, 1980; 1982). In the present study, performed with chicken and rat embryos, immunohistochemical data obtained with monoclonal antibodies to gangliosides support the above conclusion.

MATERIAL AND METHODS

Monoclonal Antibodies

The mouse monoclonal antibody AbR24, recently identified as recognizing the disialoganglioside GD3 (Pukel et al., 1982; Goldmann et al., 1984), was a gift of Dr. K.O. Lloyd (New York). A second mouse monoclonal antibody, designed as Q211, has been raised against chicken embryonic retinal membranes (Henke-Fahle, 1983). The antigen detected by Q211 was identified on TLC-plates (Brockhaus et al., 1981) in

---

Abbreviations: TLC, thin chromatography; PB, sodium phosphate buffer; IgG-B, biotirylated antimouse IgG; IgG-F, fluorescein-labeled antimouse IgG; A-F, fluorescein-labeled avidin; PGAA, polysialoganglioside-associated antigen; DRG, dorsal root ganglion; VR, ventral root; GH, hexa-sialoganglioside.

various polysialoganglioside fractions of embryonic chicken brain (Henke-Fahle, 1983). These fractions migrate on TLC-plates below GQ1b and were preliminarily characterized as representing polysialogangliosides with 4, 5 and 6 sialic acid residues (Rösner, 1981). The structure of the pentasialo-fraction has now been established by fast atom bombardment mass spectrometry according to Egge et al. (1985), Rösner et al. (in preparation). Both antibodies were used as ascites fluids, diluted up to 200 fold with 0.1 M sodium-phosphate buffer (PB), pH 7.4.

Other Materials

Biotinylated anti-mouse IgG (IgG-B) fluorescein-labeled antimouse IgG (IgG-F), and fluorescein-labeled avidin (A-F) were from Atlanta, Heidelberg. Neuraminidase from Clostridium perfringens (type VIII) and bovine serum albumin were obtained from Sigma, Munich. All other chemicals of analytical grade were from Merck, Darmstadt.

Tissue Preparation

Chicken embryos at ages E1 to E3 (corresponding to stages 9-18 of Hamburger and Hamilton, 1951) were fixed in 2.7% freshly prepared paraformaldehyde in PB at 4 °C for 1 day, embedded in polyacrylamide gels according to Hausen and Dreyer (1981), washed in PB, and frozen in isopentan cooled with liquid nitrogen. Elder chicken embryos E 4-11 as well as dissected brains from prenatal rat embryos of 16-21 days of gestation were prepared and fixed in 3.7% paraformaldehyde in PB for 1 to 3 days at 4 °C, washed and frozen as described above.

Immunohistochemistry

Antibody Binding

Cryostat sections (6-10 μm thick) were mounted on chromalaungelatin coated microscope slides and allowed to dry. All the following steps were performed at room temperature. The sections were incubated for 30-45 min with either AbR24 or Q211, diluted up to 200-fold in PB. They were then washed at least 3 times with a large excess of PB and incubated in IgG-B (0.5 mg solid/ml PB) for another 30 min. The slides were again washed with PB and finally treated with A-F (diluted 2000 fold with PB), washed again and mounted in glycerol-PB 4:1, by volume.

Alternatively IgG-F was used as second antibody instead of IgG-B and A-F. Both procedures gave identical results. Since pretreatment of the sections with 0.1 % bovin-serum-albumin in PB had no effect on the antibody binding, this step was omitted in most cases. Selected sections were incubated with PB instead of the first antibody. These controls were stained only negligibly, indicating an absence of unspecific binding of either IgG-B or IgG-F or A-F.

Neuraminidase and Ethanol Treatment

Selected sections were incubated at room temperature with neuraminidase (Sigma

VIII, 0.3 U/ml of 0.1 M acetate buffer, pH 5.5) for 10 min or for 5 min with 90 % ethanol before immunohistochemistry.

Pretreatment of AbR24 with Gangliosides

To test the specificity of AbR24, the antibody was adsorbed with the purified gangliosides GD1a, GT1b or GD3 (0.2-4mM) for 1 h at room temperature. After centrifugation at 4000 g for 20 min, the supernatant was then used for immunohistochemistry as described above.

Fluorescence Microscopy

The slides were examined by epifluorescent illumination (HBO 50 W lamp) on a Zeiss microscope with filters BP546, FT510, 580, LP520, 590 using neofluar objectives 6.3, 10, 16 and 40. Photographs were taken with Ilford HP5 (400 ASA) films at known exposure times.

RESULTS

Developmental Expression of GD3 and Polysialoganglioside Associated Antigen (PGAA) in Embryonic Chicken Brain, Detected by AbR24 and Q211, respectively

Mesencephalon

First staining with AbR24 appeared at E2-2.5 sparsely and more or less evenly distributed over the neuroectodermal wall as well as over clusters of adjacent cells within the very weakly stained mesenchyme (Fig. 1a). Due to their location these clustered cells, expressing GD3 most likely represented migrating derivatives of the neural crest. In contrast, the polysialoganglioside-associated antigen reacting with Q211 was restricted to distinct peripheral areas of the neuroepithelial wall (Fig. 1b), containing first postmitotic neurons of the oculomotor nerve nuclei (Romanoff, 1960) (Fig. 1c). Figure 1d shows that the expression of GD3 by cells of the whole mesencephalic wall had increased up to 4 days of incubation (E4). The PGAA was now stained by Q211 in a narrow peripheral layer (Fig. 1e). One day later (E5), the stained layer corresponding to the forming mantle layer was much broader (Fig. 1f). Comparison with a paraffin section at higher magnification, corresponding to the inset in Figure 1f, strongly suggests that the labeled cells represent migrating, postmitotic neurons, which have left the unstained germinal layer (Fig. 1g).

Eye and Optic Stalk

Like the mesencephalon, the retina was stained only sparsely by AbR24 up to 2.5 days. At E4, E5, however, an intensive expression of GD3 by retinal cells was indicated by high staining with the antibody (Fig. 2a, e). At this developmental step PGAA, however, was expressed only by cells of the most inner retinal layer (adjacent to the inner limiting membrane), known to contain the first postmitotic retinal ganglion cells and first postmitotic amacrine cells (Rager, 1976) (Fig. 2b, f). A cross section through the optic stalk (Fig. 2d, g) clearly indicated that the PGAA-label was restricted to the ventral half of the stalk, containing growth cones and axons of retinal ganglion cells (Rager, 1976). The epithelial cells of the stalk, however, were not stained by Q211.

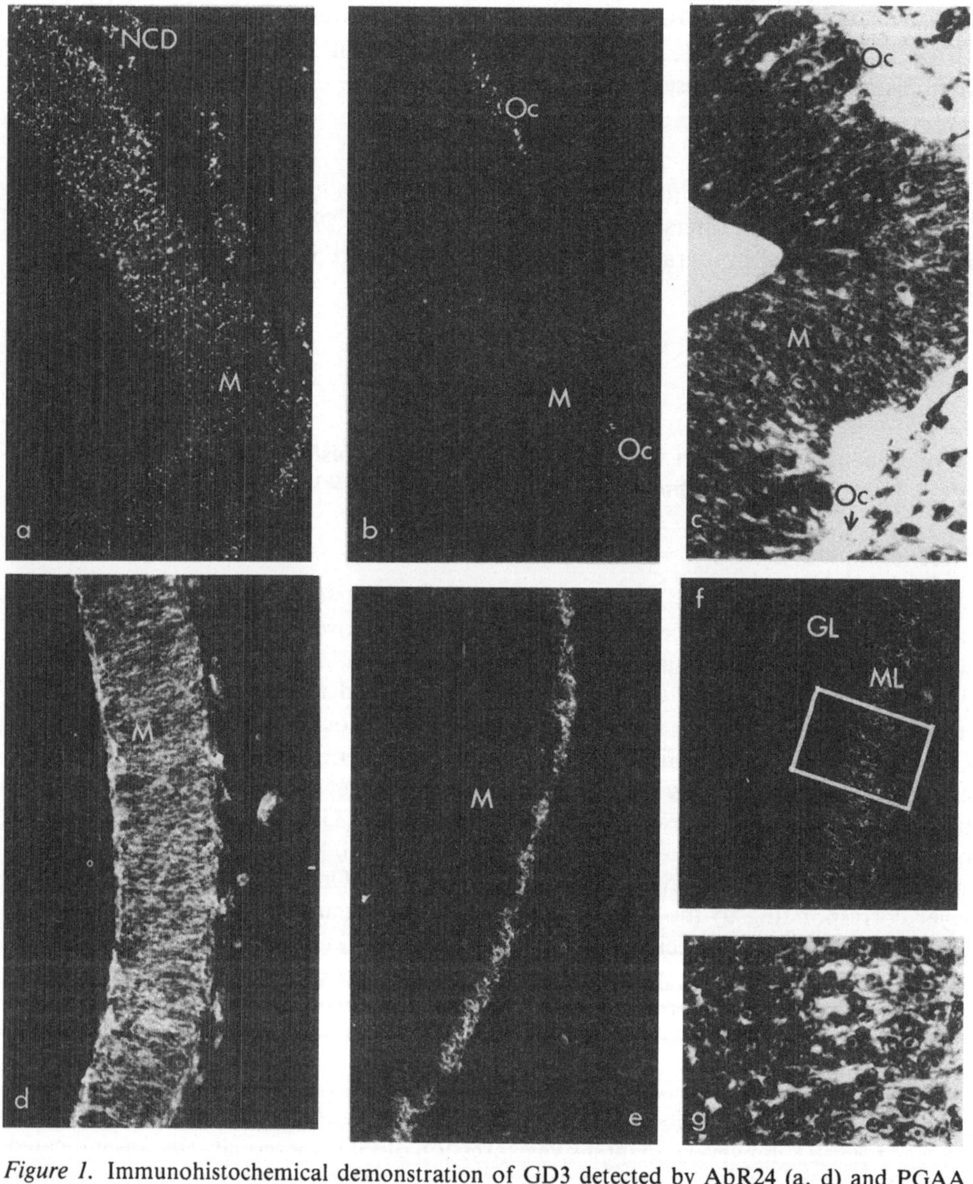

Figure 1. Immunohistochemical demonstration of GD3 detected by AbR24 (a, d) and PGAA detected by Q211 (b, e, f) on sections from the mesencephalon of an E2.5 (a, b), E4 (d, e), and E5 embryo (f); paraffin sections from the mesencephalon of an E2.5 (c) and E5 embryo (g, corresponding to the inset in Fig. 1f). GL, germinal layer; M, mesencephalon; ML, mantle layer; MES, mesenchyme; NCD, neural crest derivatives; Oc, oculomotor nerve nucleus; V, ventricle.

Diencephalon and Spinal Cord

Sections from the diencephalon of E4 embryos (Fig. 3b, c) and the spinal cord of E5 embryos (Fig. 3e, f) revealed further information about the developmental expres-

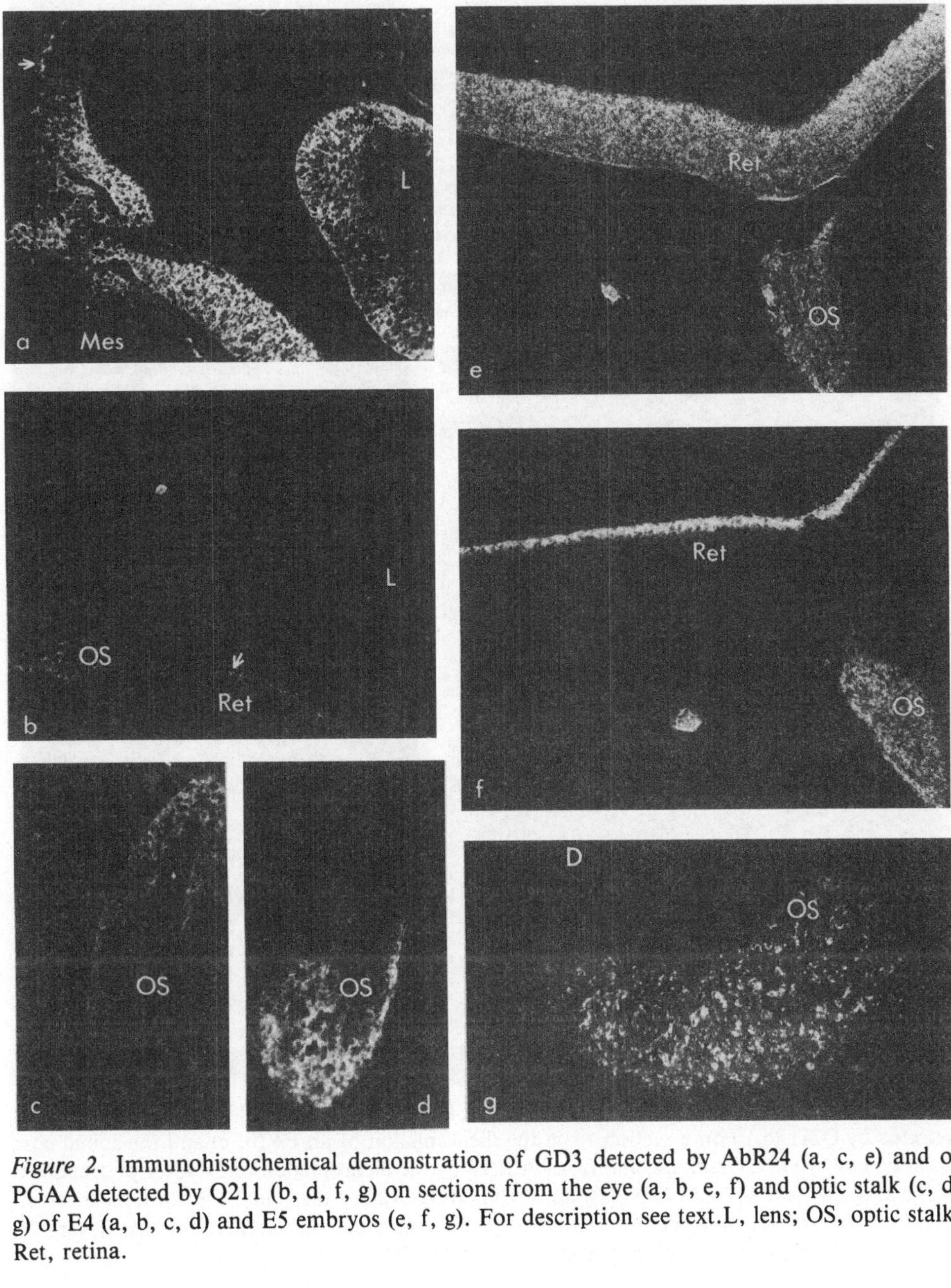

Figure 2. Immunohistochemical demonstration of GD3 detected by AbR24 (a, c, e) and of PGAA detected by Q211 (b, d, f, g) on sections from the eye (a, b, e, f) and optic stalk (c, d, g) of E4 (a, b, c, d) and E5 embryos (e, f, g). For description see text.L, lens; OS, optic stalk, Ret, retina.

sion of both ganglioside antigens. In the E4 diencephalon, the mitotically active germinal layer (GL) appeared much more intensely labeled by AbR24 than the outer mantle layer (ML) (Fig. 3b). The same was observed in the spinal cord of E5 embryos (Fig.

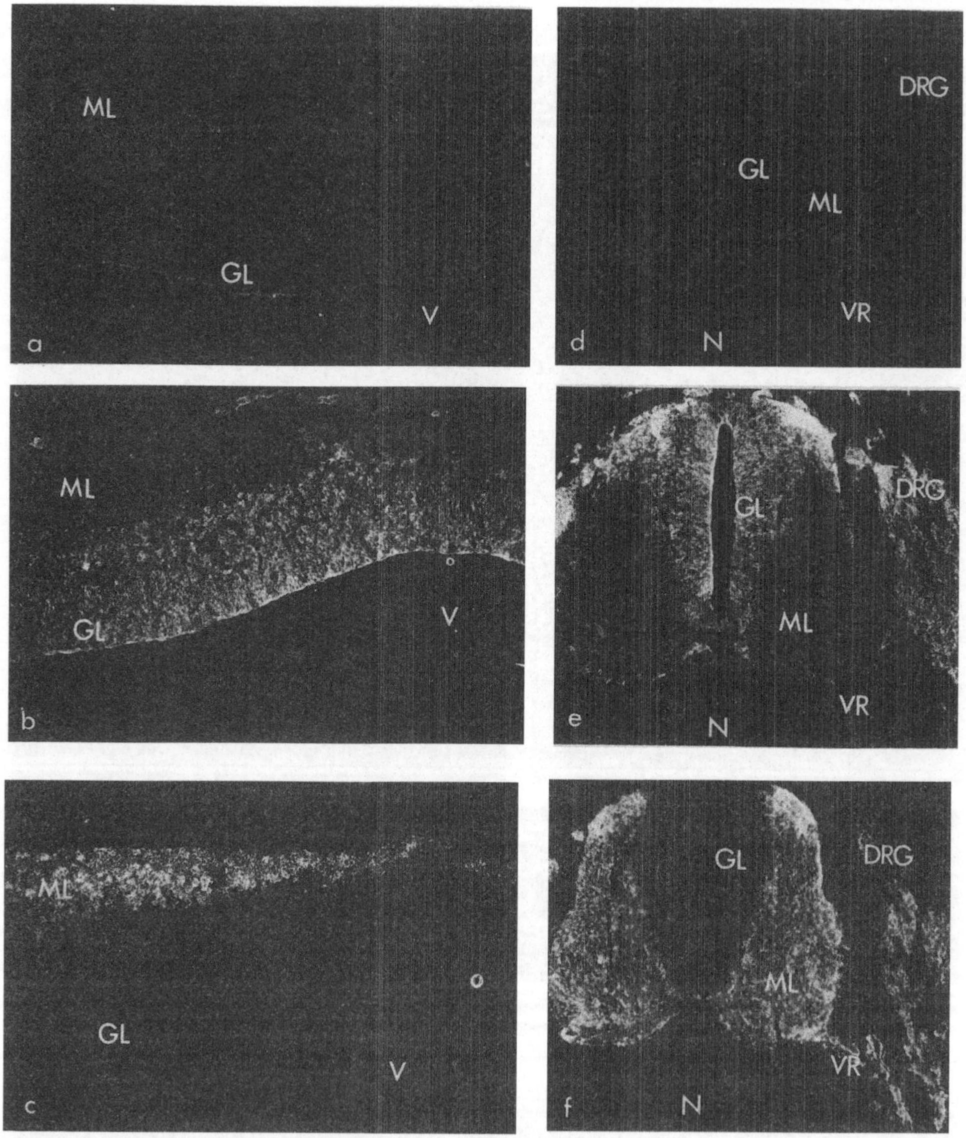

Figure 3. Immunohistochemical demonstration of GD3 detected by AbR24 (b, e) and of PGAA detected by Q211 (c, f) on a section from the diencephalon of an E4 (b, c) and the spinal cord of an E5 embryo (e, f); lack of binding of AbR24 by the diencephalon of E4 after pretreatment of the antibody with 0.2 mM GD3 (a); no binding of AbR24 by the spinal cord of E5 after treatment of the section with neuraminidase (d). GL, germinal layer; ML, mantle layer; N, notochord; DRG, dorsal root ganglion; V, ventricle; VR, ventral root.

3e), suggesting a decrease in the expression of GD3 in postmitotic neurons of the ML as compared to their precursor cells of the GL. The opposite was found for PGAA detected by Q211. This was much more intensely expressed by cells of the ML in both structures (Fig. 3c, f). Furthermore, Figures 3e, f demonstrate the presence of both antigens in dorsal root ganglion (DRG). In the ventral root (VR), however, only polysialogangliosides were stained.

Pretreatment of AbR24 Purified Gangliosides

Incubation of diluted AbR24-ascites fluids with different concentratations of the gangliosides GD3, GD1a, or GT1b (see methods) led to an inhibition of the antibody binding by all 3 gangliosides. However, GD3 was 10- to 20-fold more effective. Thus a 0.2 mM concentration of GD3 was found to cause 100% inhibition of the binding of AbR24 (Fig. 3a), whereas GD1a had to be used in 3.5 mM and GT1b in 2.3 mM concentrations to get the same effect.

Pretreatment of the Sections with Neuraminidase or Ethanol

Pretreatment of cryostat sections with neuraminidase from Clostridium perfringens (see methods) destroyed the antigenic activity for both AbR24 (Fig. 3d) and Q211 (not shown). In addition, there was no binding of either antibody to sections which had been extracted with 90% ethanol (not shown).

Developmental Expression of GD3 and PGAA in the Prenatal Rat Forebrain

At 18 days of gestation, expression of GD3 (stained by AbR24) was found over the ventricular and intermediate zones (Fig. 4a). Less immunoreactivity was seen in the superficial layer corresponding to the sublate layer, forming the cortical plate. The marginal layer was stained again more intensely. In contrast, the PGAA, detected by Q211 (Fig. 4b) seemed to be absent in the ventricular and intermediate zones. A narrow area of staining occurred at the border of the cortical plate and the superficial part of the intermediate zone. This interface, designated as sublate layer, is well described as an area of ingrowing cortical projections and early synaptogenesis (Rakič, 1972; 1974; Kostovic and Molliver, 1974). The cortical plate itself was more or less sparsely labeled and again more concentrated staining was revealed in the region of the marginal zone.

At 21 days of gestation the distribution of staining had changed remarkably (Fig. 4c, d). In the ventricular zone immuno-reactivity of GD3 was high, followed superficially by a narrow less stained and a second heavily stained zone, both corresponding to the subventricular layer (Fig. 4c). The much broader sublate layer and cortical plate were nearly unstained except for blood vessels. Intensive staining was also found over the small pia. In contrast, immunoreactivity of polysialogangliosides (PGAA) occupied all layers superficial to the unlabeled ventricular zone, with lesser staining of the sublate layer (Fig. 4d).

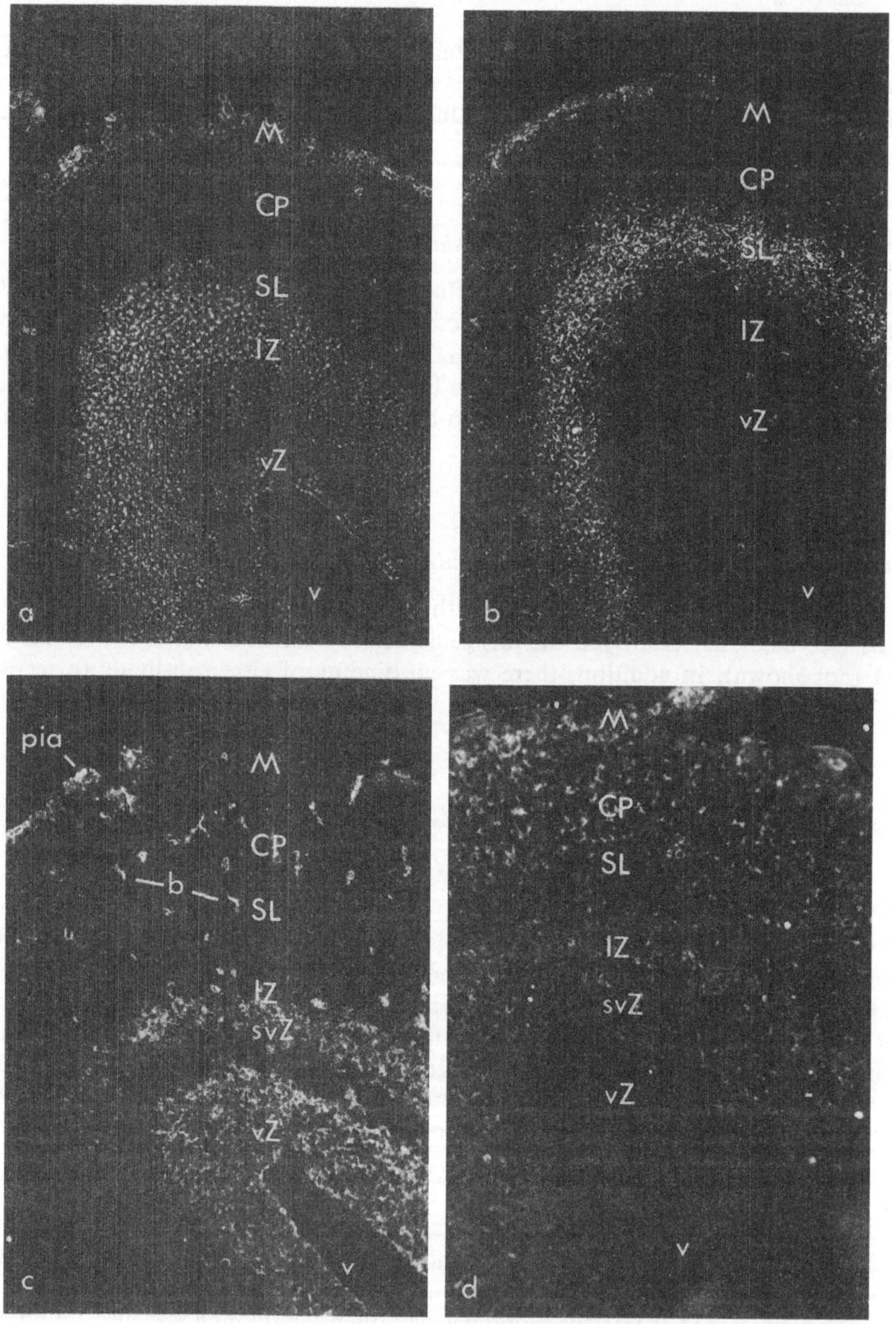

Figure 4. Immunohistochemical demonstration of GD3 detected by AbR24 (a, c) and of PGAA detected by Q211 (b, d) on sections from the neocortex of prenatal rats of 18 days (a, b) and 21 days (c, d) of gestation. M, marginal zone; CP, cortical plate; SL, sublate layer; IZ, intermediate zone, vZ, ventricular zone; svZ, subventricular zone; v, ventricle; b, blood vessels.

DISCUSSION

Specificity of the Antibodies

The present study revealed striking differences in the developmental expression by chicken embryonic tissues of two antigens which were previously found to represent different gangliosides (Pukel et al., 1982; Henke-Fahle, 1983). The ganglioside nature of both antigens was further supported in this study by the complete loss of their activity after brief neuraminidase or ethanol (not presented) treatment of cryostat sections. Furthermore, the specificity of GD3 as the antigen reacting with AbR24 was emphasized in this study by the 10- to 20-fold higher inhibitory potency of purified GD3 as compared to the gangliosides GD1a and GT1b.

Expression of GD3

The data presented demonstrate that ganglioside GD3, so far as detected by AbR24, is heavily expressed by immature, dividing cells of the neuroepithelium and drops to a much lower level when the cells differentiate to the postmitotic state. In a recently published paper, we showed that this phenomenon can be generalized to all brain areas as well as to the peripheral nervous system of the chick embryo (Rösner et al., 1985). These immunohistological data are in agreement with and substantiate well-known biochemical findings, which demonstrated GD3 to be the predominant ganglioside of the early immature chicken brain and retina which decreases during development (Rösner, 1975; 1980; 1982; Dreyfus et al., 1975; Engel et al., 1979).

A similar decrease of GD3 has been described for the developing mammalian brain (Hilbig et al., 1982; Merat and Dickerson, 1973; Sonnino et al., 1981). The data presented show that in the rat GD3 is heavily expressed by mitotically active cells of the ventricular zones, containing mostly neural precursors (Rakič, 1972; 1974) and later also by cells of the subventricular zone, containing mostly glial precursors (Raedler et al., 1980). Recently, Goldman et al. (1984) also described an intense expression of GD3 in proliferating zones of the postnatal rat cerebellum. Thus, the following generalization seems to us justified: "Intense expression of ganglioside GD3 is a common feature of mitotically active neuroectodermal cells. Upon transition to the postmitotic phase of differentiation in neurons, GD3 drops to a much lower level." The immunohistochemical staining of sections from older embryos further indicates that non-neuronal elements of the more differentiated brain (e.g. the pia), as well as muscle, endothelia of small blood vessels or phagocytes, express much more GD3 than neurons.

Expression of the Polysialoganglioside-associated Antigen

The PGAA was shown by immunoautoradiography (Henke-Fahle, 1983) to be recognized specifically by Q211 and to be present in various ganglioside fractions which were recently detected by Rösner (1980; 1981) in embryonic chicken brain. These fractions migrate on TLC plates below GQ1b and have been preliminarily characterized as polysialogangliosides possessing 4, 5 and 6 sialic acid residues (Rösner, 1981). One of these fractions has been now identified as GP1c (in preparation). The antigen is extractable by organic solvents, sensitive to neuraminidase treatment, but is not affected by protease digestion. All this supports the identification of the Q211 epitope as a lipid-bound polysialylated structure common to gangliosides of the 'c' pathway (Yu and Ando, 1980). In contrast to the expression of GD3, the PGAA, which is recognized by Q211, is absent in proliferating cells of the neuroepithelium. Since the first appearance in all regions of the brain and spinal cord of chicken is restricted to cells within the peripheral mantle layer, this antigen seems to be expressed only by neuroblasts which have completed their final mitosis. This conclusion is especially supported by the localization of the antigen in postmitotic neurons of the oculomotor nuclei in the mesencephalon (Romanoff, 1960) and by the expression of the PGAA in retinal ganglion cells correlating well with the spatio-temporal genesis of these neurons (Dütting et al., 1983; Kahn, 1974). Cross sections of the optic stalk as well as 'whole mounts' from the retina (not shown) clearly demonstrate the presence of the PGAA on the surface of growing axons, and suggest an enrichment at growth cones. Sections from peripheral nerves including neural crest derivatives, which were incubated with Q211, gave similar results and support the following general conclusion: "*A polysialoganglioside-associated antigen, specifically reacting with Q211, is expressed by young neurons immediately after withdrawl of their precursor cells from the mitotic cycle. The antigen appears over the whole cell surface including the growing processes.*"

Immunostaining of older developmental stages shows that the expression of this antigen drops during synaptogenesis and has nearly disappeared after hatching (data not shown). These presented immunohistochemical data from chicken brain are in full agreement with our previous biochemical findings (Rösner, 1982), which revealed a transient developmental accretion of those polysialogangliosides ('GQ1c', GP1c, 'GH'), which has been shown to contain the epitope reacting with Q211 (Henke-Fahle, 1983).

Interestingly, none of these gangliosides has been described yet in mammalian brains. The data presented from embryonic rat forebrain, however, clearly revealed a staining by Q211, comparable to the chicken brain. The staining was transient and restricted to those neocortical layers containing postmitotic neurons as well as ingrowing cortical projections (Rakic, 1972). As in chicken the Q211-antigen was absent in the proliferating ventricular zones.

Future studies will have to elucidate whether the antigen to Q211 in the developing rat brain is, as in chicken, of ganglioside or proteinaceous nature. In this context it is of interest to note that Finne (1982) recently reported a transient occurrence of polysialylated peptides in the developing rat brain.

REFERENCES

Brockhaus MJL, Magnani M, Blaszeyk Z, Steplewski H, Korprowski KA, Karesson G, Larsen G, Ginsburg V (1981) Monoclonal antibodies directed against the human Le^b blood group antigen. J Biol Chem 256: 13223-13225.

Dreyfus H, Urban PF, Edel-Harth S, Mandel P (1975) Developmental patterns of gangliosides and phospholipids in chick retina and brain. J Neurochem 10: 429-440.

Dütting D, Gierer A, Hansmann G (1983) Self-renewal of stem cells and differentiation of nerve cells in the developing chick retina. Develop Brain Res 10: 21-32.

Egge H, Peter-Katalinic J, Reuter G, Schauer R, Ghidoni R, Sonnino S, Tettamanti G (1985)

Analysis of gangliosides using fast atom bombardment mass spectrometry. Chem Phys Lipids. 37: 127-141.

Engel EL, Wood JG, Byrd FI (1979) Ganglioside pattern and cholera toxin-peroxidase labeling of aggregating cells from the chick optic tectum. J Neurobiol 10: 429-440.

Finne J (1982) Occurrence of unique polysialyl carbohydrate units in glycoproteins of developing brain. J Biol Chem 257: 11966-11970.

Goldman JE, Hirano M, Yu RK, Seyfried TN (1984) GD3 Ganglioside is a glycolipid characteristic of immature neuroectodermal cells. J Neurochem 7: 179-192.

Hamburger V, Hamilton H (1951) A series of normal stages in the development of the chick embryo. J Morphol 88: 49-92.

Hausen P, Dreyer C (1981) The use of polyacrylamide as an embedding medium for immunohistochemical studies of embryonic tissue. Stain Techn 56: 287-293.

Henke-Fahle S (1983) Monoclonal antibodies recognize gangliosides in the chick brain. Neurosci Lett 160.

Hilbig R, Rösner H, Merz G, Segler-Stahl K, Rahmann H (1982) Developmental profiles of gangliosides in mouse and rat cerebral cortex. Roux's Arch Develop Biol 191: 281-284.

Kahn AJ (1974) An autoradiographic analysis of the time of appearance of neurons in the developing chick neural retina. Develop Biol 38: 30-40.

Kostovic I, Molliver MW (1974) A new interpretation of the laminar development of cerebral cortex: synaptogenesis in different layers of the neopallium in the human fetus. Anat Rec 178: 1395.

Merat A, Dickerson JWT (1973) The effect of development on the gangliosides of rat and pig brain. J Neurochem 20: 873-880.

Pukel CS, Lloyd KO, Travassos LR, Deppold WG, Oettgen HF, Lloyd JO (1982) GD3, a prominent ganglioside of human melanoma. J Exp Med 155: 1133-1147.

Raedler E, Raedler A, Feldhans S (1980) Dynamical aspects of neocortical histogenesis in the rat. Anat Embryol 158: 253-269.

Rager G (1976) Morphogenesis and physiogenesis of the retinotectal connection in the chicken. I. The retinal ganglion cells and their axons. Proc Royal Soc Lond B 192: 331-352.

Rakić P (1972) Mode of cell migration to the superficial layers of the fetal monkey neocortex. J Comp Neurol 145: 61-84.

Rakić P (1974) Neurons in rhesus monkey visual cortex. Systematic relation between time of origin and eventual disposition. Science 183: 425-427.

Romanoff AL (1960) The Avian Embryo-Structural and Functional Development. Macmillan Company, New York, NY.

Rösner H (1975) Changes in the content of gangliosides and glycoproteins and in the ganglioside pattern of the chicken brain. J Neurochem 24: 815-816.

Rösner H (1980) Ganglioside changes in the chicken optic lobes and cerebrum during embryonic development. Transient occurrence of "novel" multisialogangliosides. Roux's Arch Develop Biol 188: 205-213.

Rösner H (1981) Isolation and preliminary characterization of novel polysialogangliosides from embryonic chicken brain. J Neurochem 37: 993-997.

Rösner H (1982) Ganglioside changes in the chicken optic lobes as biochemical indicators of brain development and maturation. Brain Res 236: 49-61.

Rösner H, Al-Aqtum M, Henke-Fahle S (1985) Developmental expression of GD3 and polysialogangliosides in embryonic chicken nervous tissue reacting with monoclonal antiganglioside antibodies. Dev Brain Res 18: 85-95.

Sonnino S, Ghidoni R, Masserini M, Aporti F, Tettamanti G (1981) Changes in rabbit brain cytosolic and membrane-bound gangliosides during prenatal life. J Neurochem 36: 227-232.

Yu RK, Ando S (1980) Structures of some new complex gangliosides of fish brain. Adv Exp Med Biol 125: 33-45.

Gangliosides and neuronal plasticity
G. Tettamanti, R.W. Ledeen, K. Sandhoff,
Y. Nagai, G. Toffano (eds.)
Fidia Research Series, vol. 6
Liviana Press, Padova, © 1986

GANGLIOSIDES AS DIFFERENTIAL MODULATORS OF MEMBRANE-BOUND PROTEIN KINASE SYSTEMS

Robert K. Yu[1], James R. Goldenring[1,2], John Y.H. Kim[1], Robert J. DeLorenzo[1,3]

[1]Department of Neurology, Yale University School of Medicine, New Haven, CT 06510, U S A; [2]Surgical Service (112), West Haven VA Medical Center, West Spring Street, West Haven, CT 06516; [3]Department of Neurology, Medical College of Virginia, Richmond, VA 23298

INTRODUCTION

Gangliosides are sialic acid-containing glycosphingolipids found in the plasma membrane of virtually all vertebrate tissues and are particularly abundant in the nervous system (Ledeen and Yu, 1982; Ledeen, 1983; Ando, 1983; Ledeen, 1985). Gangliosides comprise a major part of the glycoconjugate network extending from the neuronal membrane surface. Recent investigations suggest that gangliosides may exert a critical role in the regulation of cell growth and differentiation (Hakomori, 1981; 1983; Ledeen et al., 1984; Ando, 1983; Ledeen, 1984). Nevertheless, the molecular events which underlie the physiological and cellular events remain unresolved. Recent investigations have emphasized the importance of Ca^{++}-dependent kinase systems in the regulation of many dynamic cellular processes (Nestler et al., 1984; Cheung, 1982; DeLorenzo and Goldenring, 1984). In particular, the Ca^{++}/calmodulin-dependent kinase (Kennedy, 1983) and the Ca^{++}/phospholipid-dependent kinase (C-kinase) system (Nishizuka, 1984a,b) have been implicated as major regulators of cellular physiology. Gangliosides bind Ca^{++} with high affinity (Abramson et al., 1972; Probst et al., 1979; 1984; Quarles and Folch-Pi, 1965; Rahmann, 1983), and thus, their possible interaction with Ca^{++}-dependent kinases is of great interest. We present here our recent investigations which indicate that gangliosides may exert opposite effects on these Ca^{++}-dependent kinases with inhibition of C-kinase and stimulation of calmodulin-dependent kinase (Goldenring et al., 1985; Kim et al., 1986).

Abbreviations: EGTA, ethylene -bis(oxyethylenenitrile) tetraacetic acid; MBP, myelin basic protein; CaM, calmodulin; IC_{50}, concentration for 50% inhibition; SDS, sodium dodecylsulphate; PAGE, polyacrylamide gel electrophoresis.

MATERIALS AND METHODS

(γ-^{32}P)-ATP(5-10 Ci/mmol) was purchased from New England Nuclear. Calmodulin (CaM) was purified by established methods (DeLorenzo and Goldenring, 1984). Bovine brain gangliosides were prepared as previously described (Ando and Yu, 1977; Ledeen et al., 1973; Ledeen and Yu, 1982). GM1, GD1a, GD1b and GT1b were purified from bovine brain gangliosides by the methods of Ando and Yu (1977) and were homogeneous as determined by high performance thin layer chromatography (Ando at al., 1978). Gangliosides prepared by these methods had sodium as the counter ion. A calcium-salt of bovine brain gangliosides was prepared by incubation of gangliosides in the presence of a large excess of $CaCl_2$ followed by extensive dialysis. Asialo-GM1 was prepared by the method of Kasai et al., (1982). All other reagents were purchased from Sigma. Ganglioside concentrations were determined by the resorcinol-HC1 method (Svennerholm, 1957). Protein concentrations were determined by the method of Bradford (1976).

Assays of Phosphorylation

EGTA-treated rat brain membranes, prepared as previously described (DeLorenzo et al., 1981), contained both brain membrane and myelin fractions and were devoid of significant amounts of endogenous calmodulin. Highly purified myelin was prepared from rat brains through multiple discontinuous sucrose gradient centrifugation by a modification of the method of Norton and Poduslo (1973) as described previously (Kim et al., 1986). Phosphorylation in brain membrane and myelin preparations was assayed as previously described (Burke and DeLorenzo, 1982), under standard conditions of 7 mM ATP, 4 mM $MgCl_2$, 50 mM phosphate, pH 6.5 at 37° for one minute. The standard calcium ion concentration was 50 μM. Phosphorylated membrane proteins were separated on 10% SDS-PAGE.

Phosphorylation in myelin preparations was quantitated using a phosphocellulose paper absorption assay (Witt and Roskoski, 1975). Phosphorylation reactions contained myelin (protein concentration 15 μg/100 μl), 50 mM phosphate buffer (pH 7.3), 200 μM EGTA, 10 mM $MgCl_2$, 0.03% Triton X-100, and 7 μM ATP. Experimental reactions contained individual gangliosides in concentrations from 0-300 μM. The reactions were initiated with $CaCl_2$ to a final concentration of 100 μM (Waisman et al., 1981). The reaction mixtures (final volume 100 μl) were incubated at 30°C in a shaking water bath for 5 min. After incubation, 25 μl aliquots were pipetted directly onto 1 $\times$ 2 cm strips of phosphocellulose paper (Whatman P81) and processed as previously described (Kim et al., 1986).

RESULTS

The effects of gangliosides on phosphorylation in crude brain membrane preparations devoid of endogenous calmodulin are shown in Figure 1. Addition of gangliosides in the presence of Ca^{++} stimulated the phosphorylation of several bands with M_r's of approximately 45K, 50K, 60K, 80K, 140K, and 170K daltons. The ganglioside-dependent activation of phosphorylation was completely dependent on the presence of

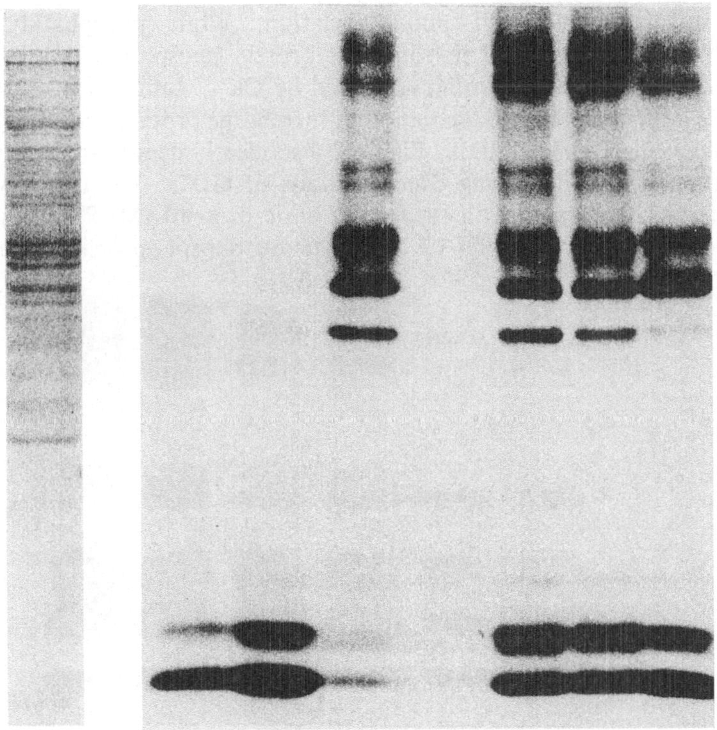

Figure 1. Calcium-dependent, ganglioside-stimulated phosphorylation in rat brain membrane. Lane 1 shows the Coomassie blue staining pattern of the brain membrane preparation. Brain membrane was phosphorylated for 1 min at 37°C with the addition of: lane 2, Mg^{++} (4 mM); lane 3, Mg^{++} and Ca^{++} (50 mM); lane 4, Mg^{++}, Ca^{++} and the sodium salt of bovine brain gangliosides (250 μM); lane 5, Mg^{++} and the sodium salt of bovine brain gangliosides (250 μM); lane 6, Mg^{++}, Ca^{++} and the calcium salt of bovine brain gangliosides (250 μM); lane 7, Mg^{++} and the calcium salt of bovine brain gangliosides (250 μM); and lane 8, Mg^{++}, Ca^{++} and calmodulin (1 μg). Autoradiographic exposure was 6 hrs. (From Goldenring et al., 1985).

calcium. Furthermore, the effects of gangliosides appeared to be directly mediated by a Ca^{++}/ganglioside complex, since a Ca^{++}-salt of bovine brain gangliosides fully stimulated phosphorylation in the absence of added Ca^{++} ions. When the pattern of Ca^{++}/ganglioside-stimulated phosphorylation was compared with that from Ca^{++}/calmodulin-stimulation of membrane phosphorylation, the patterns were qualitatively similar although the relative amounts of 50K and 60K dalton phosphoproteins were slightly different with a slightly different mobility of 50K dalton phosphoprotein.

The stimulation of phosphorylation by purified individual gangliosides was also investigated. GD1a, GD1b, and GT1b (with K_a's of 21 μM, 40 μM, 30 μM, respectively) stimulated phosphorylation at significantly lower concentrations than GM1 (K_a of 125 μM). Asialo-GM1, sulfatide, cerebroside, neuraminyl-lactose and sialic acid did not elicit any stimulation of phosphorylation. While phosphatidylserine and phosphatidylinositol stimulated myelin basic protein phosphorylation, they did not stimulate any of the phosphoproteins affected by Ca^{++}/ganglioside.

Figure 2 demonstrates the stimulation of membrane protein phosphorylation with increasing concentrations of GD1a. Figure 2 also clearly demonstrates the inhibition of three proteins with increasing concentrations of GD1a. The two low molecular weight proteins co-migrated with rat myelin basic proteins (MBP). In addition, the Ca^{++}-dependent phosphorylation of a 78K dalton phosphoprotein, separable from

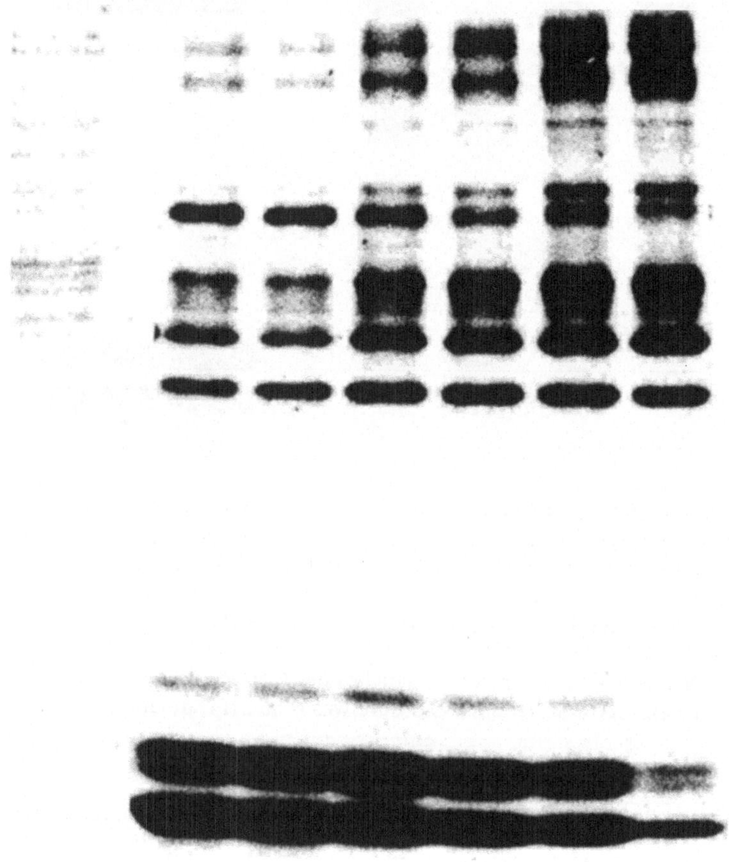

Figure 2. Effects of GD1a on calcium-dependent phosphorylation in brain membrane. Coomassie blue staining pattern (Prot) for brain membrane is shown at the left. Brain membrane was phosphorylated in the presence of Mg^{++} (4 mM), Ca^{++} (50 mM) and increasing concentrations of the sodium salt of GD1a. The autoradiograph was exposed for 24 hrs to illustrate the progressive inhibition of a 78 K dalton protein and myelin basic protein phosphorylation and the stimulation of a number of phosphoproteins by increasing concentration of gangliosides. In this figure, the 78 K dalton protein overlapped with the 80 K dalton phosphoprotein. They could be separated by a longer gel. (From Goldenring et al., 1985).

the 80 K dalton protein (synapsin I) phosphorylation stimulated by gangliosides, was also inhibited by increasing concentrations of GD1a.

Myelin basic proteins are excellent substrates for C-kinase phosphorylation in myelin preparations (Turner et al., 1982, 1984). Since gangliosides have been implicated as important modifiers of myelin basic protein function (Yohe et al., 1983), we sought to investigate MBP phosphorylation in purified myelin preparations. As reported by Turner et al. (1982; 1984), Ca^{++}-dependent MBP phosphorylation present in highly purified myelin preparations is due solely to C-kinase activity.

Phosphorylation of MBP by endogenous C-kinase in myelin was investigated in the presence of a mixture of gangliosides. As in whole brain membranes, the mixture of bovine brain gangliosides inhibited MBP phosphorylation with an IC_{50} of about 55 μM (Figure 3A). GT1b exhibited the lowest IC_{50} of approximately 40 μM (Fig. 3B) and displayed the steepest inhibition curve. GD1a and GD1b possessed equivalent IC_{50}'s of approximately 65 μM (Fig. 3C). GM1, however, required more than double the concentrations of polysialogangliosides neccessary to produce comparable effects with an approximate IC_{50} of 160 μM (Fig. 3B). Indeed, while the inhibitory effect seen with polysialoganglioside reached a plateau at approximately 75% inhibition of kinase activity, only a 50% inhibition of phosphorylation could be obtained with the monosialoganglioside GM1 (Figs. 3B, 3C). Asialo-GM1 (GA1), ceramide, and sialic acid demonstrated no inhibitory effect within a comparable concentration range (Figs. 3A, 3D). Sulfatide exhibited minor inhibitory action achieving no more than 35% inhibition at high concentrations (Fig. 3D). Ca^{++} concentrations of up to 5 mM did not alter the IC_{50}'s for gangliosides. Analysis of myelin phosphorylation on SDS-PAGE gels demonstrated that the small and large MBP's were inhibited with similar IC_{50}'s.

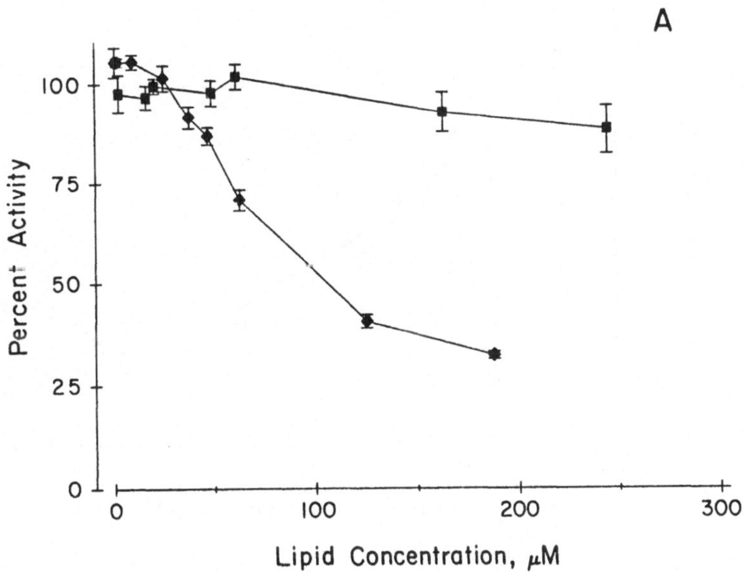

Figure 3. Continued next page.

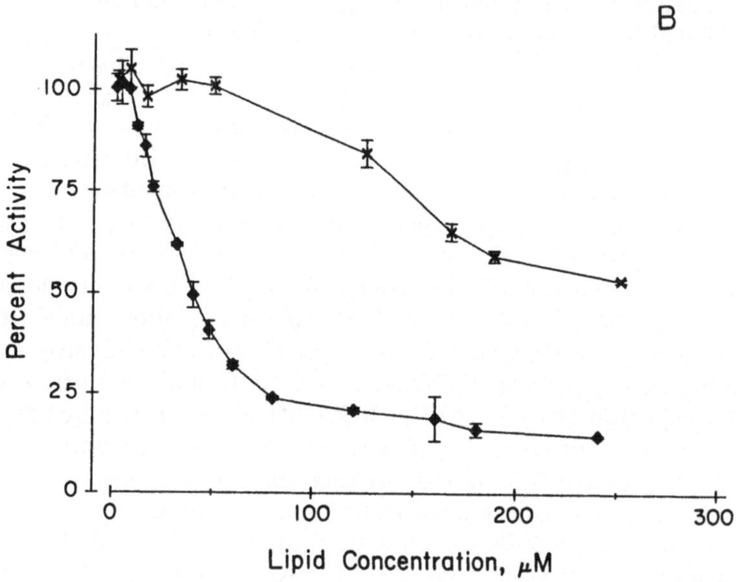

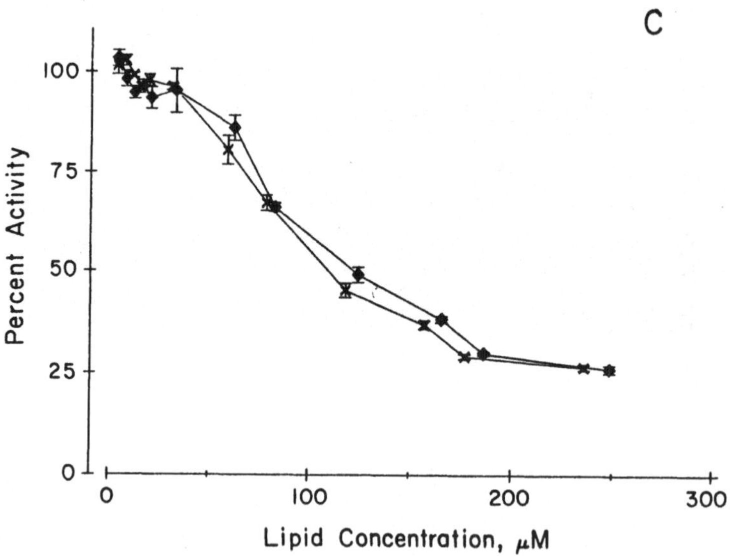

Figure 3. Continued next page.

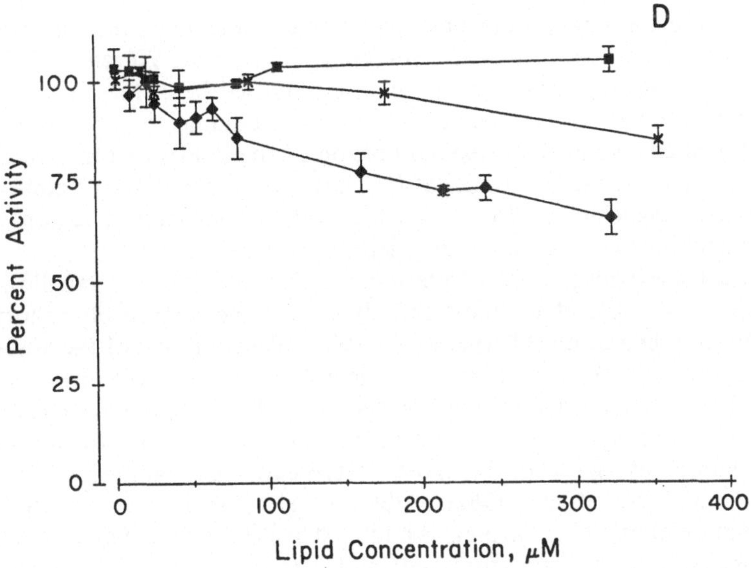

Figure 3. Effects of various exogenously added substances on phospholipid-sensitive Ca^{2+}-dependent protein kinase (C-kinase) phosphorylation of rat MBPs. The data were obtained by phosphocellulose paper absorption assay and illustrated by plotting the substance concentration against the resulting level of phosphorylation expressed as % total uninhibited kinase activity. Note the greater inhibitory potency of polysialogangliosides and the relative lack of inhibitory effects due to "non-ganglioside substances." (From Kim et al., 1986).
A) Inhibition of C-kinase phosphorylation of rat MBPs by bovine brain ganglioside mixture (◆) and asialo-ganglioside, GA1 (■).
B) Inhibition of C-kinase phosphorylation of rat MBPs by GT1b (◆) and GM1 (×).
C) Inhibition of C-kinase phosphorylation of rat MBPs by GD1a (◆) and GD1b (×).
D) Effects of N-acetylneuraminic acid (NANA) (■), ceramide (×), and sulfatide (◆) on C-kinase phosphorylation of rat MBPs.

DISCUSSION

We have demonstrated that gangliosides have complex effects on brain membrane kinase systems. Stimulation of phosphorylation by gangliosides is completely dependent on the presence of calcium ions (Goldenring et al., 1985). Similar observations have been made recently by Nagai's group (Tsuji et al., 1985). Addition of the Ca^{++}-salt of gangliosides in the absence of added Ca^{++} elicits full stimulation of phosphorylation (Yu et al., 1984; Goldenring et al., 1985). The pattern of stimulation is qualitatively similar to that for calmodulin-dependent stimulation of brain membrane phosphorylation. Thus, one hypothesis suggests that Ca^{++}-gangliosides may stimulate calmodulin-dependent kinase. Gangliosides have been shown to stimulate other calmodulin-dependent enzyme systems such as cyclic nucleotide phosphodiesterase (David and Daly, 1980) and adenylate cyclase (Partington and Daly,

1979). In addition, several groups have shown that acidic lipids such as oleic acid can stimulate calmodulin-dependent processes (Niggli et al., 1981; Vincenzi, 1982; Vincenzi et al., 1982). However, unlike the stimulation reported here, these previous reports showed stimulation in the absence of Ca^{++}. Ca^{++}-binding to gangliosides causes a conformational change in the proximal portion of the lipid structure (Maggio et al., 1980; Yu et al., 1981; Probst et al., 1984). Thus, Ca^{++} may induce a conformational change in the ganglioside molecule which reveals a hydrophobic region which can stimulate membrane-bound calmodulin-dependent enzymes.

The exact mechanism of MBP phosphorylation inhibition by gangliosides is not clear. C-kinase is responsible almost entirely for the endogenous phosphorylation of MBP in myelin preparations (Turner et al., 1982). Thus, in light of the absence of endogenous calmodulin-dependent kinase in the myelin preparation (Turner et al., 1982; Yu et al., 1985), any detectable calcium-dependent kinase activity appears to be attributable to C-kinase. In brain membrane, the inhibition of the 78K phosphoprotein phosphorylation by ganglioside likely represents the inhibition of C-kinase phosphorylation (Wu et al., 1982). Albert et al. (1984) have reported that the phosphorylation of an "87K" protein is inhibited by calmodulin. Therefore, the effects of gangliosides as both stimulators and inhibitors of protein phosphorylation may reflect a calmodulin-like action on both calmodulin-dependent kinase and C-kinase.

C-kinase is inhibited by W-7 (a calmodulin antagonist) (Schatzman et al., 1983b), vitamin E, retinoic acid (Nishizuka, 1984), acylcarnitines, adriamycin, phenotheazines (e.g. trifluoroperazine) (Wise and Kuo, 1983), local anesthetics (e.g. dibucaine and tetracaine) (Nishizuka, 1984b), polyamines (Qi et al., 1983), melittin and polymixin B (Mazzei et al., 1982). These substances do not interact with the catalytically active site of C-kinase as demonstrated by their inability to affect the active fragment yielded by limited proteolysis (Kishimoto et al., 1983). Most of the above inhibitors interact competitively with the phospholipid cofactor (Nishizuka, 1984b). Nevertheless, none of these compounds is specific for C-Kinase and all of them also inhibit calmodulin-dependent kinase in varying concentrations. Gangliosides, however, appear to stimulate calmodulin-dependent kinases, while they inhibit C-kinase activity. Thus, gangliosides represent the first agents which can pharmacologically differentiate between C-kinase and calmodulin-dependent kinase *in vitro*.

The utility of gangliosides as *in vivo* modifiers of the calcium-dependent kinase systems remains to be determined. Developmental studies have demonstrated ganglioside-induced changes in cell growth and differentiation in various system (Ledeen, 1984). Gangliosides may modulate receptor phosphorylation (Bremer et al., 1984), and thus, these glycolipids may also regulate other kinases, such as tyrosine kinases, as well. Further investigations are in progress to define the exact molecular basis of ganglioside action on protein kinases.

ACKNOWLEDGMENTS

This work was carried out with the financial assistance provided by USPHS grants NS 11853, NS-23102 and National Multiple Sclerosis Society grant RG 1289-C-4. JRG is a recipient of a Medical Scientist Training Program Fellowship.

REFERENCES

Abramson MB, Yu RK, Zaby V (1972) Biochim Biophys Acta 280: 365-372.

Albert KA, Wu WC-S, Nairn AC, Greengard P (1984) Proc Natl Acad Sci USA 818: 3622-3625.

Ando S (1983) Neurochem Int 5: 507-537.

Ando S, Yu RK (1977) J Biol Chem 252: 6247-6250.

Ando S, Chang N-C, Yu RK (1978) Anal Biochem 89: 437-450.

Bennet MK, Erondu NE, Kennedy MB (1983) J Biol Chem 258: 12735-12744.

Bradford MM (1976) Anal Biochem 72: 248-254.

Bremer EG, Hakomori S-I, Bowen-Pope DF, Raine E, Ross R (1984) J Biol Chem 259: 6818-6825.

Burke BE and DeLorenzo RJ (1982) Brain Res 236: 393-415.

Cheung WY (1982) Fed Proc 41: 2253-2257.

Cohen P (1982) Nature 296: 613-620.

Davis CW and Daly JW (1980) Proc Pharmacol 17: 206-211.

DeLorenzo RJ, Burdette S, Holderness J (1981) Science 213: 546-549.

DeLorenzo RJ and Goldenring JR (1984) In: Marangos RJ, Campbell IC, Cohen RM (eds): Brain Receptor Methodologies. Academic Press, pp. 191-207.

Hakomori S (1981) Ann Rev Biochem 50: 733-764.

Hakomori S (1983) In: Kanfer JN and Hakomori S (eds): Sphingolipid Biochemistry. Plenum Press, New York, pp. 327-379.

Kasai N, Sillerud LO, Yu RK (1982) Lipids 173: 107-110.

Kennedy MB (1983) Ann Rev Neurosci 6: 493-526.

Kim JYH, Goldenring JR, DeLorenzo RJ, Yu RK (1986) J Neurosci Res 15: 159-166.

Kishimoto A, Kajikawa N, Shiota M, Nishizuka Y (1983) J Biol Chem 258: 1156-1164.

Ledeen RW (1984) J Neurosci Res 12: 147-159.

Ledeen R (1985) TINS 8: 169-174.

Ledeen RW and Yu RK (1982) Methods Enzymol 83: 139-191.

Ledeen RW, Yu RK, Eng LF (1973) J Neurochem 218: 829-839.

Ledeen RW, Yu RK, Rapport MM, Suzuki K (eds) (1984) Ganglioside Structure, Function, and Biomedical Potential. Adv Exp Med Biol, Vol. 174, Plenum Press, New York.

Maggìo B, Cumar FA, Caputto R (1980) Biochem J 1889: 435-440.

Mazzei GJ, Katoh N, Kuo JF (1982) Biochem Biophys Res Comm 109: 1129-1133.

Nestler EJ, Walaas SI, Greengard P (1984) Science 225: 1357-1364.

Niggli V, Adunyah ES, Carafoli E (1981) J Biol Chem 256: 8588-8592.

Nishizuka Y (1984a) Science 225: 1365-1370.

Nishizuka Y (1984b) Nature 308: 693-698.

Norton WT and Poduslo SE (1973) J Neurochem 21: 749-758.

Partington CR and Daly JW (1979) Mol Pharmac 15: 484-491.

Petrali EH and Sulakhe PV (1982) Prog Brain Res 56: 125-144.

Probst W, Rosner H, Wiegandt H, Rahmann H (1979) Hoppe-Seyler's Z Physiol Chem 360: 979-986.

Probst W, Mobius D, Rahmann H (1984) Cell Molec Neurobiol 4: 157-176.

Quarles R and Folch-Pi J (1965) J Neurochem 12: 543-553.

Qi DF, Schatzman RC, Mazzei GJ, Turner RS, Raynor RL, Liao S, Kuo JF (1983) Biochem J 231: 281-288.

Rahmann H (1983) Neurochem Int 5: 539-547.

Schatzman RC, Grifo JA, Merrick WC, Kuo JF (1983a) FEBS Lett 159: 167-170.

Svennerholm L (1957) Biochim Biophys Acta 24: 604-611.

Tsuji S, Nakajima J, Sasaki T, Nagai Y (1985) J Biochem 97: 969-972.

Turner RS, Chou C-HJ, Kibler RF, Kuo JF (1982) J Neurochem 39: 1397-1404.
Turner RS, Chou C-HJ, Mazzei GJ, Dembure P, Kuo JF (1984) J Neurochem 43: 1257-1264.
Vincenzi FF (1982) Ann N Y Acad Sci 402: 368-380.
Vincenzi FF, Adunyah ES, Niggli V, Carafoli E (1982) Cell Calcium 3: 545-559.
Waisman DM, Gimble JM, Goodman DBP, Rasmussen H (1981) J Biol Chem 256: 409-414.
Wise BC and Kuo JF (1983) Biochem Pharm 32: 1259-1265.
Witt JJ and Roskoski R (1975) Anal Biochem 66: 253-258.
Wu WC-S, Walaas SI, Nairn AC, Greengard P (1982) Proc Natl Acad Sci U S A 79: 5249-5253.
Yohe HC, Jacobson RI, Yu RK (1983) J Neurosci Res 9: 401-412.
Yu RK, Sillerud LO, Schafer DE, Prestegard JH, Konigsberg W (1981) Trans Amer Soc Neurochem 12: 78.
Yu RK, Goldenring JR, DeLorenzo RJ (1984) INSERM 1984 126: 335-354.

Gangliosides and neuronal plasticity
G. Tettamanti, R.W. Ledeen, K. Sandhoff,
Y. Nagai, G. Toffano (eds.)
Fidia Research Series, vol. 6
Liviana Press, Padova, © 1986

MEMBRANE AGING OF THE BRAIN SYNAPTOSOMES WITH SPECIAL REFERENCE TO GANGLIOSIDES

Susumu Ando, Yasukazu Tanaka, Kazuo Kon

Department of Biochemistry, Tokyo Metropolitan Institute of Gerontology, Sakaecho, Itabashi-ku, Tokyo-173, Japan

INTRODUCTION

Giacobini (1982) has proposed an idea of synaptic aging, according to which synapses are functionally developed to the maximal state in the developmental stage, and then may deteriorate in their function in senescence. Giacobini emphasized the relevance of reduced synaptic transmission to malfunction of the aging brain. Age-related alterations of synaptic functions have been reported in terms of the levels, synthetic rates and releasing activities of neurotransmitters. Gibson and Peterson (1981a) reported the reduced synthetic rate of acetylcholine in aged mice with little change in acetylcholine content. Gibson and Peterson (1981b) also pointed out that the release of acetylcholine is severely affected with aging. We assumed that possible alterations of synaptic membranes would be responsible for the decrease in neurotransmission. We, therefore, have isolated synaptosomes from different age groups of mice and rats, and attempted to correlate the membrane changes with altered synaptic functions such as acetylcholine release.

EXPERIMENTAL

Different age groups of C57BL/6 mice and Wistar rats were supplied from the aging farm of our institute. The animals were maintained with ordinary chow diet under a specific pathogen-free condition. Synaptosomal fractions were prepared from cerebral cortices by using isotonic discontinuous media according to Booth and Clark (1978).

Total lipids were extracted from an aliquot of the synaptosomal fraction (0.5 - 1.0 mg protein) with $CHCl_3$ / MeOH / water (1:2:0.8, v/v; total volume, 3.8 ml) (Bligh

---

Abbreviations: TLC, thin layer chromatography; ESR, electron spin resonance.

and Dyer, 1959). Aliquots of the extract were analyzed for total phosphorus (Bartlett, 1959) and free cholesterol (cholesterol oxidase method using an assay kit, Determiner[R] FC-555, Kyowa Medics Co., Tokyo). The rest of the total extract was applied to a DEAE-Toyopearl column (DEAE-Toyopearl 650M, Toyo Soda Manufacturing Co., column volume, 0.5 ml). The procedure for the elution of gangliosides was principally based on the original method of Ledeen et al. (1973). Neutral lipids were eluted with 3 ml of $CHCl_3$ / MeOH / water (30:60:8), and acidic lipids were quantitatively obtained by elution with 5 ml of $CHCl_3$ / MeOH / aqueous 1.0 M sodium acetate (30:60:8). The acidic lipid fraction was mixed with a small amount of aqueous 10 N NaOH (final concentration, 0.2 N), and warmed at 37°C for 2 h. The fraction was freed from organic solvents and redissolved in 1.0 ml water. The solution was applied to a Bio-Gel P-6DG column (column volume, 10 ml), and total gangliosides were obtained in the effluent of 3.5 ml following the void volume elution, based on the method of Ueno et al. (1978). Gangliosides were quantitated as sialic acid by Svennerholm's method (1957) modified by Miettinen and Takki-Luukainen (1959).

Membrane microviscosity was monitored at 37°C as fluorescence polarization (P value = $(I_{\parallel} - I_{\perp}) / (I_{\parallel} + I_{\perp})$) by using 1,6-diphenyl-1,3,5-hexatriene (DPH) according to Shinitzky and Barenholz (1978). P values were directly read with an Elcint Al fluorospectrometer.

The release of acetylcholine was measured as follows. Freshly prepared synaptosomes were incubated with [^{14}C]-choline in Krebs-Ringer salt solution containing physostigmine for 30 min at 37°C, and washed with the same solution three times, followed by preincubation for 15 min to reach a steady level of the spontaneous release. The labeled synaptosomes were incubated for 5 min at 37° with or without stimulation (veratridine or high potassium). After centrifugation, the radioactivities of the supernatant and pellet were measured.

RESULTS AND DISCUSSION

Membrane Compositional Changes of Synaptosomes

Figure 1 shows the age-related changes in rat synaptosomal lipids. Total lipid phosphorus decreases slightly with advancing age, while the cholesterol level remains almost constant. Consequently, the molar ratio of cholesterol to lipid phosphorus (C/P) shows a gradually increasing tendency. A similar increase of C/P ratio with age was found in mouse myelin (Ando et al., 1984). The ganglioside content rapidly decreases in the young adult stage and then the changes slow down in the adult and senescent stages. It was reported by Suzuki (1965) and Vanier et al. (1971) that rat brain total gangliosides rapidly increase during 1 - 2 weeks of postnatal age, and then decrease. Their findings may correspond to the decrease in synaptosomal gangliosides during the young adult period. On the other hand, myelin gangliosides were previously reported to increase progressively until senescence (Ando et al., 1984). Thus, age-related changes of gangliosides may occur differently in each structural component of the central nervous system.

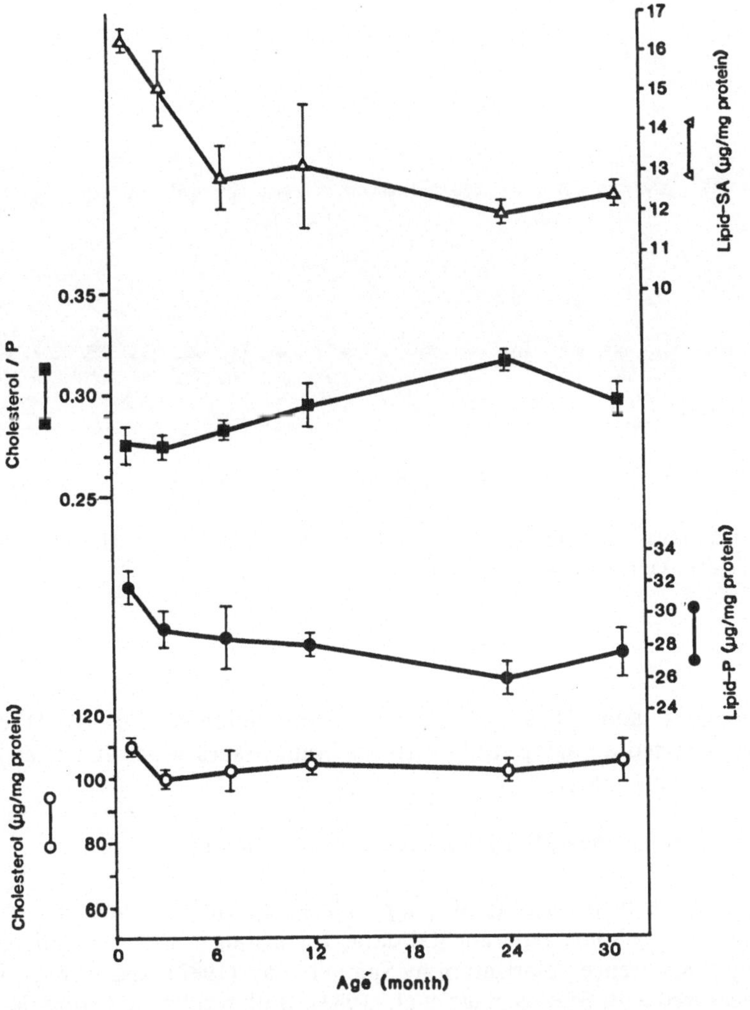

Figure 1. Age-related changes in the ganglioside, phospholipid and cholesterol contents of rat brain synaptosomes.

Rodent synaptosomes were revealed to have a unique set of ganglioside species. As shown in Figure 2, GD1a, GT1b and GQ1b were observed as the major species in mouse synaptosomes, and only a trace amount of GM1 was detected by ordinary TLC. Dreyfus et al. (1980) reported the presence of high amounts of polysialogangliosides, and of low amounts of monosialo species in cultured neuronal cells isolated from chick embryo hemispheres. Dreyfus et al. (1980) also observed a high increase of GD1a during synapse formation. Their findings seem to be supportive of ours. Therefore, GD1a, GT1b and GQ1b may be assigned as synaptosomal gangliosides proper. Relative distributions of these gangliosides appear to shift with aging, that is, GD1a (a series)

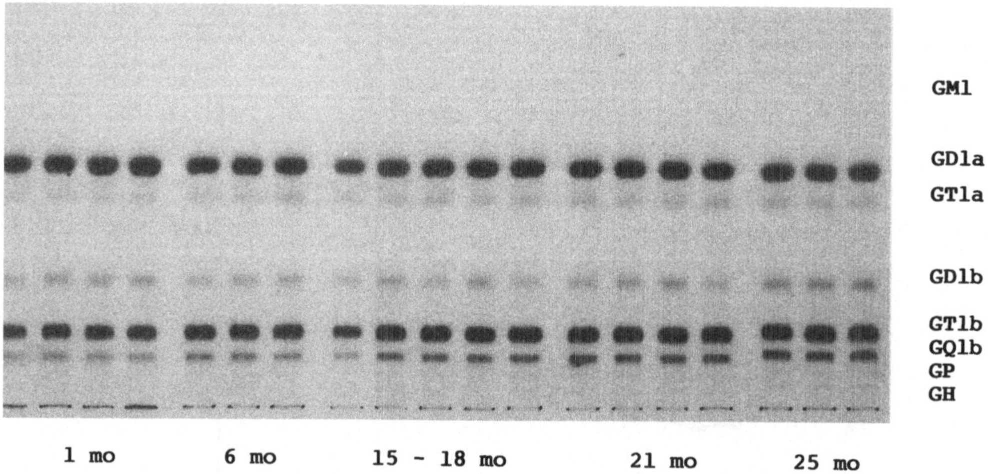

Figure 2. Thin-layer chromatogram of mouse synaptosomal gangliosides. GP, pentasialoganglioside; GH, hexasialoganglioside.

steadily decreases, and GT1b and GQ1b (b series) increase (Fig. 3). Hilbig et al. (1983/1984) reported an aging study with rat brain tissues which indicated a similar trend as seen in this study.

Changes in the Membrane Microviscosity of Synaptosomes

The membrane microviscosity of synapses has been alleged to increase with aging. This is supported by some experimental data, for instance, increased microviscosity measured by fluorescence polarization by Samuel et al. (1982), and increased correlation time measured with ESR by Nagy et al. (1983). Both results imply that the synaptic membrane microviscosity apparently increases during aging in a linear fashion, simply because only two or three age points were monitored in those studies. We have tested more age points with rats by using the fluorescence polarization method (Shinitzky and Barenholz, 1978). The whole course of the fluorescence polarization changes appears to be a non-linear curve throughout the life span of rat, as shown in Figure 4. It is noticed that there are definite discrepancies between the changes in the fluorescence polarization and in the C/P ratio (Fig. 1). This seems to be unexpected and unusual, because the membrane microviscosity is thought to be principally determined by the C/P ratio, even though other factors such as fatty acid composition, contents of sphingomyelin and glycokalyx, etc., may somewhat affect the microviscosity too (Stubbs and Smith, 1984). The changes of fluorescence polarization (Fig. 4) appear to be roughly a mirror image of the changes of ganglioside contents (Fig. 1), suggesting an inverse relationship between the ganglioside content and the membrane microviscosity.

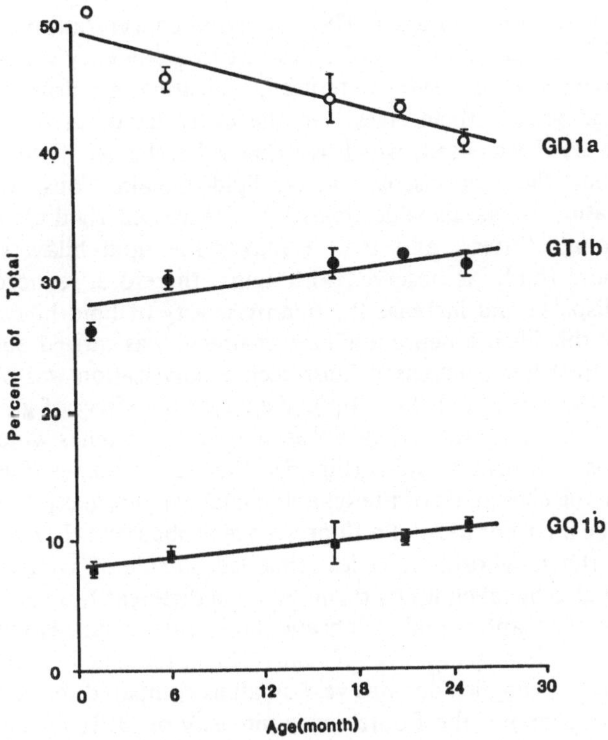

Figure 3. Age-related compositional changes of the mouse synaptosomal ganglioside species.

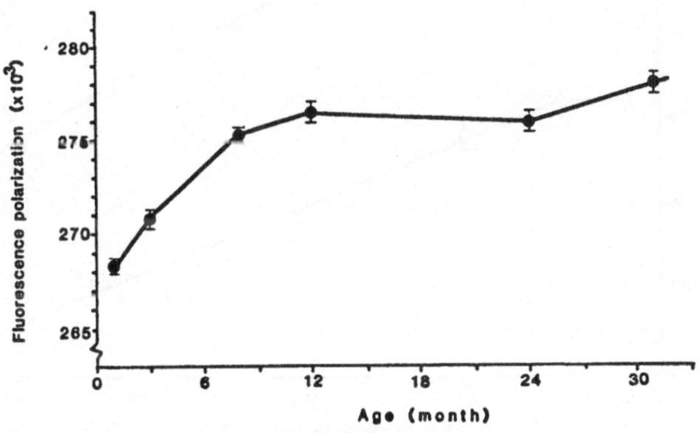

Figure 4. Age-related changes in the fluorescence polarization of DPH incorporated in the rat synaptosomal membranes.

According to Sharom and Grant (1978), increased concentration of gangliosides decreases the mobility of ganglioside molecules due to a tendency of aggregation. In their system, spin-labeled gangliosides were used, indicating primarily the mobility of the ganglioside head groups themselves. On the other hand, in our experiment a hydrophobic probe DPH was used, which is supposed to be deeply localized in lipid bilayers, and to report the microviscosity of the lipid domain. Thus, we assume that increased concentration of gangliosides causes the increased rigidity of ganglioside molecules by forming clusters, and that it makes the lipid bilayers depleted of gangliosides and more fluid. As observed with aging, therefore, reduced amounts of gangliosides may disperse and increase the microviscosity of lipid bilayers.

In order to test this idea, a neuraminidase treatment was carried out with synaptosomes. After the treatment, increased fluorescence polarization was observed, that is, from 272.8 ± 0.3 to 274.0 ± 1.2 ($\times 10^{-3}$). We tested the effect of ganglioside supplementation on the microviscosity of synaptosomes and liposomes which were made of the total synaptosomal neutral lipids (Fig. 5). The synaptosomes (0.6 mg protein) were incubated with micelles made of the synaptosomal gangliosides (2-10 μg / mg protein), followed by the incorporation of a fluorescence probe DPH. It was assumed that if some portion of the ganglioside micelles remained outside of synaptosomal membranes, DPH might also be taken up by them, giving a different fluorescence polarization value from that of synaptosomal membranes and a false result. However, the contribution from ganglioside micelles, even if half of them remained, should be negligible, because the mass of gangliosides was very small as compared to the total lipids in the membranes. Furthermore, the fluorescence intensity of DPH in the micelles was half as much as that in the membranes. The possible contribution was thus estimated to be less than 0.4% of the fluorescence polarization value observed. As shown in Figure 5, fluorescence polarization was linearly decreasing in both the membranes and

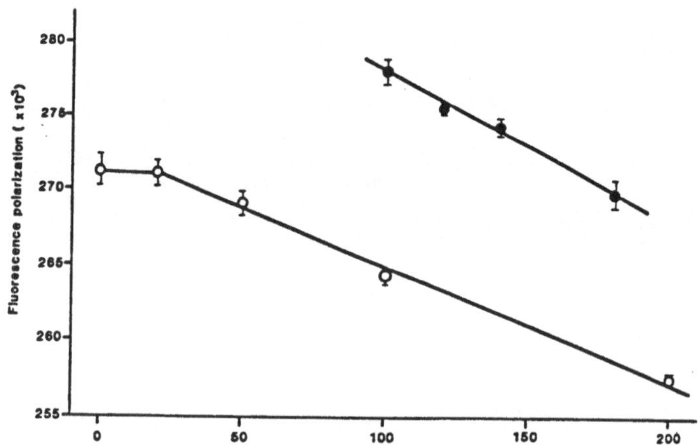

Figure 5. Effect of ganglioside supplementation on the fluorescence polarization of DPH in the synaptosomes (—•—•—) isolated from 24 month-old rats and liposomes (—○—○—) made of the neutral synaptosomal lipids.

liposomes supplemented with gangliosides. The synaptosomes isolated from 24 month old rats were modified by 80% supplementation with gangliosides to show about the same microviscosity as those from 3 month old rats (Fig. 4). Thus, it is supposed that the microviscosity of lipid bilayers is partially regulated by the ganglioside concentration. It seems likely that the decreased ganglioside content increases the membrane microviscosity of synaptosomes in the aging brain. Hintzemann and Harris (1984) also reported the effect of gangliosides on microviscosity in a liposome system, but they obtained confused results; that is, increased microviscosity was reported by trimethylamino-DPH as a probe and decreased microviscosity by DPH when gangliosides were added.

How is the Synaptic Function Affected by Aging?

Acetylcholine release, which was evoked by veratridine, was measured with synaptosomes isolated from different age groups of rats. Figure 6 reveals that acetylcholine release is the highest in the developmental stage, and rapidly decreases to the lowest level during the adult stage. The releasing activity then rebounds somewhat in the late adult stage, and remains almost constant during senescence. The profile of the age-related changes in the activity seems to be different from that reported by Gibson and Peterson (1981b), who found a linearly decreasing change in the release activity with aging. The rapidly decreasing phase in the young adult stage (Fig. 6) appears to coincide with that of gangliosides (Fig. 1).

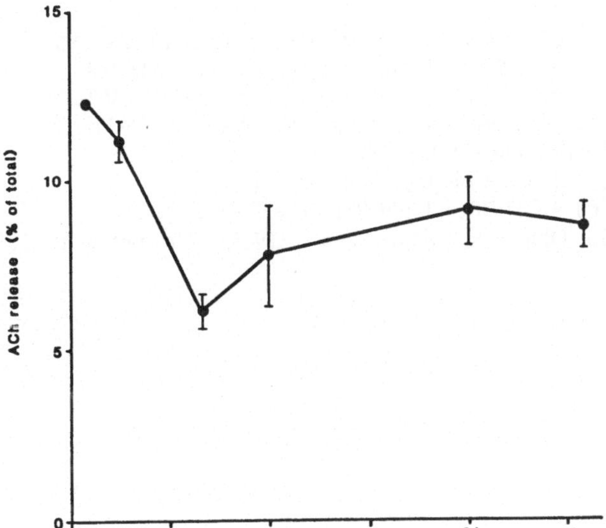

Figure 6. Age-related changes in the acetylcholine release of the rat synaptosomes. The release was evoked by veratridine (0.1 mM).

The relationship between acetylcholine release and ganglioside content was examined by neuraminidase treatment. The release, when evoked by high potassium, was reduced from 15.8 ± 1.0 to 14.3 ± 0.3 (percent of the total radioactivity) after neuraminidase treatment with the synaptosomes isolated from 12 month old rats. This decrease may partially explain the in vivo alterations in synaptic function with aging.

REFERENCES

Ando S, Tanaka Y, Ono Y, Kon K (1984) in: Ledeen RW, Yu RK, Rapport MM, Suzuki K (eds): Ganglioside Structure, Function, and Biomedical Potential, Plenum, New York, pp. 241-248.

Bartlett GR (1959) J Biol Chem 234: 466-468.

Bligh EG and Dyer WJ (1959) Can J Biochem Physiol 37: 911-917.

Booth RFG and Clark JB (1978) Biochem J 176: 365-370.

Dreyfus H, Louis JC, Harth S, Mandel P (1980) Neurosci 5: 1647-1655.

Gibson GE and Peterson C (1981a) Science 213: 674-676.

Gibson GE and Peterson C (1981b) J Neurochem 37: 978-987.

Hilbig R, Lauke G, Rahmann H (1983/1984) Dev Neurosci 6: 260-270.

Hintzemann RJ and Harris A (1984) Dev Brain Res 14: 113-120.

Ledeen RW, Yu RK, Eng LF (1973) J Neurochem 21: 829-839.

Miettinen T and Takki-Luukkainen IT (1959) Acta Chem Scand 13: 856-858.

Nagy K, Simon P, Zs.-Nagy I (1983) Biochem Biophys Res Commun 117: 688-694.

Samuel D, Heron DS, Hershkowitz M, Shinitzky M (1982) The Aging Brain: Cellular and Molecular Mechanisms of Aging in the Nervous System. Raven Press, New York, pp. 93-97.

Sharom FJ and Grant CWM (1978) Biochim Biophys Acta 507: 280-293.

Shinitzky M and Inbar M (1976) Biochim Biophys Acta 433: 133-149.

Shinitzky M and Barenholz Y (1978) Biochim Biophys Acta 515: 367-394.

Stubbs CD and Smith AD (1984) Biochim Biophys Acta 779: 89-137.

Suzuki K (1965) J Neurochem 12: 969-979.

Svennerholm L (1957) Biochim Biophys Acta 24: 604-611.

Ueno K., Ando S, Yu RK (1978) J Lipid Res 19: 863-871.

Vanier MT, Holm M, Öhman R, Svennerholm L (1971) J Neurochem 18: 581-592.

Gangliosides and neuronal plasticity
G. Tettamanti, R.W. Ledeen, K. Sandhoff,
Y. Nagai, G. Toffano (eds.)
Fidia Research Series, vol. 6
Liviana Press, Padova, © 1986

POSSIBLE ASSOCIATION OF DEGENERATIVE MOTOR NEURON DISEASE (ALS) WITH ABNORMAL GANGLIOSIDE METABOLISM: ABNORMAL GANGLIOSIDES IN ALS SPINAL CORD AND ALS-LIKE SYMPTOMS IN A UNIQUE PARTIAL HexB DEFICIENCY SYNDROME

G. Dawson, L.W. Hancock, A.L. Horwitz, R. Wollman, N. Cashman, J. Antel

Joseph P. Kennedy Jr. Mental Retardation Research Center,
Departments of Pediatrtics, Neurology and Pathology, University of Chicago,
Chicago, Illinois, USA

INTRODUCTION

Two major ganglioside storage diseases are known in man, GM2 (II^{3}NeuAcGgOse$_3$Cer)-Gangliosidosis (Tay-Sachs disease and its β-hexosaminidase-deficient variants) and GM1-gangliosidosis (β-Galactosidase deficiency) (O'Brien, 1983). Both diseases encompass many genotypic variants and many clinical phenotypes, but the major pathological characteristic is the lysosomal accumulation of gangliosides in neurons (O'Brien, 1983). Ultrastructural studies of neurons in humans and animals with GM2- and GM1-gangliosidosis have revealed aberrant sprouting of neurites from axon hillocks (meganeurites) (Purpura and Suzuki, 1976) prior to neuronal death and this has led to much of the current interest in gangliosides as neuritogenic agents. Much of this Symposium will be devoted to the stimulatory, neurite regenerative capacity of gangliosides, so it is important to remember the neuropathological origins of the "ganglioside effect".

One can also use these accidents of nature to study the relative susceptibility of neuronal populations to abnormal ganglioside catabolism. This is most clearly seen in patients who have partial enzyme deficiencies and slower onset of symptoms. For example, ten or more patients with an abnormality of the α-chain locus which results in a profound loss of β-hexosaminidase activity (HexA) toward GM2 ganglioside have

---

Abbreviations: ALS, amyotrophic lateral sclerosis; SDS, sodium dodecylsulphate; EM, electron microscopy; 4MU (4MU-), 4 methyl umbelliferone (4-methylumbelliferyl-); PDGF, platelet derived growth factor; EGT, epithelial cells growth factor; PAGE polyacrylamide gel electrophoresis; TLC, thin layer chromatography.

been described (Johnson, 1981; Argon and Navon, 1984; Frisch et al., 1984). Such patients have cerebellar atrophy and motor neuron disease, suggesting that these populations of neurons either metabolize gangliosides more rapidly than those, say, of the visual system (retinal ganglia) or are more susceptible to ganglioside accumulation.

We have recently described a patient with a slow onset form of amyotrophic lateral sclerosis and a complete deficiency of β-hexosaminidase B (Hancock et al., 1985; Cashman et al., 1986). A complete deficiency of HexB with normal β-hexosaminidase A (HexA) activity has not been previously described. Surprisingly, despite the presence of substantial amounts of HexA activity toward synthetic substrates, this patient had a marked decrease in ability to hydrolyze GM2. The focus of the disease on upper and lower motor neurons suggests that these neurons have a unique ganglioside metabolism. Together with our recent findings of abnormal Hex*N*Ac-containing gangliosides in ALS (Dawson and Stefansson, 1984) it appeared that a study of the patient would yield important insights into human ganglioside metabolism in vivo.

CLINICAL DESCRIPTION

The female patient (KL) of German-Irish ancestry was initially seen at the University of Chicago's Joseph P. Kennedy, Jr. Mental Retardation Research Center at the age of 7 years with speech problems and evidence of neurodegenerative disease. At age 26, she has a progressive upper and lower motor neuron syndrome resembling amyotrophic lateral sclerosis. There is little evidence of tremor, sensory abnormalities or cerebellar degeneration (in contrast to reported HexA deficiencies). Her major clinical problems are widespread fasciculations, peripheral muscle atrophy, and weakness in the limbs with little evidence of intellectual or visual impairment.

BIOCHEMICAL STUDIES

Enzymology

Total β-Hexosaminidase (as measured with the 4-methylumbelliferyl substrate) (Dawson and Tsay, 1977), was markedly reduced, with all activity appearing to be HexA. Her mother and one sibling exhibited evidence of heterozygote status for HexB (Table 1), whereas her father and one other sibling had normal ratios of HexA and HexB and normal total β-Hex. KL fibroblasts, grown as monolayers in Dulbecco's modified Eagle's medium supplemented with 10% fetal calf serum, showed total β-Hex activity in the normal range, but more than 90% of activity was lost after incubating at 50°C for 2 hr. The low (almost undetectable) level of HexB was confirmed by cellogel electrophoresis (Fig. 1). Mixing studies with extracts of normal fibroblasts and both α- and β-chain locus mutants revealed no evidence of inhibitors. The residual HexA activity in the patient's fibroblasts behaved as normal HexA on DEAE Sephadex ion exchange chromatography and thermal inactivation studies revealed a normal rate of inactivation.

Subsequent studies on the patient's urinary β-Hex by Drs. Y.T. Li and S.C. Li

(Tulane University, New Orleans, LA) confirmed the total absence of HexB, and suggested the presence of normal β-Hex activator protein.

Table 1. *Tissue Hexosaminidase Levels in β-Hex Deficiencies*

	Total β Hex (% of control)	% HexA*
Plasma		
Patient KL	15	98
Mother	50	80
Father	95	65
Sibling 1	110	70
Sibling 2	85	50
Tay-Sachs	110	0
Juvenile spinal muscular atrophy	50	10
Leukocytes		
Patient KL	10	100
Mother	40	90
Father	80	70
Tay-Sachs	100	0
Juvenile spinal muscular atrophy	120	10
Fibroblasts		
Patient KL	80	96
Tay-Sachs	230	0
Sandhoff dis.	0	0
Adult GM2 gangliosidosis	100	10
GM1-gangliosidosis	150	45

* Determined by thermal inactivation (Hancock et al., 1985; Myerowitz et al., 1983) confirmed by Cellogel electrophoresis.

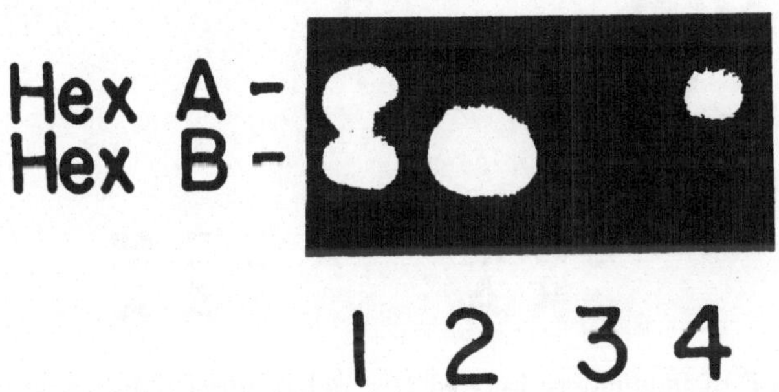

Figure 1. Cellogel electrophoresis profile of fibroblast homogenates following 4MU released from 4MU-β-GlcNAc as described in text. Lane 1, normal control demonstrating both Hex A and Hex B activity; Lane 2, Tay-Sachs disease with absent Hex A; Lane 3, Sandhoff disease with absent Hex A and Hex B; Lane 4, Patient A with absent Hex B and diminished Hex A. Tay-Sachs and Sandhoff disease fibroblasts obtained from the cell collection of the Kennedy Mental Retardation Research Center of the University of Chicago.

Processing of α- and β- Chains in Normal and KL Fibroblasts

Fibroblasts were pulse-labelled with [^{3}H]leucine by the method of Hasilik and Neufeld (1980) and β-Hex immunoprecipitated with a polyclonal antibody to HexB ($\beta_A\beta_B$) generously donated by Drs. Proia (NIH, Bethesda, MD) and E Neufeld (Dept. Biochem., UCLA, Los Angeles, CA). SDS 9% polyacrylamide gel electrophoresis showed that both control and KL fibroblasts synthesized mature α-chains (54KDa) and β-chains (29KDa, 25KDa, etc.), but the α/β ratio was substantially higher in patient KL. When fibroblasts were treated with NH_4Cl to prevent acidification and cause secretion of newly synthesized β-Hex precursors into the culture medium (Hasilik and Neufeld, 1980; Myerowitz et al., 1983) both normal and KL showed α-precursor (67KDa) and β-precursor (63KDa) (Hasilik and Neufeld, 1980; Myerowitz et al., 1983). However, the relative amount of β-chain precursor was greatly diminished in KL (Fig. 2), a ratio of $\alpha/\beta=2$ compared to 1 in controls.

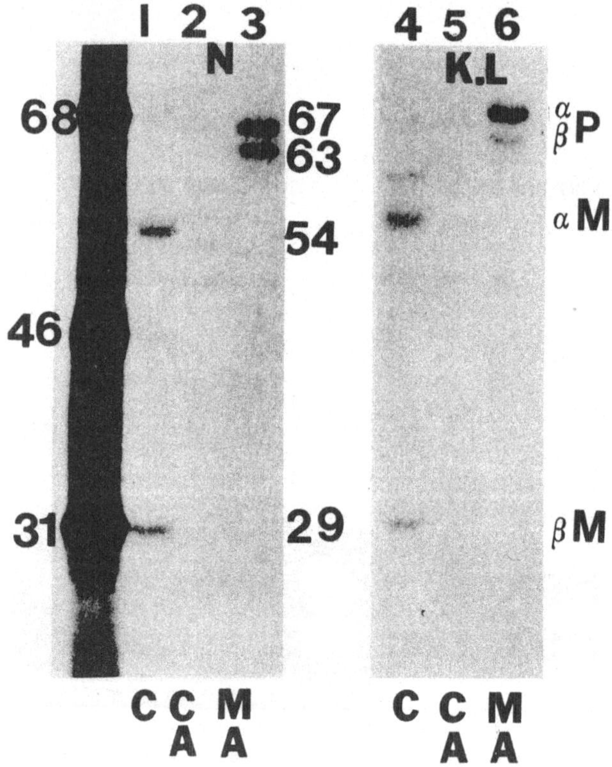

Figure 2. SDS-PAGE of immunoprecipitated Hex polypeptides (labelled with [^{3}H]Leu):
Lane 1, Normal cells
Lane 2, Normal cells treated with 10 m*M* NH_4Cl
Lane 3, Medium from NH_4Cl-treated normal cells
Lane 4, Patient K.L. cells
Lane 5, Patient K.L. cells treated with 10 m*M* NH_4Cl
Lane 6, Medium from Patient K.L. cells treated with NH_4Cl
M = mature polypeptide chain; P = precursor polypeptide chain

This suggested that a partial failure of β-chain synthesis was responsible for the apparent HexB deficiency in KL, with all available β-chains combining with α-chains to form HexA. It is also possible that some β-locus defect in KL causes a failure of β-β chain self-association or premature degradation of β_A (acidic) or β_B (basic) chains (O'Dowd et al., 1985).

Evidence for Abnormalities in Ganglioside Metabolism in Patient KL

Since β-Hex abnormalities have been described in patients who have no clinical symptoms (O'Brien, 1983), it was important to relate the apparent defect in ganglioside catabolism to actual evidence for pathological ganglioside accumulation. Because of the limited availability of biopsy material, storage was confirmed in a rectal biopsy by electron microscopy (EM) (Fig. 3) and in cultured skin fibroblasts following GM2 loading (Hancock et al., 1986).

A rectal suction biopsy obtained at age 24 showed the presence of cells enriched in granules which had the staining characteristics of lysosomes at the light microscopic level. EM study revealed numerous concentric, lamellar inclusions in ganglion cells (Fig. 3). The demonstration of the membrano-cytoplasmic storage bodies in myenteric plexus is considered pathognomic for a glycolipid storage disease (O'Brien, 1983).

Chemical Identification of Storage Material

Analysis of gangliosides and neutral glycolipids from cultured fibroblasts revealed a slight increase in endogenous GM2, as is observed in fibroblasts cultured from patients with Tay-Sachs disease (Fig. 4). Thin-layer chromatography of neutral glycolipids revealed some elevation of asialo-GM2 ($GgOse_3Cer$) comparable to that in Tay-Sachs disease, but not elevation of globoside ($GbOse_3Cer$) as observed in Sandhoff fibroblasts (total β-Hex deficiency) (Fig. 5).

Analysis of possible oligosaccharide storage revealed no accumulation in control, Tay-Sachs or KL fibroblasts in contrast to the accumulation of a hepta- or hexa- saccharide (GlcNAc(2→3)Man_3GlcNAc) in Sandhoff fibroblasts (Hancock et al., 1986). The storage of globoside and GlcNAc (glycoprotein-derived) oligosaccharides is generally associated with HexB deficiency (since it occurs in Sandhoff tissue, but not in Tay-Sachs tissue). It was, therefore, surprising that the absence of HexB in KL did not lead to such storage, suggesting that residual enzyme with the biochemical properties of HexA was able to hydrolyze both $GbOse_4Cer$ and oligosaccharide - or that the apparent HexB deficiency was an artifact of using a synthetic substrate (4MUβGlcNAc).

To test the ability of fibroblasts to degrade GM2, [^{3}H]GM2 (labelled in the terminal GalNAc residue by the galactose oxidase-NaB^3H_4 procedure) (Kolodny and Raghavan, 1983) was fed to fibroblasts for 7 days, and gangliosides isolated from the cells. Accumulation of [^{3}H] GM2 was observed in Tay-Sachs, Sandhoff and KL fibroblasts (the amount of GM2 storage in KL was approximately half that observed in Tay-Sachs), but not in normals (Fig. 6). This partial inability of KL fibroblasts to degrade GM2 was confirmed in collaboration with Dr. Y.T. Li who showed that urinary β-Hex from patient KL was only 25% as effective in degrading GM2 (in the presence of normal or Kl activator protein) as was the enzyme isolated from normal urine.

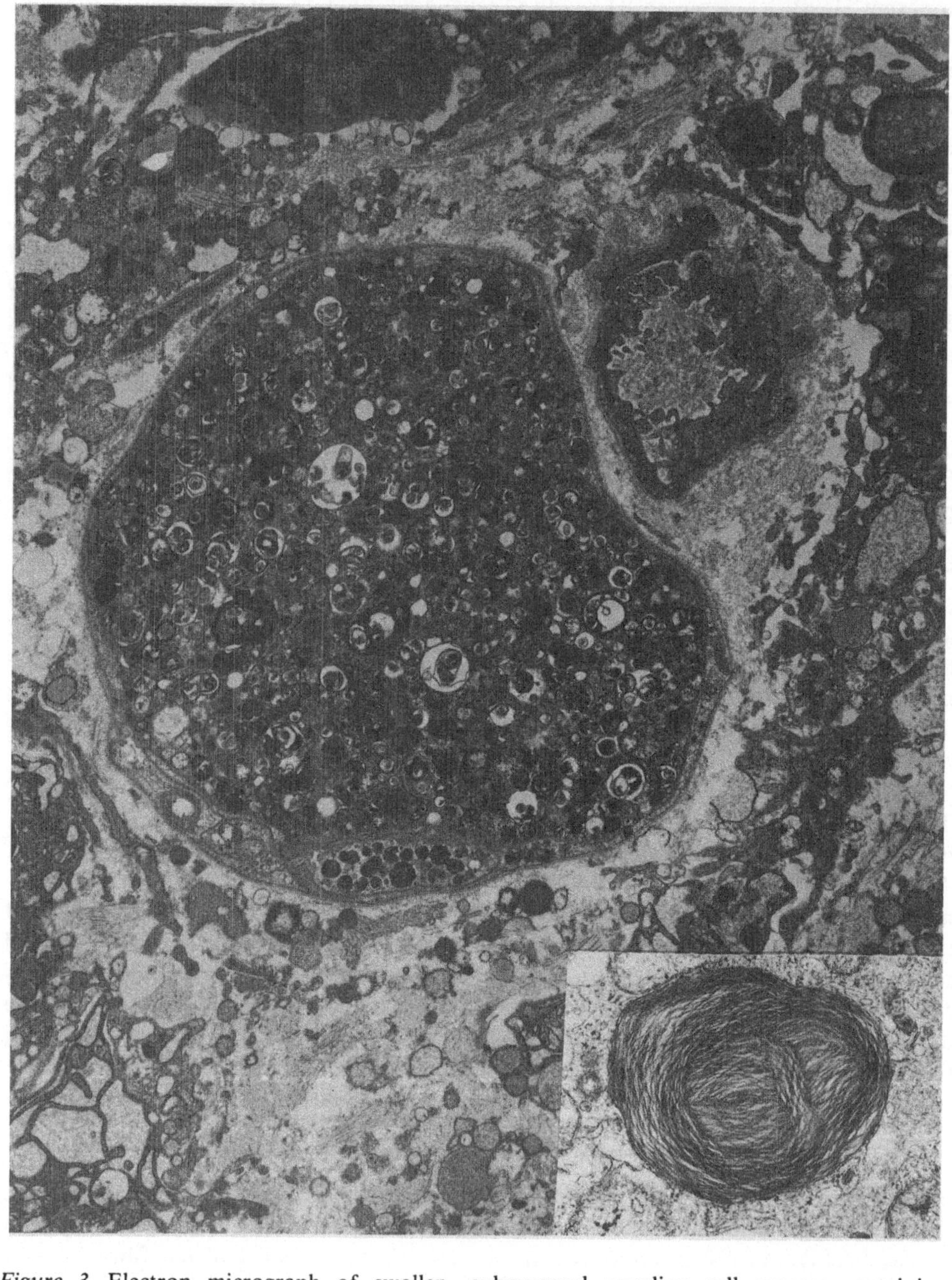

Figure 3. Electron micrograph of swollen, submucosal ganglion cell process containing numerous secondary lysosomes filled with membranous lamellar bodies. Suction rectal biopsy, 6875x.

INSET: Electron micrograph of "membranous cytoplasmic bodies" characteristic of the lysosomal inclusions found in the GM2 gangliosides. Suction rectal biopsy, 20625x

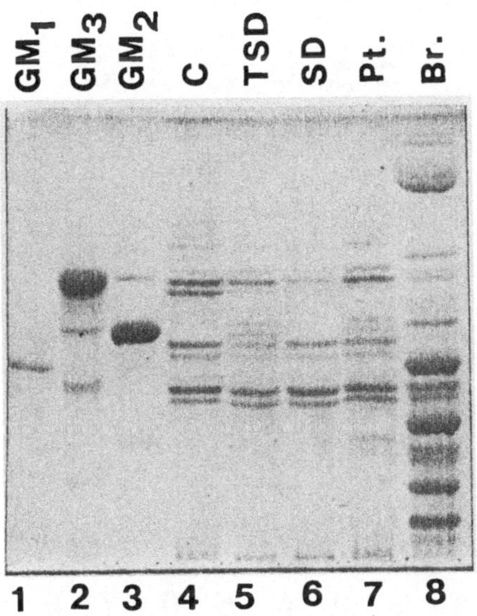

Figure 4. Thin-layer chromatogram of human skin fibroblast gangliosides, developed twice in $CHCl_3$: CH_3OH:0.02% aqueous $CaCl_2$(60:40:9v/v). Lane 1, GM1; Lane 2, GM3; Lane 3, GM2; Lane 4, normal cells; Lane 5, Tay-Sachs disease; Lane 6, Sandhoff disease; Lane 7, Patient K.L.; Lane 8, normal human brain.

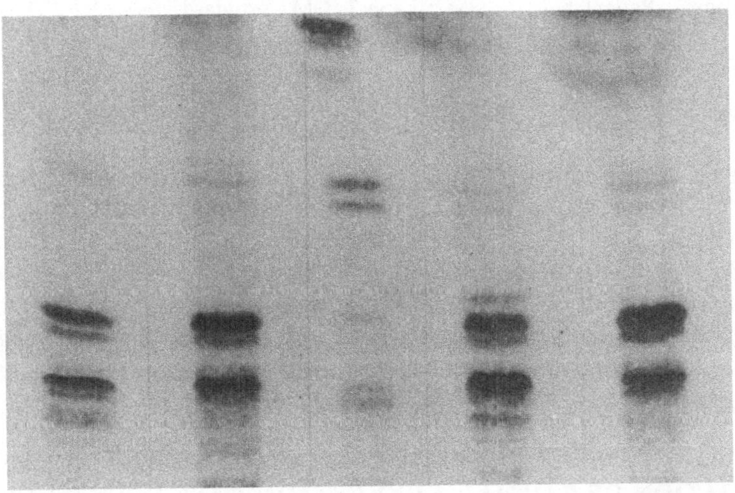

1 2 3 4 5

Figure 5. Thin-layer chromatogram of human skin fibroblast glycolipids developed in chloroform-methanol-water (100-42-6 v/v). Lane 1, control; Lane 2, Patient K.L.; Lane 3, human spleen glycolipids (major doublet is lactosylceramide); Lane 4, Sandhoff's Disease; Lane 5, Tay-Sachs disease.

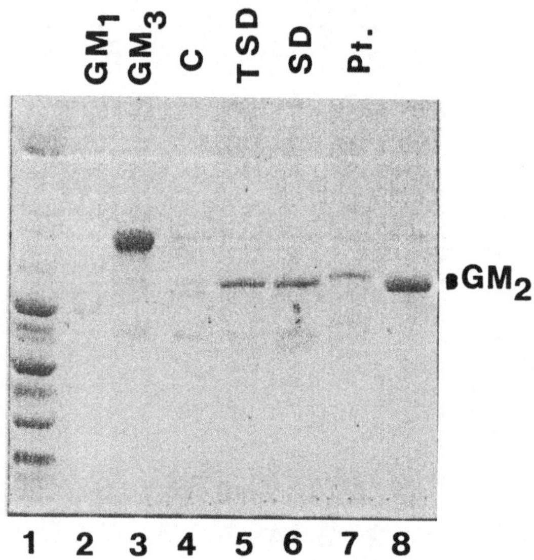

Figure 6. Thin-layer chromatogram of human skin fibroblast gangliosides following feeding with [^{3}H]GM2; Lane 1, human brain gangliosides; Lane 2, GM1 standard; Lane 3, GM3 standard; Lane 4, normal cells + GM2; Lane 5, Tay-Sachs cells + GM2; Lane 6, Sandhoff cells + GM2; Lane 7, Patient K.L. + GM2; Lane 8, GM2 standard.

Conclusions

These results are consistent with a partial deficiency of functional HexA activity in patient KL despite the apparent presence of normal amounts of HexA in this patient. Studies on β-Hex processing revealed a reduced amount of β-chains, sufficient only to combine with α-chains to form the $\alpha\beta_A\beta_B$, which behaved like HexA according to electrophoretic mobility, thermal stability and charge characteristics. If the defect does lie in the β-chain, then studies with activator protein suggest that the β-chains are involved in forming a stable complex between $\alpha\beta_A\beta_B$ ganglioside (GM2) substrate, and activator protein.

ASSOCIATION OF PARTIAL β-HEXOSAMINIDASE DEFICIENCY WITH SPINAL MUSCULAR ATROPHY AND ALS

A number of neurological syndromes have been associated with partial HexA deficiency, most notably the cases of spinal muscular atrophy or ALS-like syndromes first described by Johnson et al. (Johnson, 1980; Johnson et al., 1982) and Mitsumoto et al. (1985). In general, the symptoms resemble those in our case of apparent HexB deficiency and GM2 degradation rates in fibroblasts were also comparably slow (Argov and Navon, 1984; Frisch et al., 1984). The occurrence of muscle atrophy in HexA-deficient patients (either with or without an encephalopathic course) and the fact that spinal or bulbar neurons in 4 such patients with delayed-onset GM2 gangliosidosis had membranous cytoplasmic bodies raises the question of involvment of gangliosides in typical, sporadic amyotrophic lateral sclerosis.

GANGLIOSIDE ABNORMALITIES IN ALS

Rapport et al. (1985) compared the ganglioside pattern in cortex of middle frontal gyrus from postmortem brain of 16 ALS and 11 non-ALS patients and found a relative increase in simple gangliosides (GM2 and GM3) and a relative decrease in GD1b, GT1b and GQ1b, although the total ganglioside levels remained unchanged. We have examined spinal cord from 9 ALS and 9 control patients and find a relative increase in faster moving (low molecular weight) gangliosides tentatively identified as GM2, II^3 $NeuAcLnOse_3Cer$, LM1 ganglioside (NeuAc(2→3)Gal-GlcNAc-Gal-Glc-Ceramide) and sialosylglobotetraosylceramide (NeuAc(2→3)GalNAc-Gal-Gal-GlcCer) (Dawson and Stefansson, 1984), the latter being a major abnormal component of ALS muscle (Kundu et al., 1984) (Fig. 7).

POSSIBLE PATHOLOGICAL ROLE OF GANGLIOSIDES IN ALS

There is no neuropathological evidence to suggest that ALS is a lysosomal storage disorder. However, it is possible that ganglioside metabolism is specifically affected in ALS motor neurons and results in the accumulation of the same Hex*N*Ac-rich gangliosides which accumulate in patients with partial HexA or HexB deficiency and late-onset symptoms of motor neuron disease. Abnormal ganglioside metabolism in ALS motor neurons could be the result of viral invasion (since DNA and RNA viruses inhibit GM3:GalNAc transferase and GM2:gal transferase in rodent cells) (Brady and Fishman, 1984). Alternatively, a retrovirus could cause translocation/inactivation of the 5q13 → qter region of 50% of chromosome 5 where the β-chain locus of β-Hex is localized) (Fox et al., 1984) or the observed abnormalities could be the result of an autoimmune process. Of interest, we have followed two sisters, 25-30 years old, with ALS preceded by onset of Graves disease, a disorder in which anti-ganglioside antibodies are claimed to circulate. At present, we have no consistent evidence for the presence of anti-ganglioside or neutral glycolipid antibodies in serum from patients with ALS, but this question needs to be re-examined with more sensitive techniques. We would anticipate that the finding of anti-glycolipid antibodies in ALS could be of particular significance since in an analogous sensory system, mouse dorsal root ganglia, express a variety of unique Gal*N*Ac glycolipids (e.g., anti-$GbOse_4$ (SSEA3) on 10-15% of DRG^s) (Dodd et al., 1984). It seems likely that motor neurons will also express specific Glc*N*Ac glycolipids, but this has not yet been determined. However, evidence from fetal pathology of Tay Sachs disease is consistent with unusually rapid ganglioside turnover in motor neurons, making them highly susceptible to any impairment in metabolism.

It is, therefore, clear that while gangliosides may be important cell surface stimulators of neuronal differentiation and neurite growth, they can be toxic when ac-

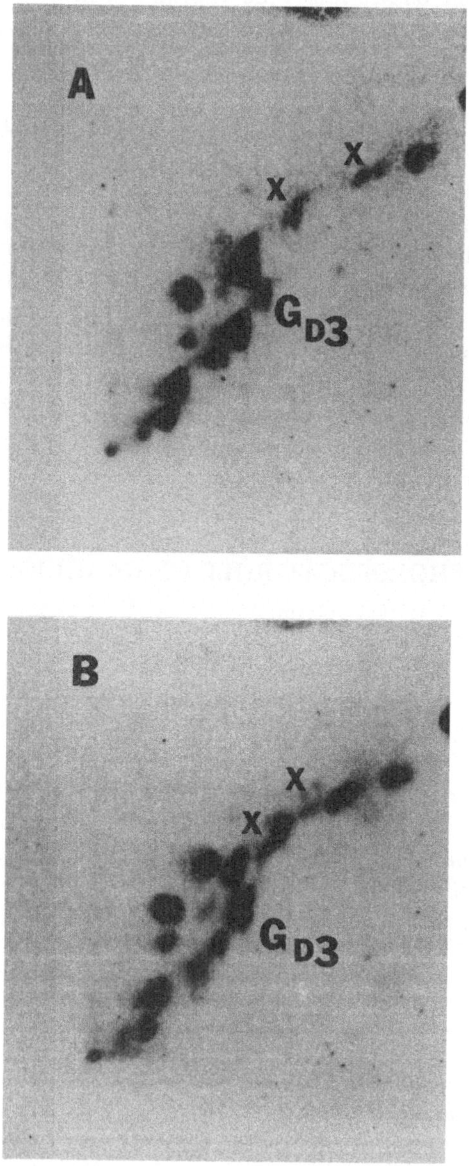

Figure 7. Two-dimensional thin-layer chromatogram of gangliosides from ALS motor cortex (A) and spinal cord (B) showing abnormal gangliosides. TLC plate is developed in $CHCl_3$:CH_3OH:0.02% aqueous $CaCl_2$ (60:45:9 v/v) followed by $CHCl_3$:CH_3OH:2.5N aqueous NH_4OH (65:35:8 v/v).

cumulating in lysosomes or perhaps even when abnormal gangliosides are generated through failures in normal anabolism or catabolism. We know very little about the biological role of gangliosides, but it has been suggested that gangliosides could modulate growth hormone action (GM1 inhibits PDGF stimulation of protein (tyrosine) kinase, while GM3 inhibits EGF stimulation) (this book; and Ledeen, 1985) convert low-affinity serotonin binding sites to high affinity binding sites (Berry-Kravis & Dawson, 1984), facilitate ligand coupling to adenylate cyclase (e.g., cholera toxin - GM1) (Cuatrecasas, 1973) and are highly antigenic (tumor antigens) (Ledeen, 1985). Gangliosides also have the non-specific ability to chelate divalent cations and form rigid regions in cell membranes (Yamakawa and Nagai, 1978), suggesting that an abnormality in cellular ganglioside metabolism could have a profound effect on the ability of that cell to function — perhaps leading to the type of early death of anterior horn cells which is observed in ALS.

ACKNOWLEDGMENTS

This work was supported in part by USPHS Grants HD-06426, HD-04583, HD-09402 and NS-21442. NRC is a fellow of the Amyotrophic Lateral Sclerosis Association. GD is a Joseph P. Kennedy, Jr. Scholar.

REFERENCES

Argov Z, Navon R (1984) Clinical and genetic variations in the syndrome of adult GM2 gangliosidosis resulting from hexosaminidase A deficiency. Ann Neurol 16: 14-20.

Berry-Kravis L, Dawson G (1984) Gangliosides as modulators of the coupling of neurotransmitters to adenylate cyclase. In: Ledeen RW, Yu RK, Rapport MM, Suzuki K (eds): Ganglioside Structure, Function and Biomedical Potential. Plenum press, New York, pp. 341-353.

Brady RO, Fishman P (1974) Biosynthesis of glycolipids in virus-transformed cells. Biochim Biophys Acta 335: 121-148.

Cashman NR, Antel JP, Hancock LW, Dawson G, Horwitz AL, Johnson WG, Huttenlocher PR, Wollmann RL (1985) Ann Neurol 19: 568-572.

Cuatrecasas P (1973) Interaction of vibrio cholerae enterotoxin with cell membranes. Biochemistry 12: 3547-3582.

Dawson G, Stefansson K (1984) Gangliosides of human spinal cord: Aberrant composition of cords from patients with amyotrophic lateral sclerosis. J Neurosci Res 12: 213-220.

Dawson G, Tsay G (1977) Substrate specificity of human α-L-fucosidase. Arch Biochem Biophys 184: 12-23.

Dodd J, Solter D, Jessel T (1984) Monoclonal antibodies against carbohydrate differentiated antigens identify subsets of primary sensory neurones. Nature 311: 469-472.

Fox MF, Dutoit DL, Warnich L, Retief AE (1984) Regional localization of α-galactosidase (GAL) to Xpter → q22, hexosaminidase B (HexB) to 5q13 → qter and arylsulfatase B (ARSB) to 5 pter → q13. Cytogenet Cell Genet 38: 45-49.

Frisch A, Baram D, Navon R (1984) Hexosaminidase A deficiency adults: presence of α-chain precursor in cultured skin fibroblasts. Biochem Biophys Res Commun. 119: 101-107.

Hancock LW, Horwitz AL, Cashman NR, Antel JP, Dawson G (1985) *N*-acetyl-*β*-hexosaminidase deficiency in cultured fibroblasts from a patient with progressive motor disease. Biochem Biophys Res Commun 130: 1185-1192.

Hancock L, Li YT, Dawson GD (1985) Abnormal glycolipid metabolism in a patient with HexB deficiency. J Biol Chem, in press.

Hasilik A and Neufeld EF (1980) Biosynthesis of lysosomal enzymes in fibroblasts synthesis as precursors of higher molecular weight. J Biol Chem 255: 4937-4945.

Johnson W (1981) The clinical spectrum of hexosaminidase deficiency diseases. Neurology 31: 1453-1456.

Johnson WG, Wigger HJ, Karp HR (1982) Juvenile spinal muscular atrophy: A new hexosaminidase phenotype. Ann Neurol 11: 11-16.

Kolodny EH, Raghavan SS (1983) GM2-Gangliosidosis. Trends in Neurol Sci 6: 16-20.

Kundu SK, Harati Y, Misra LK (1984) Sialosylglobotetraoscylceramide: A marker for amyotrophic lateral sclerosis. Biochem Biophys Res Commun 118: 82-89.

Ledeen RW (1985) Gangliosides of the neuron. Trends in Neurol Sci 9: 169-174.

Mitsumoto H, Sliman RJ, Schafer IA, Sternick CS, Kaufman B, Wilbourn A, Horwitz SJ (1985) Motor neuron disease and adult hexosaminidase A deficiency in two families: Evidence for multisystem degeneration. Ann Neurol 17: 378-385.

Myerowitz R, Robins AR, Proia RL, Sahagian GG, Puchalski CM and Neufeld EF (1983) *N*-acetyl-*β*-D-hexosaminidase. Method Enzymol 96: 729-736.

O'Brien JS (1983) Gangliosides. In: Stanbury JB, Wyngaarden JB, Fredrickson DS, Goldstein JL, Brown MS (eds): The Metabolic Basis of Inherited Disease. McGraw-Hill, New York, pp. 945-969.

O'Dowd B, Quan F, Willard HF, Lamhonwah AM, Korneluk RG, Lowden JA, Gravel RA, Mahuren DJ (1985) Isolation of cDNA clones encoding the *β*-hexosaminidase gene. Proc Natl Acad Sci USA 82: 1184-1188.

Purpura DP, Suzuki K (1976) Distortion of neuronal geometry and formation of aberrant synapses in neuronal storage diseases. Brain Res 116: 1-21.

Rapport MM, Donnenfeld H, Brunner W, Hungund B, Bartfeld H (1985) Ganglioside patterns in amyotrophic lateral sclerosis brain regions. Neurology, in press.

Yamakawa T, Nagai Y (1978) Biological role of glycolipids. Trends Biochem Sci 3: 128-131.

Gangliosides and neuronal plasticity
G. Tettamanti, R.W. Ledeen, K. Sandhoff,
Y. Nagai, G. Toffano (eds.)
Fidia Research Series, vol. 6
Liviana Press, Padova, © 1986

ULTRASTRUCTURAL LOCALIZATION OF CALCIUM AT SYNAPSES AND MODULATORY INTERACTIONS WITH GANGLIOSIDES

Hinrich Rahmann and Wolfgang Probst

Institute of Zoology, University of Stuttgart-Hohenheim,
D 7000 Stuttgart 70 (Hohenheim), Federal Republic of Germany

INTRODUCTION

In 1975/76 we published the functional hypothesis of an involvement of sialo-glycomacromolecules, especially gangliosides, in the process of synaptic transmission, including memory formation (Rahmann, 1976; Rahmann et al., 1975; 1976). In our proposal it had been discussed that due to the ability of ganglioside-bound negatively charged sialic acids to form labile complexes together with Ca^{2+}-ions, these glycosphingolipids are assumed to act as modulatory compounds for the Ca^{2+}-dependent release of transmitter substances at synapses. Since 1975 we have been able to add to this hypothesis extensive experimental evidence (Rahmann, 1983; 1984; Rahmann et al., 1982; Probst et al., 1984). In the meantime it also had been supported by several other authors (Svennerholm, 1980; Tettamanti et al., 1980; Maggio et al., 1981; Veh and Sander, 1981; Leskawa and Rosenberg, 1981). The essential basis for our model is the well-known fact that almost every stage of neuronal activity, especially electrical responsiveness, depends on the presence of extra-cellular Ca^{2+}. With regard to this, during recent years extensive experimental efforts were undertaken to establish subcellular Ca^{2+}-deposits in the synaptic terminal. Rough endoplasmic reticulum, mitochondria and vesicles were determined as intracellular Ca^{2+}-storage structures (McGraw et al., 1980; Chan et al., 1983). The electronmicroscopical methods used until now, however, failed to establish extra-cellular deposits of Ca^{2+}.

By means of biochemical investigations during the last few years, Ca^{2+}-binding molecules, such as phospho-inositides, especially PIP_2 (Nishizuka, 1984) or calmodulin (Stoclet, 1981), were found to be involved in intra-cellular synaptic functions. However, the initiation of the electro-chemical transmission process in synapses

---

Abbreviations: CNS, central nervous system; PIP2, phosphoinositide biphosphate (or triphosphoinositide); PC, phosphatidylcholine; PS, phosphatidylserine; EGTA, ethylen glycol bis (β-aminoethylether)-N,N,N',N',-tetra acetic acid.

requires first of all modulatory changes at the outer surface of the synaptic membrane, by which the influx of Ca^{2+} from its extra-cellular storage sites into the synapse is induced.

This is the reason why our research interest during the last few years has been focussed mainly on three major questions: a) on the establishment of any possible extra-cellular Ca^{2+}-storage sites within the synaptic cleft; b) on the investigation of any specific physico-chemical properties of gangliosides in their interaction with Ca^{2+} (Fig. 1), which do not exist in other membrane lipids, and finally c) on the demonstration of physiologically relevant data concerning changes in brain ganglioside content and composition (Table 1). Over all, our research aims at elucidating a possible causal mechanism demonstrating the functional role of gangliosides as neuromodulatory substances in the process of synaptic transmission (Rahmann, 1983).

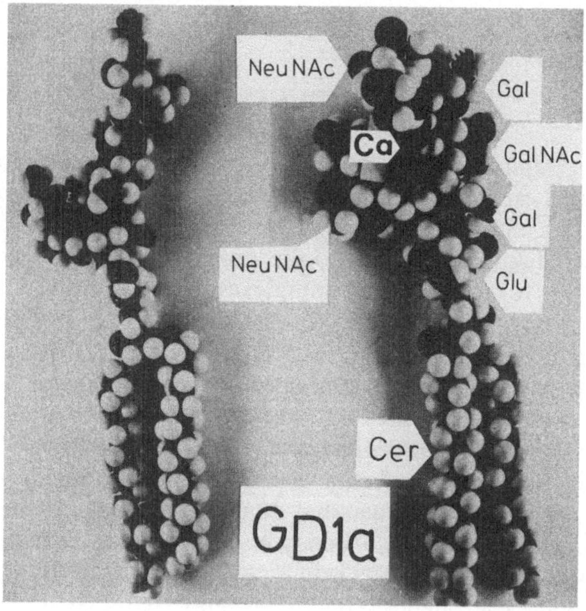

Figure 1. Space-filling Corey-Pauling-Koltun molecular model of the disialoganglioside GD_{1a} without and with bound Ca^{2+}, demonstrating the change of molecular conformation by intramolecular Ca^{2+}-binding.

MATERIALS AND METHODS

The electronmicroscopical demonstration of Ca^{2+} at the subcellular level was performed as calcium-osmium-phosphate deposits by the use of a $K_2Cr_2O_7/OsO_4$-mixture for postfixation of glutaraldehyde/phosphate prefixed tissue (Probst, 1985).

Table 1. *Review of Literature References Concerning Physico-Chemical and Physiological Peculiarities of Gangliosides as Compared to Other Membrane Lipids*

PHYSICO-CHEMICAL AND PHYSIOLOGICAL PECULIARITIES OF GANGLIOSIDES AS COMPARED TO OTHER MEMBRANE LIPIDS	REFERENCES
A. SPECIFIC PHYSICO-CHEMICAL PROPERTIES	
1. CHANGE FROM HYDRO- TO LIPOPHILIC SOLUBILITY FOLLOWING Ca^{2+}-ADDITION	SVENNERHOLM 1956; QUARLES & FOLCH-PI 1965; HAYASHI & KATAGIRI 1974; RAHMANN ET AL. 1976
2. DISCONTINUITY OF Ca^{2+}-BINDING	ABRAMSON ET AL. 1972; BEHR & LEHN 1973; PROBST ET AL. 1979; HAYASHI ET AL. 1984; MÜHLEISEN ET AL. 1983
3. Ca^{2+}-RELEASE FROM Ca^{2+}-GANGLIOSIDE-COMPLEXES BY MEANS OF	
A) METALLIC CATIONS (K^+, Na^+, Li^+, Mg^{2+}, Ca^{2+})	MÜHLEISEN ET AL. 1979; PROBST ET AL. 1979; HAYASHI ET AL. 1984
B) NEURO-TRANSMITTERS (ACH, 5-HT)	MÜHLEISEN ET AL. 1979
C) NEUROTOXIN (CURARINE)	MÜHLEISEN ET AL. 1979
D) TEMPERATURE CHANGES	PROBST & RAHMANN 1980
4. CHANGES OF MEMBRANE BEHAVIOUR OF GANGLIOSIDES (MOLECULAR SURFACE REQUIREMENT, SURFACE POTENTIAL, "VISCOSITY") BY MEANS OF	
A) Ca^{2+}	MAGGIO ET AL. 1980; PROBST ET AL. 1984
B) TEMPERATURE	PROBST ET AL. 1984
C) INTERACTION WITH OTHER LIPIDS	PROBST ET AL. 1984; PETERS ET AL. 1984
B. PHYSIOLOGICAL DATA BEING CORRELATED WITH PHYSICO-CHEMICAL FINDINGS	
1. STIMULATION OF MEMBRANE-BOUND ENZYMES BY MEANS OF Ca^{2+}-GANGLIOSIDE-COMPLEXES	GOLDENRING ET AL. 1985; TSUJI ET AL. 1985
2. CHANGES IN NEURONAL GANGLIOSIDE-COMPOSITION (≙ POLARITY) IN DEPENDENCE OF	
A) PHYLOGENETIC LEVEL OF NERVOUS ORGANIZATION	RAHMANN & HILBIG 1983
B) ONTOGENETIC NERVE DIFFERENTIATION	HILBIG ET AL. 1982
C) NEUROPATHY	SUZUKI 1984
D) THERMAL ADAPTATION	RAHMANN 1981
E) PHYSIOLOGICAL STIMULATION	RAHMANN 1979

Ca^{2+}-ganglioside interactions were investigated on the basis of monolayer techniques using a teflon-trough (Fig. 2) enabling simultaneous measurements of the surface requirement of molecules, the surface potential and the "viscosity". The "viscosity" was measured as the damping constant of a standing wave contactlessly generated at the buffer surface by means of a razor blade connected with a frequency generator. The waves were registered by laser light deflection. The other measurements were done as described previously (Probst et al., 1984).

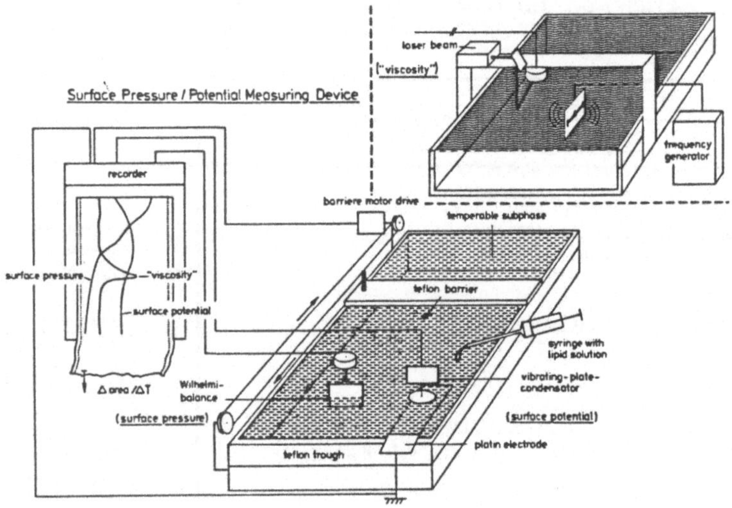

Figure 2. Teflon trough for measuring surface pressure/molecular area-isotherms (space requirement), surface potential and surface "viscosity" of lipid monolayers.

RESULTS

Proof of Extra-cellular Ca^{2+}-Deposits in the Synaptic Cleft

With the new technique developed by us it was possible to substantiate, in addition to the well known subcellular Ca^{2+}-storage sites in nerve terminals (rough endoplasmic reticulum, mitochondria and storage vesicles), extra-cellular Ca^{2+}-accumulations in the synaptic cleft, especially at the very local zone of synaptic contact (Fig. 3a). The concentration of Ca^{2+} in the synaptic cleft was about three times as high as in the extracellular space of the very close vicinity of the synapse. In fish synapses it was much more pronounced than in those of mammals, which correlates with the fact that the extracellular Ca^{2+}-concentration in the CNS of bonyfish is about three times higher than that of mammals (3 mM versus 1.2 mM; Veh and Sander, 1981). The proof of Ca^{2+} specificity had been established by pre-treatment of the ultra thin slices with the specific Ca^{2+}-chelator EGTA (Fig. 3b). Summarizing, these electron-microscopical data provide the first definite proof for the assumption that the extra-cellular space of the very local zone of synaptic contact is a preferential site for the storage of Ca^{2+}; from which it is well known that it is essentially needed for the process of synaptic transmission of information (Rahmann, 1983).

Physico-chemical Peculiarities of Ca^{2+}-Ganglioside Interactions

Since gangliosides have the ability to complex strongly and specifically with Ca^{2+}, such complexes in the outer leaflet of the synaptic membrane are assumed to

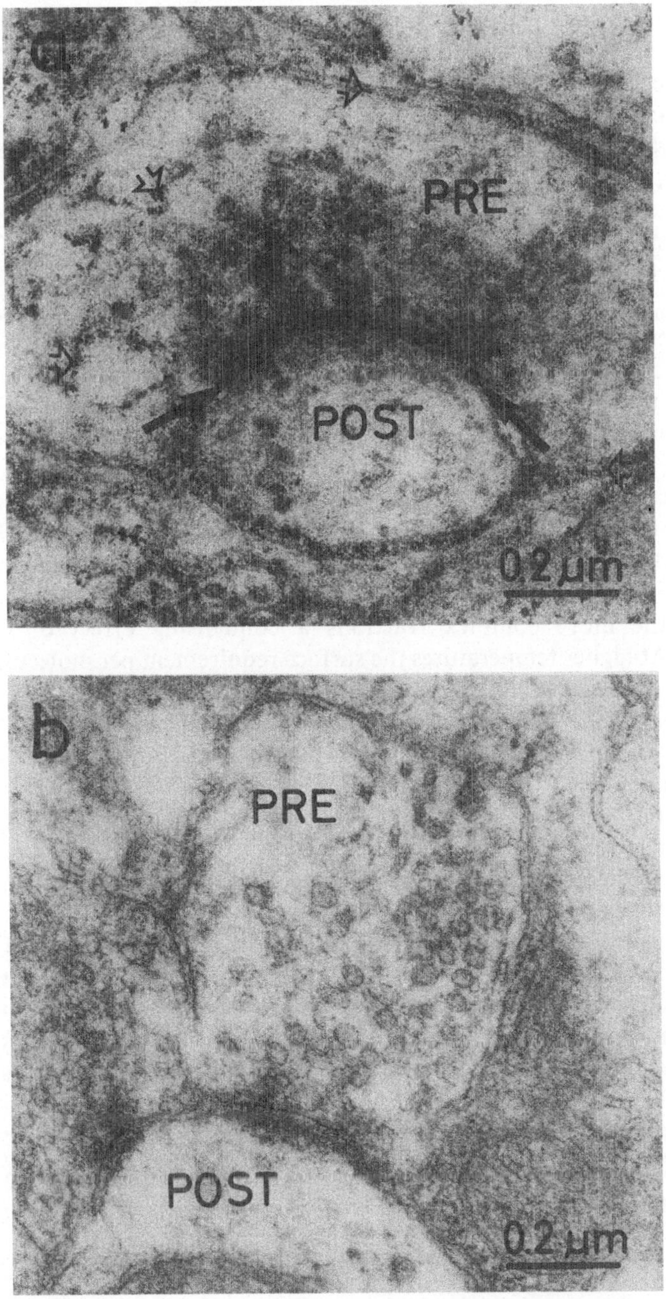

Figure 3. Extra-cellular localization of Calcium in the synaptic cleft of a synapse from the optic tectum of a carp. Calcium deposits are represented as calcium-osmium-phosphate particles (a); removal of calcium following incubation of the ultrathin section by means of the calcium-specific chelator EGTA (b).

reflect an efficient molecular mechanism for modulating the membrane-mediated processes in the course of synaptic transmission and by this of functional synaptic plasticity (Rahmann 1983). As can be followed from the review in Table 1, in previous studies specific physico-chemical properties of Ca^{2+}-ganglioside interactions were found which are almost unknown for other competing membrane lipids. These refer, for instance, to characteristic changes from hydro - to lipophilic solubility, discontinuity in Ca^{2+}-binding, high lability of Ca^{2+}-binding as against metallic cations, neurotransmitters, neurotoxins and temperature changes.

In addition to these findings, a peculiar surface pressure behaviour of gangliosides had been described (Probst et al., 1984) by means of monolayer studies. In continuation of these experiments we now measured surface pressure-area-isotherms of monolayers being composed of different ganglioside species (GM1, GD1a, GD1b, GT1b) and compared them with those of phosphatidylcholine (PC) and phosphatidylserine (PS). These experiments were undertaken under the influence of changes in Ca^{2+}-content (0; 0,01 and 0,5 mM in the subphase) and variation in temperature (11°, 20° and 37° C).

The surface requirement of PC and even that of the negatively charged PS remained almost uninfluenced by changes in Ca^{2+}-content and temperature (Fig. 4 a, b). The changes of the isotherms of gangliosides, however, were considerable in the sequence GD1a > GT1b > GD1b > GM1 = PS > PC. In other words, the ganglioside data revealed for the more complex fractions a condensing effect of Ca^{2+} at low temperatures. At higher temperatures the surface requirement per molecule is partly increased in the presence of Ca^{2+} (Fig. 4 c - f).

In addition to these specific Ca^{2+}/temperature influences on surface behaviour of gangliosides it can be inferred from Figure 4 that only in the case of gangliosides, during the transition from liquid-expanded to liquid-condensed phase, do characteristic conformational changes in the inner-molecular organisation occur (formation of shoulder in isotherms), which is unknown for most of the other lipids. The onset of the conformational changes of gangliosides in a first preliminary study in the case of GM1 versus PS, for example, was shown to be correlated with remarkable changes in the surface potential (Fig. 5). In another set of experiments the surface "viscosity" of GD1b-mono-layers was measured without Ca^{2+}-addition. The astonishing result was that dramatic changes in the viscosity of the GD1b-film occurred during transition from liquid-expanded to liquid-condensed phase. Extrapolating from the condensing effect of Ca^{2+} (see also Figure 4), the course of "viscosity" should also change as indicated by the dotted line in Figure 6. The transitional changes in the increase in rigidity precede the conformational changes in the arrangement of the ganglioside molecules, thus indicating possible correlative interactions between these two parameters.

DISCUSSION

The electron-microscopical proof of a special extra-cellular Ca^{2+}-storage at the very local zone of synaptic contact is regarded as an essential presupposition for our functional hypothesis of an involvement of gangliosides in the process of synaptic transmission (Rahmann, 1976; 1983; Rahmann et al., 1975; 1976; 1982). The different

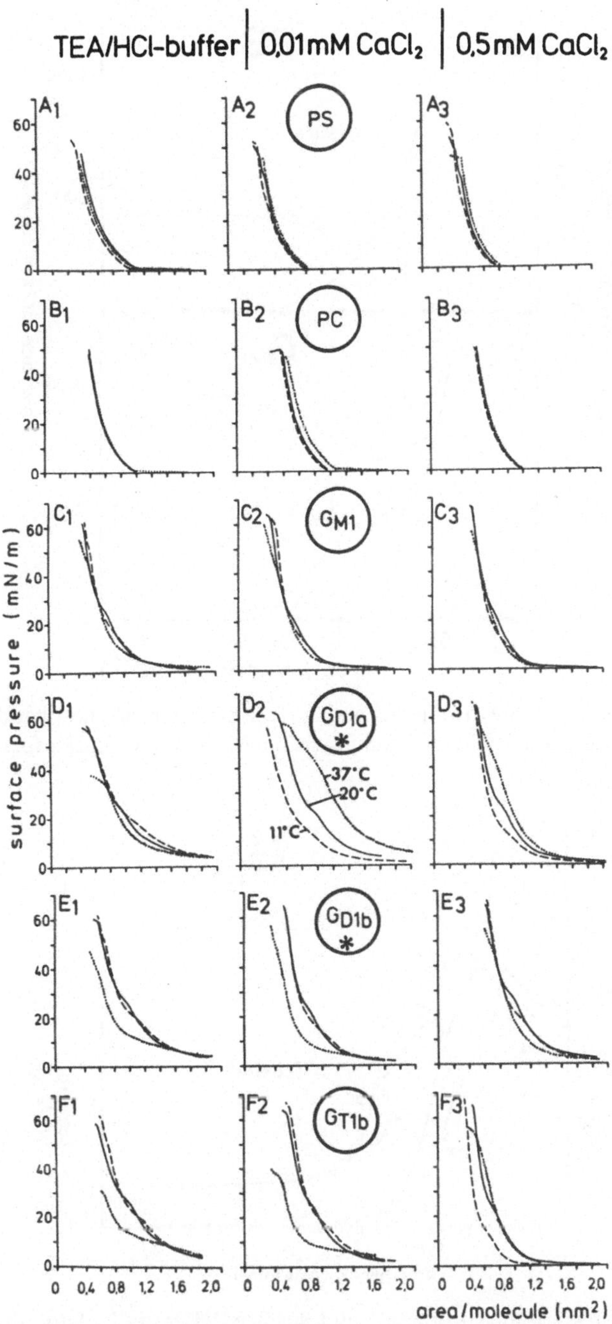

Figure 4. Surface pressure/area isotherms of PC, PS, and the gangliosides GM1, GD1a, GD1b and GT1b on 5 mM triethanolamine (TEA)/HCl-buffer (pH 7.4) as subphase at different temperatures (11°, 20°, 37° C) and Ca^{2+}-concentrations (0; 0,01 and 0,5 mM). Note shoulder formation in gangliosides only during transition from liquid-expanded to liquid-condensed phase.

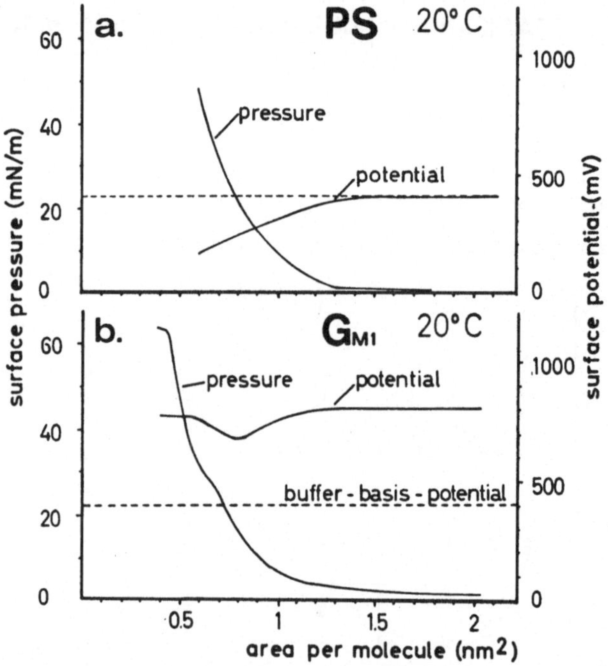

Figure 5. Surface pressure/area isotherms and surface potential of phosphatidylserine (PS, a) and the mono-sialoganglioside GM1 (b). Subphase: 5 mM TEA/HCl-buffer (pH 7.4); measuring temperature: 20° C.

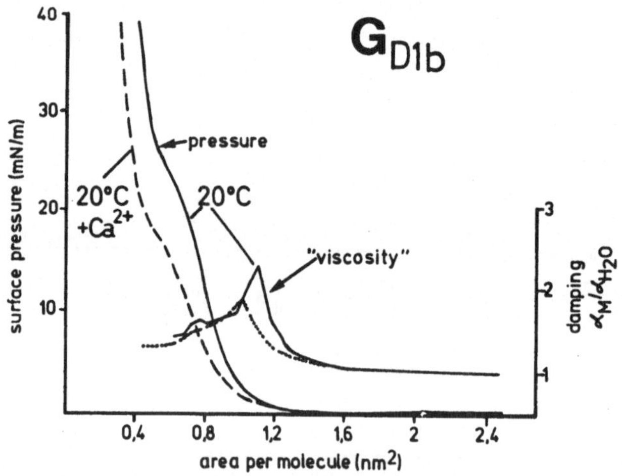

Figure 6. Surface pressure/area isotherms and surface-"viscosity" of the disialoganglioside GD1b. The "viscosity" was measured as the quotient of the damping constant of a standing wave generated contactlessly at water surface with and without lipid monolayer. Subphase: Aqua bidest without and with 0.01 mM $CaCl_2$ (in case of surface pressure/area-isotherms only); dotted line was extrapolated from Ca^{2+}-effect on surface pressure/area-isotherm. Measuring temperature: 20° C.

amounts of Ca^{2+}-accumulation in the synaptic cleft of lower and higher vertebrates is in agreement with the different Ca^{2+}-concentrations in the extra-cellular space in the animal kingdom (Veh and Sander, 1981). All told, the histo-chemical evidence of extra-cellular Ca^{2+}-storage in synapses is in full conformity with the various physico-chemical and physiological peculiarities of gangliosides as compared to other membrane lipids, as summarized in Table 1.

The very specific physico-chemical interactions of gangliosides with Ca^{2+} (change in solubility; discontinuity in binding curve; lability of Ca^{2+}-ganglioside binding as against metallic ions, neuro-transmitters, neuro-toxins and temperature; changes in monolayer behaviour concerning surface requirement, surface potential or surface "viscosity") are correlated with (and probably responsible for?) various physiological peculiarities of these glycosphingolipids. On the one hand, GD1a - and GQ1b - Ca^{2+} complexes were found to influence specifically the activity of membrane-bound protein kinases in synapses, thus giving strong evidence for the assumption that gangliosides may promote or initiate intra-cellular metabolic changes following neuronal stimulation within the synapse. On the other hand, the dramatic variations in ganglioside content and composition depending on biologically relevant parameters (e.g. phylogenetic level of nervous organisation, ontogenetic differentiation of nerve cells, neuropathy, thermal adaptation, physiological stimulation) also demonstrate indirectly that these neuronal membrane constituents are most likely responsibly involved in basic neuronal functions. With regard to synaptic transmission, gangliosides in combination with Ca^{2+}-ions seem to fulfill the demand of being neuromodulators, being able to adapt under the influence of physical (e.g. temperature, pressure) and chemical (e.g. ion environment, enzyme activities) factors thus enabling an undisturbed synaptic transmission process despite environment changes.

ACKNOWLEDGMENTS

This work was supported by the Deutsche Forschungsgemeinschaft (Grant: Ra 166/13-6).

REFERENCES

Abramson MB, Yu RK, Zaby V (1972) Ionic properties of beef brain gangliosides. Biochim Biophys Acta 280: 365-372.

Behr JP, Lehn JM (1972) Stability constants for the complexation of alkali and alkaline-earth cations by N-acetyl-neuraminic acid. FEBS Lett 22: 178-180.

Chan SY, Ochs S, Jersild RA jr (1984) Localization of calcium on nerve fibers. J Neurobiol. 15: 89-108.

Goldenring JR, Otis LC, Yu RK, DeLorenzo RJ (1985) Calcium/ganglioside-dependent protein kinase activity in rat brain membrane. J Neurochem 44: 1229-1234.

Hayashi K, Katagiri A (1974) Studies on the interaction between gangliosides, protein and divalent cations. Biochem Biophys Acta 337: 107-117.

Hayashi K, Mühleisen M, Probst W, Rahmann H (1984) Binding of Ca^{2+} to phosphoinositols, phosphatidyl serines and gangliosides. Chem Phys Lipids 34: 317-322.

Hilbig R, Rösner H, Merz G, Segler-Stahl K, Rhamann H (1982) Developmental profiles of gangliosides in mouse and rat cerebral cortex. Roux's Archives Dev. Biol 191: 281-284.

Leskawa KC, Rosenberg A (1981) The organization of gangliosides and other lipid components in synaptosomal plasma membranes and modifying effects of calcium ions. Cell Molec Neurobiol 1: 373-388.

Maggio B, Cumar FA, Caputto R (1980) Configuration and interactions of the polar head group in gangliosides. Biochem J 189: 435-440.

Maggio B, Cumar FA, Caputto R (1981) Molecular behaviour of glycosphingolipids in interfaces. Possible participation in some properties of nerve membranes. Biochim Biophys Acta 650: 69-87.

McGraw CF, Somlyo AV, Blaustein MP (1980) Localization of calcium in presynaptic nerve terminals. An ultrastructural and electron microprobe analysis. J Cell Biol 85: 228-241.

Mühleisen M, Probst W, Wiegandt H, Rahmann H (1979) In vitro studies on the influence of cations, neurotransmitters and tubocurarine on calcium ganglioside interactions. Life Sci 25: 791-796.

Mühleisen M, Probst W, Hayashi K, Rahmann H (1983) Calcium binding to liposomes composed of negatively charged lipid moieties. Japan J. exper Med 53: 103-107.

Nishizuka Y (1984) Phospholipid turnover in signal transduction. Proceedings 25 th Intern Conf Biochem Lipids, Antwerp, pp. 27-28.

Peters MW, Barber KR, Grant CWM (1984) Lateral distribution of gangliosides in bilayer membranes: lipids and ionic effects. J of Neurosci Res 12: 343-353.

Probst W (1986) Electron-microscopical evidence for extra-cellular calcium deposits in the synaptic cleft. Histochemistry, in press.

Probst W, Rösner H, Wiegandt H, Rahmann H (1979) Das Komplexations-vermögen von Gangliosiden für Ca^{2+}. Hoppe-Seyler's Z Physiol. Chem. 360: 979-986.

Probst W, Rahmann H (1980) Influence of temperature changes on the ability of gangliosides to complex with Ca^{2+}. J therm Biol. 5: 243-247.

Probst W, Möbius D, Rahmann H (1984) Modulatory effects of different temperatures and Ca^{2+}-concentrations on gangliosides and phospholipids in monolayers at air/water interfaces and their possible functional role. Cell Mol Neurobiol 4: 157-176.

Quarles R, Folch-Pi J (1965) Some effects of physiological cations on the behaviour of gangliosides in a chloroform-methanol-water biphase system. J Neurochem 12: 543-553.

Rahmann H (1976) Neurobiologie. UTB-Verlag Ulmer, Stuttgart, p. 273.

Rahmann H (1979) The possible functional role of gangliosides for synaptic transmission and memory formation. In: Matthies H, Krug M, Popov N (eds): Biological aspects of learning, memory formation and ontogeny of the CNS. Akad Wiss DDR, Berlin, pp. 83-110.

Rahmann H (1981) Die Bedeutung der Hirnganglioside bei der Temperatur-adaptation der Vertebraten. Zool Jb Physiol 85: 209-248.

Rahmann H (1983) Functional implication of gangliosides in synaptic transmission. Neurochem Intern 5: 539-547.

Rahmann H (1984) Lernen und Gedächtnis vom Standpunkt der Neurobiologie. Therapiewoche 34: 7139-7154.

Rahmann H, Hilbig R (1983) Phylogenetical aspects of brain gangliosides in vertebrates. J Comp Physiol 151: 215-224.

Rahmann H, Rösner H, Breer H (1975) Sialo macromolecules in synaptic transmission and memory formation. IRCS Med Sci Forum 3: 110-112.

Rahmann H, Rösner H, Breer H (1976) A functional model of sialoglycomacromolecules in synaptic transmission and memory formation. J Theor Biol 57: 231-337.

Rahmann H, Probst W, Mühleisen M (1982) Gangliosides and synaptic transmission. Japan J Exper Med: 275-286.

Stoclet JC (1981) An ubiquitous protein which regulates calcium-dependent cellular functions and calcium movements. Biochem. Pharmacol. 30: 1723-1729.

Suzuki K (1984) Gangliosides and disease: A review. In: Ledeen RW, Yu RK, Rapport MW, Suzuki K (eds): Ganglioside structure, function, and biomedical potential. Adv Exp Med Biol 174: 407-418.

Svennerholm L (1956) The quantitative estimation of cerebrosides in nervous tissue. J Neurochem 1: 42-53.

Svennerholm L (1980) Gangliosides and synaptic transmission. In: Svennerholm L, Mandel P, Dreyfus H, Urban PF (eds): Structure and Function of gangliosides. Adv Exp Med Biol, Plenum Press, New York, pp. 533-544.

Tettamanti G, Preti A, Cestaro B, Masserini M, Sonnino S, Ghidoni R (1980) Gangliosides and associated enzymes at the nerve-ending membranes. In: Sweeley CC (ed): Cell Surface Glycolipids. ACS Symp Ser 128 pp. 321-343.

Tsuji S, Nakajima J, Sasaki T, Nagai Y (1985) Bioactive gangliosides: IV. Ganglioside GQ1b/Ca^2 dependent protein kinase activity exists in the plasma membrane fraction of neuroblastoma cell line GOTO. J Biochem 97: 969-972.

Veh RW, Sander M (1981) Differentiation between gangliosides and sialyllactose sialidases in human tissue. Perspect Inher Met Dis 4: 71-109.

Gangliosides and neuronal plasticity
G. Tettamanti, R.W. Ledeen, K. Sandhoff,
Y. Nagai, G. Toffano (eds.)
Fidia Research Series, vol. 6
Liviana Press, Padova, © 1986

INVOLVEMENT OF GANGLIOSIDES IN THE SYNAPTIC TRANSMISSION IN THE HIPPOCAMPUS AND STRIATUM OF THE RAT BRAIN

A. Wieraszko and W. Seifert

Max-Planck-Institute für biophysikalische Chemie, Department of Neurobiology, Laboratory of Molecular Neurobiology, D-3400 Göttingen, Federal Republic of Germany

INTRODUCTION

For a long time it has been suggested that gangliosides can somehow modulate synaptic transmission (Svennerholm, 1980; Rahmann, 1983). However, the experimental proof is still missing. It was demonstrated over 20 years ago that gangliosides help neuronal tissue to recover from the transient exposure to cold (Balakrishman and McIlwain, 1961; McIlwain, 1961). In the last 10 years functional studies on gangliosides concentrated mainly on the neurotrophic effects of gangliosides in nerve cell cultures (Morgan and Seifert, 1979; Seifert, 1981) and on the potential influence of gangliosides on the reinnervation processes in the nervous system (Ledeen, 1984). Surprisingly, very little effort has been made to test the hypothesis put forward for the first time by Svennerholm (1963, 1980) and later in more detailed form by Rahmann et al. (1976) concerning participation of gangliosides in neurotransmission. Our study is addressed directly to this problem.

It is beyond the scope of this article to describe in detail the molecular structures of gangliosides; however, some general comments concerning peculiar features of gangliosides which make them potential candidates for modulation of synaptic events at the molecular level are necessary.

Each ganglioside molecule consists of a hydrophobic ceramide part consisting of sphingosine and fatty acid. This portion is embedded in the membrane. Outside the membrane protrudes the hydrophilic, oligosaccharide part of the ganglioside molecule containing sialic acid. Although gangliosides are very ubiquitous in vertebrate

---

Abbreviations: NVCh, Neuraminidase from Vibrio Cholerae; NAu, Neuraminidase from Arthrobacter Ureafaciens; RR, Ruthenium Red; ChT, Cholera Toxin; TT, Tetanus Toxin; anti-GM1, Antiserum against ganglioside GM1; AP, 3,4-diamino-pyridine; LTP, Long Term Potentiation.

organisms, their high concentration in neurons has been known for a long time. Their possible presence at the synapse has received special consideration (Avrova et al., 1973; Hansson et al., 1977; Ledeen, 1978). Sialic acid, which is negatively charged, seems to be crucial in all events in which gangliosides are supposed to participate at the cell surface. The concentration of gangliosides in a given region of the membrane is not stable and depends on calcium concentration. In the presence of calcium, gangliosides tend to concentrate into clusters, which makes the membrane more rigid (Bertoli et al., 1981; Sharom and Grant, 1977). The calcium-ganglioside interaction can thereby change membrane fluidity and thus influence several processes taking place at the cell surface. Neurotransmission could be one of them. The efficiency of neurotransmission could be directly or indirectly dependent on the ganglioside-calcium interaction. In fact it has been demonstrated that gangliosides influence neurotransmission related processes such as release of neurotransmitters (Cumar et al., 1978), uptake of their precursors (Schulze and Rommelspacher, 1978), calcium binding (Behr and Lehn, 1973; Hayashi and Katagiri, 1974; Probst et al., 1979) and binding of neurotransmitters to specific proteins (Tamir et al., 1980). Also, some enzymes functionally linked to neurotransmission are influenced by gangliosides (Daly, 1981; Jeserich et al., 1981).

In 1979 Römer and Rahmann demonstrated that injection of Neuraminidase from Vibrio Cholerae (NVCh), an enzyme which splits off sialic acid from its conjugates, influenced neuronal activity in the fish optic tectum and frog spinal cord. Similar results were observed when NVCh was injected into Aplysia neurons (Hipp et al., 1980; Simonneau et al., 1980; Tauc and Hinzen, 1974). However, the results are difficult to interpret since the presence of sialic acid in Aplysia has been recently questioned (Segler et al., 1978). To our knowledge there are no data which would try to correlate the electrical activity in the mammalian brain with treatments which modify endogenous gangliosides, or in which the level of endogenous gangliosides would be changed by exogenous ganglioside application. It was our goal to throw more light on this problem.

Therefore we tested the influence of different factors which modify endogenous gangliosides (and sometimes other sialoconjugates) on the efficiency of synaptic activity in the sliced brain tissue. Brain slices are an especially convenient model for this kind of study (Lynch and Schubert, 1980; Weiler et al., 1982). The tissue is kept at constant temperature (33°C in our case) in a proper ionic environment with the supply of CO_2/O_2. The spontaneous and evoked activity of the neurons is preserved and can be monitored with conventional techniques (Lynch and Schubert, 1980; Weiler et al., 1982). The investigator has an easy access to the tissue and addition of different compounds directly to the bathing medium is possible. In this way one can follow the electrical activity of the slices and its change evoked by a given treatment.

It has been reported that the ganglioside content in synaptic plasma membranes obtained from the "cholinergic" subfraction of brain homogenate contained more gangliosides than the "noncholinergic" subfraction of the brain homogenate. The difference was in the 5-fold range (Avrova et al., 1973). It has also been demonstrated in human brain that the dominance of one of the ganglioside biosynthesis pathways ("a" versus "b") leading to biosynthesis of different gangliosides depends on the brain structure (Kracun et al., 1984). Finally, different proportions of gangliosides in the cell membrane have been observed in mouse cerebellar cells (Seyfried et al., 1983). Having that in mind we chose for our study structures such as striatum and hippocampus where the efficiency of two different neurotransmitter systems can be monitored with electrophysiological techniques.

In striatal slices one can follow with the extracellularly recorded evoked potentials the efficiency of cholinergic neurotransmission (Misgeld and Bak, 1979). The localization of stimulating and recording electrodes is not defined by the structure of striatum. Both electrodes were placed at a distance of about 2 mm from each other until the evoked potential in the range of 1-2 mV was found. In the hippocampal slices stimulation of Schaffer collaterals (electrode S1, Fig. 4) resulted in a discharge of pyramidal neurons. This discharge was followed with the recording electrode localized in the CA1 hippocampal field (R1, Fig. 4). In this way the efficiency of glutamergic synapses (Wieraszko, 1983) could be monitored.

The ganglioside influencing factors used in the present investigations are listed in Figure 1. To make our experimental approach more reasonable, we will briefly describe the action of each of these compounds. They can be divided into two groups. Those belonging to the first group bind to gangliosides, probably leaving their structures intact (toxins, anti-GM1, Ruthenium Red). Those from the second group modify gangliosides by modification or liberation of sialic acid (enzymes, $NAIO_4$).

The compound which should influence all gangliosides is sodium periodate. This compound attacks sialic acid (Gahmberg and Anderson, 1977) oxidizing it to the aldehyde which cannot, in contrast to intact sialic acid, bind calcium any longer (Jaques et al., 1980). Another general treatment which also attacks sialic acid, hydrolyzing it from sialoglycoproteins and gangliosides (except GM1), is Neuraminidase from Vibrio Cholerae (NVCh) (Hansson et al., 1977). As a result of the action of NVCh, polysialogangliosides are transformed to monosialoganglioside GM1. Another enzyme, Neuraminidase Arthrobacter Ureafaciens (NAu) liberates sialic acid from glycoproteins and gangliosides including GM1 (Sugano et al., 1978). Thus, using these two different enzymatic dissections, one can evaluate the potential role of GM1 in processes at the molecular level.

Two bacterial toxins which specifically bind to different gangliosides were used in the present investigations. Cholera Toxin (ChT) is very specifically bound by GM1

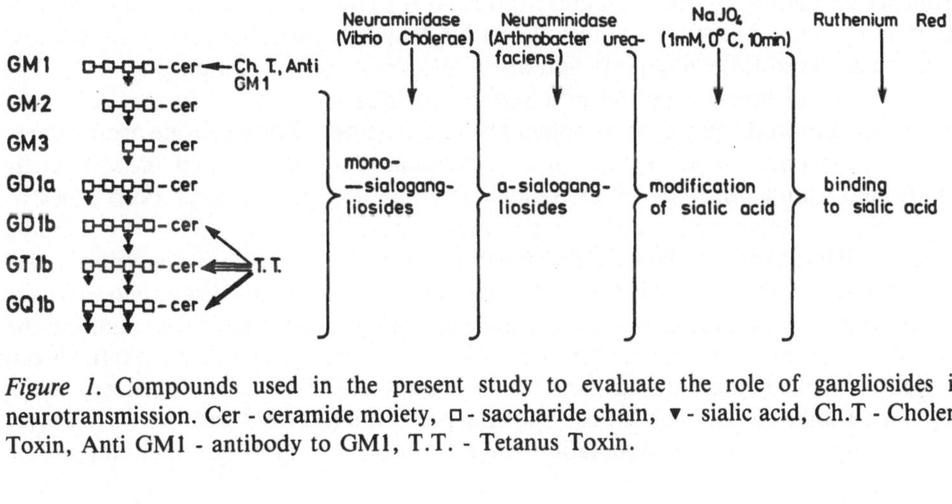

Figure 1. Compounds used in the present study to evaluate the role of gangliosides in neurotransmission. Cer - ceramide moiety, □ - saccharide chain, ▾ - sialic acid, Ch.T - Cholera Toxin, Anti GM1 - antibody to GM1, T.T. - Tetanus Toxin.

(Hansson et al., 1977) while Tetanus Toxin (TT) binds to polysialogangliosides, most strongly to GT1b (Rogers and Snyder, 1981). Another compound, Ruthenium Red (RR), an inorganic dye, is very often used as a "sialic acid binding compound" (Simonneau et al., 1980). However, this statement had never been supported experimentally. We demonstrate some preliminary data which show that RR is really bound by sialic acid. There are still some other ganglioside or sialic acid interacting compounds which were not used in our study but should be mentioned here. Lectins are short peptides of animal origin which can bind sialic acid in a very specific way (Sharon and Lis, 1972). However, as some of them can also bind calcium (Sharon and Lis, 1972), which is necessary for neurotransmission, lectins were not used in our study. Moreover, there are three other toxins which were not used by us but which use gangliosides as their receptors. Botulinum toxin binds to GT1b and GQ1b (Kitamura et al., 1980). Staphylococcal α-toxin binds to N-acetylglucosamine-containing ganglioside (Kato and Naiki, 1976) and toxin from Vibrio parahaemoliticus binds to GT1 gangliosides (Takeda et al. 1976).

In all ganglioside-toxin interactions sialic acid is of great importance (Kato and Naiki, 1976; Roger and Snyder, 1981; Yavin and Habig, 1984). We have mainly used the two neuraminidases, cholera toxin and tetanus toxin in different experimental approaches. The general conclusion is drawn that GM1 is crucial for neurotransmission in the hippocampal glutamergic synapses, while higher polysialogangliosides might be involved in GABA-ergic hippocampal neurotransmission and in cholinergic neurotransmission in the striatum.

RESULTS AND DISCUSSION

Chemical Treatments

Hippocampal slices prepared and maintained as described previously (Wieraszko and Lynch, 1979) were exposed to sodium periodate at three different concentrations: 1 mM, 10 mM and 60 mM. Sodium periodate at the concentration of 1 mM had no influence on the size of the population spike. At the concentrations of 10 mM sodium periodate abolished the population spike within minutes, but full recovery of potential could be achieved following perfusion of the slices with fresh solution (Fig. 2 A). The situation was different when 60 mM sodium periodate was used. In this case both ortho- and antidromically evoked potentials were followed. The electrode arrangement is depicted in Figure 4 A. Orthodromic stimulation (S1) was followed 80 ms later by antidromic stimulation (52) of pyramidal neurons. The record after both types of stimulation was taken with the same recording electrode (R1) localized in the pyramidal cell layer. The shapes of the recorded potentials are demonstrated in the upper panel of Figure 2B. In these records the first potential seen with 80 ms delay represents the pyramidal cell discharge activated antidromically. After sodium periodate addition the population spike is completely suppressed, while antidromically evoked potential is only slightly affected (compare "a" and "b" — Figure 2B). Only partial recovery of the population spike could be observed following perfusion (see "b" and "c"). However, almost full recovery of the population spike was observed after 3,4-diaminopyridine

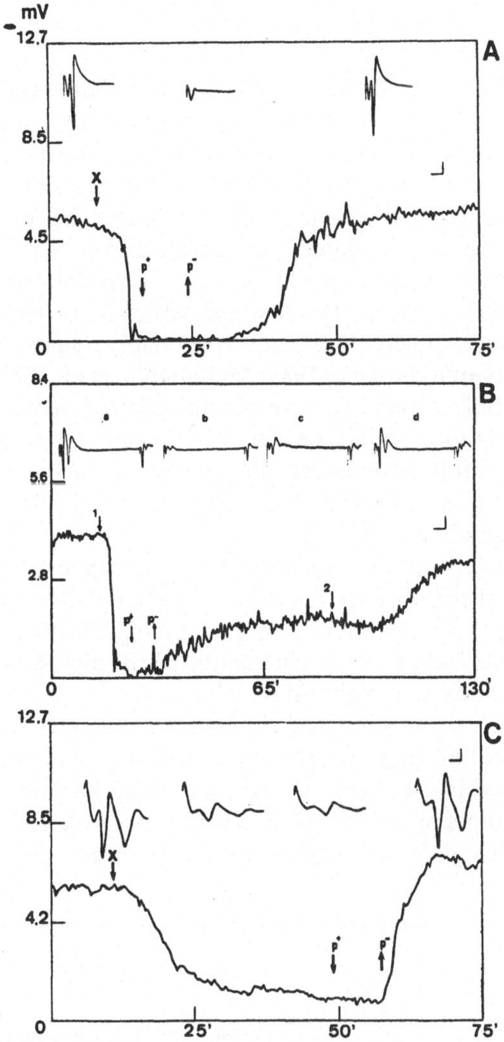

Figure 2. The influence of sodium periodate and sodium perchlorate on the evoked potentials recorded from hippocampal slices. A - sodium periodate (final conc. 10.0 mM) was added to the chamber ("X") when the size of the population spike was already stabilized. The potential disappeared, but full recovery could be observed following a 7 minute perfusion of the slices ("p^+" and "p^-" - start and end of the perfusion, respectively) with fresh solution. Calibration: 5 ms, 1 mV. B - sodium periodate (final conc. 60.0 mM) was added to the chamber at the time marked by "1". In this experiment both orthodromically evoked potential (population spike) and antidromically driven cell discharge (80 ms delay) were followed. The electrode arrangement was as shown in Figure 4. After $NaIO_4$ addition only the population spike disappeared, while the antidromically evoked potential remained almost unchanged (compare "a" and "b"). Only partial recovery of the population spike could be seen following perfusion (compare "b" and "c"). Full recovery of the population spike could be observed after AP (71 μM) addition at the time marked by "2" (compare "c" and "d"). Calibration: 5 ms, 1.2 mV. C - sodium perchlorate was added to the chamber ("X") to achieve a final concentration of 60.0 mM. Otherwise as in "A". Note the full recovery of the population spike following perfusion alone. Calibration: 5 ms, 1.5 mV.

(AP) addition (see "c" and "d"), the drug known to facilitate calcium transport into the nerve cells (Thesleff, 1980). These data suggest that $NaIO_4$ specifically influences synaptic transmission, as cell excitability (indicated by antidromically evoked potentials) remained almost unchanged. As a control for the experiments we used sodium perchlorate which is chemically similar to sodium periodate, but does not modify sialic acid. As can be seen in Figure 2C, sodium perchlorate also evokes a reversible decrease in the size of the hippocampal population spike. Our data suggest that sodium periodate has a dual mode of action. The first one, which is exerted at lower concentration, is common with sodium perchlorate and is probably due to chaotropic action of both these compounds (Gomolla et al., 1983; McLaughlin et al., 1975) on the cell membrane. This action is fully reversible. The second type of action, demonstrated by $NaIO_4$ at higher concentration, might be due to sialic acid modification. That action of sodium periodate probably disturbs calcium transport ability of the nerve terminals, as the sodium periodate effect could be reversed by increasing calcium penetration into the cell with diamino-pyridine (AP).

Ruthenium Red (RR) is an inorganic dye which is supposed to bind to negative charges at the biological membranes due to its highly positive charge (Luft, 1971). Sialic acid is considered as a potential site for RR binding. If this is the case, RR could be considered as an additional "tool" in elucidation of the biological role of sialic acid-containing compounds such as gangliosides.

Ruthenium Red (0.71 mM) abolished the hippocampal population spike after 40 min following its application (Fig. 3). The time necessary to abolish the potential by RR is markedly increased in the slices in which sialic acid has been partially liberated by NVCh (Fig. 3). It points to sialic acid as a target molecule for RR. The action of RR, similarly as in the case of sodium periodate, was specifically directed to

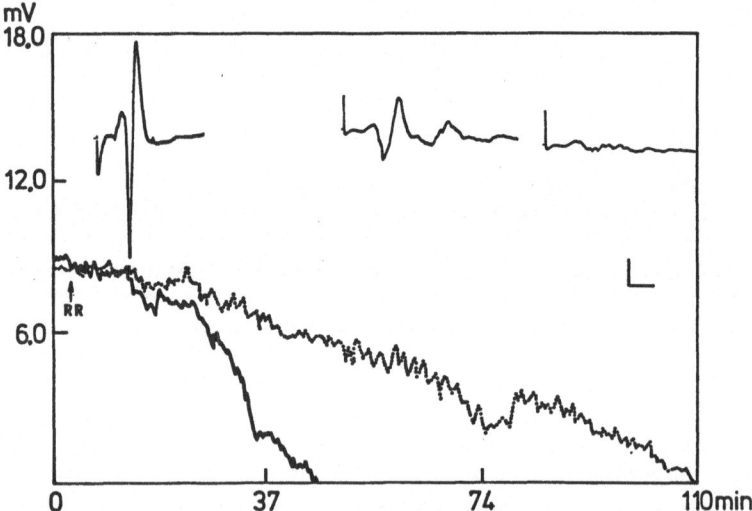

Figure 3. The influence of Ruthenium Red (0.71 mM) on the size of the potential recorded from control hippocampal slices (solid line) and from the slices incubated for 3 hrs with NVch - 0.14 U/ml (broken line). The upper panel represents the shape of the potential corresponding to the slice incubated with NVCh. Calibration: 5 ms, 1 mV.

neurotransmission as antidromically driven potentials remained unchanged (Wieraszko, submitted). Moreover, the depressing effect of RR on the population spike could be reversed as in the case of $NaIO_4$ by AP application (Wieraszko, 1985). This finding supports the view (Kamino et al., 1976) that RR binds to sialic acid residues and disturbs Ca^{2+} binding to this molecule.

Effect of Toxins

Cholera toxin at the concentration of 0.8 U/ml evoked a slight decrease in the size of the population spike in the hippocampal slice. However, at ten times higher concentration ChT blocked the potential (Wieraszko, Janigro and Gorio, unpublished). We interpret the ChT action as a result of interaction of this toxin with GM1. Tetanus toxin had no effect on the size of the population spike recorded from hippocampal pyramidal layer during 4 hrs incubation period. But we have found an effect of this toxin on the efficiency of recurrent inhibition in the hippocampus (Wieraszko, submitted). The experimental arrangement is shown in Figure 4. The orthodromically evoked population

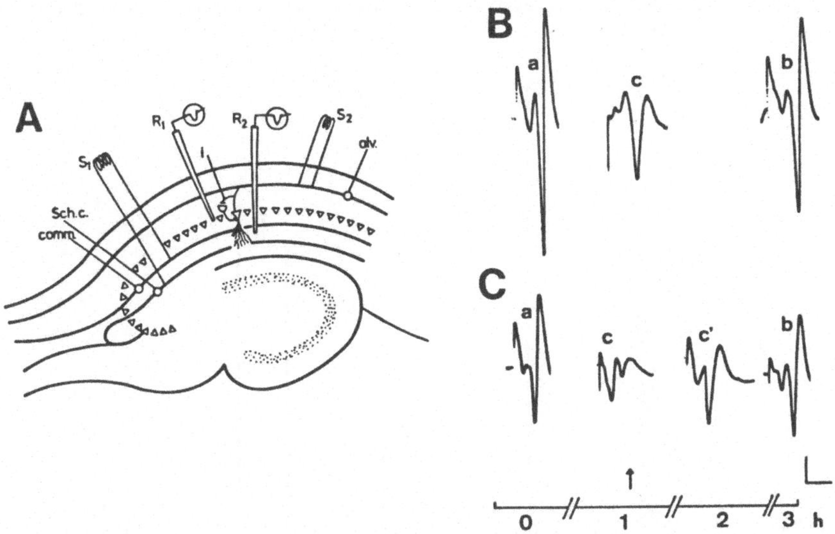

Figure 4. The influence of Tetanus Toxin (TT) on the recurrent inhibition in the hippocampal slices. A - diagram of the hippocampal slice, S1 - the electrode orthodromically stimulating Schaffer collaterals-commissural fibres (Sch. c. comm.), R1 - the electrode monitoring extracellularly the discharge of the pyramidal neurons, S2 - the electrode placed on alveus (alv) antidromically stimulating pyramidal neurons and inhibitory loop (i), R2 - the electrode placed in the dendritic layer recording the changes in the size of EPSP; B and C - the size of the potential at the beginning "a" and at the end of the experiment "b". In both time points ("a" and "b") the electrode stimulating pyramidal neurons antidromically was switched off. "c" represents the potential when the antidromically stimulating electrode was switched on. In control experiment (B) the size of the potential reduced by recurrent inhibition (potential "c") remained stable during the 2.5 hrs of duration of the experiment. When T.T. was added at the time marked by an arrow (C) the potential, reduced in its previous value, slowly increased with time (compare "c" and "c' "). Calibration: 5 ms, 1 mV.

spike (electrode S1) is markedly reduced in its size when it is preceded (20ms) by antidromic activation (electrode S2) of the pyramidal neurons by the electrode placed on alveus. Having this experimental paradigm, we checked the influence of TT on the efficiency of the inhibitory loop. TT at the concentration of 10^6 mouse LD_{50}/ml reduced the strength of inhibition (Fig. 5). The size of the population spike, initially reduced to 45% of its control value by activated inhibitory neurons, increased after TT addition to 75% (Fig. 5). During the whole experiment the strength of both ortho-and antidromic stimulation had not been changed. The effect of TT was no longer observed in the slices incubated for 3 hrs with NVCh (0.14 U/ml). After such a treatment the level of GT1b, by which TT is mainly bound (Rogers and Snyder, 1981) was reduced twofold.

The recurrent inhibitory loop activated by us antidromically uses GABA as a neurotransmitter of the basket neurons (Ribak et al., 1978). It has also been demonstrated that TT decreases the GABA level released from synaptosomes (Osborne and Bradford, 1973) and from hippocampal slices (Collingridge et al., 1981). Thus attenuation of inhibition in hippocampal slices by TT can be explanined as being a result of diminishing of GABA release evoked by stimulation. TT can no longer exert its ac-

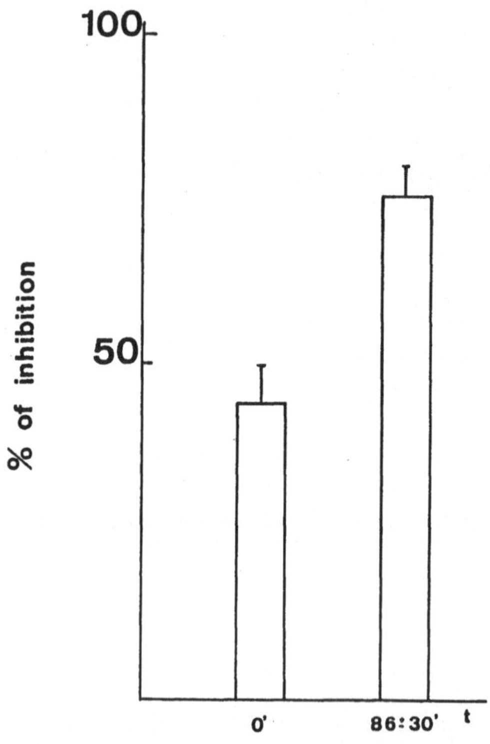

Figure 5. The average attenuation of the inhibition in the hippocampal slices by T.T. Experimental paradigm as depicted in Figure 4. Time "0" shows the inhibition at the moment of T.T. addition. As 100% the average size of the potential was taken at the beginning and at the end of the experiment, when the antidromically stimulating electrode was switched off.

tion when the level of polysialogangliosides (especially GT1b) decreases below a certain level after NVCh treatment. It would suggest that GABA-ergic neurons are richer in polysialogangliosides than glutamergic nerve endings. This is why Tetanus toxin can be bound by these nerve endings. Another explanation is also possible as well. The level of polysialogangliosides could be the same in both types of neurons; however, polysialogangliosides would be more important in GABA- than in glutamate-mediated neurotransmission.

Enzyme Treatments

Neuraminidase from Vibrio Cholerae was added to the striatal slices containing chamber at the concentration of 0.14U/ml. The extracellular record taken from striatal slices following electrical stimulation consists usually of two spikes (Fig. 6A). The first spike results from antidromic cell activation (Misgeld and Bak, 1979). The second, which is the result of the discharge of synaptically driven cells (Misgeld and Bak, 1979), was abolished by the enzyme treatment. It has been demonstrated (Misgeld and Bak, 1979) and confirmed in our own study (Wieraszko and Seifert, 1984) that the striatal cells from which the extracellular potential is recorded use acetylcholine as a neurotransmitter. As a result of the enzyme action the level of GM1 was doubled while polysialogangliosides, mainly GD1a and GT1b, were reduced. GD1b appeared to be relatively resistant to the enzyme action. Subsequently, the level of free sialic acid in the enzyme treated tissue increased tenfold.

The results were different when hippocampal slices were incubated with the same concentration of NVCh. No changes in the population spike evoked by low frequency

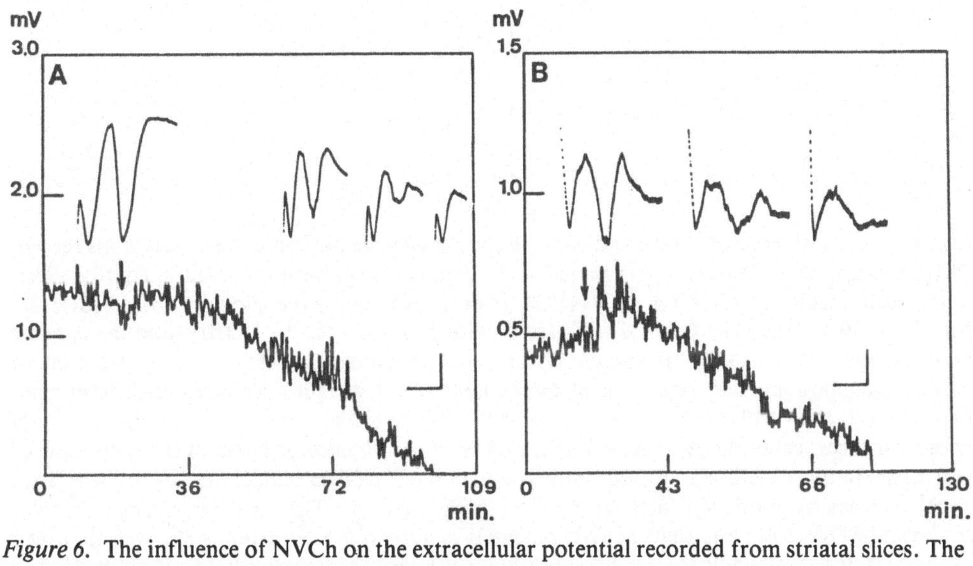

Figure 6. The influence of NVCh on the extracellular potential recorded from striatal slices. The arrow marks the enzyme addition (0.14 U/ml). Note that only the second spike (see "A") which is a result of synaptically driven cell discharge, is abolished by the enzyme treatment. Calibration: 2 ms, 0.2 mV.

(0.06 Hz) stimulation could be observed up to 8 hrs of incubation with the enzyme (Fig. 7). The ganglioside pattern had been severely changed (Fig. 7). The amount of GM1, which is about 13-15% of the total ganglioside content in the hippocampus, increased to over 70% after 8 hrs of incubation (Fig. 8). Polysialogangliosides were reduced 2.3-fold, except GD1a which was reduced by 20% only. However, when the level of GM1 was also reduced by treatment with the other neuraminidase (NAu) (Fig. 7), the

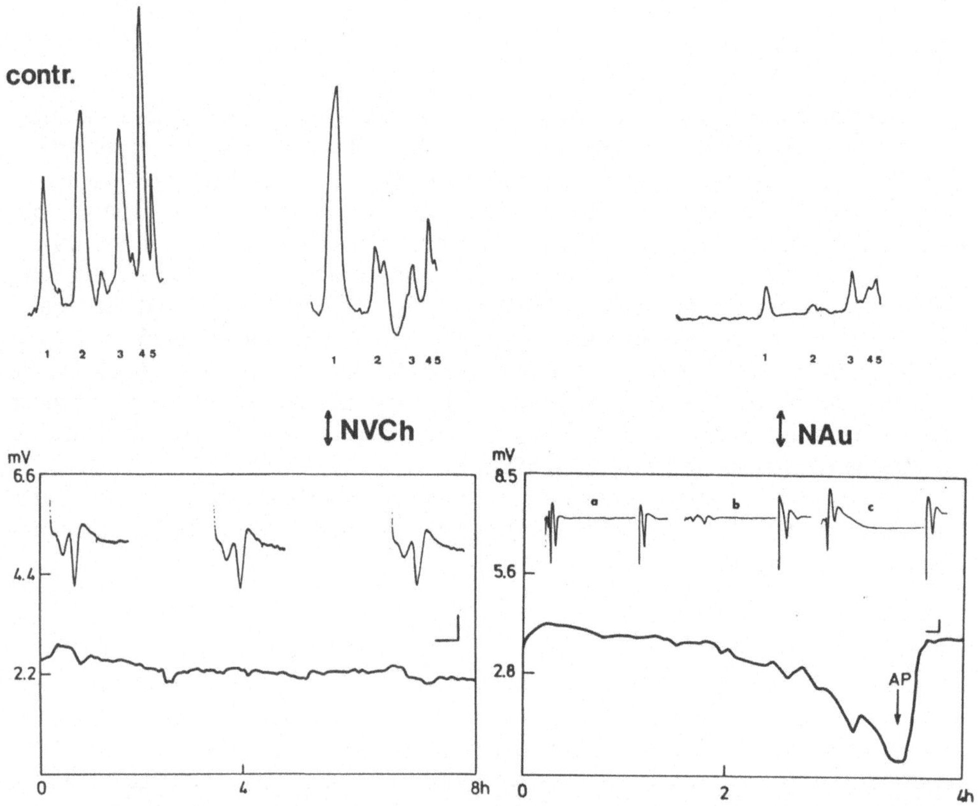

Figure 7. The influence of NVCh and NAu on the ganglioside pattern and the potentials recorded from hippocampal slices. Upper panel — the ganglioside pattern observed in control slices, in the slices incubated for 8 hrs with NVCh (0.14 U/ml) and in the slices incubated for 5 hrs with NAu, (0.5 U/ml); 1 - GM1, 2 - GD1a, 3 - GD1b, 4 - GT1b, 5 - GQ1b. Note the increase in GM1 level after NVCh treatment, which is accompanied by reduction in the level of polysialogangliosides, and reduction in the content of all gangliosides after NAu treatment. Lower panel: left - lack of the influence of NVCh on the evoked potential recorded from hippocampal slices; calibration: 5 ms, 0.5 mV; right — the influence of NAu on the hippocampal potentials. In this experiment orthodromic stimulation of Schaffer collaterals was followed with an 80 ms delay by antidromic activation of pyramidal neurons. The electrode arrangement was as shown in Figure 4. Note that the first potential representing population spike was abolished by the enzyme action, while the second antidromically evoked potential was only slightly affected (compare "a" and "b"). The recovery of the population spike could be observed following AP (71 μM) application (compare "b" and "c"). Calibration: 5 ms, 1 mV. In the case of both enzymes the recording of the potential started after enzyme addition.

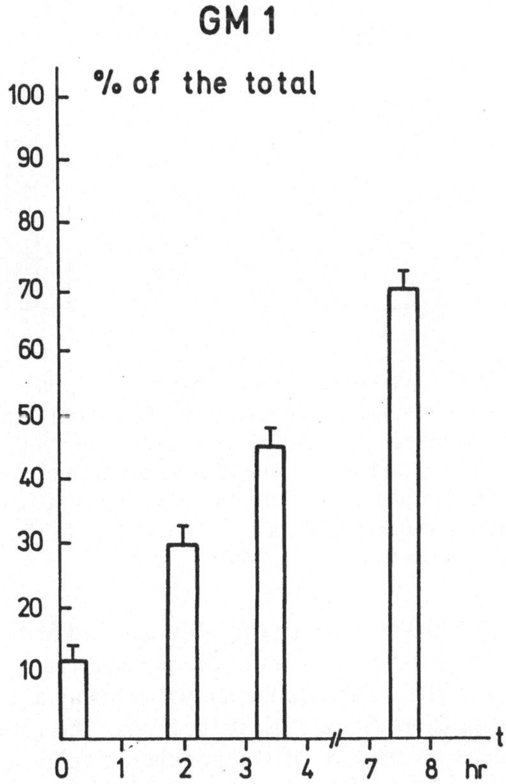

Figure 8. Increase in the GM1 content in the hippocampal slices incubated for different times with NVCh (0.14 U/ml). Each bar represents the mean of 4 experiments ± SEM. The amount of GM1 is expressed as a % of the total ganglioside content in the extract from 10 hippocampal slices.

size of the population spike was reduced (Fig. 7) or completely abolished. This would suggest that a certain critical level of GM1 is necessary to maintain glutamergic neurotransmission in the hippocampus.

However, the situation was different when high frequency stimulation was applied to the same hippocampal pathway. It is well known that this brief high frequency stimulation (in the range of hundreds of Hz) permanently increases the efficiency of stimulated pathways (Bliss and Lomo, 1973). This phenomenon, called "Long Term Potentiation" (LTP), is one of the most intriguing phenomena in neurobiology, having implications to memory processes (Lynch and Baudry, 1984). We found (Wieraszko and Seifert, 1985) that hippocampal slices in which the level of GM1 was increased by treatment with NVCh can potentiate better then their matched controls. Better potentiation could be observed both 2 and 15 min after high frequency stimulation (Fig. 9). A similar increase in the size of potentiation was observed in experiments in which slices were enriched with GM1 by incubation with this ganglioside (71.4 μM). Such treatment also should increase the GM1 level in the membranes as GM1 and other

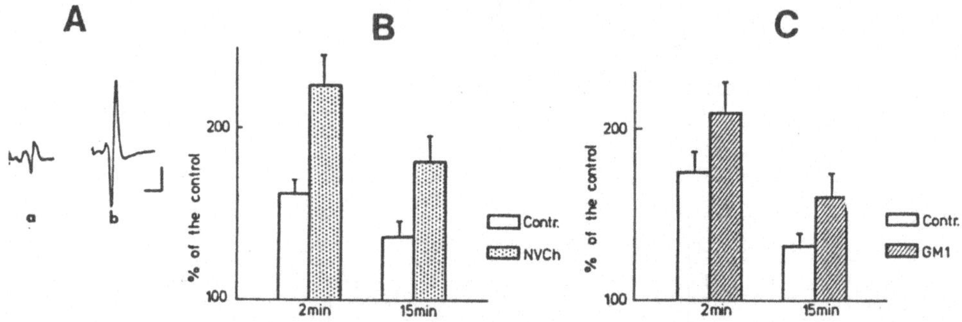

Figure 9. The influence of NVCh treatment and incubation of the slices with GM1 on the size of the potentiation. A - typical example of the increase in the size of the potential from the control value "a" to the value obtained after high frequency stimulation "b". In this particular case the potentiation exceeded 300%; calibration: 5 ms, 1 mV. B - the magnitude of potentiation in control slices and in the slices incubated for 3 hrs with NVCh (0.14 U/ml), n = 44 pairs, $p < 0.001$. C - the magnitude of potentiation in control slices and in the slices incubated for 3 hrs with GM1 (71.4 μM/ml), n = 29 pairs, $p < 0.002$.

gangliosides incorporate in the nerve cell easily (Morgan and Seifert 1979, Toffano et al., 1980). These data imply that enrichment of the cell membrane with GM1 improves cell excitability. Thus in a glutamergic hippocampal pathway a decrease in the GM1 level disturbs neurotransmission. An increase in the level of this ganglioside in the same pathway has no influence on the size of the population spike when low frequency stimulation (0.06 Hz) is employed. The influence of GM1 is seen only when the triggering of higher efficiency by high frequency stimulation takes place.

It is interesting to note here that calcium/ganglioside-dependent kinase activity has been described recently in the rat brain (Goldenring et al., 1985). On the other hand protein phosphorylation is a phenomenon which accompanies LTP (Bär et al., 1982; Browning et al., 1980; Routtenberg and Lovinger, 1985) and which is probably responsible for an increase of glutamate receptor activity following high frequency stimulation (Lynch and Baudry, 1984). Recent data from our laboratory (Hollmann and Seifert, 1985) have demonstrated that gangliosides can activate the glutamate receptor in a calcium-dependent manner. All these findings point again to the importance of the proper level of GM1 in the plastic adjustements of the nerve ending's efficiency in response to a new working condition.

Effect of Antibodies to GM1

Antiserum to GM1 was kindly provided by Dr.S. Karpiak (New York) in an amount which allowed us to conduct two experiments with hippocampal slices. In both of them abolishing of the potential could be observed while addition of a control serum had no effect. One of the experiments is shown on Figure 10.

This result is in agreement with the effect of arthrobacter neuraminidase treatment (Fig. 7) where degradation of GM1 leads to abolishing of the population spike, and with the data reporting strong behavioral effects of antiserum to GM1 in vivo (Karpiak

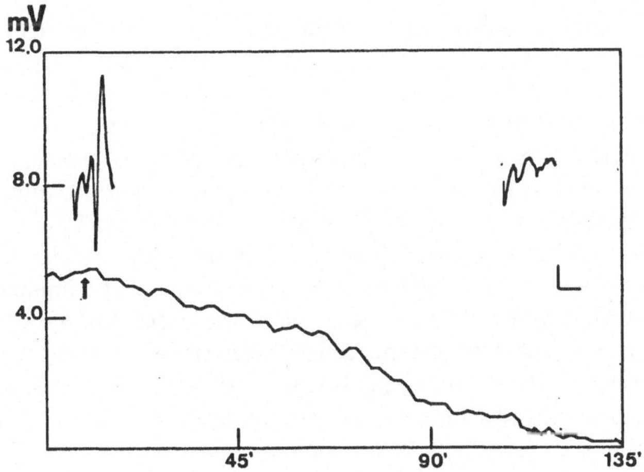

Figure 10. The influence of antibody to GM1 on the hippocampal potential. The time of antibody addition marked by an arrow. Calibration: 5 ms, 0.8 mV.

et al., 1976; Karpiak et al., 1978). Thus, both enzymatic degradation of GM1 or blocking of GM1 by specific antibodies results in the same effect on the population spike of this hippocampal pathway. Therefore we conclude that in the glutamergic hippocampal pathways the ganglioside GM1 plays a crucial role for synaptic transmission.

CONCLUDING REMARKS

We have demonstrated in our study that factors which in some way modify endogenous gangliosides can influence synaptic transmission, but not cell excitability. Moreover, our data suggest that different types of gangliosides could be involved in neurotransmission in the synapses using different types of neurotransmitters. Our experiments point to GM1 as being involved in glutamergic neurotransmission in the hippocampus. This "involvement" shows two aspects. At first it seems that a certain level of GM1 is crucial to maintain neurotransmission at the proper level. The second point is that enrichment of the hippocampal membranes in GM1 improves the ability of these synapses to demonstrate plastic properties. A similar role played by GM1 in glutamergic synapses could be filled by polysialogangliosides in GABA-ergic and cholinergic synapses. This suggestion has support from our experiments performed with tetanus toxin on hippocampal slices and with NVch on striatal slices. The mechanism through which gangliosides could influence synaptic transmission is discussed by us in detail elsewhere (Wieraszko and Seifert, 1984; Wieraszko and Seifert, 1985; Wieraszko and Seifert, 1985). In general we would like to support the hypothesis (Rahmann, 1983; Svennerholm, 1980) that GM1-Ca^{2+} interaction is mainly responsible for the observed effects. It is known that gangliosides can bind calcium not only with sialic acid, but also with an oxygen rich region of the molecule present in some of them ("calcium pocket") (Tettamanti et al., 1985). However, other explanations are plausible as well. The research on gangliosides in the last ten years has demonstrated that gangliosides can influence the biochemistry of nerve cells in several different ways (Daly, 1981; Jeserich et al., 1981; Seifert, 1981). Apart from direct or indirect influences on membrane-bound enzyme systems, gangliosides may also function as receptors or receptor modulators (Morgan and Seifert, 1979). This indeed has been demonstrated recently in our laboratory for glutamate receptor binding with isolated synaptic plasma membranes from rat cortex and hippocampus which is stimulated by individual gangliosides (Hollmann and Seifert, 1986).

We are still far from a detailed understanding of the mechanism of action of gangliosides in the nervous system. However, our experiments indicate that gangliosides play an important functional role in the process of synaptic transmission and therefore seem to substantiate this old hypothesis. Moreover, our results lead us to conclude that certain individual gangliosides play such functional roles at different types of synapse, — as shown here for glutamergic, GABA-ergic and cholinergic synaptic transmission.

REFERENCES

Avrova NF, Chenykaeva E Yu, Abukhova EL (1973) J Neurochem 20: 977-1004.

Balakrishman S and McIlwain H (1961) Biochem J 81: 72-78.

Bär PR, Tielen AM, Lopes da Silva FH, Zwiers H, Gispen WH (1982) Brain Research 245: 69-79.

Behr J-P and Lehn J-M (1973) FEBS Lett 31: 297-300.

Bertoli E, Masserini M, Sonnino S, Ghidoni R, Cestaro B, Tettamanti G (1981) Biochim Biophys Acta 467: 196-202.

Bliss TVP and Lomo T (1973) J Physiol (Lond) 232: 331-356.

Browning M, Dunwiddie T, Bennet W, Gispen W, Lynch G (1980) Science 203: 60-62.

Collingridge GL, Thompson PA, Davies J, Mellanby J (1981) J Neurochem 37: 1039-1041.

Cumar FA, Maggio B, Caputto R (1978) Biochim Biophys Res Comm 84: 65-69.

Daly JW (1981) in: Rapport, MM and Gorio A (eds): Gangliosides in neurological and neuromuscolar function, development and repair, Raven Press, New York, pp. 55-56.

Gahmberg CG and Anderson LC (1977) J Biol Chem 252: 5888-5894.

Goldenring JR, Otis LC, Yu RK, de Lorenzo RJ (1985) J Neurochem 44: 1229-1234.

Gomolla M, Gottschalk G, Luttgan H Ch (1983) J Physiol (Lond) 343: 197-214.

Hansson H-A, Holmgren J, Svennerholm L (1977) Proc Natl Acad Sci USA 74: 3782-3786.

Hayashi K and Katagiri A (1974) Biochim Biophys Acta 337: 107-117.

Hipp FX, Gielen W, Margaret AD, Hinzen DH (1980) Pflügers Arch 385: 45-50.

Hollmann M, Seifert W (1986) Neurosci Lett 65: 133-138.

Jaques LW, Riesco BF, Weltner W Jr (1980) Carbohydrate Res 83: 21-32.

Jeserich G, Breer H, Duvel M (1981) Neurochem Res 6: 465-474.

Kamino K, Ogawa M, Uyesaka N, Inouye A (1976) J Membr Biol 26: 345-356.

Kato I and Naiki M (1976) Infect Immunol 13: 289-291.

Karpiak SE, Graf L, Rapport MM (1976) Science 194: 735-737.

Karpiak SE, Graf L, Rapport MM (1978) Brain Research 151: 637-640.

Kitamura M, Iwamori M, Nagai Y (1980) Biochim Biophys Acta 628: 328-335.

Kracun I, Rösner H, Cosovic C, Stavljenic A (1984) J Neurochem 43: 979-989.

Ledeen RW (1978) J Supramolec Str 8: 1-17.

Ledeen RW (1984) J Neurosci Res 12: 147-159.

Lloyd CW (1975) Biol Rev 50: 325-350.

Luft JH (1971) Anat Rec 171: 369-416.
Lynch G and Schubert P (1980) Ann Rev Neurosci 3: 1-22.
Lynch G and Baudry M (1984) Science 224: 1057-1063.
McIlwain H (1961) Biochem J 78: 24-32.
McLaughlin S, Bruder A, Chen S, Moser C (1975) Biochim Biophys Acta 394: 304-313.
Misgeld U and Bak J (1979) Neurosci Lett 12: 277-282.
Morgan J and Seifert W (1979) J Supramolec Str 10: 111-124.
Osborne RH and Bradford HF (1973) Nature (New Biology) 244: 157-158.
Probst W, Rösner H, Wiegandt H, Rahmann H (1979) Physiol Chem 360: 979-986.
Rahmann H, Rösner H, Breer H (1976) J Theor Biol 57: 231-237.
Rahmann H (1983) Neurochem Int 5: 539-547.
Ribak CE, Vaughn JE, Saito H (1978) Brain Research 140: 315-332.
Rogers TB and Snyder SH (1981) J Biol Chem 256: 2402-2407.
Römer H and Rahmann H (1970) Exp Brain Res 34: 49-58.
Routtenberg A and Lovinger DM (1985) Behavioral and Neuronal Biol 43: 3-11.
Schulze G and Rommelspacher H (1979) in: Abstr. 11th Collegium Int Neuro-Psychopharmacologicum Congress, Vienna, p. 427.
Segler K, Rahmann H, Rösner H (1978) Bioch Systematics and Ecology 6: 87-93.
Seifert W (1981) in: Rapport, M and Gorio A, (eds): Gangliosides in neurological and neuromuscular function, development and repair, Raven Press, New York, pp. 99-117.
Seyfried TN, Miyazawa N, Yu RK (1983) J Neurochem 41: 491-505.
Sharon N and Lis H (1972) Science 177: 949-959.
Sharom FJ and Grant CWM (1977) Biochem Biophys Res Comm 74: 1039-1045.
Simonneau M, Baux G, Tauc L (1980) J Physiologie (Paris) 76: 427-433.
Sugano K, Saito M, Nagai Y (1978) FEBS Lett 89: 321-325.
Svennerholm L (1980) in: Svennerholm L, Mandel P, Dreyfus H, Urban PF (eds): Structure and function of gangliosides, Plenum Press, New York and London, pp. 533-544.
Takeda Y, Takeda T, Honda T, Miwatami T (1976) Infect Immunity 14: 1-5.
Tamir H, Brunner W, Casper D, Rapport MM (1980) J Neurochem 34: 1719-1724.
Tauc L and Hinzen DH (1974) Brain Research 80: 340-344.
Tettamanti G, Sonnino S, Ghidoni R, Masserini M, Venerando B (1985) in: V De Giorgio, M Corti (eds): Physics of Amphiphiles: Micelles, vesicles and microemulsions, Soc Italiana di Fisica, Bologna, Italy, pp. 607-635.
Thesleff S (1980) Neuroscience 5: 1413-1419.
Toffano G, Benvegnu D, Bonetti AC, Facci L, Leon A, Orlando P, Ghidoni R, Tettamanti G (1980) J Neurochem 35: 861-866.
Veh RW and Sander M (1981) Perspectives in Inherited Metabolic Diseases 4: 71-109.
Weiler MH, Misgeld U, Jenden DJ (1982) in: Hanin I and Goldberg M (eds): Progress in Chol Biol: Model of cholinergic synapses, Raven Press, New York, pp. 271-287.
Wieraszko A (1983) in: Seifert W (ed): Neurobiology of the Hippocampus, Academic Press, London, pp. 175-196.
Wieraszko A (1985) Life Sci 37: 2059-2065.
Wieraszko A and Lynch G (1979) Brain Research 160: 372-376.
Wieraszko A and Seifert W (1984) Neurosci Lett 52: 123-128.
Wieraszko A and Seifert W (1985) Brain Research 345: 159-164.
Wieraszko A and Seifert W (1986) Brain Research 371: 305-313.
Wieraszko A (1986) Brain Research, in press.
Yavin E and Habing WH (1984) J Neurochem 42: 1313-1320.

Gangliosides and neuronal plasticity
G. Tettamanti, R.W. Ledeen, K. Sandhoff,
Y. Nagai, G. Toffano (eds.)
Fidia Research Series, vol. 6
Liviana Press, Padova, © 1986

ANTIBODY TO GLYCOLIPID IN A PATIENT WITH IgM PARAPROTEINEMIA AND POLYRADICULONEUROPATHY

Hiroko Baba, Nobuyuki Miyatani, Shuzo Sato, Kyoko Nakamura[1], Tatsuhiko Yuasa, Tadashi Miyatake

Department of Neurology, Brain Research Institute, Niigata University, Asahimachi-dori, Niigata 951, Japan and [1] Department of Biochemistry and Metabolism, The Tokyo Metropolitan Institute of Medical Science, Tokyo, Japan

INTRODUCTION

Although peripheral neuropathy is a well-known complication of myeloma and other paraproteinemias, it has been uncertain whether these paraproteins play a primary role in the pathogenesis of neuropathy or not. Recently Braun et al. (1982) demonstrated that IgM monoclonal antibodies from patients with IgM paraproteinemia and polyneuropathy reacted with myelin-associated glycoprotein (MAG). Ilyas et al. (1984) reported that IgM from some of these patients reacted with both MAG and an acidic glycolipid in human peripheral nerve ganglioside fraction, and suggested that a carbohydrate determinant which reacted with these patients' IgM was shared between MAG and this acidic glycolipid. We reported previously (Baba et al., 1985) a patient with IgM paraproteinemia and polyradiculoneuropathy whose serum only reacted with a lipid component in human peripheral nerve ganglioside fraction but not with MAG. We demonstrate here further studies on this glycolipid.

PATIENT

A 65 year old Japanese male was admitted because of paresthesias and weakness of the limbs for three months. Muscular weakness and atrophy were found particularly in the distal parts of the limbs. Tendon reflexes were absent. There was sensory loss to touch, pain and temperature in a glove-and-stocking pattern. Sensation to vibration was markedly diminished in the distal parts of the limbs. Serum IgM was 1167 mg/dl,

Abbreviations: MAG, myelin associated glycoprotein; SDS, sodium dodecylsulphate; PAP, peroxidase anti-peroxidase; BSA, bovine serum albumin; DAB, diaminobenzidine; TLC, thin layer chromatography; SLPG, sialosyllactosaminyl paragloboside.

and serum protein electrophoresis showed a small peak in the gamma range which was identified as IgM kappa on immunoelectrophoresis. The cold agglutinin titer was over × 4096. After treatment with prednisolone and cyclophosphamide for three months, there was marked improvement in the strength and sensation of the limbs. He died two years and nine months later. On post-mortem examination he was diagnosed as affected by malignant lymphoma.

MATERIALS AND METHODS

Human brain myelin was prepared following the method of Norton and Poduslo (1973). MAG was isolated from the myelin by the lithium diiodosalicylate-phenol method and further purified by gel filtration on a Sepharose CL-6B column (Pharmacia) equilibrated with 0.5% (w/v) sodium dodecyl sulfate (SDS)/1 mM dithiothreitol/10 mM Tris-HCl pH 7.4 and then dialyzed against distilled water. Anti-human MAG antiserum was obtained from a Japanese white rabbit immunized by purified MAG, as Sato et al. (1983) described previously.

Myelin proteins (20 μg) were electrophoresed on 10% SDS-polyacrylamide slab gel and transferred to nitrocellulose paper by the method of Towbin et al. (1979). For the immunostaining of the blots, the peroxidase anti-peroxidase (PAP) method was used. Briefly, the blots were incubated with 3% BSA/0.9% NaCl/10 mM Tris-HCl, pH 7.4 (BSA-saline) at 40°C for 2 hours. Then the blots were incubated at 25°C for an hour with the first antibody diluted by BSA-saline, and rinsed with washing buffer (0.9% NaCl/10 mM Tris-HCl, pH 7.4) for an hour with changes of the buffer every ten minutes. The same incubation procedure was performed for the bindings of the second and third antibodies, and PAP. A substrate solution containing 0.5 mg of diaminobenzidine (DAB) per ml/0.01% (v/v) hydrogen peroxide/20 mM Tris-HCl pH 7.4 was prepared and the blots were developed at room temperature in the substrate solution, and were then rinsed well to stop the reaction.

Gangliosides were isolated from the human lumbar plexus following the method of Nakamura et al. (1983b). Briefly, total lipids of the peripheral nerve were extracted with chloroform-methanol (2:1, v/v). The simple lipids were removed by a column of Iatrobeads (Iatron, Tokyo) and crude gangliosides were separated by a column of DEAE-Sephadex A-25 (Pharmacia) and alkali treatment, then desalted by Sephadex LH-20 (Pharmacia) column and finally applied to a Iatrobeads column to remove sulfatide. Crude gangliosides were then applied to a column of DEAE-Sephadex A-25 (acetate form) and eluted with a gradient of ammonium acetate in methanol. The fractions of mono-, di-, and trisialogangliosides were obtained and desalted.

The total human gray matter gangliosides were used as the standard. Sialosylparagloboside and Sialosyllactosaminylparagloboside (SLPG) were purified from hog skeletal muscles described previously by Nakamura et al. (1983a).

The gangliosides were chromatographed on high performance-TLC plate (aluminium sheets silica gel 60; Merck) with the following solvent systems: A, chloroform/methanol/0.22% $CaCl_2 \cdot 2H_2O$ (50:45:10, by volume); B, chloroform/methanol/5N NH_4OH/0.4% $CaCl_2 \cdot 2H_2O$ (60:40:4:5, volume). After chromatography, the dried plates were dipped for one minute in 0.0025% polyisobutylmethacrylate (Polyscience) in n-hexane and immunostained with the same PAP method as for blots.

RESULTS

As shown in Figure 1, the patient's serum IgM bound to a minor component of the peripheral nerve ganglioside fraction which migrated between GM1 and GD1a.

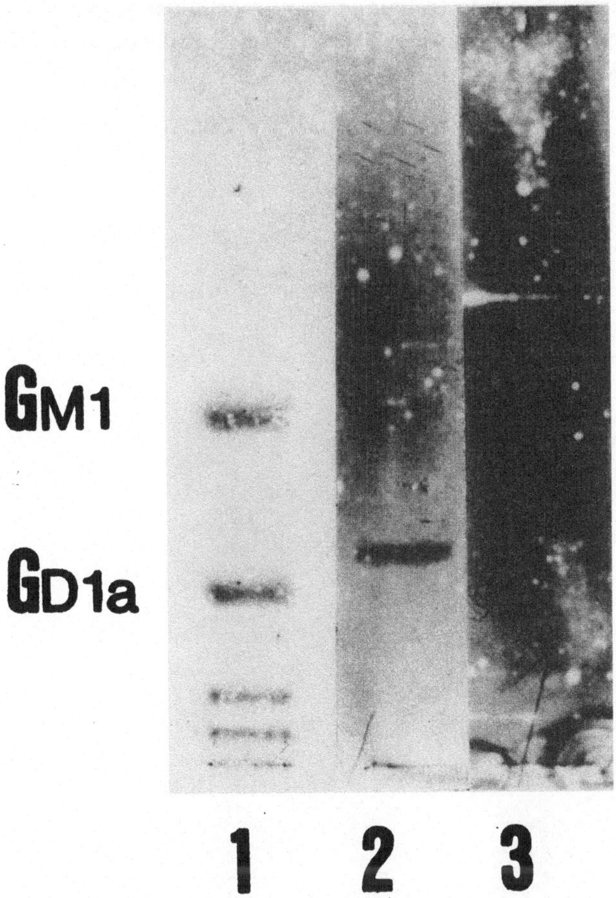

Figure 1. Immunostaining of gangliosides of human nervous system. Gangliosides isolated from human brain (Lane 1) and peripheral nerve (Lane 2, 3), containing 2 μg of sialic acid per lane, were separated by TLC. Lane 1 was examined with orcinol reagent. On lane 2 and 3, each dried thin layer plate was dipped for 1 min in n-hexane dissolving polyisobutylmethacrylate prior to incubation with antibodies. Lane 2 was immunostained with a 1:100 dilution of the patient's serum as the first antibody, a 1:200 dilution of anti-human IgM (rabbit) (Cappel) as the second antibody, a 1:40 dilution of anti-rabbit IgG (goat) (Cappel) as the third antibody, and then a 1:80 dilution of peroxidase anti-peroxidase complex (rabbit) (DAKO). Lane 3 was immunostained with a 1:100 dilution of anti-human MAG antiserum as the first antibody.

There was no distinct band corresponding to this glycolipid when examined with the orcinol reagent. Anti-MAG antiserum produced in a rabbit also bound to a peripheral nerve glycolipid between GM1 and GD1a, but this glycolipid moved slightly faster than the component which reacted with the patient's serum on TLC.

After the fractionation with anion exchange column chromatography, the component that reacted with our patient's serum was found in the monosialo-ganglioside fraction (not shown), and the peripheral nerve component that reacted with this serum had the same Rf values as SLPG in two solvent systems as shown in Figures 2 and 3. The patient's serum also bound to purified SLPG.

Using the immunoblotting technique on nitrocellulose paper, the patient's serum did not react with MAG while anti-MAG antiserum produced in a rabbit reacted with both MAG and MAG derivative (Fig. 4).

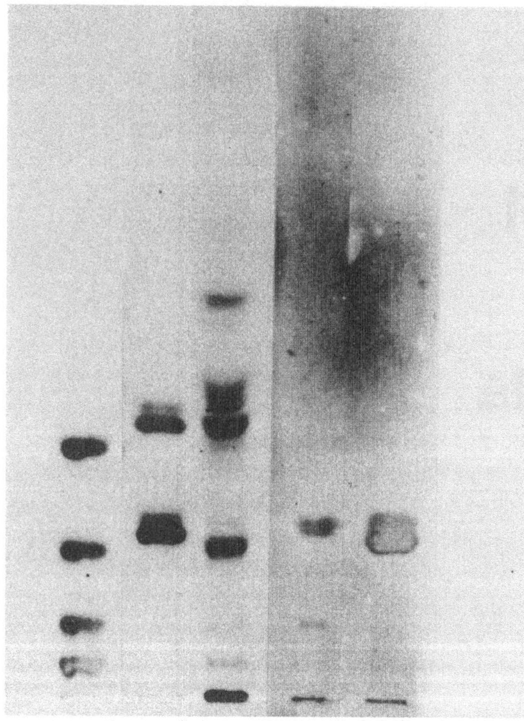

Figure 2. Binding of the patient's serum to gangliosides from human peripheral nerve. The plate was developed with solvent system A, chloroform/methanol/0.2% $CaCl_2 \cdot 2H_2O$ (55:45:10, by volume). Lane 1; gangliosides from human brain gray matter, lane 3 and 4; gangliosides from human peripheral nerve, lane 2 and 5; sialosyparagloboside (upper) and sialosyllactosaminylparagloboside (lower), lane 1, 2 and 3; gangliosides were stained with the orcinol reagent, lane 4 and 5; gangliosides were immunostained with the patient's serum.

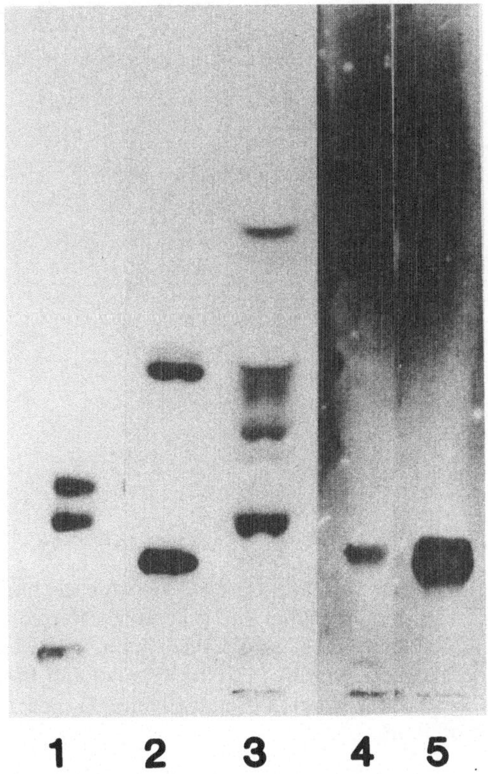

Figure 3. Binding of the patient's serum to human peripheral nerve gangliosides. The plate was developed with solvent system B, chloroform/methanol/5N NH_4OH/0.4% $CaCl_2{\cdot}2H_2O$ (60:40:4:5, by volume). Lane 1 to 5; the same as Fig. 2.

DISCUSSION

The glycolipid that reacted with the patient serum was a minor component of peripheral nerve monosialo-ganglioside fraction. This component had the same Rf values as SLPG using two different solvent systems and this serum also reacted with purified SLPG itself. SLPG is a glucosamine containing ganglioside of neolacto-series, and recently Chou et al. (1985b) reported that SLPG was present in rat sciatic nerve myelin. These data suggested that the peripheral nerve component which reacts with this serum is SLPG.

Our patient's serum did not bind to purified human MAG, and the Rf value of the glycolipid which bound anti-MAG antiserum was different from SLPG. Recently Chou et al. (1985a) reported that an acidic glycolipid which reacted with anti-MAG

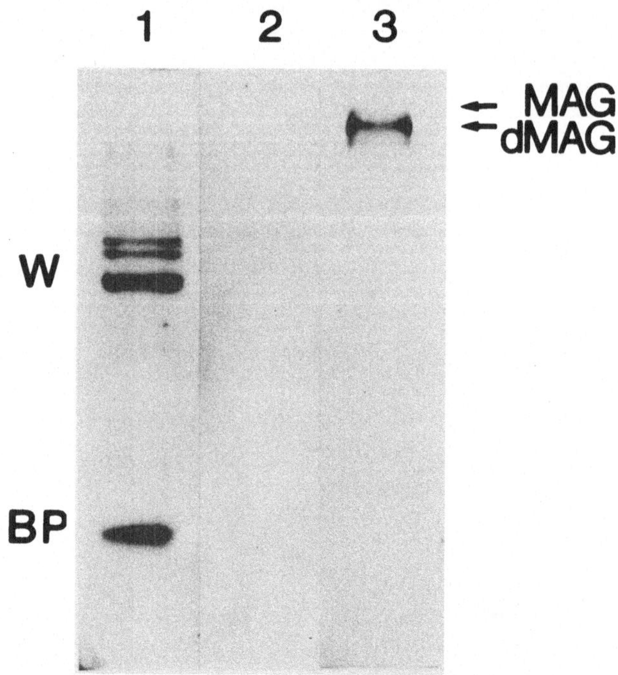

Figure 4. Immunoblotting of human myelin proteins. The two-cycle human myelin (20 μg) was electrophoresed on SDS-polyacrylamide (10%) slab gel and transferred to nitrocellulose sheets. Lane 1 was stained with amido black. Lane 2 was stained immunologically by the same method as Fig. 1. Lane 3 was also immunostained with a 1:100 dilution of anti-MAG antiserum produced in a rabbit as the first antibody, a 1:40 dilution of anti-rabbit IgG (goat) as the second antibody and finally a 1:80 dilution of PAP.

IgM from the patients with polyneuropathy had a glucuronic acid and an extra negatively charged group, probably, sulfate.

Sherman et al. (1983) and Nemni et al. (1983) demonstrated that IgM M proteins from the patients with axonal polyneuropathy are bound to chondroitin sulfate C, a constituent of axons and connective tissue. The present study demonstrates that the serum from a patient with IgM paraproteinemia and polyneuropathy recognizes SLPG in peripheral nerve but not MAG or other myelin proteins. These findings suggest that autoantibodies to various peripheral nerve constituents including MAG, glycolipid and chondroitin sulfate C may play a role in the pathogenesis of neuropathy in IgM paraproteinemia.

It is uncertain whether these circulating antibodies to peripheral nerve antigens play a primary role in the pathogenesis of neuropathy, or are produced secondarily as the result of the peripheral nerve damage. Hays et al. (1983) reported that intraneural injection of serum from a patient with polyneuropathy and monoclonal IgM which reacted with MAG produced focal demyelination in the sciatic nerve of cats and they suggested that neuropathy in patients with anti-MAG is an antibody-mediated autoimmune disorder. It may therefore be suggested that antibody to SLPG in our case may also cause neuropathy.

REFERENCES

Baba H, Miyatani N, Sato S, Yuasa T, Miyatake T (1985) Antibody to glycolipid in a patient with IgM paraproteinemia and polyradiculoneuropathy. Acta Neurol Scand 72: 218-221.

Braun PE, Frail DE, Latov N (1982) Myelin-associated glycoprotein is the antigen for a monoclonal IgM in polyneuropathy. J Neurochem 39:1261-1265.

Chou KH, Ilyas AA, Evans JE, Quarles RH, Jungalwala FB (1985a) Structure of a glycolipid reacting with monoclonal IgM in neuropathy and with HNK-1. Biochem Biophys Res Commun 128: 383-388.

Chou KH, Nolan CE, Jungalwala FB (1985b) Subcellular fractionation of rat sciatic nerve and specific localization of ganglioside LM1 in rat nerve myelin. J Neurochem 44: 1898-1912.

Hays AP, Takatsu M, Latov N et al. (1983) Focal demyelination of cat sciatic nerve induced by intraneural injection of serum from patients with polyneuropathy and monoclonal IgM reactive with myelin-associated glycoprotein. J Neuropath Exp Neurol 42: 349.

Ilyas AA, Quarles RH, MacIntosh TD, Dobersen MJ, Trapp BD, Dalakas MC, Brady RO (1984) IgM in a human neuropathy related to paraproteinemia binds to a carbohydrate determinant in the myelin-associated glycoprotein and to a ganglioside. Proc Natl Acad Sci USA 81: 1225-1229.

Nakamura K, Nagashima M, Sekine M, Igarashi M, Ariga T, Atsumi T, Miyataka T, Suzuki A, Yamakawa T (1983a) Gangliosides of hog skeletal muscle. Biochim Biophys Acta 752: 291-300.

Nakamura K, Ariga T, Yahagi T, Miyatake T, Suzuki A, Yamakawa T (1983b) Interspecies comparison of muscle gangliosides by two-dimentional thin-layer chromatography. J Biochem (Tokyo) 94: 1359-1365.

Nemni R, Galassi G, Latov N, Sherman WH, Olarte MR, Hays AP (1983) Polyneuropathy in nonmalignant IgM plasma cell dyscrasia: A morphological study. Ann Neurol 14: 43-54.

Norton WT and Poduslo SE (1973) Myelination in rat brain: Method of myelin isolation. J Neurochem 21: 749-757.

Sato S, Baba H, Tanaka M et al. (1983) Antigenic determinant shared between myelin-associated glycoprotein from human brain and natural killer cells. Biomed Res 4: 489-494.

Sherman WH, Latov N, Hays AP, Takatsu M, Nemni R, Galassi G, Osserman EF (1983) Monoclonal IgM_k antibody precipitating with chondroitin sulfate C from patients with axonal polyneuropathy and epidermolysis. Neurology (NY) 33: 192-201.

Towbin H, Staehelin T, Gordon J (1979) Electrophoretic transfer of proteins from polyacrylamide gels to nitrocellulose sheets: Procedure and some applications. Proc Natl Acad Sci USA 76: 4350-4354.

Gangliosides and neuronal plasticity
G. Tettamanti, R.W. Ledeen, K. Sandhoff,
Y. Nagai, G. Toffano (eds.)
Fidia Research Series, vol. 6
Liviana Press, Padova, © 1986

AXONAL TRANSPORT OF GANGLIOSIDES AND NEUTRAL GLYCOLIPIDS IN THE PERIPHERAL NERVOUS SYSTEM. IDENTIFICATION OF GANGLIOSIDE TYPES IN MOTONEURONS OF THE PNS

D.A. Aquino, M.A. Bisby[1], R.W. Ledeen

Departments of Neurology and Biochemistry, Albert Einstein College of Medicine, Bronx, New York 10461;
[1]Department of Medical Physiology, University of Calgary, Alberta T2N 4N1, Canada

INTRODUCTION

The highly compartmentalized nature of ganglioside biosynthesis in cells (Yusef et al., 1983) together with their widespread distribution over the cell surface (Ledeen, 1978) has suggested the presence of efficient transport/transfer mechanisms. One aspect of these processes has been elucidated in the central nervous system (CNS) through demostration of fast axonal transport as the primary mechanism for conveyance of gangliosides to axonal and nerve-ending membranes (Forman and Ledeen, 1972; Rosner et al., 1973; Landa et al., 1979; Ledeen et al., 1981; Gammon et al., 1985). Recent studies with sciatic nerve of rabbit (Yates et al., 1984) and rat (Aquino et al., 1985a) have demonstrated similar anterograde flow of gangliosides in the PNS as well. The above rat system, involving motoneurons, also revealed another aspect of ganglioside movement in the neuron, i.e., retrograde axonal transport.

More recently our attention has focused on the sensory component of rat sciatic nerve as a means of studying glycolipid movement in a different neuronal system (Aquino et al., 1985b). This study has revealed the same bidirectional flow of gangliosides as we had previously found for motoneurons of this nerve, but in addition a parallel movement of neutral glycosphingolipids (GSLs). We have thus found sciatic nerve a convenient system not only for the study of axonal transport of these glycoconjugates but also for depicting the unique glycolipid contents of the different neuronal types. This is based on the fact that among the neuronal/glial mixture of glycolipids extracted from whole nerve segments, only the neuronal species will be radiolabeled in the axonal transport paradigm. The results of this approach are presented here.

Abbreviations: CNS, central nervous system; PNS, peripheral nervous system; GSL, glycosphingolipid; DRG, dorsal root ganglia; C, chloroform; M, methanol; W, water.

MATERIALS AND METHODS

Labeling of Glycolipids in Rat Sciatic Sensory Neurons

Adult Sprague-Dawley rats (250-350 g) were anesthetized and D[6-^{3}H]glucosamine hydrochloride (200-400 μCi) in 2 μl Ringer's solution was injected at several sites into the L4 and L5 dorsal root ganglia (DRG) (Bisby, 1977). The specific activity of the labeled precursor (Amersham, Oakville, On., Canada) was 20 Ci/mmole. In some studies 200 μCi of [^{14}C]glucose (50-60 mCi/mmol; American Radiochemical Inc., St. Louis, MO) was injected in 2 μl Ringer's solution. Injections were made unilaterally, so that each animal could serve as its own control for blood-borne labeling. At various time intervals after injection (4-70 h) the animals were reanesthetized and the sciatic nerve exposed. Two ligatures were tied approximately 25 and 35 mm from the DRG. After an additional 2 h to allow transported materials to accumulate at the collection ligatures, nerves were excised and dissected into five segments (Fig. 1). The accumulation segment for anterograde transport was the 2 mm segment just proximal to the proximal ligature, while that for retrograde transport was the 2 mm segment just distal to the distal ligature. Lipids were extracted from these segments, and gangliosides and neutral GSLs separately isolated (see below).

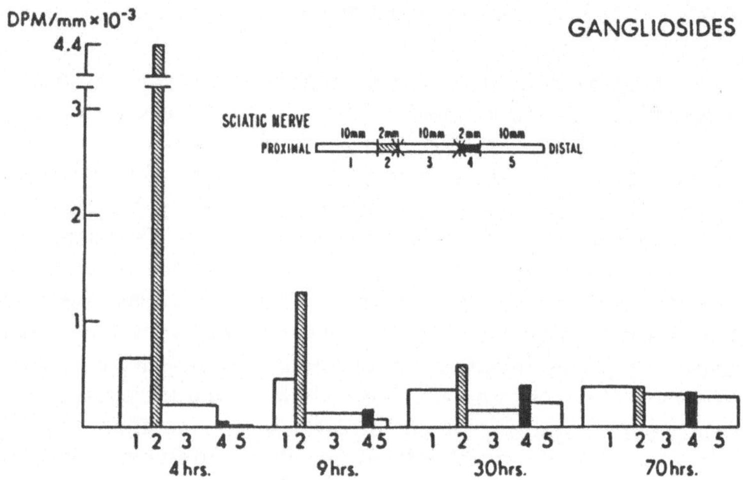

Figure 1. Profiles of ganglioside transported radioactivity in sensory neurons of sciatic nerve. Ligatures were tied (upper diagram) at the times indicated following [^{3}H]glucosamine injection into DRG and accumulation progressed for 2 h; nerves were then dissected into the five segments shown. Gangliosides were isolated from each segment as described.
Radioactivity accumulating at segments 2 and 4 represented anterograde and retrograde transport, respectively. Anterograde transport was discernable at the earliest time studied (4 h), while retrograde transport was first seen at 9 h and became more evident at 30 h. At 70 h there was no longer appreciable accumulation in either direction. Comparable levels of radioactivity (relative to the extent of radiolabel incorporation) were determined from both accumulation segments of 4 additional nerves.

Isolation of Gangliosides and Neutral Glycosphingolipids

Gangliosides and neutral GSLs were extracted and isolated from individual nerve segments by a procedure designed to optimize yield and purity from such tissues (Byrne et al., 1985). Each nerve segment was homogenized in 2 ml chloroform (C)-methanol (M)-water (W) (5:5:1), and the mixture was subsequently brought to 0.1 N H^+ (overall) by addition of a small volume of concentrated HC1. After standing 15-20 min the tissue extract was centrifuged and the resulting supernatant applied to a column of Sephadex LH-20, packed in C-M-W (5:5:1). Lipids eluted as a single fraction, which was then divided into neutral and acidic fractions on a DEAE-Sephadex column.

The resulting two fractions were subjected to mild akaline methanolysis and chromatographed sequentially on Sephadex LH-20 and silica gel (Iatrobeads). In the case of neutral GSLs, the latter chromatography was carried out according to Vance and Sweeley (1967). Additional purification was achieved in some runs by reverse-phase chromatography according to Williams and McCluer (1980), but employing Sepralyte (Analytichem International, Harbor City, CA 90710) in place of Sep-Pak. Radioactivity was determined by dissolving the sample in 1 ml water + 10 ml Hydrofluor (National Diagnostics, Sommerville, NJ) and counting with a Packard Tri-Carb 300 Liquid Scintillation Spectrometer.

Fractionation of Gangliosides

Purified gangliosides from anterograde accumulation segments were pooled and fractionated by FPLC on a Mono Q anion exchange resin (Pharmacia, Piscataway, NJ) into mono-, di-, tri- and polysialogangliosides. These were eluted with a discontinuous gradient of 0.05, 0.15, 0.20 and 0.23 M potassium acetate in methanol, respectively (Mansson et al., 1985). Radiolabeled rabbit brain gangliosides were fractionated for comparison.

Structure Determination

Gangliosides of motoneurons of rat sciatic nerve were radiolabeled with [^{3}H]glucosamine and allowed to undergo axonal transport as previously described (Aquino et al., 1985a). Segments 1, 3 and 5 (Fig. 1), approximately 20 h after precursor injection, were pooled and gangliosides extracted and purified as described (Byrne et al., 1985). These were fractionated by FPLC on a Pharmacia Mono Q column (see above) and the resulting fractions desalted by dialysis, followed by sequential chromatography on Sephadex LH-20 and silica gel (Iatrobeads) columns. Each was then treated with neuraminidase (C1. Perfringens, Type V, Sigma, St. Louis, MO) as previously described (Byrne et al., 1983). The resulting products were fractionated on a DEAE-Sephadex column into neutral (F1) and acidic (F2) products (see above). The former of these were chromatographed in sequence on Sephadex LH-20 (see above) and silica gel (Iatrobeads) according to Vance and Sweeley (1967), while the latter fractions were desalted with a Sepralyte column (see above) and subjected to TLC with the system commonly-employed for gangliosides (Byrne et al., 1985). The radiolabeled samples were co-chromatographed with unlabeled bovine brain ganglioside mixture to provide markers for the scraping and counting of individual fractions.

RESULTS

Bidirectional Transport of Glycolipids in Sensory Axons

Sensory axons of rat sciatic nerve received labeled glycoconjugates via axonal transport after injection of [^{3}H]glucosamine into L4 and L5 dorsal root ganglia, as evidenced in accumulation of transported radiolabeled materials at ligatures tied at various time intervals following precursor injection. Collection of transported materials ensued for 2 h. Background labeling, resulting from blood-borne precursor, was ignored since control (uninjected) nerves from the same animals contained only 7-8% as much radioactivity as injected nerves. It was found that both gangliosides (Fig. 1) and neutral GSLs (Fig. 2) underwent fast bidirectional transport in these neurons. Glycoproteins radiolabeled with glucosamine exhibited the same phenomenon (data not shown).

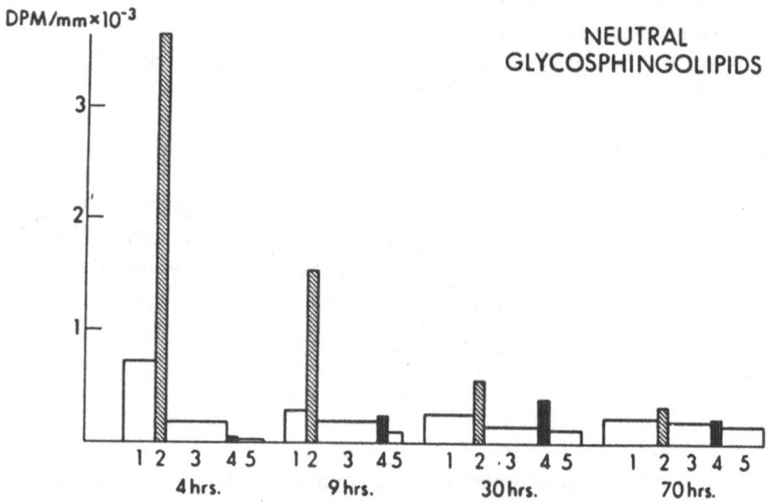

Figure 2. Profiles of neutral glycosphingolipid transported radioactivity in sensory neurons of sciatic nerve. See legend to Figure 1. As with gangliosides, retrograde transport became evident at 9 h and more so at 30 h.

Anterograde transport determined as radioactivity accumulating proximal to the proximal ligature (segment 2), appeared as early as 4-6 h after precursor injection (ligatures were tied at 4 h, collection then ensuing for 2 h). Velocity of the wave front of [^{3}H]glucosamine-labeled materials was calculated as 15 ± 3.6 (SEM) mm/hr with an extrapolated intercept on the time axis of 0.27 hrs, representing processing time in the cell body (data not shown).

Accumulation of radioactivity distal to the distal ligature (segment 4), indicating retrograde transport, was first perceptible at 9 h and rose to a higher level by 30 h. It was not possible to estimate retrograde transport velocity since distal nerve fibers undergo extensive branching, resulting in large variations in retrograde transport distances. At 70 h there was no longer appreciable accumulation of radioactivity at either ligature.

The presence of neutral GSLs among the axonally transported glycoconjugates was indicated by the method of isolation, following [^{3}H]glucosamine injection (Fig. 2). More definitive evidence came from TLC radioautography (Fig. 3) of isolated neutral GSLs following injection of [^{14}C]glucose into the DRG. Several radiolabeled components were detected in this fraction, the major ones migrating in the general vicinity of globoside, paragloboside and asialo GM1 (GA1). However, identification of these substances must await more rigorous methods of structural analysis.

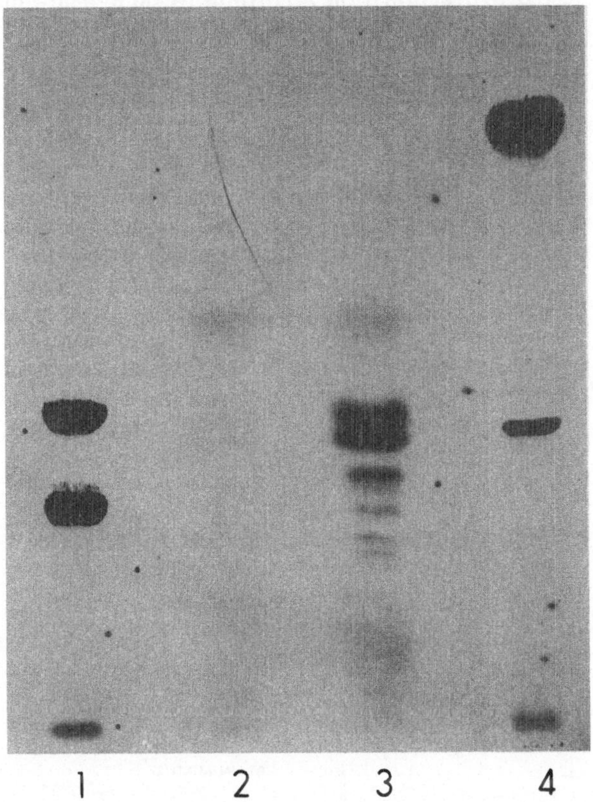

Figure 3. Radioautogram of neutral glycosphingolipids from sensory neurons of rat sciatic nerve. Dorsal root ganglia of several rats were injected with [^{14}C]glucose and anterograde transported material collected at the proximal ligature. Neutral GSLs were extracted and purified as described and applied (225 DPM) to a silica gel 60 TLC plate (lane 3). A similarly-prepared extract of motoneurons (following injection of [^{14}C]glucose into the ventral horn of the lumbosacral spinal cord) was applied to lane 2. Standards: globoside (upper) and GA1 (lower), lane 1. lactosylceramide (upper) and paragloboside (lower), lane 4. The plate was developed in chloroform-methanol-aq. KC1 (0.25%) 65:35:8, dried, and exposed to XAR-5 Kodak X-ray film for approximately 4-5 months. Standards were revealed by iodine vapor.

Purified gangliosides from anterograde accumulation segments of sensory axons were fractionated by FPLC on a Mono Q anion exchange resin. The distributional pattern was similar to that previously observed for motor axons of the same nerve and not radically different from that of brain (Fig. 4). The percentages of mono-, di-, tri- and polysialogangliosides for sensory axons were approximately 22 ± 1, 48 ± 0.3, 21 ± 1, and 9 ± 0.4% (n = 3), respectively.

Structure Determination

Gangliosides of motoneurons that had been labeled in the cell bodies with [^{3}H]glucosamine and allowed to undergo transport were extracted from nerve segments and subjected to fractionation according to sialic acid number on a Pharmacia Mono Q column, as described. Following neuraminidase treatment of each frac-

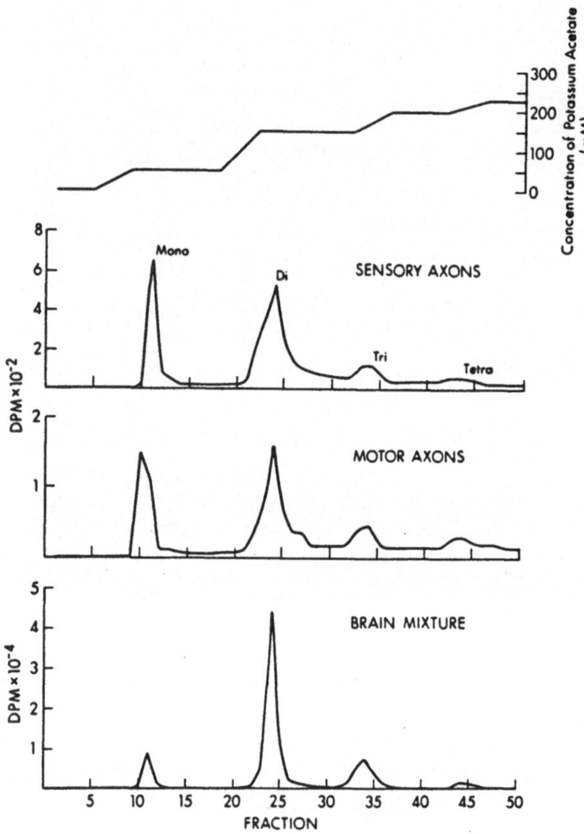

Figure 4. Chromatographic separation of gangliosides according to sialic acid content by FPLC on Mono Q anion exchange resin. Purified gangliosides from anterograde accumulation segments of sensory and motor axons are compared with radiolabeled brain mixture (rabbit).

tion, the majority of counts were then eluted from a DEAE-Sephadex column in the acidic (F2) components while lesser amounts of radioactivity were eluted with the neutral components (F1) (Table 1). The latter included little if any neutral GSL, as determined by chromatography on a silica gel column according to Vance and Sweeley (1967); few if any counts were obtained in the acetone-methanol (9:1) eluent.

Table 1. *Study of Ganglioside Structures in Rat Sciatic Motoneurons Following Axonal Transport*

	Mono	Di	Tri	Poly
Subjected to Neuraminidase	428	835	630	780
DEAE-Sephadex F1 (Neutral GSLs)	125	210	244	251
DEAE-Sephadex F2 (Acidic GSLs)	257	529	396	476
Reverse Phase Chromatography				
M/W 1:1	85	200	210	230
C/M 1:1	130	260	120	200
resin	42	69	66	46
Overall recovery	89%	89%	102%	93%

Gangliosides were labeled in neuronal cell bodies by injection of [^{3}H]glucosamine into the lumbosacral spinal cord. Axonal transport was allowed to proceed and gangliosides were then isolated from pooled tissues representing segments 1, 3 and 5 (Aquino et al., 1985a). The gangliosides were fractionated according to sialic acid number on a Pharmacia Mono Q column, and each fraction was treated with neuraminidase followed by chromatography on DEAE-Sephadex. The neutral eluent (FI) was counted and then subjected to chromatography on Sephadex LH-20 and finally silica gel according to Vance and Sweeley (1967); none of the counts eluted in the fraction which would contain neutral GSL (acetone-methanol 9:1).
Fractionation of the acidic substances (F2) on a reverse-phase (Sepralyte) column produced substantial radioactivity in the fraction that would contain ganglioside (C-M 1:1). TLC analysis of the latter (not shown) revealed radioactivity to comigrate with GM1 for each of the original ganglioside types, indicating structures in the ganglio-series. Numbers represent DPM in each fraction.

The acidic (F2) components were chromatographed on a reverse phase (Sepralyte) column and a significant number of counts were found in the C-M (1:1) eluent in each case (Table 1). This is the glycolipid-containing fraction, and subsequent TLC with the ganglioside solvent system revealed virtually all recovered counts to comigrate with GM1. This applied whether the original fraction had been mono-, di-, tri- or polysialoganglioside. The substantial radioactivity detected in the M-W 1:1 fractions, representing low molecular weight materials, undoubtedly included free sialic acid generated by neuraminidase and possibly some low molecular weight contaminants. Counts sticking to the resin could have been residual glycopeptides. Overall recoveries of radioactivity were 89% or better.

Fractionation on TLC of the Mono Q-separated fractions (prior to neuraminidase) revealed concentration of counts where expected for "brain-type" gangliosides (i.e., ganglio-series). Thus, the monosialo fraction had significant counts comigrating with GM1 (46% of counts in the ganglioside region) while the disialo fraction had a majority (66%) of counts comigrating with GD1a and GD1b. The tri- and polysialo fractions had 60% and 46% of counts in the ganglioside region comigrating with GT1b and GQ1b, respectively. Considerable counts were also detected above and below the ganglioside region of the TLC, indicating some remaining radiolabeled contaminants despite the extensive purification. These experiments pointed to the presence of "brain type" gangliosides as the major species in this component of the PNS.

DISCUSSION

Anterograde and Retrograde Flow of Glycosphingolipids of Sensory Neurons of PNS

This study has demonstrated bidirectional transport of gangliosides in the sensory neurons of rat sciatic nerve, analogous to the previous results with motoneurons of the same nerve (Aquino et al., 1985a). Unlike the latter system, however, appreciable levels of neutral GSLs were also detected among the transported components of the sensory neurons. These substances, while not yet structurally identified, were shown to belong to this glycoconjugate category by virtue of their (a) having incorporated radiolabeled glucosamine, (b) method of isolation and (c) radioautographic pattern following TLC. Their level of labeling was somewhat less than that of gangliosides in the transported components of sensory neurons. To our knowledge neutral GSLs have not yet been observed in mature neurons of the CNS, although there is some evidence they occur at earlier stages of development (Ledeen et al., 1985; Yamamoto et al., 1985).

The transport characteristics of these glycolipids closely resembled those of gangliosides in both the anterograde and retrograde modes. The estimated anterograde transport velocity of 360 $\pm$ 86 mm/day agrees well with the value (284-446 mm/day) we reported for ganglioside movement in motoneurons (Aquino et al., 1985a), and the velocity (428 mm/day) previously reported for leucine-labeled proteins in this system (Bisby, 1976). The velocity of retrograde transport could not be directly calculated, but the early return of these glycoconjugates seemed consistent with the relatively rapid velocities (equivalent to 1/2 anterograde flow) previously estimated for other substances (Bisby, 1980; Bulger and Bisby, 1978).

Fractionation of purified gangliosides from anterograde collection sites on the basis of sialic acid content revealed a distributional pattern which was similar to that of motoneurons of the same nerve. Comparison with gangliosides from retrograde collection sites as well as examination of individual species are in progress.

Fast axonal transport appears to be the primary mechanism through which axonal and synaptic membranes receive glycoconjugates and other membrane components from the cell body, the primary locus of synthesis. The present report, in conjunction with our previous study of sciatic nerve motoneurons, has demonstrated that glycolipids of both neutral and ganglioside types display this phenomenon of bidirectional flow, thus providing at least a partial answer to the question of "turnover" mechanism of these substances in axonal and nerve-ending membranes.

Use of the Axonal Transport Paradigm to Study Neuron-Specific Gangliosides of the PNS

The fact that radiolabeled gangliosides migrate considerable distances down the nerve from their site of synthesis in the cell body provides an opportunity to identify neuron-specific gangliosides among the mixture extracted from whole nerve segments. Applying this technique to rat sciatic motoneurons, we have found evidence for the presence of the ganglio-series, similar to the "brain-type" gangliosides already so well-characterized in the CNS (Ledeen, 1983; 1985). This evidence consisted mainly in the TLC identification of GM1 as the major glycolipid product of neuraminidase. The relative proportions of mono-, di-, tri- and polysialogangliosides were not greatly different from those found in brain. Based on the apparent absence of any neutral GSL products from neuraminidase, we tentatively conclude that gangliosides of the globo-, lacto- or neolacto-types are scarce or absent from these particular neurons.

It would be wrong, however, to extrapolate this findings to PNS neurons in general. Similar experiments now being carried out with sensory neurons of rat sciatic nerve suggest that these may have a somewhat different glycolipid profile, including a mixture of neutral GSLs (see above) and perhaps some gangliosides which are not part of the ganglio-series. In due course it should be possible to exploit the axonal transport paradigm to more fully elucidate the relatively unknown area of glycolipid composition of PNS neurons.

ACKNOWLEDGMENTS

This project was supported by PHS grants NS 04834 and NS 03356 (to RL) and Medical Research Council of Canada Grant MT 5198 (to MB). D.A.A. received assistance from NIH training grant MH 15788.

REFERENCES

Aquino DA, Bisby MA, Ledeen RW (1985a) J Neurochem 45: 1262-1267.

Aquino DA, Ledeen RW, Bisby MA (1985b) Trans Amer Soc Neurochem 16: 134.

Bisby MA (1976) Exp Neurol 50: 628-640.

Bisby MA (1977) J Neurochem 28: 249-251.

Bisby MA (1980) In: Federoff S and Herz L (eds): Advances in Cellular Neurobiology. Academic Press, New York, pp. 69-117.

Bulger VT and Bisby MA (1978) J Neurochem 31: 1411-1418.

Byrne MC, Ledeen RW, Roisen FJ, Yorke G, Sclafani JR (1983) J Neurochem 41: 1214-1222.

Byrne MC, Sbaschnig-Agler M, Aquino DA, Sclafani JR, Ledeen RW (1985) Anal Biochem 148: 163-173.

Forman DS and Ledeen RW (1972) Science 177: 630-633.

Gammon GM, Goodrum JF, Toews AD, Okabe A, Morell P (1985) J Neurochem 44: 376-387.

Landa CA, Maccioni HJF, Caputto R (1979) J Neurochem 33: 825-838.

Ledeen RW (1978) J Supramolec Struct 8: 1-17.

Ledeen RW (1983) In: Lajtha A (ed): Handbook of Neurochemistry. Vol. 3, Plenum Press, New York, pp. 41-90.

Ledeen RW (1985) TINS 8: 169-174.

Ledeen RW, Sbaschnig-Agler M, Pfenninger K (1985) Trans Amer Soc Neurochem 16: 266.

Ledeen RW, Skrivanek JA, Nunez J, Sclafani JR, Norton WT, Farooq M (1981) In: Rapport MM and Gorio A (eds): Gangliosides in Neurological and Neuromuscular Function, Development, and Repair. Raven Press, New York, pp. 211-223.

Mansson J-E, Rosengren B, Svennerholm L (1985) J Chromatogr 322: 465-472.

Rosner H, Wiegandt H, Rahmann H (1973) J Neurochem 21: 655-665.

Vance DE and Sweeley CC (1967) J Lipid Res 8: 621-630.

Williams MA and McCluer RH (1980) J Neurochem 35: 266-269.

Yamamoto M, Boyer AM, Schwarting G (1985) Proc Natl Acad Sci USA 82: 3045-3049.

Yates AJ, Tipnis UR, Hofteig JH, Warner JK (1984) In: Ledeen R, Yu RK, Rapport MM, Suzuki K (eds): Ganglioside Structure, Function and Biomedical Potential. Plenum Press, New York, pp. 155-168.

Yusuf HKM, Pohlentz G, Schwarzmann G, Sandhoff K (1984) J Neurosci Res 12: 161-178.

Gangliosides and neuronal plasticity
G. Tettamanti, R.W. Ledeen, K. Sandhoff,
Y. Nagai, G. Toffano (eds.)
Fidia Research Series, vol. 6
Liviana Press, Padova, © 1986

CONCEPTS OF GANGLIOSIDE METABOLISM

U. Hinrichs, S. Sonderfeld, G. Schwarzmann, E. Conzelmann, K. Sandhoff

Institut f. Organische Chemie und Biochemie, Universität Bonn,
Gerhard Domagkstr. 1, 53 Bonn, Federal Republic of Germany

INTRODUCTION

Gangliosides (Fig. 1), although characteristic components of all mammalian plasma membranes, are highly abundant in nervous tissues only. They are arranged asymmetrically in the outer leaflet of cellular membranes, their carbohydrate chains facing the extracellular space. Little is known about their function and about the regulation of their metabolism, the elucidation of which might be helpful in understanding the pathogenetic mechanisms underlying the clinical symptoms of ganglioside storage diseases.

SCHEME OF GANGLIOSIDE METABOLISM

The main steps of ganglioside metabolism are shown in Figure 2. They are synthesized by membrane-bound glycosyltransferases mainly in the lumen of the Golgi apparatus, starting from their ceramide portion, to which activated sugars are bound in a stepwise manner. Water soluble sugar nucleotides formed in the cytosol are transported across the Golgi membrane by carrier proteins. For some activated sugars this uptake can be inhibited by tunicamycin (Yusuf et al., 1983 a,b; Sommers and Hirschberg, 1982).

After synthesis gangliosides reach the plasma membranes presumably by vesicular membrane flow. Here oligosialogangliosides can partially be degraded to the monosialoganglioside GM1 by a membrane-bound sialidase, which is in the same membrane, as has been found by in vitro studies (Scheel et al., 1982) (Fig. 3). The enzyme-substrate interaction depends on the lipid bilayer's physical properties. Membrane perturbing agents like primary alcohols or general anaesthetics which increase membrane fluidity also enhance degradation of oligosialogangliosides without stimulating the en-

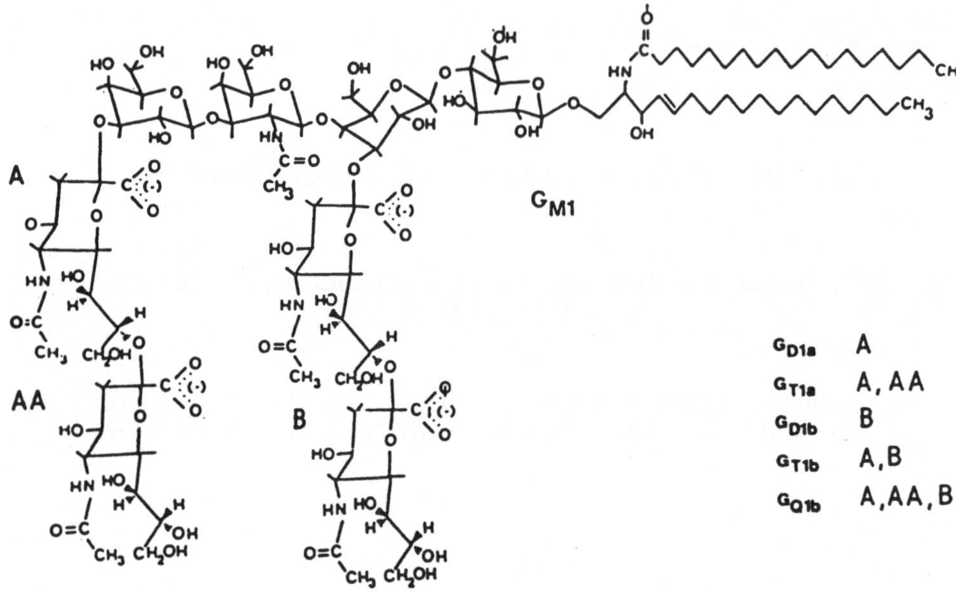

Figure 1. Mammalian brain gangliosides with GM1 as their basic structure. GM1: basic structure; GD1a: basic structure and residue A; GT1a: basic structure and residues A, AA; GD1b: basic structure and residue B; GT1b: basic structure and residues A, B; GQ1b: basic structure and residues A, AA and B.

zyme itself (Scheel et al., 1985). Those agents are thought to facilitate the interaction between membrane-bound substrate and enzyme, probably by enhancing the lateral diffusion of both the reaction partners and/or by melting membrane domains containing glycolipids.

Gangliosides are completely degraded by lysosomal enzymes. Starting from the non-reducing end of the carbohydrate chain, several exoglycosidases clip off individual sugars in a stepwise manner up to the ceramide moiety (Sandhoff and Christomanou, 1979) which is subsequently catabolized to sphingosine and a fatty acid. Sphingosine as well as the fatty acids can be utilized in other subcellular compartments for biosynthetic routes (Schwarzmann et al., 1984). The degradation of membrane-bound gangliosides by water soluble enzymes is made possible by activator proteins. These acid glycoproteins extract specific gangliosides from the lysosomal membranes, the resulting ganglioside-activator complex then serves as the sugar hydrolases' true substrate (Sandhoff, 1984). Activator proteins have been described for the degradation of sulfatides by arylsulfatase A (Mehl and Jatzkewitz, 1964), of ganglioside GM1 by β-galactosidase (Li and Li, 1976) and of ganglioside GM2 by hexosaminidase A (Conzelmann and Sandhoff, 1978; 1979). Figure 4 depicts the lysosomal degradation of a membrane-bound ganglioside (for example GM2) by a water-soluble enzyme. The high affinity of the GM2-activator protein for ganglioside GM2 has been demonstrated by affinity labeling with N-bromoacetyl-[^{3}-H]-lyso GM2, which binds covalently to the activator protein (Neuenhofer and Sandhoff, 1985).

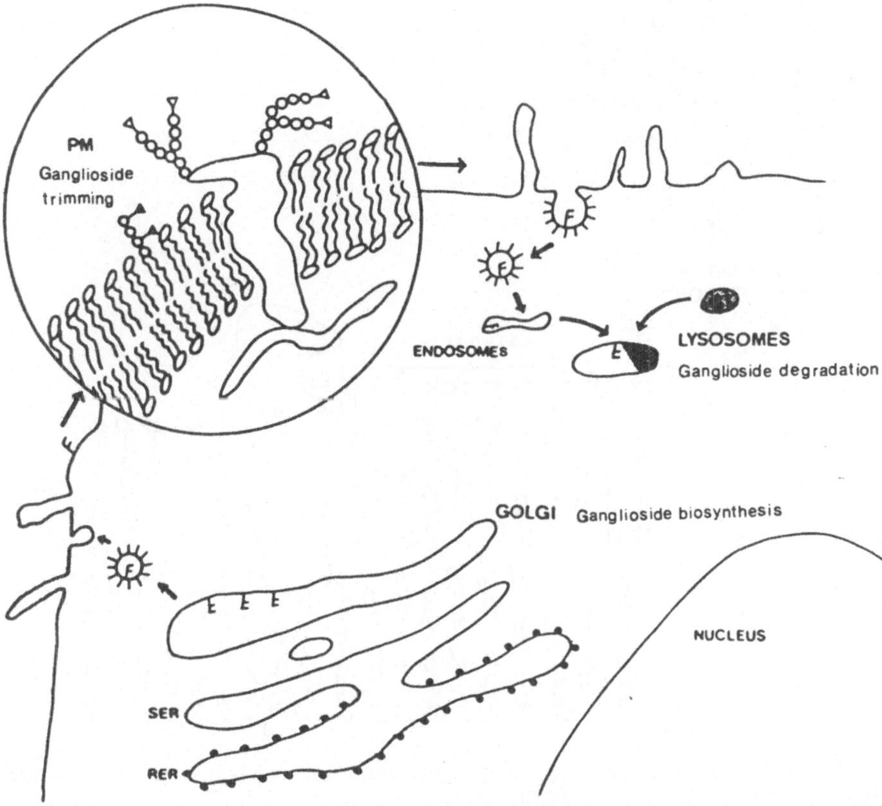

Figure 2. Main steps of ganglioside metabolism. Gangliosides are synthesized by membrane bound glycosyltransferases in the lumen of Golgi vesicles. In plasma membranes (PM) oligosialogangliosides can be degraded to the monosialoganglioside GM1 by a membrane bound sialidase. Final catabolism is achieved in lysosomes by exoglycosidases and activator proteins. SER, smooth endoplasmic reticulum; RER, rough endoplasmic reticulum.

METABOLISM OF GM2 IN SKIN FIBROBLASTS

In order to get closer insight into the interplay of the various metabolic routes, we have investigated the metabolism of exogenous gangliosides in cultured cells. Electron spin resonance studies have already shown that a small fraction of exogenously added gangliosides is fully incorporated into the plasma membranes, whereas a much larger portion is only attached to the cellular surfaces in a trypsin-removable fashion (Schwarzmann and Sandhoff, 1983).

[^{3}H]-GM2 and [^{3}H]- GD1a labeled in their sphingoid moiety (Schwarzmann, 1978) were added to the culture media of normal and mutant skin fibroblasts at a concentration of 5×10^{-5}M. As ganglioside insertion depended strongly on the media's serum content, serum concentration was reduced to 0,3% (Sonderfeld et al., 1985).

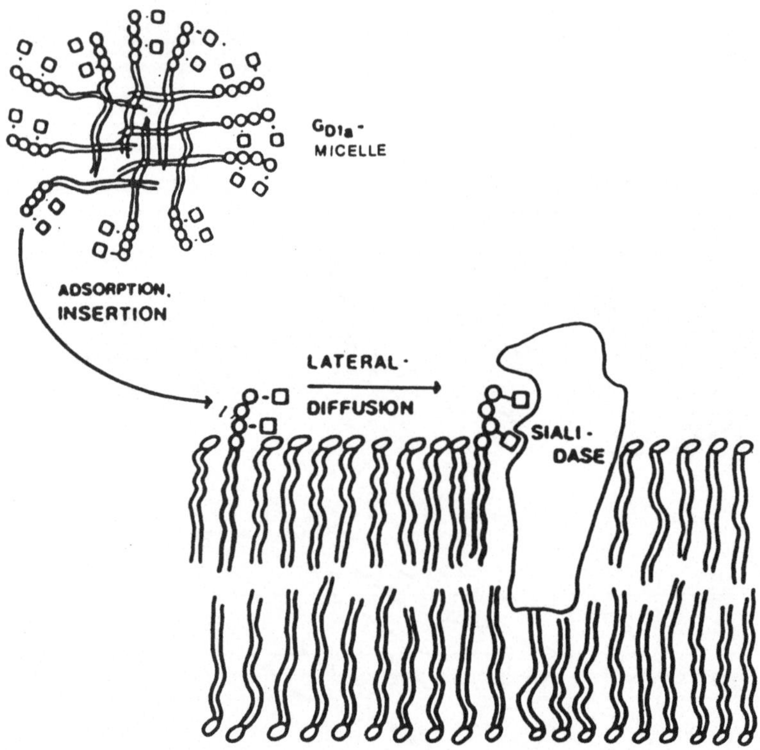

Figure 3. Interaction of exogenous ganglioside GD1a and membrane bound sialidase (Sandhoff and Pallmann, 1978; Scheel et al., 1985). In water GD1a forms micelles which partly adsorb to the sialidase-containing neuronal membranes. A fraction of exogenous gangliosides is inserted into the membrane as shown by electron spin resonance spectroscopy. The inserted molecules can diffuse laterally to the enzyme and can subsequently be degraded.

After various incubation times cells were harvested by trypsinization and lipid extracts were analyzed by thin layer chromatography either directly or after separation according to charge by ion exchange chromatography. Individual radioactive bands were visualized by fluorography. According to kinetic studies skin fibroblasts from normal probands as well as from patients with GM2-gangliosidosis variant 0 incorporated 13-14 nmol GM2/mg protein after 92h. Apart from a somewhat more rapid initial phase, the amount of ganglioside taken up by cells in a trypsin resistant fashion increased almost linearly with time (Sonderfeld et al., 1985). Similar results were obtained after [^{3}H]-GM3 administration to normal fibroblasts and to cells of patients with various glycolipid storage diseases (Schwarzmann and Sandhoff, 1983; Schwarzmann et al., 1984). In normal fibroblasts exogenous [^{3}H]-GM2 was catabolized to [^{3}H]-GM3, [^{3}H]-LacCer, [^{3}H]-GlcCer and [^{3}H]-ceramide, whereas in GM2-gangliosidosis variant B and variant 0 cells only traces of degradation products, mainly [^{3}H]-GA2,

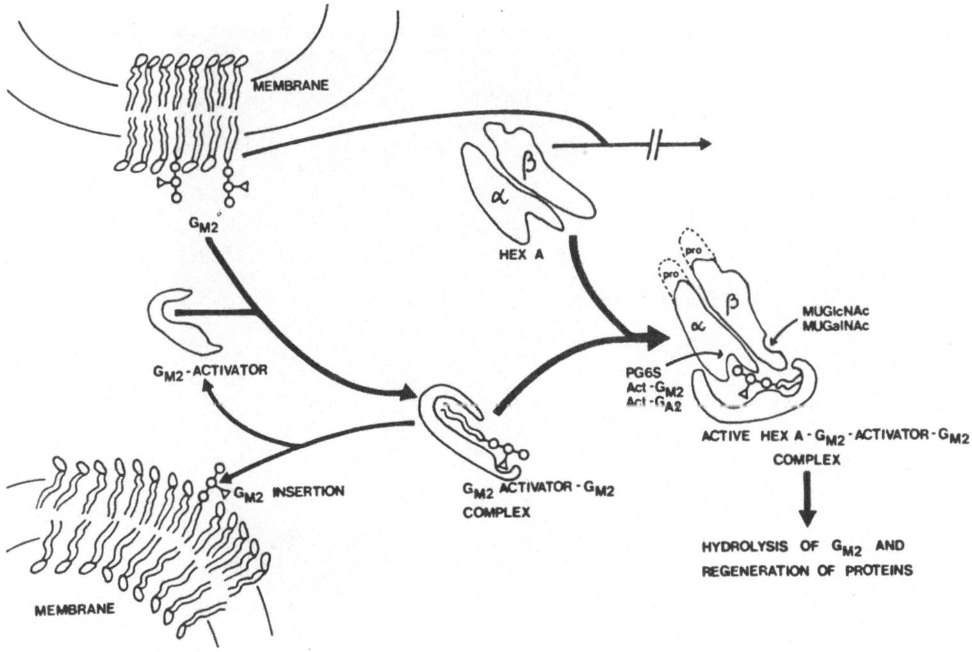

Figure 4. Function of GM2-activator protein as glycolipid transferprotein and as stimulator for GM2 degradation by hexosaminidase A. GM2-activator protein acts as glycolipid transfer protein and as stimulator for the GM2 degradation by hexosaminidase A (Sandhoff, 1984). Hexosaminidase A is a heteropolymer composed of subunits, named α and β. Each subunit has an active site. The active site of the β-subunit degrades mainly water soluble N-acetylglucosaminides and N-acetylgalactosaminides; the active site of the α-subunit mainly degrades a sulfated N-acetylglucosaminide (PG 6 S) and the glycolipids GM2 and GA2 when bound to the GM2-activator protein. MU GlcNAc, 4-methylumbelliferyl-N-acetilglucosaminide, MUGalNAc, 4 methylumbelliferyl-N-acetylgalactosaminide.

were found (Fig. 5). Administration of the GM2- activator molecule to GM2-gangliosidosis variant AB cells, which lack this protein, fully restored catabolic activity to the cultures (Fig. 6). Apart from the above mentioned catabolic metabolites biosynthetic compounds, i.e. [^{3}H]-sphingomyelin, [^{3}H]-phosphatidyl choline, [^{3}H]-GM1 and [^{3}H]-GD1a, were detected in normal cells. The latter two occured in similar amounts in normal cells as well as in mutant fibroblasts (Figs. 5 and 6). Feeding of variant AB cells with both [^{3}H]-GM2 and the activator protein did not increase the production of [^{3}H]-GD1a significantly (Fig. 6). These data indicate that formation of [^{3}H]-GD1a results from direct glycosylation of intact [^{3}H]-GM2 in the Golgi compartment (Sonderfeld et al., 1985) besides reutilisation of a lysosomal catabolite as has been found for [^{3}H]-GD1a formation from [^{3}H]-GM3 in normal cells (Schwarzmann et al., 1984).

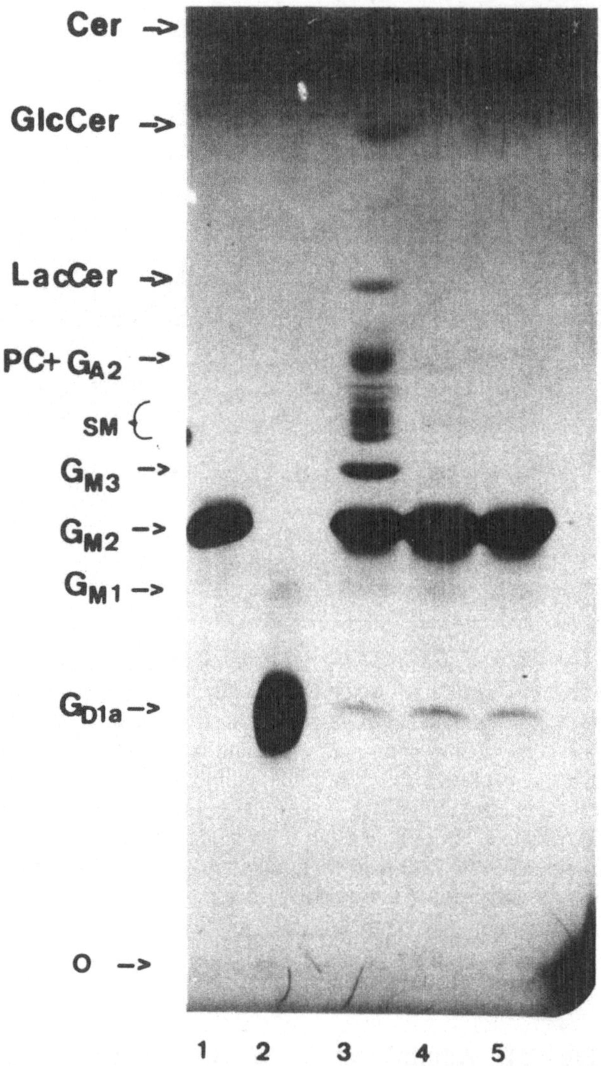

Figure 5. Metabolism of exogenously added [^{3}H]-GM2 in fibrobasts. Skin fibroblasts from normal controls and from patients with GM2-gangliosidosis, variant B and variant 0, were fed with [^{3}H]GM2 for 26h, then harvested and extracted. The lipids were separated by thin layer chromatography and the radioactive spots were visualized by fluorography. Lane 1: standard ganglioside GM2; lane 2: standard ganglioside GD1a; lane 3: normal controls; lane 4: GM2-gangliosidosis variant 0 cells; 0: origin; SM: sphingomyelin; PC: phosphatidyl choline; Glc Cer: glucosylceramide; Lac Cer: lactosylceramide (Sonderfeld et al., 1985).

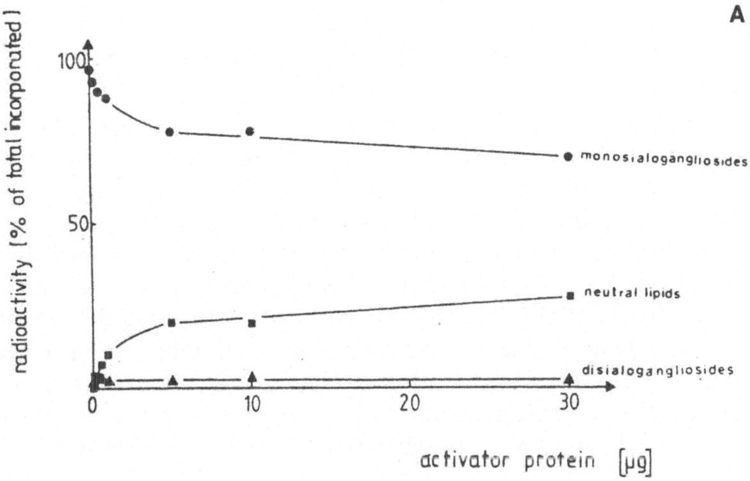

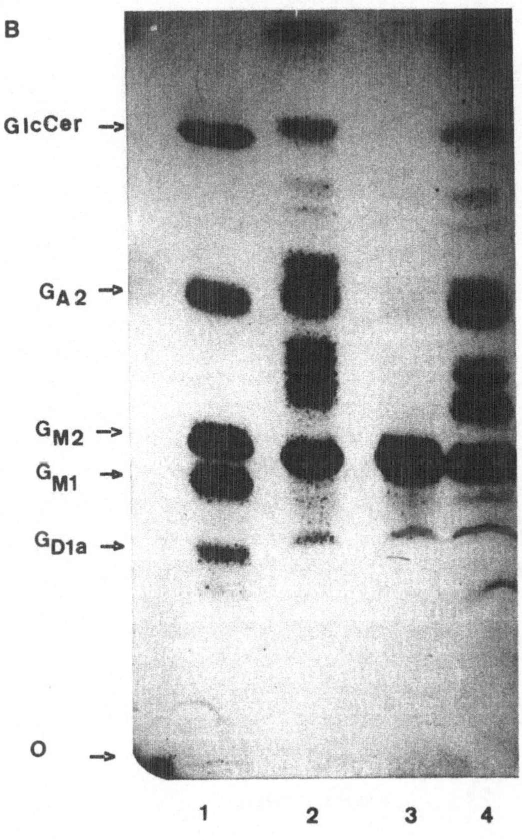

Figure 6. Metabolism of [^{3}H]-GM2 in cells from GM2-gangliosidosis, variant AB, as function of exogenously added GM2-activator protein. The fibroblasts were fed with [^{3}H]-GM2 (50 μM) and the indicated amounts of purified GM2-activator protein. After 70h the cells were harvested and extracted. A: Quantitative distribution of radioactivity between different lipid classes. The lipids were separated on DEAE-Sepharose Cl 6 B into a neutral lipid fraction (■), monosialoganglioside fraction (•) and disialoganglioside fraction (▲). The radioactivity of each fraction was measured by scintillation counting. B: Separation of total lipid extracts by thin layer chromatography. Lane 1: Standards GD1a, GM1, GM2, GA2, glycosylceramide (GlcCer); lane 2: total lipid extracts of normal cells; lane 3: total lipid extracts of cell from GM2-gangliosidosis, variant AB; lane 4: total lipid extracts of cells from GM2-gangliosidosis, variant AB after feeding 30 μg GM2-activator protein for 70h; 0: origin (Sonderfeld et al., 1985).

ROUTES OF INTRACELLULAR TRANSPORT

Routes of intracellular ganglioside flow were further investigated by subcellular fractionation studies. When GM2-gangliosidosis (infantile variant 0) cells were fed with [^{3}H]-GM2, total radioactivity increased quite similarly with incubation time in both lysosomal and light membrane fraction. Expressed as a percentage of total radioactivity taken up by the cells the share found in the lysosomal fraction remained almost constant at a fifty percent level (12-48h) after a rapid initial increase (data not shown).

After application of [^{3}H]-GD1a to GM2-gangliosidosis (infantile variant B) cells [^{3}H]-GM2 did not only accumulate in the lysosomal fraction but was also detected in substantial amounts in the light membrane fraction (Schwarzmann et al., 1984; Sonderfeld et al., 1985). These data are not compatible with the hypothesis of an unidirectional intracellular flow of gangliosides [^{3}H]-GM2 and [^{3}H]-GD1a from the plasma membrane to the lysosomes and subsequent total degradation.

Instead, the results suggest that the exogenous gangliosides as well as their degradation products were transported bidirectionally between the two subcellular compartments (Fig. 7).

A

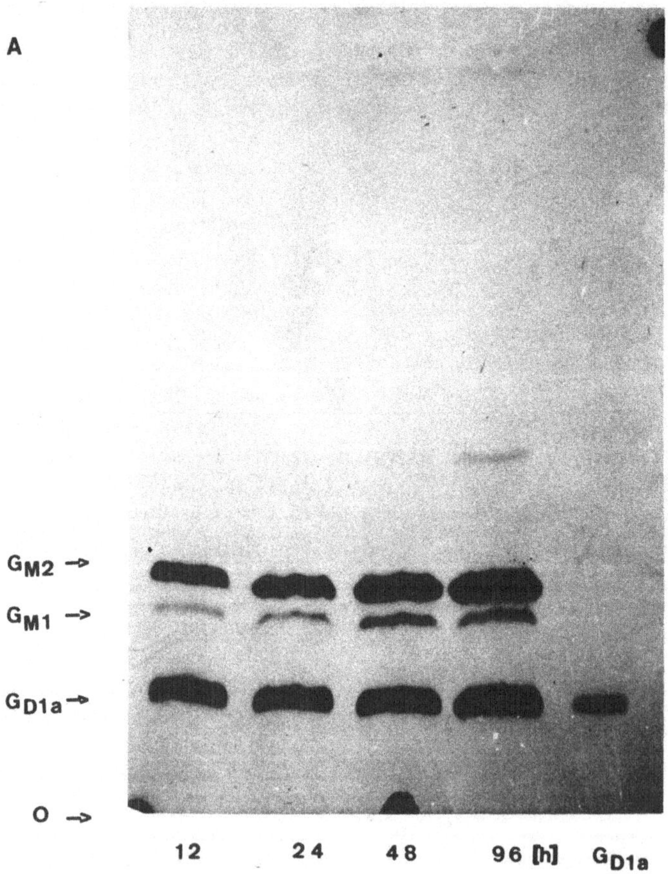

Figure 7a. See caption on opposite page.

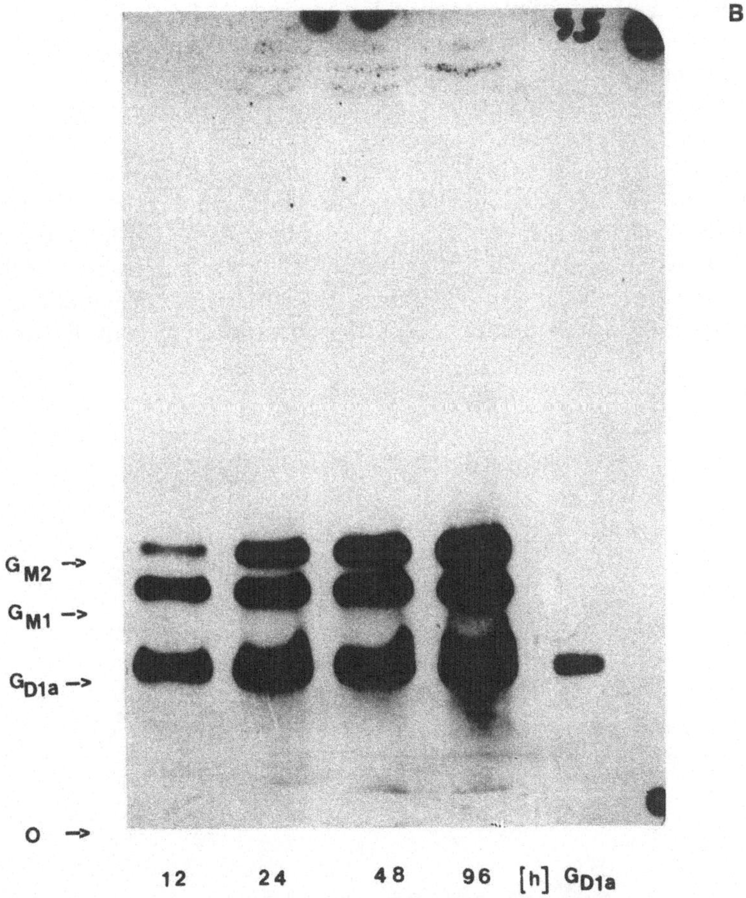

Figure 7. Radioactive lipids in lysosomal and light membrane fractions of cells from GM2-gangliosidosis, variant B. Cells were incubated at 37°C with 50μM [^{3}H]-GD1a for the tissues indicated in the figures, harvested and fractioned. Lipid extracts of lysosomal (A) and light membrane fractions (B) were subjected to thin layer chromatography and the radioactive spots were visualized by fluorography. 0: origin (Sonderfeld et al., 1985).

GANGLIOSIDE METABOLISM IN BRAIN CELLS

Glycolipid metabolism of primary murine cerebellar cultures derived from 6 day old animals (Hatten and Liem, 1981) and — for comparison — of human skin fibroblasts were determined by feeding cultures with [^{14}C]-galactose for 48h. Further experimental conditions corresponded to those described above for [^{3}H]-GM2 incorporation. After 4, 12 and 18 days in vitro cerebellar cultures exhibited a significantly different glycolipid pattern than did fibroblasts (Fig. 8). Brain cells mainly contained the more complex gangliosides [^{14}C]-GD1a and [^{14}C]-GT1b, whereas in fibroblasts neutral glycolipids and monosialogangliosides dominated. Cerebellar glycolipid profiles were fully developed after 4 days in vitro (Fig. 8) and corresponded largely to those found in the cerebelli of 6 and 7 day old mice in vivo (data not shown). Reduction of serum concentration from 10% (standard conditions) to 0,3% did not significantly influence glycolipid synthesis (data not shown).

To investigate the metabolism of exogenous gangliosides both cell types were incubated with 1×10^{-5}M[^{3}H]-GM2 and [^{3}H]-GD1a, labeled in their sphingoid base, for 48h, the media's serum concentration again amounting to 0,3%. Both cell types mainly degraded the exogenous gangliosides, significant differences in metabolic patterns being obvious. Thus, cerebellar cells contained less [^{3}H]-glycosylceramide and [^{3}H]-GM3 after [^{3}H]-GM2 as well as after [^{3}H]-GD1a incorporation than did fibroblasts.

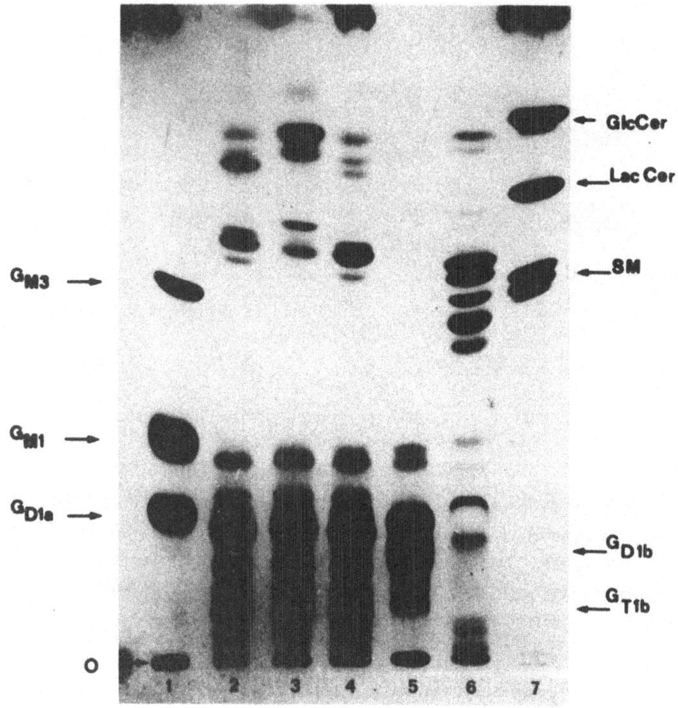

Figure 8. Incorporation of [^{14}C]-galactose into the glycolipids of cerebellar and fibroblast cultures. Cells were incubated with $4{,}2 \times 10^{-5}$M [^{14}C]-galactose for 48h, harvested and processed as described for Figure 5. Lane 1: standard gangliosides GD1a, GM1, GM3; lanes 2, 3 and 4: total lipid extracts of cerebellar cells after 4, 11 and 18 days in vitro, respectively; lane 5: standard gangliosides GT1b, GD1b, GD1a, GM1; lane 6: total lipid extracts of confluent fibroblast cultures; lane 7: standard sphingolipids sphingomyelin, lactosylceramide (LacCer), glycosylceramide (GlcCer). For symbols see Figure 5.

In contrast, feeding with [^{3}H]-GD1a yielded significantly more [^{3}H]-GM1 in brain cultures than it did in fibroblasts (Fig. 9). As for biosynthetic products, cells metabolized exogenous [^{3}H]-GM2 to [^{3}H]-sphingomyelin, [^{3}H]-GM1, [^{3}H]-GD1a (see above) and [^{3}H]-GT1b and exogenous [^{3}H]-GD1a to [^{3}H]-sphingomyelin and [^{3}H]-GD1a. In both cases [^{3}H]-GT1b was only formed by cerebellar cells (Fig. 9). These data again

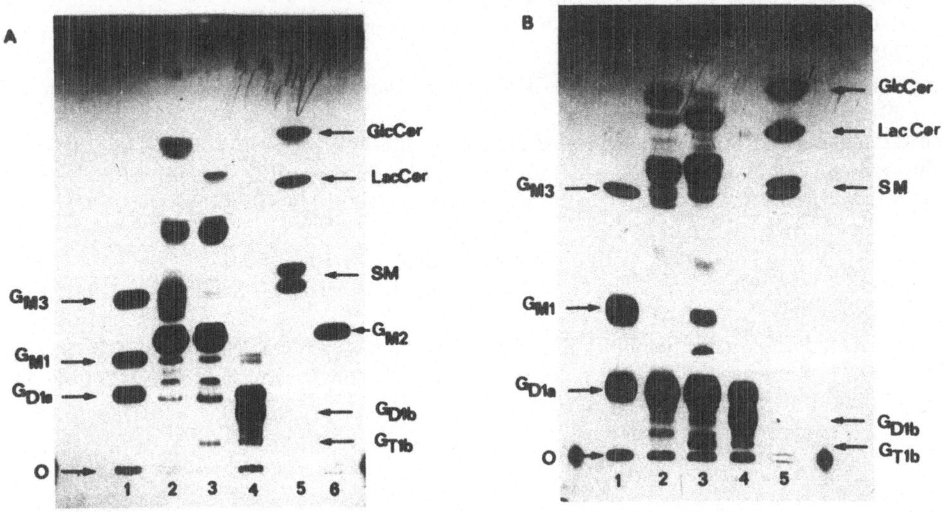

Figure 9. Metabolism of exogenously added [^{3}H]-GM2 and ^{3}H-[GD1a] in cerebellar cells and in fibroblasts. Cells were incubated with 1×10^{-5}M [^{3}H]-GM2 (A) and [^{3}H]-GD1a (B) for 48h, harvested and processed as described for Figure 5. A: lane 1: standard gangliosides GD1a, GM1, GM3; lane 2: total lipid extracts of fibroblasts; lane 3: total lipid extracts of cerebellar cells; lane 4: standard gangliosides GT1b, GD1b, GD1a, GM1; lane 5: standard sphingolipids sphingomyelin, lactosylceramide, glucosylceramide; lane 6: standard ganglioside GM2. B: lanes 1, 2, 3: see Figure 9 A; lane 4: standard gangliosides GT1b, GD1b, GD1a; lane 5: see Figure 9 A. For symbols see Figure 5.

indicate that, after insertion into the plasma membrane, exogenous gangliosides, are not exclusively transported to and fully degraded in the lysosomes, but that also the intact molecules as well as their metabolic products can enter the glycolipid biosynthetic pathways in the Golgi. The routes of intracellular ganglioside flow in cerebellar cells as described above for fibroblasts as well as the regulation of ganglioside metabolism remain to be investigated.

REFERENCES

Conzelmann E and Sandhoff K (1978) Proc Natl Acad Sci USA 75: 3979-3983.
Conzelmann E and Sandhoff K (1979) Hoppe-Seyler's Z Physiol Chem 360: 1837-1849.
Hatten ME and Liem RKH (1981) J Cell Biol 90: 622-630.

Li S-C and Li Y-T (1976) J Biol Chem 251: 1159-1163.
Mehl E and Jatzkewitz H (1964) Hoppe-Seyler's Z Physiol Chem 339: 260-276.
Neuenhofer S and Sandhoff K (1985) FEBS Lett 185: 112-114.
Sandhoff K and Pallmann B (1978) Proc Natl Acad Sci USA 75: 122-126.
Sandhoff K and Christomanou H (1979) Hum Genet 50: 107-143.
Sandhoff K (1984) in: Barranger JA and Brady RO (eds): The Molecular Basis of Storage Disorders. Academic Press, Washington, pp. 19-49.
Scheel G, Acevedo E, Conzelmann E, Nehrkorn H, Sandhoff K (1982) Eur J Biochem 127: 245-253.
Scheel G, Schwarzmann G, Hoffmann-Bleihauer P, Sandhoff K (1985) Eur J Biochem 153: 29-35.
Schwarzmann G (1978) Biochim Biophys Acta 529: 106-114.
Schwarzmann G and Sandhoff K (1983) in: Chester MA, Heinegård D, Lundbland A, Svensson S (eds): Proceedings of the 7th International Symposium on Glycoconjugates. University of Lund, Sweden, pp. 238-239.
Sommers CW and Hirschberg CD (1982) J Biol Chem 257: 10811-10817.
Sonderfeld S, Conzelmann E, Schwarzmann G, Burg J, Hinrichs U, Sandhoff K (1985) Eur J Biochem 149: 247-255.
Yusuf HKM, Pohlentz G, Sandhoff K (1983a) Proc Natl Acad Sci USA 80: 7075-7079.
Yusuf HKM, Pohlentz G, Schwarzmann G, Sandhoff K (1983b) Eur J Biochem 134: 47-54.

Gangliosides and neuronal plasticity
G. Tettamanti, R.W. Ledeen, K. Sandhoff,
Y. Nagai, G. Toffano (eds.)
Fidia Research Series, vol. 6
Liviana Press, Padova, © 1986

METABOLISM OF EXOGENOUS GM1 AND RELATED GLYCOLIPIDS IN THE RAT

R. Ghidoni, M. Trinchera, B. Venerando, A. Fiorilli, G. Tettamanti

Study Center for the Functional Biochemistry of Brain Lipids, Department of Medical Chemistry and Biochemistry, The Medical School, University of Milan, via Saldini 50, 20133, Milan, Italy

INTRODUCTION

Knowledge of the fate and metabolic pathways run by gangliosides administered to animals is interesting from both the pharmacological and biochemical points of view. Ceccarelli et al. (1976) and Obata et al. (1977) first discovered that exogenous gangliosides facilitate development of neuromuscular junctions in *in vivo* and *in vitro* systems. Since then many reports definitely established the ability of exogenous gangliosides to induce the formation of neurite-like processes in a number of neural cells cultivated *in vitro*, and to facilitate functional recovery of damaged peripheral and central nervous system by promoting neuronal survival, nerve regeneration and reinnervation (for review see Ledeen, 1985). The molecular mechanisms underlying these effects are unknown and possibly connected with the physiological performances of the endogenous gangliosides.

Studies aimed at exploring the above mechanisms must be based on definite assessment of some basic points, such as: (a) do exogenous gangliosides reach the different tissues after injection into animals; (b) how exogenous gangliosides interact with cells in *in vivo* and *in vitro* systems; (c) do exogenous gangliosides penetrate into cells; (d) are exogenous gangliosides metabolized by cells after uptake and internalization. Most of these points have been cleared. It is known for instance that exogenous gangliosides bind to cells in culture (Radsak et al., 1982) and become incorporated into the external lipid leaflet of the cell membrane (Moss et al., 1976; Schwarzmann et al., 1983), where they can mimick the activity of endogenous gangliosides (Moss et al., 1976). As well, gangliosides injected into animals were shown to be absorbed and distributed to different organs, and to bind to plasma and intracellular membranes (Tettamanti et al.,

---

Abbreviations: DPQ, dichloro-diciano benzo quinone; HPLC, high performance liquid chromatography; Sph, spningosine; EDTA, ethylene diamino tetracetic acid; t.l.c., thin layer chromatography.

1981; Lang, 1981). Moreover exogenous gangliosides were proved to undergo metabolic processing when either added to cells cultivated in vitro (Kinders et al., 1982; Sonderfeld et al., 1985) or administered to animals (Ghidoni et al., 1983; Tettamanti et al., 1984). These evidences from one side prove that exogenous ganglioside reach enzymatic machineries that are intracellularly located, and, from the other, provide an exceptionally useful approach for studying ganglioside metabolism, especially at the subcellular level.

With the present work we focused attention on the following situations: (a) definite assessment that injected gangliosides reach the central nervous system; (b) recognition of the pattern of metabolic processing of injected gangliosides with particular emphasis to the subcellular aspects of the process; (c) the organ specificity of the metabolic processing of injected gangliosides. The reported data refer to experiments carried out on rats, at the level of brain and liver.

BASIC EXPERIMENTAL APPROACHES

The used experimental approach consists in the administration of isotopically labelled ganglioside to rats, followed by: (a) determination of the distribution and uptake of radioactivity in tissues, with particular reference to brain; and (b) recognition of the formed metabolites in different tissues (liver and brain) and at different subcellular levels.

Liver and brain were chosen for the following reasons. Liver is known (Tettamanti et al., 1981) to be capable to take up the largest amount of administered gangliosides, and is particularly suitable for preparing subcellular fractions in adequate amounts; brain is a crucial organ, with the highest content of endogenous gangliosides and with a direct involvement in the major effects produced by exogenous gangliosides.

Labelling procedures. Ganglioside GM1, prepared in pure form (over 99%) from calf brain (Tettamanti et al., 1973), was isotopically tritium labelled at the level of terminal galactose by the galactose-oxidase/$NaBH_4$ method (Ghidoni et al., 1977), or at the C-3 of the long chain base by the DDQ/$NaBH_4$ method (Ghidoni et al., 1981) followed by removal of the threostereoisomers by HPLC (Sonnino et al., 1984a). The specific radioactivity was 1.5 Ci/mmol for (Gal-^{3}H)GM1 and 1.3 Ci/mmol for (Sph-^{3}H)GM1. The radiochemical purity was better than 99% for both compounds. Glucosylceramide and galactosylceramide, supplied by Sigma Chem. Co. (St. Louis, Mo, USA), were tritiated according to the DDQ/$NaBH_4$ procedure as described by Iwamori et al. (1975).

Animals. Adult male Wistar rats (mean weight, 150 g), provided by Charles River, were employed.

Intravenous injection. The animals were intravenously injected in the tail with 50 μCi of tritiated GM1, dissolved in 100 μL of physiological saline solution.

Intracisternal injection. When brain metabolism of exogenous GM1 was investigated, the animals were intracisternally injected with 8 μl (2 μl/min) of physiological saline solution containing 50 μCi of tritiated GM1, in the absence or presence of L-arabinose (final concentration, 1.6 M), that increases the permeability of blood-brain barrier (Barranger et al., 1979).

Brain treatment. Animals were killed by heart perfusion with a physiological saline solution and the brain removed, weighed, washed and homogenized with 5 volumes of 0.25 M sucrose solution containing 1mM phosphate buffer and 0.1 mM EDTA, pH 7.2, and then centrifuged at 150,000 g for 1 h. The supernatant was accurately syphoned off and the pellet, washed once more with 2 ml of sucrose solution, centrifuged again and the new supernatant added to the previous one. The final pellet, dissolved with redistilled water, constituted the total particulate fraction. The pooled supernatants constituted the "soluble" fraction. Brain subcellular fractionation and preparation of the "plasma membrane" fraction were accomplished by the method of Morgan et al. (1971). Brain microvessels were prepared according to Hjelle et al. (1978).

Liver treatment. Animals were anaesthetized and killed by liver perfusion with a heparin-containing Krebs buffer. Liver was removed, weighed, washed and: a) lyophilized and submitted to chemical analysis; or b) homogenized with buffered isotonic sucrose solution and submitted to preparation of the total particulate and soluble fractions (see above) or to systematic subcellular fractionation. In particular a highly purified "lysosomal" fraction and "Golgi apparatus" fraction were prepared. These fractions were obtained by the methods of Sawant et al. (1964) and of Sandberg et al. (1980), respectively.

Lipid and ganglioside analysis. Specimens from liver and brain were submitted to lipid extraction and fractionation into individual entities (Tettamanti et al., 1973; Stoffel and Sticht, 1967). The individual gangliosides and non-gangliosidic lipids (neutral glycolipids; ceramide; sphingomyelin) were counted for incorporated radioactivity (for details see Ghidoni et al., 1983).

Determination of radioactivity. Radioactivity was determined by liquid scintillation counting, radiochromatoscanning, and fluorography (for details see Ghidoni et al., 1983).

Calculation of Relative Specific Radioactivity (RSR). The relative enrichment of radioactive materials in any subcellular fraction was assessed by determining the Relative Specific Radioactivity. This was referred to the whole homogenate and was calculated similarly to the Relative Specific Activity (RSA) of enzymes, with the enzyme activity being substituted with the radioactivity carried by the individual lipids.

UPTAKE OF EXOGENOUS GM1 IN LIVER AND BRAIN

The distribution and uptake of injected radioactivity in the liver and brain of rats essentially followed the trend described in mice (Tettamanti et al., 1981).After intravenous injection of (Sph-^{3}H)GM1, liver retained a much higher amount of radioactivity (18% of total administered at 4hr and 10% at 40hr after administration) than brain (1.6% at 4h and 0.44% at 40hr after administration). When (Gal-^{3}H)GM1 was employed approximatively the same proportions of radioactivity uptake were maintained between the two organs but the absolute uptake was 10-30%, depending on the time after injection, of that recorded with (Sph-^{3}H)GM1. This difference reflects the different metabolism and much greater turnover of ^{3}H-galactose than ^{3}H-sphingosine. It should be emphasized that the radioactivity measured in the two organs is authentic tissue-linked radioactivity and not due to blood contamination. Infact animals were killed by heart or liver perfusion in order to remove blood and, anyhow, the amount of radioactivity per gr (or ml) unit was definitely much greater in liver and brain than in blood (Table 1). Volatile radioactivity (mostly tritiated water) began being produced soon after injection, indicating the early occurrence of metabolic processing of exogenous GM1. Much more volatile radioactivity was formed after injection of (Gal-^{3}H)GM1 than (Sph-^{3}H)GM1, confirming the different metabolic involvement of ^{3}H-galactose and ^{3}H-sphingosine. With regards to production of tritiated water from (Gal-^{3}H)GM1, liver displayed a different behaviour than brain (Table 2). In the liver radioactivity was proportionately higher in the particulate fraction (that does not carry volatile radioactivity), especially at the longer times after injection, while in the brain the radioactivity was preponderant in the soluble fraction (that is rich in volatile radioactivity). However the soluble fraction from liver contained almost only volatile radioactivity, while that prepared from brain contained a substantial amount (3-10%, depending on the time after injection) of non-volatile radioactivity corresponding to unmodified tritiated GM1. This may suggest that in brain the metabolic pool of cytosolic or soluble gangliosides (Sonnino et al., 1984b) is much greater than in the liver. On the other hand (Table 3) the persistence of radioactive GM1 in brain is much longer than in liver. Therefore it should be inferred that GM1 remains associated to the tissue structures longer in brain than in liver; however, once it has been introduced into the metabolic machinery of the tissue it is catabolized to water faster in brain than in liver.

The penetration of injected and taken up GM1 into cells has been assessed by subcellular studies. These showed that in rats, as well as in mice (Tettamanti et al., 1981), and in both liver and brain, radioactivity is bound to a number of subcellular fractions. With particular reference to brain (Table 4) the soluble fraction carried the highest specific radioactivity, in terms of both total radioactivity (volatile plus non volatile) and non-volatile radioactivity, followed by the plasma membrane fraction that contained only non-volatile radioactivity. Non-volatile specific radioactivity in the soluble fraction remained somewhat higher than that of the plasma membrane fraction (7,690 DPM/mg protein versus 5,600). Noteworthy, a purified preparation of brain microvessels carried a substantial amount of radioactivity all of which in the non-volatile form (specific radioactivity, 6,050 DPM/mg protein). Since the specific radioactivity of blood cells under the same experimental conditions was several fold lower than that of brain microvessels, the radioactivity associated to microvessels likely corresponded to GM1 molecules tighly bound to the vessel walls.

All these findings lead to conclude that injected GM1 is taken up by the brain, binds to the capillary network, penetrates into neural cells, associates to both plasma

Table 1. *Radioactivity linked to rat brain and blood after 40 hr from a single injection of (Gal-$_3$H) GM1 (50 μCi/animal). The data refer to 1 g of fresh tissue, or ml of blood, and are the mean values of 3 experiments. Standard deviation was less than 10% of the mean value.*

TISSUE	DPM/G FRESH TISSUE
Liver	308,000
Brain	204,000
Blood	150,500
plasma	124,500
cells	26,000

Table 2. *Time course of total radioactivity in rat liver and brain following intravenous injection of (Gal-$_3$H) GM1 (50 μCi/animal). The percent distribution of radioactivity in the "total particulate" and "soluble" fractions is shown. The "soluble" fraction contains practically all the volatile radioactivity present in the starting tissue. The data refer to 1 mg of fresh tissue and are the mean values of 3 experiments. Standard deviation was less than 10% of the mean value.*

	LIVER		BRAIN	
	4h	40h	4h	40h
Total DPM	1,620	308	761	204
Particulate fraction (%)	55	58	19	10
Soluble fraction (%)	45	42	87	90
Volatile (%)	44	40.5	78	87
Non volatile (%)	1	1.5	9	3

Table 3. *Persistence of GM1 in rat brain and liver after a single intravenous injection of (Gal-^{3}H) GM1 (50 μCi/animal). The data indicate the percentage of radioactivity linked to GM1 over total non volatile radioactivity. The data are mean of 3 independent experiments. Standard deviation was less than 10% of mean value.*

TIME AFTER INJECTION (hr)	BRAIN	LIVER
4	69.1	31.2
8	54.4	21.6
16	42.5	12.8
24	32.0	10.3
40	20.9	4.1

Table 4. *Subcellular distribution of radioactivity in rat brain after 4 hr from intravenous injection of (Gal-^{3}H) GM1 (50 μCi/animal). Protein recovery after subfractionation ranged from 84 to 90%. The data are referred to 1 g starting fresh tissue. Data on isolated brain capillaries are also given. Values are mean of 3 independent experiments. Standard deviation was less than 10% of the mean value.*

FRACTION	TOTAL DPM	% (*)	SPECIFIC RADIO-ACTIVITY (**)
Homogenate	761,600	—	17,505
Nuclear fraction$^{(+)}$	17,515	2.3	2,250
Mitochondrial fraction (P2)	34,270	4.5	3,200
Microsomal fraction$^{(+)}$	47,980	6.3	5,100
Soluble fraction			
Total radioactivity	616,895	81.0	61,520
Non-volatile radio-activity	77,110	—	7,690
Recovery	716,660	94.1	—
Plasma membranes$^{(+)}$	—	—	5,600
Brain microvessels$^{(+)}$	—	—	6,050

(*) Referred to homogenate and calculated on total radioactivity
(**) DPM/mg protein
(+) The radioactivity associated to this fraction is almost completely non volatile.

and intracellular membranes and partly resides in soluble form in the cytosol or/and the interstitial fluid. Although the quantity of brain-linked exogenous GM1 is relatively small as compared to that taken up by liver, GM1 tends to remain longer in brain that in liver. On the other hand the velocity of GM1 metabolic processing or, more precisely, of GM1 degradation seems to be higher in brain than in liver, provided that the ganglioside has reached the intracellular site responsible for the process.

It should be reminded that detailed studies on the metabolic processing of exogenous ganglioside in brain require the presence of sufficiently high amounts of radioactivity in this tissue. To this purpose (Table 5), the use of intracisternal injection of radiolabelled ganglioside, especially under conditions that render more permeable the blood brain barrier (addition of 1.6 M arabinose), seems adequate. In fact using this route of administration the amount of bound radioactivity can increase up to 7-fold (from 480 DPM/mg fresh tissue to 3,510) with (Sph-^{3}H)GM1. Of course these findings clearly indicate that the blood-brain barrier plays a very selective role in the penetration of exogenous ganglioside into brain.

THE USE OF EXOGENOUS GANGLIOSIDE FOR INSPECTING THE INTRACELLULAR ASPECTS OF GANGLIOSIDE METABOLISM

Ganglioside metabolism has been mainly inspected by enzymological investigations and incorporation studies that used radioactive precursors (Tettamanti, 1984). According to the results obtained in many tissues gangliosides appear to be biosyn-

Table 5. *Incorporation of radioactivity in rat brain after 40 hr from intravenous or intracisternal injection of (Sph-^{3}H) GM1 or (Gal-^{3}H) GM1 (in each case, 50 μCi/animal). The data refer to 1 mg of fresh tissue and are the mean of three experiments. Standard deviation was less than 10% of the mean value. The total radioactivity incorporated in whole brain is also given as percentage of total injected radioactivity.*

MODE OF INJECTION	PRECURSOR	INCORPORATED RADIOACTIVITY DPM/MG FRESH TISSUE	INCORPORATED RADIOACTIVITY WHOLE BRAIN/ TOTAL INJECTED, %
Intravenous	(Gal-^{3}H) GM1	204	0.18
Intravenous	(Sph-^{3}H) GM1	480	0.44
Intracisternal	(Sph-^{3}H) GM1	1,114	1.03
Intracisternal (plus 1.6 M arabinose)	(Sph-^{3}H) GM1	3,510	3.18

thesized at the level of the Golgi apparatus and then transported to the plasma membrane, which is the main site of their cellular location. From the plasma membrane they move to the lysosomes where they are degraded. The process of biosynthesis of gangliosides consists of sequential additions of saccharide units to a starting precursor, under the catalysis of glycosyltransferases. The degradation of gangliosides consists of sequential removal of saccharide units, starting from the non reducing terminus (Gatt, 1970). The process is catalyzed by specific glycohydrolases which eventually produce ceramide, split by ceramidase into long chain base and fatty acid (Gatt, 1970).

The mechanism by which gangliosides are transported to and from the plasma membrane is largely unknown. It was recently demonstrated (Miller-Podraza and Fishman, 1982) that in cultured neurotumor cells newly synthesized gangliosides remain inside the cell for about 20 minutes before they appear at the cell surface. These intracellular forms of gangliosides are associated with membranes and unavailable to sialidase. This suggests the possible occurrence of ganglioside-carrying vesicles with the gangliosides associated to the inner surface of the vesicle membrane. Fusion of the transport vesicle with the plasma membrane would expose gangliosides on the outer surface of the same membrane. With regard to the plasma membrane-lysosome relationships, it can be suggested that some surface events would lead a patch of the plasma membrane (carrying gangliosides) to internalize giving rise to an endosome. This fuses with a primary lysosome and forms a secondary lysosome where the degradation of gangliosides takes place. Of course other possibilities of transport, for instance mediation by specific transport proteins, can be postulated.

The approach based on precursor incorporation studies and enzymological investigations left several unsolved questions, the most relevant of which are the following: (a) possible occurrence of synchronization between Golgi-located biosynthetic and lysosomes-located degradative processes; (b) direct verification of the presence of "multiglycosyltransferases systems" for ganglioside biosynthesis and the mutual connections of the different transferases in the system; (c) the mechanism of ganglioside traffic between plasma membranes and intracellular organelles; (d) the importance of compartmentation in the regulation of ganglioside metabolism.

Some answers to these questions have been just provided by experiments in which radioactive GM1 was administered to animals and the labelled metabolites produced in various organs or tissues recognized. After injection of labelled GM1 a total lipid extract was prepared from the tissues. This extract was found to carry all the radioactivity contained in the starting tissue. The total lipid extract, obtained from whole tissue or subcellular preparations, was then fractionated into: (a) a dialysed aqueous phase, which contained gangliosides; (b) a dialysate, that contained tritiated water and diffusable substances of low molecular weight; and (c) an organic phase, which contained lipid materials of non-ganglioside nature (neutral glycolipids, neutral lipids, phospholipids etc.). The use of (Gal-^{3}H)GM1 enabled to detect any direct involvement of GM1 in the biosynthesis of more complex gangliosides; the use of (Sph-^{3}H)GM1 enabled to follow the complete pattern of GM1 degradation and of reutilization of by-products for biosynthesis purposes. The data obtained with rat liver and brain shall be presented and discussed.

Verification of the Occurrence of Ganglioside Degradation in the Lysosomes

According to enzymological studies (Gatt, 1970) ganglioside GM1 should be degraded via the sequence (Fig. 1): GM1 → GM2 (removal of terminal galactose by β-galactosidase), → GM3 (removal of N-acetylgalactosamine by β-hexosaminidase), → lactosylceramide (removal of sialic acid by sialidase), → glucosylceramide (removal of internal galactose by β-galactosidase), → ceramide (removal of ceramide-linked glucose by β-glucosidase), → long chain base and fatty acid (breakdown of the carboamidic bond by ceramidase). All the enzymes involved in the process have a lysosomal localization. Among the radioactive compounds produced after administration of (Sph-^{3}H)GM1 in rat liver (intravenous injection) and brain (intracisternal injection) it was possible to identify GM2, GM3, lactosylceramide, glucosylceramide and ceramide (Fig. 2). These exactly correspond to the expected products of GM1 degradation. As shown in Table 6 differences in the relative concentrations of the various products are present in liver and brain, reflecting tissue specificity in metabolic processing. The only catabolite which could not be recognized in this investigation was the free long chain base. However substantial amounts of volatile radioactivity were detected in the dialysable fraction (likely tritiated water) indicating the occurrence of complete degradation of the long chain base. Therefore, it seems likely that the long chain base was formed, but immediately degraded and removed with a fast-rate process. This kept its concentration so low to be undetectable under the used experimental conditions. Subcellular fractionation studies, performed on the liver, clearly indicated a primary involvement of lysosomes in the degradation of GM1. In fact (Fig. 3A) lysosomes incorporated large amounts of radioactivity, with a several-fold increase of specific radioactivity compared to that of the starting homogenate. This radioactivity was distributed mainly in the dried organic phase, which contains GM1 catabolites lacking sialic acid (neutral glycolipids and ceramide). Moreover, the highest RSR values for the above catabolites and for GM2 and GM3 were found in the lysosomal fraction (Fig. 3B) indicating a net enrichment of these compounds in the same fraction.

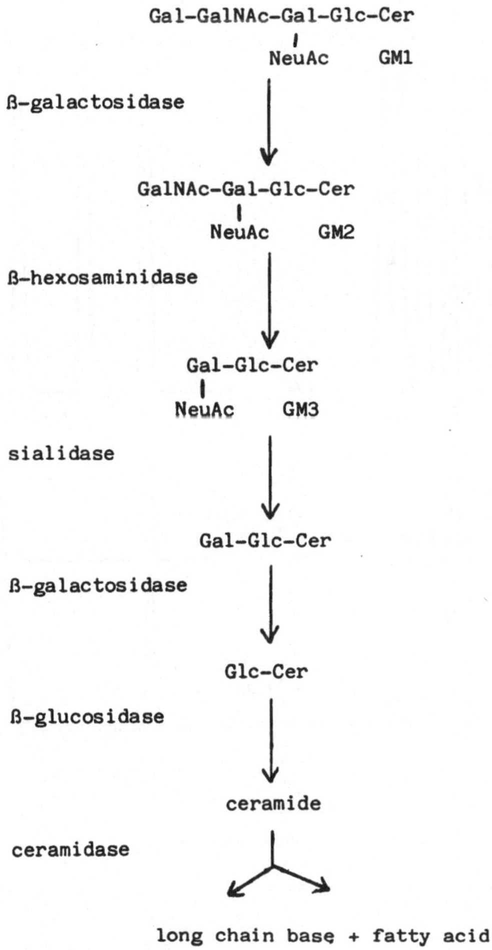

Figure 1. Route for degradation of ganglioside GM1.

Ganglioside Biosynthesis in the Golgi Apparatus: Occurrence of Two Different Mechanisms

The biosynthetic routes for the most abundant gangliosides in brain and liver are schematically illustrated in Figure 4. This scheme collects the results of enzymological and incorporation studies (Roseman, 1970; Fishman and Brady, 1976; Tettamanti, 1984). It is a consolidated evidence that all the gangliosides residing in the plasma membrane appear to be biosynthesized with similar rates. This led to suggest (Roseman, 1970; Caputto et al., 1976) that each ganglioside is biosynthesized by a multienzyme system composed of membrane-bound glycosyltransferases and located in the Golgi apparatus. In this system the glycolipid product of one glycosyltransferase (that can be a ganglioside) serves as substrate for the next enzyme. The transient intermediates,

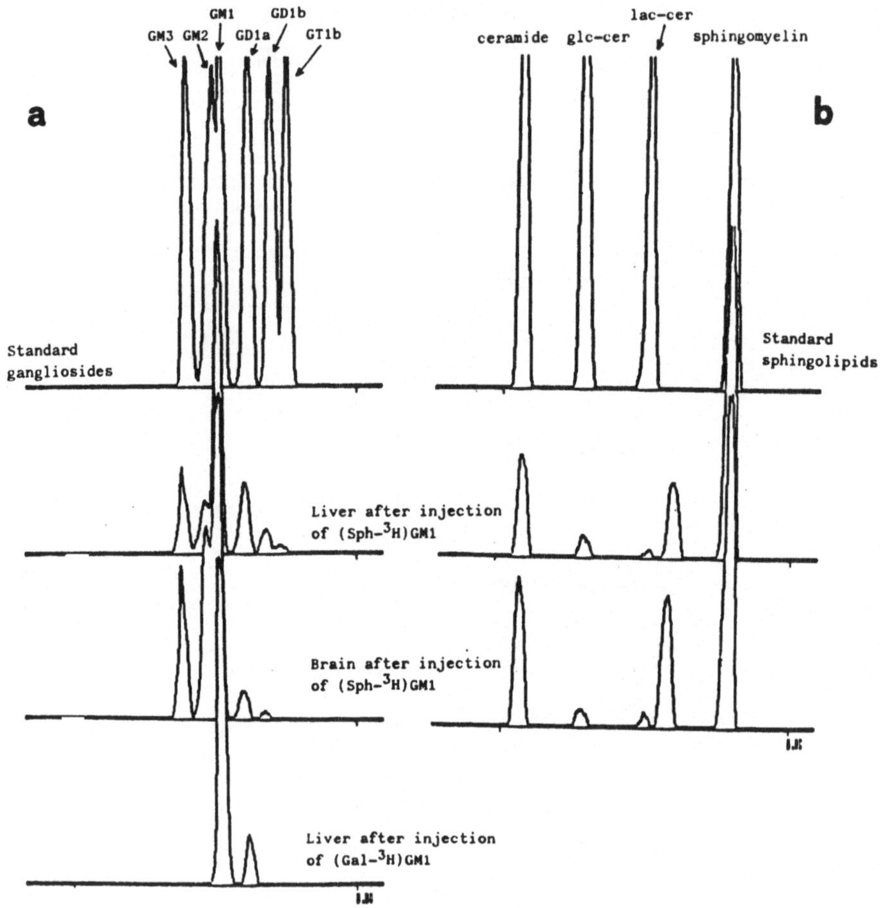

Figure 2. Distribution of radioactivity in the individual liver and brain lipids present in the dialyzed aqueous phase (gangliosides) (a) and in the dired organic phase (non-ganglioside lipids) (b), obtained after injection of 50 μCi of (Sph-^{3}H) GM1 or (Gal-^{3}H)GM1.
The radioactive lipids were separated by t.l.c. and the plates were submitted to radiochromatographic scanning.

most of them gangliosides, are tighly bound to the system, which releases only the final product to be delivered to the plasma membrane. Several multienzyme systems are supposed to occur, their number equalling that of the gangliosides residing on the plasma membrane. Therefore two distinct pools of gangliosides would exist: a minor one, constituted by transient precursors, located in the Golgi apparatus; a larger one, constituted by the final products, located on the plasma membrane. Of course according to this concept no direct metabolic correlation is expected to occur between the gangliosides sitting in the plasma membrane and the transient gangliosides linked to the biosynthetic machinery of the Golgi apparatus.

After injection of (Sph-^{3}H)GM1 in both liver and brain radioactive gangliosides

Table 6. *Distribution of the radioactivity incorporated in different lipids of liver and brain 40 hours after injection of tritiated GM1 (50 μCi/animal). The data refer to 1 g fresh tissue and are the mean of three independent experiments.*

	Incorporatd radioactivity DPM/mg fresh tissue		
	Liver		Brain
	($Sph-^3H$) GM1	($Gal-^3H$) (GM1	($Sph-^3H$) GM1
	Intravenous injection		Intracisternal injection
Ceramide	307.2	—	186.4
Glc-cer	55.1	—	29.1
Lac-cer	21.0	—	30.2
Sphingomyclin	1190.4	—	525.0
GM3	48.0	—	79.9
GM2	33.9	—	87.3
GD1a	44.2	14.4	12.1
GD1b	12.5	—	2.5
GT1b	3.1	—	tr

ÇD1a, GD1b, and GT1b could be isolated (Fig. 2a; Table 6). Since these gangliosides are more complex than GM1, their formation clearly indicate that they are biosynthetic products starting from by-products of GM1 degradation or from unmodified GM1. After injection of ($Gal-^3H$)GM1 the formation of radioactive GD1a was clearly observed in both liver and brain (Fig. 2; Table 6) (data on brain not shown); however no radioactive GD1b or GT1b could be detected. In other words ($Sph-^3H$)GM1 produced tritiated GD1a, GD1b and GT1b, while ($Gal-^3H$)GM1 only GD1a. A time course study on the formation of these complex gangliosides showed (Fig. 5) that the radioactivity/mg fresh tissue, incorporated into GD1a, GD1b and GT1b tended to increase with time after injection of labelled GM1. At all times the radioactivity linked to GD1a was significantly higher (more than two-fold) using ($Sph-^3H$) GM1 than equimolar amounts of ($Gal-^3H$)GM1, carrying the same radioactivity. Since the degradation of GM1 begins with release of galactose, in the case of injected ($Gal-^3H$)GM1 the only route for GD1a formation consists in direct sialylation of GM1. In the case of injected ($Sph-^3H$)GM1 this route is operating and the higher production of labelled GD1a indicates that part of GD1a is produced via a by-product of GM1 degradation. Therefore the biosynthesis of these gangliosides appear to proceed along two pathways: direct sialylation, which applies only to GD1a, and multiple-step glycosylation of a simple precursor (originated from GM1 degradation) that applies to GD1a, GD1b and GT1b.

Subcellular fractionation studies carried out on the liver, clearly indicated that both pathways for ganglioside biosynthesis take place in the Golgi apparatus. In fact (Fig. 3A) the Golgi-enriched fraction incorporated large amounts of radioactivity with a several-fold increase of specific radioactivity compared to that of the starting homogenate. In addition the highest RSR values for GD1a, GD1b, and GT1b were found in the Golgi apparatus fraction (Fig. 3C), indicating a net enrichment of these

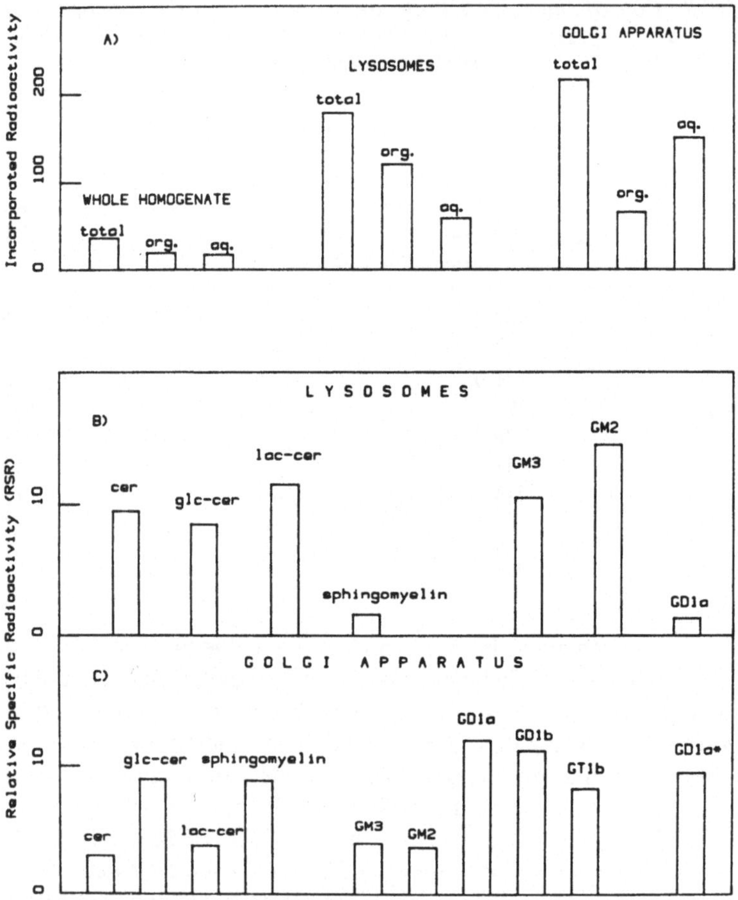

Figure 3. Distribution of radioactivity, fractionated in organic and aqueous phase (A) and RSR values (B and C) of individual liver lipids in the 'lysosome' and 'Golgi apparatus' fractions. The data refers to 1 mg of fresh tissue and are obtained 4 h after injection of 50 μCi of (Sph-^{3}H)GM1, with the only exception of GD1a, that is obtained after injection of 50 μCi of (Gal-^{3}H)GM1.

compounds in the same fraction. It should be noted that other compounds, like GM3, GM2 and glucosylceramide were also present in the Golgi apparatus fraction with RSR values (especially that of glucosylceramide) that unequivocally indicated their substantial enrichment in this fraction. A possible explanation is that these compounds, that can be produced by GM1 degradation, can also be generated through a biosynthetic process utilizing by-products of GM1 degradation.

The observed occurrence of direct sialylation of GM1 to GD1a but not to GD1b and GT1b deserves a comment. A sialyltransferase was described (Fishman and Brady, 1976) that catalyzes sialylation of GM1 to GD1a, while no sialyltransferase is known that can introduce sialic acid on GM1 to produce GD1b and GT1b (Fig. 4). If we

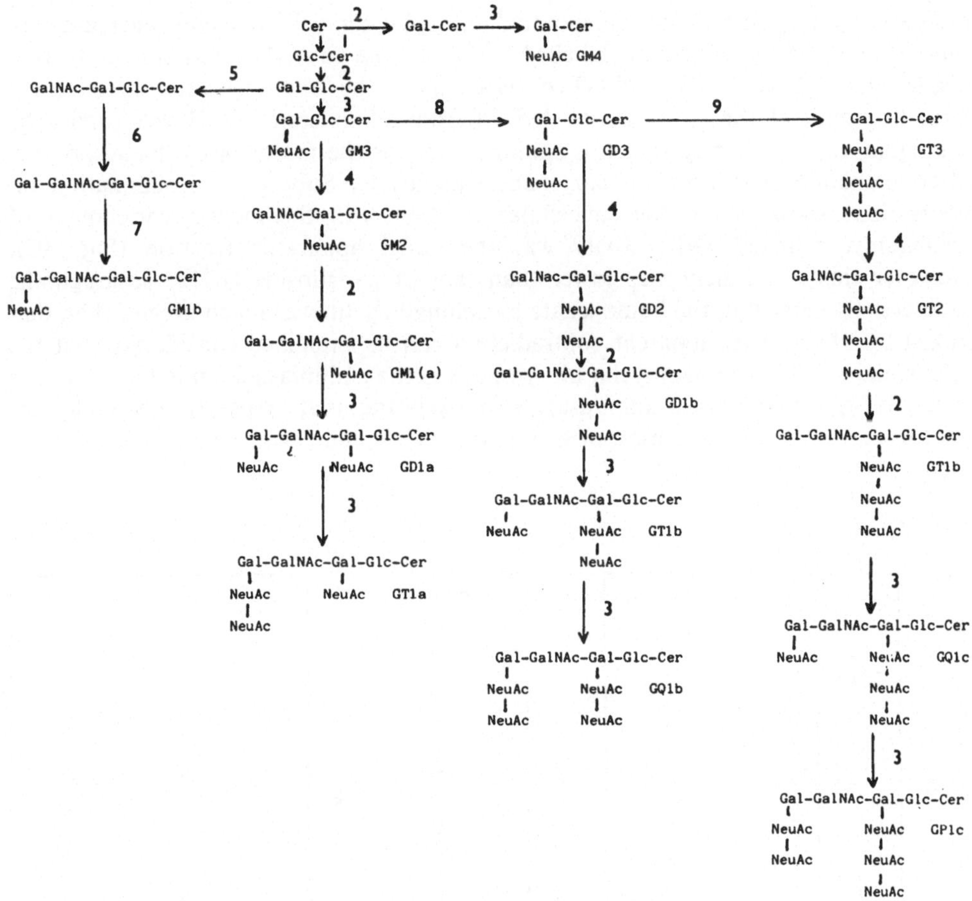

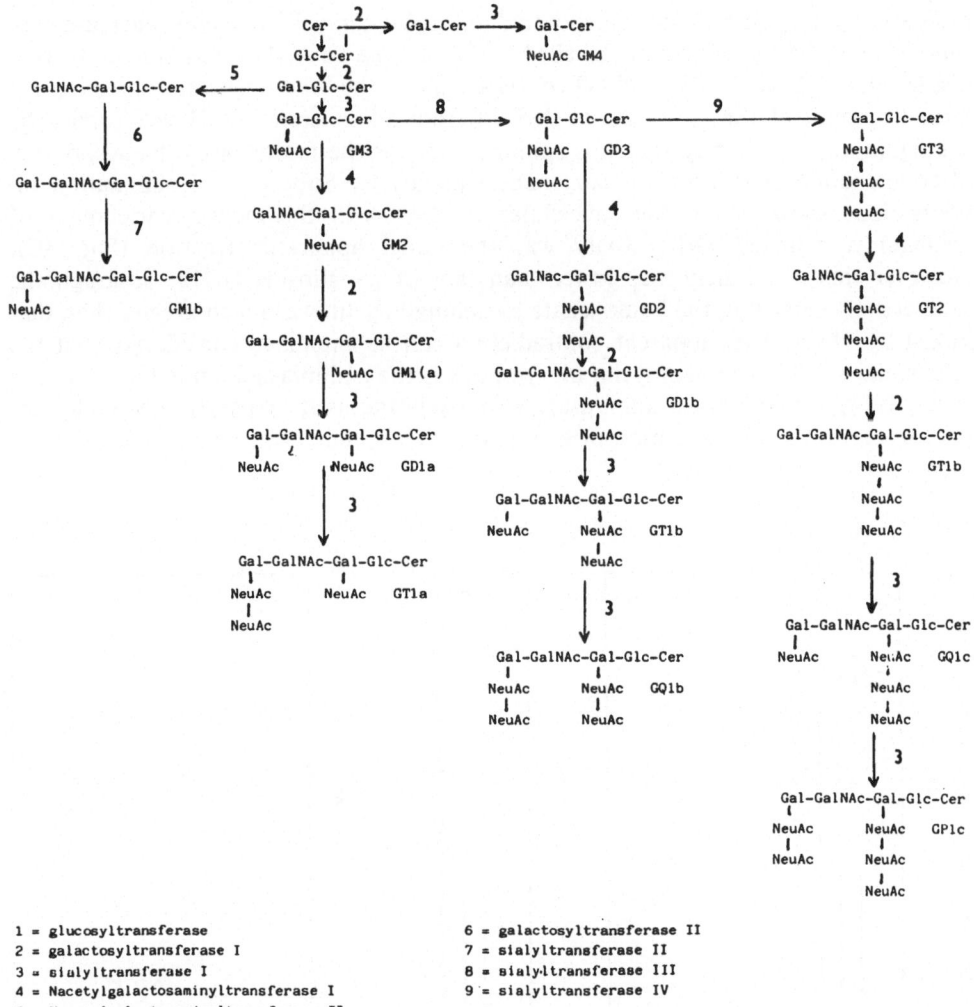

1 = glucosyltransferase
2 = galactosyltransferase I
3 = sialyltransferase I
4 = Nacetylgalactosaminyltransferase I
5 = Nacetylgalactosaminyltransferase II
6 = galactosyltransferase II
7 = sialyltransferase II
8 = sialyltransferase III
9 = sialyltransferase IV

Figure 4. Major routes for the biosynthesis of gangliosides.

assume that this sialyltransferase is part of the multienzyme system responsible for the biosynthesis of GM1 it might be suggested that the multienzyme system can accept as an 'exogenous' precursor one of its own intermediate metabolites and convert it to the final product.

Biosynthesis of Sphingomyelin from a By-Product of GM1 Degradation

Sphingomyelin was found in both liver and brain to carry a substantial portion of incorporated radioactivity after administration of (Sph-^{3}H)GM1 (Table 6). A time course study on the formation of labeled sphingomyelin showed (Fig. 5) that the radioactivity/mg fresh tissue incorporated into sphingomyelin tended to increase with time after injection of tritium labelled GM1, a behaviour that was similar to that displayed by GD1a, GD1b, and GT1b. Especially at the long times after injection the largest amount of the radioactivity derived from administered GM1 was carried by sphingomyelin, indicating that this sphingolipid was the major biosynthetic product after metabolic processing of exogenous ganglioside. Subcellular fractionation investigations showed that the subcellular fraction with the highest enrichment of sphingomyelin-linked radioactivity was the Golgi apparatus fraction (Fig. 3C). However this enrichment was lower than that of ganglioside GD1a, although the radioactivity carried in the homogenate by sphingomyelin was much greater than that linked to GD1a. This apparent contradiction can be explained considering that the subcellular site for sphingomyelin biosynthesis is the endoplasmic reticulum (Fleisher et al., 1974), not the Golgi apparatus. Very likely the used preparation of Golgi apparatus carried a contamination of endoplasmic reticulum, which contributed for the

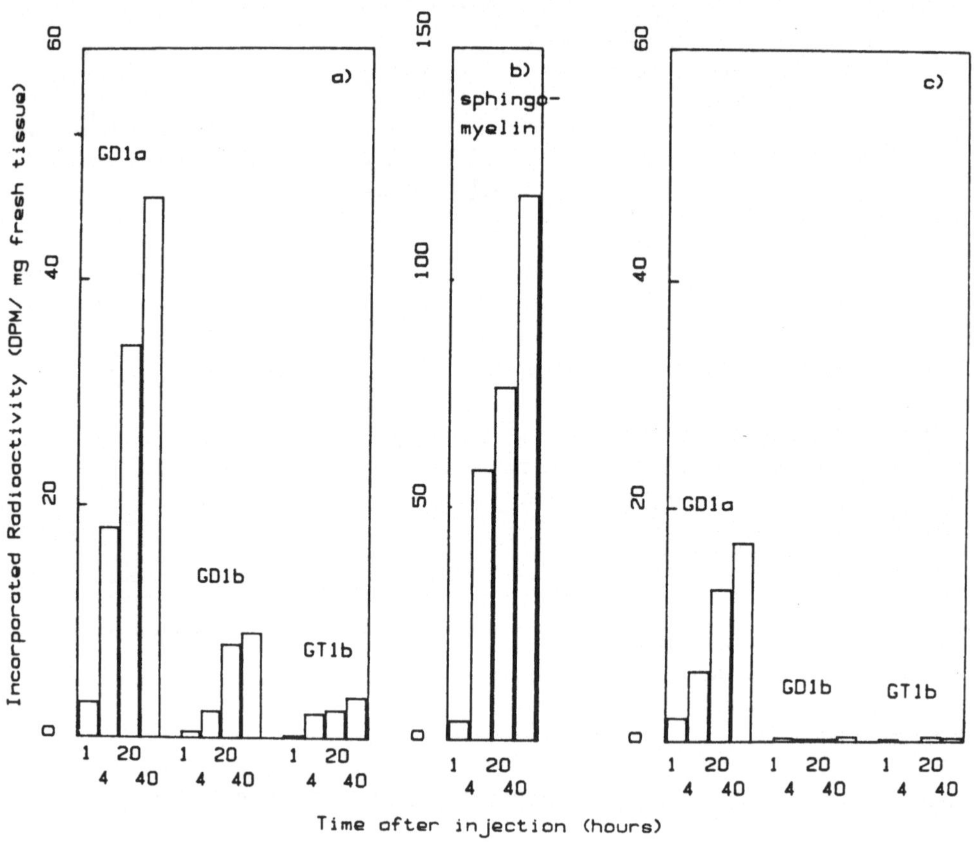

Figure 5. Time course of radioactivity distribution in the liver oligosialogangliosides and sphingomyelin, derived from tritiated GM1. a) and b): injection with 50 μCi of (Sph-^{3}H)GM1; c): injection with 50 μCi of (Gal-^{3}H)GM1.

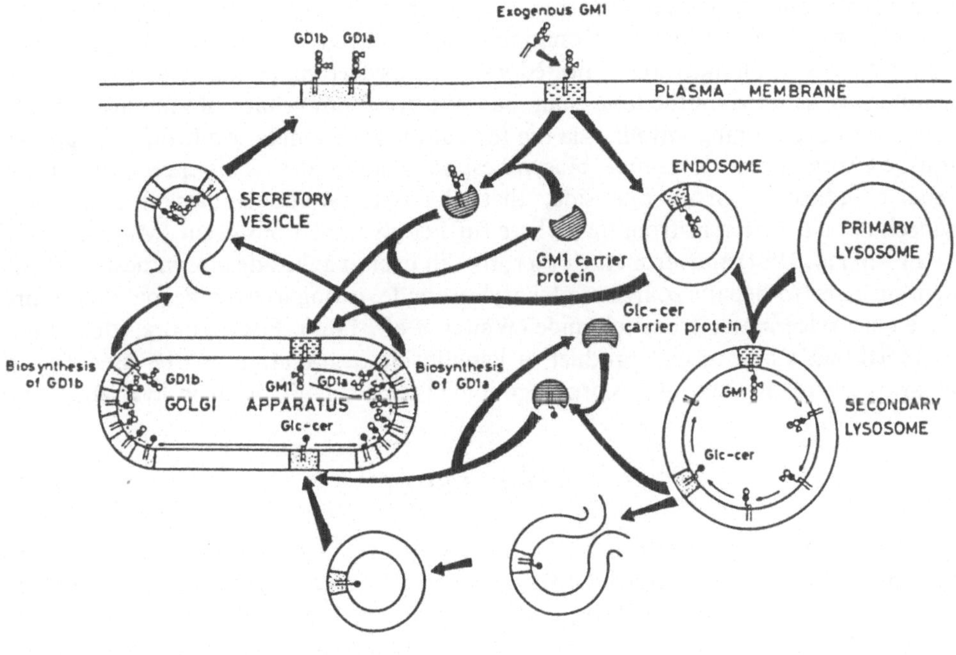

Figure 6. Proposed metabolic pathways of exogenous gangliosides GM1 in rat liver.

sphingomyelin-bound radioactivity. An endoplasmic reticulum enriched preparation would have provided much higher values of RSR for sphingomyelin than the Golgi apparatus fraction.

Synchronization between Lysosomes and Golgi Apparatus in Ganglioside Biosynthesis: Role of Glucosylceramide

Since the degradation of exogenous GM1 takes place in the lysosomes, while the biosynthesis of gangliosides occurs in the Golgi apparatus, it must be assumed that lysosomes release an end- or by-product of GM1 degradation, which is taken up and recycled by the Golgi system. This should be true also for sphingomyelin biosynthesis, with the assumption that in this case the endoplasmic reticulum is involved. Since free long chain base was not detected among the products of GM1 degradation, the simplest catabolite that could be involved in sphingomyelin biosynthesis is ceramide (cer + CDP-coline → SM + CMP). With regards to gangliosides the by-product produced in the lysosomes and utilized for their biosynthesis in the Golgi apparatus cannot be easily ascertained. A first indication can be obtained by analyzing the RSR values of different lipids in the lysosomes and Golgi apparatus fractions. As shown in Figure 3, glucosylceramide displayed RSR values that are high and almost identical in the Golgi apparatus and the lysosomal fraction, while other catabolites (GM2, GM3, lactosylceramide, ceramide) have much higher RSR values in the lysosomal fraction than in the Golgi apparatus one. This may suggest that glucosylceramide is the above key intermediate. A more direct confirmation was given by an experiment in which two different groups of animals were injected with glucosylceramide and galactosylceramide isotopically tritiated on the long chain base moiety. Both groups of animals were able to biosynthesize sphingomyelin, via the formation of ceramide, while only the animals injected with glucosylceramide biosynthesized gangliosides. This meets with recent reports indicating, from one side, that glucosylceramide is biosynthesized at a subcellular site different from that where further glycosylations occur (Miller-Podraza and Fishman, 1984) and, from the other, that in liver ganglioside biosynthesis is dependent on an extra-hepatic source of glucosylceramide, owing to liver incapability to produce glucosylceramide from ceramide (Walter et al., 1983). Furthermore, it cannot be excluded that a by-, or end-product of ganglioside degradation can be used for the biosynthesis of the neutral glycosphingolipids that normally occur in liver and brain.

CONCLUSION

The investigations summarized in this report provided the following indications: (a) intravenously administered GM1 is extensively taken up by liver, a small but definite amount being also incorporated into brain; presumably all the events encountered by incorporated GM1 in these two organs follow a similar trend; (b) exogenous GM1 is internalized into cells and faces two different fates: the first consists in a degradation process taking place in the lysosomes, the second in a sialylation process, occurring in the Golgi apparatus; (c) lysosomes can operate complete degradation of GM1 ganglioside, as suggested by the formation of water as end-product. A fast-rate processing seems to involve free long-chain base, removed immediately after being produced. Anyway some catabolites can escape full degradation and be delivered out from lysosomes; (d) at least two of these catabolites appear to be re-utilized for biosynthetic purposes: ceramide, employed for sphingomyelin biosynthesis, and glucosylceramide, used as precursor in ganglioside biosynthesis. The latter event occurs at the level of the Golgi apparatus; (e) following the concept of multiglycosyltransferase system governing the biosynthesis of gangliosides, it seems that the system producing GD1a can accept exogenous GM1 as a biosynthetic intermediate.

In other words, the experimental approach discussed in this report gave direct demonstration that a ganglioside is degraded in the lysosomes, that gangliosides are biosynthesized in the Golgi apparatus and that a direct (possibly functional) connection exists between the degradation machinery and the biosynthetic system. The next goal which seems to be pursued using the same experimental approach, is clarification of the mechanism for ganglioside intracellular transport. Since exogenous ganglioside can be stably inserted into the plasma membranes of isolated cells *in vitro*, it is reasonable to conceive that a similar phenomenon occurs also *in vivo* after intravenous administration of ganglioside. The prompt lysosomal degradation of exogenous GM1 together with the concomitant process of sialylation in the Golgi apparatus, suggests the hypothesis that a portion of the plasma membrane, carrying ganglioside, generates an endosome that can either fuse with lysosomes or reach the Golgi apparatus (Fig. 6). Whether this is the mechanism and whether the process is mediated by clathrin coated pits remain to be explored. By analogy, the release from lysosomes of catabolites and their transport to the Golgi apparatus might also follow a vesiculation mechanism. Of course, other possibilities of internalization and of intracellular vehiculation of ganglioside from and within the compartments of metabolic processing cannot be excluded. These include ganglioside transfer to specific cytosoluble 'carrier proteins' followed by the interaction of the formed complexes with various intracellular membranes. These are also fields for future investigations that have been opened by the experimental approach discussed here.

ACKNOWLEDGMENTS

The authors wish to thank Prof. Sandro Sonnino for valuable discussion. This work was supported in part by a grant from the Consiglio Nazionale delle Ricerche (CNR), Rome, Italy (Progetto finalizzato Medicina preventiva e riabilitativa, grant n° 85.00791.56).

REFERENCES

Barranger JA, Rapoport SI, Fredericks WR, Pentchev PG, MacDermot KD, Steusing JK, Brady RO (1979) Modification of the blood-brain barrier: Increased concentration and fate of enzymes entering the brain. Proc Natl Acad Sci USA 76: 481-485.

Caputto R, Maccioni HJ, Arce A, Cumar FA (1976) Biosynthesis of brain gangliosides. Adv Exptl Med Biol 71: 27-44.

Ceccarelli B, Aporti F, Finesso M (1976) Effects of brain gangliosides in experimental regeneration and reinnervation. Adv Exptl Med Biol 71: 275-293.

Fishman PH, Brady RO (1976) Biosynthesis and function of gangliosides. Science 194: 906-910.

Fleisher B, Zambrano F, Fleisher S (1974) Biochemical characterization of Golgi complex of mammalian cells. J Supram Struct. 2: 737-750.

Gatt S (1970) Enzymatic aspects of sphingolipid degradation. Chem Phys Lipids 5: 235-242.

Ghidoni R, Tettamanti G, Zambotti V (1977) Labeling of natural substrates for the radiochemical assay of enzymes involved in lipid storage diseases: a general procedure for tritiation of gangliosides. Bioch Exptl Biol 13: 61-69.

Ghidoni R, Sonnino S, Masserini M, Orlando P, Tettamanti G (1981) Specific tritium labelling of gangliosides at the 3-position of sphingosines. J Lipid Res 22: 1286-1295.

Ghidoni R, Sonnino S, Chigorno V, Venerando B, Tettamanti G (1983) Occurrence of glycosylation and deglycosylation of exogenously administered ganglioside GM1 in mouse liver. Biochem J 213: 321-329

Hjelle JT, Baird-Lambert J, Cardinale G, Spector S, Udenfriend S (1978) Isolated microvessels: the blood-brain barrier in vitro. Proc Natl Acad Sci USA 75: 4544-4548.

Iwamori M, Moser HW, McCluer RH, Kishimoto Y (1975) Specific tritium labeling of cerebrosides at the 3-position of erythro-sphingosine and threo-sphingosine. J Lipid Res 16: 332-336.

Kinders RJ, Rintoul DA, Johnson TC (1982) Ganglioside GM1 sensitizes tumor cells to growth inhibitory glycopeptides. Biochem Biophys Res Common 107: 663-669.

Lang W (1981) Pharmacokinetic studies with ^{3}H-labeled exogenous gangliosides injected intramuscularly into rats. In: Rapport MM and Gorio A (eds): Gangliosides in neurological and neuromuscular function, development and repair. Raven Press, New York, pp. 241-251.

Ledeen RW (1985) Gangliosides of the neuron. TINS 8: 169-174.
Miller-Podraza H and Fishman PH (1982) Translocation of newly synthesized gangliosides to the cell surface. Biochemistry 21: 3265-3270.
Miller-Podraza H and Fishman PH (1984) Effects of drugs and temperature on biosynthesis and transport of glycosphingolipids in cultured neurotumors cell. Biochim Biophys Acta 804: 44-51.
Morgan IG, Wolfe LS, Mandel P, Gombos G (1971) Isolation of plasma membranes from rat brain. Biochim Biophys Acta 241: 737-751.
Moss J, Fishman PH, Manganiello VC, Vaughan M, Brady RO (1976) Functional incorporation of gangliosides into intact cells: induction of choleragen responsiveness. Proc Natl Acad Sci USA 73: 1034-1037.
Obata K, Oide M, Handa S (1977) Effects of glycolipids on in vitro development of neuromuscular junctions. Nature 266: 369-371.
Radsak K, Schwarzmann G, Wiegandt H (1982) Studies on the cell association of exogenously added sialoglycolipids. Hoppe Seyler Z Physiol Chem 363: 263-272.
Roseman S (1970) The synthesis of complex carbohydrates by multi-glycosyltransferase systems and their potential function in intercellular adhesion. Chem Phys Lipids 5: 270-281.
Sandberg PO, Marzella L, Glausmann H (1980) A method for rapid isolation of rough and smooth microsomes and Golgi apparatus from rat liver in the same sucrose gradient. Exptl Cell Res 130: 393-400.
Sawant PL, Shibko S, Kumta US, Tappel AL (1964) Isolation of rat liver lysosomes and their general properties. Bioch Biophys Acta 85: 82-92.
Schwarzmann G, Hoffmann-Bleihauer P, Schubert J, Sandhoff K, Marsh D (1983) Incorporation of ganglioside analogues in fibroblast cell membranes. Biochemistry 22: 5041-5048.
Sonderfeld S, Conzelmann E, Schwarzmann G, Burg J, Hinrichs U, Sandhoff K (1984) Incorporation and metabolism of ganglioside GM2 in skin fibroblasts from normal and GM2 gangliosidosis subjects. Eur J Biochem 149: 247-255.
Sonnino S, Ghidoni R, Gazzotti G, Kirschner G, Galli G, Tettamanti G (1984a) High performance liquid chromatography preparation of the molecular species of GM1 and GD1a gangliosides with homogeneous long chain base composition. J Lipid Res 25: 620-629.
Sonnino S, Ghidoni R, Fiorilli A, Venerando B, Tettamanti G (1984b) Cytosolic ganglioside of rat brain: their fractionation into protein-bound complexes of different ganglioside composition. J Neurosc Res 12; 193-203.
Stoffel W, Sticht G (1967) Metabolism of sphingosine bases, I Hoppe-Seyler's Z Physiol Chem 348: 941-943.
Tettamanti G (1984) An outline of ganglioside metabolism. Adv Exptl Med Biol 174: 197-211.
Tettamanti G, Bonali F, Marchesini S, Zambotti V (1973) A new procedure for the extraction, purification and fractionation of brain gangliosides. Biochim Biophys Acta 296: 160-170.
Tettamanti G, Venerando B, Roberti S, Chigorno V, Sonnino S, Ghidoni R, Orlando P, Massari P (1981) The fate of exogenously administered brain gangliosides. In: Rapport MM and Gorio A (eds): Gangliosides in neurological and neuromuscular function, developmental and repair: Raven Press, New York, pp. 225-240.
Tettamanti G, Ghidoni R, Sonnino S, Chigorno V, Venerando B, Giuliani A, Fiorilli A (1984) Adv Exptl Med Biol 174: 273-284.
Walter VP, Sweeney K, Morrè DJ (1983) Neutral lipid precursors for gangliosides are not formed by rat liver homogenates or by purified cell fractions. Biochim Biophys Acta 750: 346-352.

Ganglioside and neuronal plasticity
G. Tettamanti, R.W. Ledeen, K. Sandhoff,
Y. Nagai, G. Toffano (eds.)
Fidia Research Series, vol. 6
Liviana Press, Padova, © 1986

GANGLIOSIDE-MEDIATED MODULATION OF GROWTH FACTOR RECEPTOR FUNCTION

Sen-itiroh Hakomori, Eric Bremer*, Yoshio Okada**

Program of Biochemical Oncology/Membrane Research,
Fred Hutchinson Cancer Research Center and Departments of
Pathobiology, Microbiology and Immunology,
University of Washington, Seattle, WA 98104 USA

INTRODUCTION

Two classes of glycosphingolipids can be distinguished in cells; one class with long and complex carbohydrates characterizes types of cells and may play an important role in cell-cell recognition (cell social function), while the other class is composed of a few basic structures, such as GM3, lactosylceramide, and glucosylceramide, common to many types of cells. The latter class of glycosphingolipids and even some of the former class may play a basic role in regulating the function of intrinsic membrane proteins such as receptors and transporters. There have been a few lines of evidence that GM3 or GM1 ganglioside may affect and modulate the function of the receptors for fibroblast growth factor (FGF), platelet-derived growth factor (PDGF), and epidermal growth factor (EGF). There is a study that has suggested a possible association of a tumor-associated glycolipid antigen (gangliotriaosylceramida, Gg3Cer) and the transferrin receptor in mouse lymphoma L5178Y. Furthermore, some data indicate that cell adhesion could be mediated by gangliosides, although much of the evidence for this possibility is still fragmentary and requires extensive further study. I will try to summarize our observations regarding the possible role of gangliosides in the regulation of receptor function.

EVIDENCE THAT GANGLIOSIDES MAY REGULATE CELL PROLIFERATION

Several lines of evidence that gangliosides may regulate cell proliferation have accumulated in the past decade and are listed in Table 1. Contact inhibition of cell growth

Abbreviations: FGF, fibroblast growth factor; PDGF platelet-derived growth factor; EGF, epidermal growth factor; FCS, fetal calf serum; DME, Dulbecco modified Eagle Medium.

Present address: * Dept of Microbiology, Rush University, Chicago, Ill, 60612, USA; ** Dept of Medicine, Okayama University, Okayama, Japan.

Table 1. *Evidence that Glycolipids May Regulate Cell Proliferation*

1. Contact inhibition of cell growth accompanies change of glycolipid synthesis (Hakomori, 1970; Sakiyama et al., 1972; Critchley and MacPherson, 1973; Yogeeswaran and Hakomori, 1975; Kijimoto and Hakomori, 1971, Langenbach and Kennedy, 1978).
2. Cell cycle-dependent change of glycolipid organization: Exposure at Gl or G0 phase (Gahmberg and Hakomori, 1975, Lingwood and Hakomori, 1977).
3. Butyrate induces cell growth inhibition and enhances GM3 synthesis (Fishman et al., 1974; Simmons et al., 1975).
4. Retinoids induce contact inhibition, enhance GM3 synthesis and glycolipid response (Patt et al., 1978).
5. Antibodies to GM3 but not to globoside inhibit 3T3 and NIL cell growth and enhance GM3 synthesis (Lingwood and Hakomori, 1977).
6. Exogenous addition of glycolipids incorporated into cell membranes inhibit cell growth through extension of the Gl phase (Laine and Hakomori, 1973; Keenan et al., 1975).

accompanies changes of ganglioside synthesis, and a loss of glycolipid response on cell contact has been observed to be associated with a loss of contact inhibition in many transformed cells (Hakomori, 1970; Sakiyama et al., 1972; Critchley and Mac Pherson, 1973; Yogeeswaran and Hakomori, 1971; Kijimoto and Hakomori, 1971; Langenbach and Kennedy, 1978). Cell cycle-dependent changes of glycolipid organization in membranes (Gahmberg and Hakomori, 1975; Lingwood and Hakomori, 1977) have been observed in synchronized cells. Butyrate enhances GM3 synthesis and induces cell growth inhibition (Fishman et al., 1974; Simmons et al., 1975); retinoids increase glycolipid response on cell contact (Patt et al., 1978). Antibodies to GM3 but not those to globoside (Gb4Cer) inhibit 3T3 and NIL cell growth and enhance GM3 synthesis (Simmons et al., 1975). Exogenous addition of glycolipids incorporated into cell membranes inhibits cell growth through extension of the Gl phase of the cell cycle (Laine and Hakomori, 1973; Keenan et al., 1975). The growth inhibition induced by exogenous ganglioside addition can be observed more clearly in chemically-defined media than in serum-containing media (Bremer and Hakomori, 1982), and this approach has been applied to observe ganglioside-mediated cell growth inhibition induced by FGF, PDGF, and EGF, as described below.

INHIBITION OF FGF-DEPENDENT BHK GROWTH BY EXOGENOUS ADDITION OF GM3

Since the growth of BHK cells in chemically-defined medium has been well established, and the cells require insulin, transferrin, hydrocortisone, and FGF (Barnes and Sato, 1980; Maciag et al., 1980), we have studied the effect of gangliosides added to the chemically-defined medium. We found that GM3 but not GM1 induced a state refractory to cell growth stimulation by FGF (Bremer und Hakomori, 1982). The data that suggest that GM3 may affect FGF receptor function are listed in Table 2. Since

Table 2. *Suggestive Data that FGF Receptor Function in BHK Fibroblasts is Influenced by GM3 (Bremer and Hakomori, 1982)*

1. BHK cells can be grown in insulin, transferrin, and hydrocortisone (1 μg/ml each) and 100 ng/ml of FGF but do not require EGF or PDGF. The growth is specifically inhibited when 10-100 μg/ml of GM3 but not GM1 or other glycolipids are included in the medium.

2. Both GM3 and GM1 added in culture media are equally incorporated into cell membranes; however, only GM3-fed cells become refractory to stimulation by FGF.

3. Those cells whose growth is inhibited by culturing in the presence of GM3 but not in the presence of other gangliosides accumulate a large quantity of the ^{125}I-labeled FGF when cells were added with ^{125}I-FGF. ^{125}I-labeled transferrin is not accumulated in such cells.

4. GM3 does not interact with FGF directly.

our knowledge of the FGF receptor has been obscure, we were unable to study the relationship between GM3 and FGF receptor function in detail. Very recently, however, Neufeld and Gospodarowicz (1985) demonstrated the FGF receptors of BHK cells as 145 and 125K membrane proteins by FGF-affinity labeling technique. It is now therefore possible to further study a possible interaction between GM3 and the FGF receptor.

THE EFFECT OF GM1 AND GM3 ON THE PDGF RECEPTOR IN SWISS 3T3 CELLS

Swiss 3T3 cell growth in chemically-defined media is dependent on PDGF, and to a lesser extent on EGF, and the cell growth (both cell number increase and DNA synthesis) in chemically-defined media is preferentially inhibited by addition of GM1, to a lesser extent by GM3, but not by NeuAcnLc4Cer nor by Gb4Cer (Bremer et al., 1984). All these glycolipids exogenously added to culture medium are incorporated equally well into cell membranes. The effect of gangliosides on mitogen-stimulated DNA synthesis (thymidine incorporation), on the concentration-dependent mitogen binding to cells, and on PDGF-dependent phosphorylation have revealed three important results: (i) A relative specificity was found between mitogens and ganglioside inhibitors, although mitogens do not interact directly with gangliosides, i.e., PDGF-dependent [^{3}H]-thymidine incorporation was preferentially inhibited by GM1 and to a lesser extent by GM3, while EGF-dependent [^{3}H]-thymidine incorporation was preferentially inhibited by GM3 and only very weakly by GM1. NeuAcnLc4Cer did not inhibit [^{3}H]-thymidine incorporation. (ii) Gangliosides were able to affect the kinetic properties of PDGF receptor interactions. Preincubation of 3T3 cells with GM1 or GM3 altered the K_D of ^{125}I-PDGF binding without altering the number of receptors. Thus, the reduced mitogenic potential of PDGF in the presence of gangliosides is consistent with possible "qualitative" changes in the PDGF-binding properties of the PDGF receptor. (iii) The ganglioside levels in membranes may affect PDGF receptor phosphorylation. This is indicated by the fact that the PDGF-dependent tyrosine phosphorylation of the PDGF receptor with a molecular weight of 170,000 was inhibited by GM1 and GM3 gangliosides but not by NeuAcnLc4Cer or the neutral glycolipid Gb4Cer (Fig. 1). Acidic or neutral detergents did not inhibit the tyrosine phosphorylation of the 170,000 receptor protein. Some detergents, e.g. sodium deoxytaurocholate and Triton X-100, rather enhanced the phosphorylation at higher concentrations. The 170,000 phosphoprotein in 3T3 cells has been proposed to be a PDGF receptor since it is phosphorylated on the tyrosine residue in response to PDGF stimulation (Bremer et al., 1984; Glenn et al., 1982) and corresponds in size to the PDGF-binding protein as demonstrated by affinity cross-linking studies. The data suggesting that membrane gangliosides could affect PDGF receptor function in Swiss 3T3 cells are summarized in Table 3.

A feasible explanation for the inhibition of mitogenic stimulation by gangliosides is that exogenous gangliosides may alter the conformation of the growth factor receptor, and hence the affinity of the receptor to their mitogens. A decrease of the K_D for PDGF binding, i.e., increase of binding affinity, without altering the number of receptors is analogous to the effect of phorbol esters, phospholipase C (Shoyab and Todaro, 1981), dexamethasone (Baker et al., 1978), and vasopressin on EGF binding; all these factors increase the K_D for EGF binding without altering the number of receptors and appear to stimulate cell growth. Our 4°C binding studies indicate that growth factor binding is affected by both GM1 and GM3 gangliosides. At subsaturating concentrations, more PDGF was bound to growth-inhibited cells than to non-inhibited cells. A change in binding properties of the receptor does not directly explain the decrease in thymidine incorporation, but it does suggest a qualitative change in receptor function. On the other hand, a possible interaction between gangliosides and PDGF itself can be eliminated by the absence of direct interaction of PDGF with ganglioside liposomes.

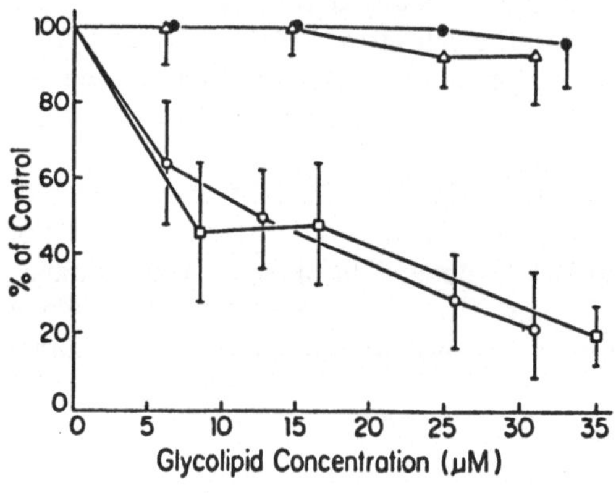

Figure 1. Effect of glycolipids on PDGF-dependent protein phosphorylation. Swiss 3T3 membranes were incubated with 10 nM [^{32}P]ATP and 60 nmols PDGF with glycolipid at the indicated concentrations for 30 min at 30°C, and proteins were separated by polyacrylamide gel electrophoresis. After visualization of [^{32}P]-labeled protein by autoradiography, the Mr-170,000 protein was cut from the gel and counted in a liquid scintillation counter. Results are expressed as % of maximum response (glycolipid concentration 0) vs. concentration of glycolipid added. Data shown is the average ± SE of at least three determinations. ○, GM1; □, GM3; •, Gb4Cer; and Δ, NeuAcnLc4Cer.

Table 3. *The Effect of GM1 or GM3 on PDGF-Dependent Cell Growth and PDGF Receptor Phosphorylation in Swiss 3T3 Cells (Bremer et al., 1984)*

1. Swiss 3T3 cell growth in chemically-defined media is dependent on PDGF, and the cell growth (cell number increase) in chemically-defined media is specifically inhibited by GM1, to a lesser extent by GM3, but not by NeuAcnLc4Cer. All these gangliosides are equally incorporated into cell membranes.

2. GM2 inhibits both PDGF- and EGF-stimulated mitogenesis determined by thymidine incorporation, while GM1 can only inhibit PDGF-stimulated mitogenesis. NeuAcnLc4Cer has no effect on mitogenesis.

3. The concentration-dependent ^{125}I-PDGF binding to cells indicates that cells whose growth is inhibited by GM1 or GM3 show an increased affinity for PDGF as compared to cells grown without addition of ganglioside, while the total number of receptors stay the same.

4. No direct interaction can be observed between gangliosides and growth factors as evidenced by the lack of competition by ganglioside-containing liposomes for cellular binding of ^{125}I-growth factors.

5. GM1 and GM3 but not NeuAcnLc4Cer nor Gb4Cer inhibit the PDGF-stimulated tyrosine phosphorylation by membrane preparations of a 120K protein, which is identified as PDGF receptor.

The change in cellular affinity of PDGF binding and simultaneous growth inhibition induced by specific types of gangliosides (GM1 and GM3) are supported by the inhibition of PDGF-dependent tyrosine phosphorylation of the receptor protein by GM1 and GM3 but not by other gangliosides and detergents. Similar results on the effect of GM3 on EGF binding to the EGF receptor in A431 and KB cells have been observed and will be discussed in the subsequent section.

EFFECT OF GM3 ON EGF-DEPENDENT MITOGENESIS AND EGF RECEPTOR PHOSPHORYLATION

The growth of human oral epidermoid carcinoma KB and ovarial epidermoid carcinoma A431 cells was shown to be inhibited by exogenous addition of GM3 and to a lesser extent by addition of GM1, and EGF-dependent tyrosine phosphorylation of the EGF receptor was inhibited in both types of cells (Bremer et al., 1986). The crucial experimental data are shown in Figs. 2-4, and the data are summarized in Table 4. A specific ganglioside, GM3, and to a lesser degree GM1, but not other glycolipids, inhibited both KB and A431 cell growth (Fig. 2). EGF-dependent mitogenesis of KB cells was inhibited by GM3 (50 nmole/ml), to a lesser extent by GM1 (Fig. 3). The cell growth inhibition on addition of GM3 was associated with inhibition of tyrosine phosphorylation of the EGF receptor *in vitro* as well as *in situ*. No other glycolipid showed the same degree of effect on EGF receptor phosphorylation in membrane preparations *in vitro* (Fig. 4). This phenomenon is analogous to our previous observations described in the foregoing sections that GM3 and GM1 but not other glycolipids inhibited tyrosine phosphorylation of the PDGF receptor *in vitro* and inhibited PDGF-dependent cell growth and DNA synthesis (Bremer et al., 1984). The EGF-dependent

Table 4. *Effect of GM3 on EGF-Dependent Mitogenesis and EGF Receptor Phosphorylation*

1. Exogenous addition of GM3 or GM1 inhibits EGF-dependent cell growth of human epidermoid carcinoma cell lines KB and A431, although GM3 shows a much stronger inhibitory effect on both KB and A431 cell growth than GM1.

2. Neither GM3 nor GM1 have any effect on the binding of ^{125}I-EGF to its cell surface receptor. GM3, and to a lesser extent GM1, are capable of inhibiting EGF-stimulated tyrosine-phosphorylation of the EGF receptor in membrane preparations of both KB and A431 cells.

3. EGF-dependent tyrosine phosphorylation and its inhibition by GM3 have also been demonstrated on isolated EGF receptors after adsorption on the anti-receptor-antibody-Sepharose complex.

4. Inhibition of tyrosine phosphorylation of the EGF receptor by GM3 was further confirmed by phospho-amino acid analysis on two-dimensional electrophoresis.

5. The inhibitory effect of GM3 on EGF-dependent receptor phosphorylation can be reproduced in membranes isolated from A431 cells that have been cultured in medium containing 50 nmoles/ml GM3 to effect cell growth inhibition.

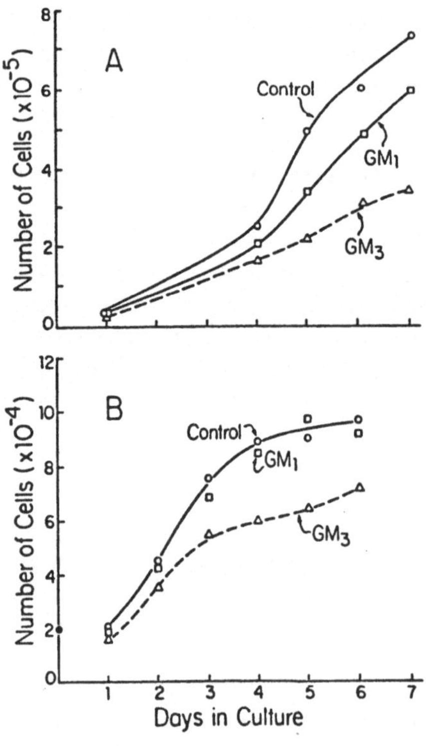

Figure 2. Effect of exogenous glycolipids added in culture media on growth of human epidermoid carcinoma A431 cells (A) and KB cells (B). The growth curve of both A431 and KB cells was measured in DME supplemented with 5% FCS and 50 nmoles/ml of glycolipid. Cells were seeded at day 0,2 × 10^4 cells/well (Falcon 24-well plate), and at day 1 the medium was changed to those with or without glycolipid. The cell numbers were counted at the indicated time. The value is the arithmetic mean of three determinations.

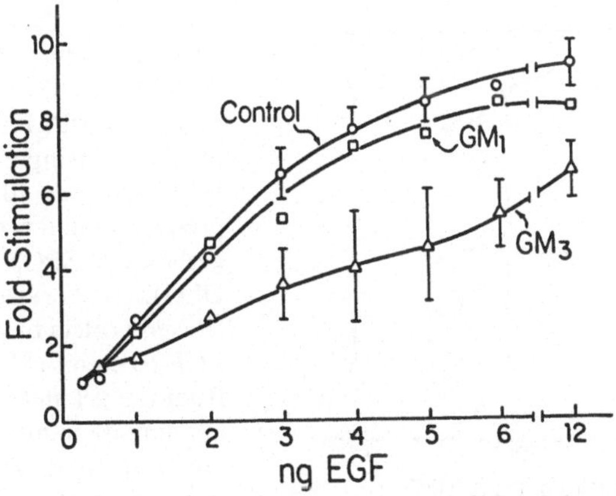

Figure 3. EGF-dependent mitogenesis of KB cells and inhibition of mitogenesis in the presence of exogenous addition of GM3. KB cells were grown for 3 days in DME containing 5% FCS with or without addition of 50 moles/ml of GM3 or GM1 ganglioside. The media were then replaced with DME without FCS containing 1% human plasma-derived serum with various amounts of EGF with or without 50 nmoles/ml of ganglioside, and cells were cultured for 18 hours followed by labeling with 1 μCi/ml of [^{3}H]thymidine for 2 hours. Cells were washed with PBS and extracted with 5% trichloroacetic acid. The insoluble residue filtered on a Millipore filter was counted.

tyrosine phosphorylation of the EGF receptor and its inhibition by GM3 were not only demonstrated on membrane preparations but also on the isolated EGF receptor after adsorption on the EGF receptor antibody bound to Sepharose (Table 5 and Fig. 5). In this case, the EGF receptor phosphorylation was enhanced on addition of phosphatidylethanolamine, indicating that the EGF receptor function is greatly affected by the membrane lipid environment. The characterization of GM3-sensitive receptor phosphorylation was performed in A431 cells, which have a higher content of the EGF receptor. The following results were of particular interest: (i) EGF-dependent tyrosine phosphorylation of the EGF receptor and its inhibition by GM3 were demonstrated on the isolated EGF receptor by immune precipitation. (ii) That the reduction of EGF receptor phosphorylation in the presence of GM3 was entirely due to the tyrosine phosphorylation was demonstrated by phosphoamino acid analysis with two-dimensional electrophoresis. Only a spot for tyrosine phosphate, but not those for serine or threonine phosphates, were reduced in the presence of GM3. (iii) The inhibitory effect of GM3 on EGF-dependent receptor phosphorylation could be reproduced in membranes isolated from A431 cells that had been cultured in medium containing 50 nmoles/ml GM3 to effect cell growth inhibition. The membrane fraction isolated from such growth-arrested cells was found to be less responsive to EGF-stimulated receptor phosphorylation. These results suggest that membrane lipids, especially GM3, can modulate EGF receptor phosphorylation *in vitro* as well as *in situ*.

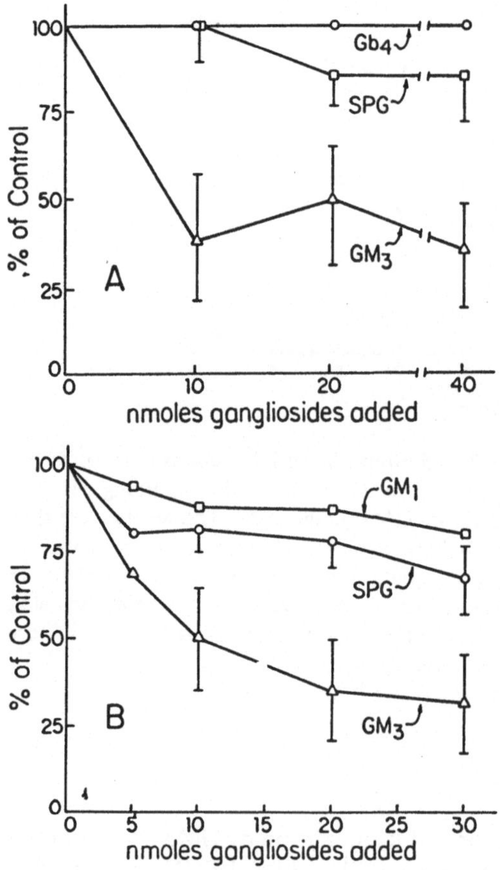

Figure 4. Reduction in EGF-receptor phosphorylation by GM3 ganglioside. A: In A431 cells. A431 membranes were incubated with 100 ng EGF for 20 min at room temperature and 10 nM [γ^{32}P]ATP for 10 min at 0°C in the presence or absence of glycolipids (Gb4, globoside; SPG, sialylparagloboside; GM3). After visualization of [^{32}P]-labeled protein by autoradiography, the EGF-receptor (Mr = 170,000) was cut from the gel and radioactivity detected in a liquid scintillation counter. Results are expressed as % of maximum response (glycolipid concentration 0) *versus* concentration of glycolipid added. Data shown is the average + S.E. of at least three determinations. B: In KB cells. KB cell membranes were incubated with EGF followed by [γ^{32}P]ATP in the presence of absence of various amounts of glycolipids (GM3; GM1; SPG, sialylparagloboside). After visualization of [^{32}P]-labeled protein by autoradiography, the activity of the EGF receptor area was determined as in Panel A.

Table 5. *Effect of Gangliosides and Phospholipids on EGF-Dependent Phosphorylation of the Immunoadsorbed EGF Receptor**

Glicolipids added (total volume of the reaction mixture, 40 μl)	100 ng EGF	[^{32}P] Recovered from EGF Receptor (fold stimulation)†		
		No phospholipid added	17 μg phosphatidylethanolamine	17 μg phosphatidylcholine
—	—	0	3.1	1.3
—	+	4.2	17.2 (5.5)†	4.9
7 nmoles GM1	+	2.4	14.9 (4.8)	4.4
14 nmoles GM1	+	2.1	13.5 (4.4)	4.2
7 nmoles GM3	—	1.1	6.5 (2.1)	2.4
14 nmoles GM3	—	1.0	6.8 (2.2)	2.3

* Phosphorylation pattern of EGF receptor was observed after the receptor was adsorbed on the antibody (29.1)-protein A-Sepharose complex. Phosphorylation was performed in the presence or absence of 100 ng EGF, or on addition of 7-14 nmoles of gangliosides, or 17 μg phospholipid in 40 μl of reaction mixture as described in the text.

† Fold stimulation over non-EGF stimulated [^{32}P] incorporation (no EGF, no phospholipid, no glycolipid).

† Values in parentheses are fold stimulation over non-EGF-stimulated [^{32}P] incorporaton, but with 0.6 μM phosphatidylethanolamine.

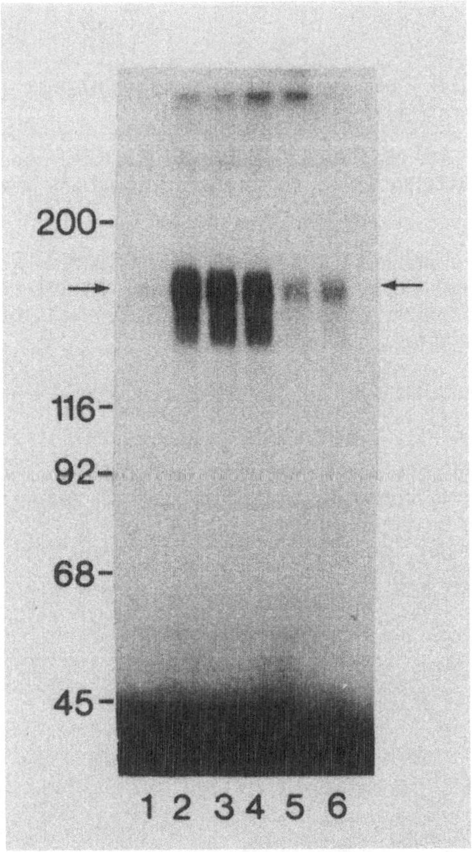

Figure 5. EGF-dependent phosphorylation of the immunoprecipitated EGF-receptor. The EGF receptor was immunoprecipitated from solubilized A431 cell membranes. The immunoprecipitated material was then assayed for EGF-stimulated phosphorylation. Glycolipids were included in the reaction mixture as indicated below. After phosphorylation, the samples were subjected to gel electrophoresis, dried, and exposed to X-ray film. The resulting autoradiogram is shown above. Arrow indicates the location of the EGF receptor. Molecular weight standards are shown $X10^{-3}$. Lane 1, no EGF, no glycolipid; Lane 2, 100 ng EGF, no glycolipid; Lane 3, 100 ng EGF, 7.0 nmoles GM1; Lane 4, 100 ng EGF, 14 nmoles GM1; Lane 5, 100 ng EGF, 7 nmoles GM3; Lane 6, 100 ng EGF, 14 nmoles GM3.

FUNCTIONAL ASSOCIATION BETWEEN GANGLIOTRIAOSYLCERAMIDE (ASIALO GM2; Gg3Cer) AND TRANSFERRIN RECEPTOR IN MOUSE LYMPHOMA L5178Y

Mouse lymphoma L5178Y has the unusual property of being able to grow in basic RPMI medium without the addition of any growth factors, such as insulin, EGF, PDGF, or FGF, which are required by most animal cells to grow. The only requirement is transferrin, which transports iron into cells. Addition of insulin, however, promotes growth, but is not essential. One of the clones, AA12, expresses a high quantity of gangliotriaosylceramide (Gg3Cer) at the cell surface, and another clone, 27AV, does not express Gg3Cer. Therefore, these cells are suitable for studying the functional role of Gg3Cer in cell growth. We previously established monoclonal antibody 2D4 directed to Gg3Cer (Young et al., 1979) and the antibody itself does not affect cell growth. When biotinyl 2D4 and avidin were added to clone AA12, cell growth was completely inhibited, although biotinyl 2D4 or avidin alone did not affect cell growth. Growth of 27AV cells was, however, not affected by biotinyl 2D4 and avidin. When growth of AA12 cells was inhibited in the presence of biotinyl antibody and avidin, cells were refractory to transferrin stimulation, i.e., transferrin-dependent cell growth was inhibited (Fig. 6). In such growth-arrested cells, internalization of ^{125}I-labeled transfer-

Table 6. *Possible Functional Association between Gangliotriaosylceramide and Transferrin Receptor in Mouse Lymphoma L5178Y (26)*

1. Mouse lymphoma L5178Y cells express both gangliotriaosylceramide (Gg3Cer) and the transferrin receptor. Cells can be grown in chemically-defined medium without any growth factor except transferrin.

2. Transferrin-dependent cell growth can be inhibited by biotinyl anti-Gg3Cer antibody plus avidin, although the cell growth is not inhibited by the addition of biotinyl-anti-Gg3Cer alone, avidin alone, or the biotinyl derivative of an IgM antibody directed to N-acetyllactosamine (IB2) cross-linked with avidin.

3. The growth inhibition induced by biotinyl-anti-Gg3Cer and avidin is associated with a block of transferrin internalization.

4. Growth-inhibited cells under the above conditions show a capping of Gg3Cer antigen, while transferrin receptors, the function of which is blocked by biotinyl-anti-Gg3Cer and avidin, are distributed independently from Gg3Cer at the cell surface.

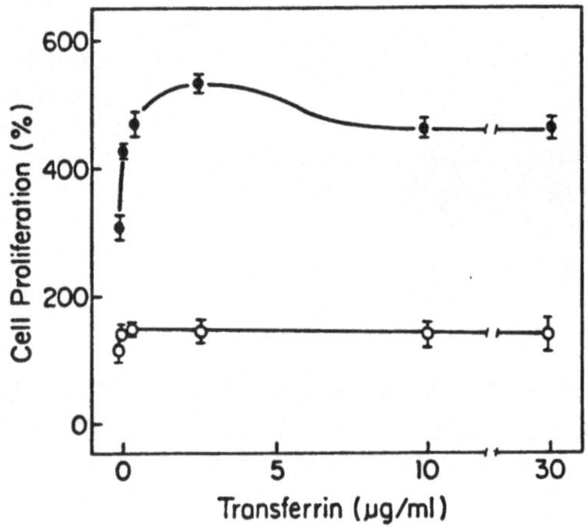

Figure 6. The effect of concentration of transferrin on cell proliferation of AA12 cells. AA12 cells were cultured in the presence of various amounts of transferrin in chemically defined medium containing biotinylated 2D4 antibody alone (•) or biotinylated 2D4 antibody and avidin (○). After 48 hr of culture, the number of cells was counted and percent cell proliferation was calculated. Results are expressed as the mean ± S.D. of three determinations.

rin was inhibited (Figs. 7 and 8). Transferrin-dependent growth stimulation in 27AV cells was not inhibited by biotinyl 2D4 and avidin (Okada et al., 1985).

Growth-arrested AA12 cells grown in the presence of biotinyl 2D4 and avidin showed a remarkable clustering of Gg3Cer at one side of the cells, while the transferrin receptor was uniformly distributed, i.e. Gg3Cer and the transferrin receptor are independently distributed at the cell surface. This is in striking contrast to non-growth-arrested cells, in which the transferrin receptor and Gg3Cer are co-distributed. These results indicate that transferrin receptors in AA12 cells are closely associated with Gg3Cer and that their function is regulated by Gg3Cer. In contrast, the transferrin receptors in 27AV cells, which lack Gg3Cer, function independently of Gg3Cer.

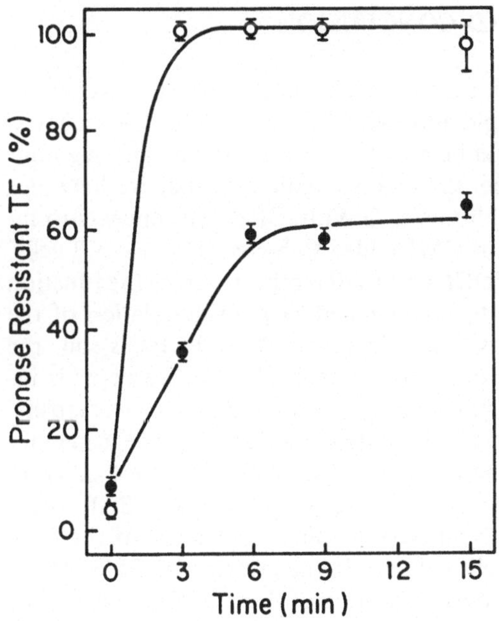

Figure 7. Internalization of transferrin. AA12 cells were cultured in the presence of 20μg/ml of biotinylated 2D4 antibody (○) or 20 μg/ml of biotinylated 2D4 antibody and 50 μg/ml of avidin (•) for 24 hr. Unbound biotinylated antibody and avidin was washed off and cells were incubated with a trace amount of [^{125}I]-transferrin at 0°C for 45 minutes. After washing unbound transferrin off, cells were incubated at 37°C for indicated time and released transferrin was washed off. Cells were digested with 10 μg/ml pronase at 0°C for 60 min and pronase-resistant radioactivity was determined. Percent pronase resistant transferrin was a fraction of pronase radioactivity in total cell-associated radioactivity at a given time. Results are expressed as the mean ± S.D. of three determinations.

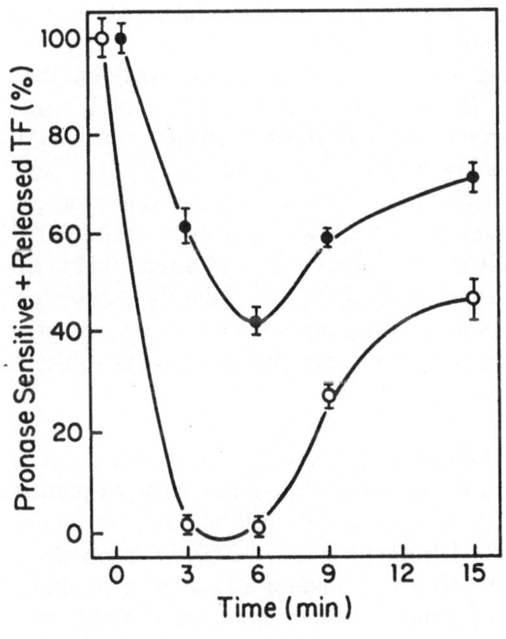

Figure 8. Internalization and externalization of transferrin. AA12 cells were cultured in the presence of 20 μg/ml of biotinylated 2D4 antibody (○) or in the presence of 20 μg/ml of biotinylated 2D4 antibody and 50 μg/ml of avidin (•) for 24 hr. After washing, cells were incubated with a trace amount of [^{125}I]-transferrin at 0°C for 45 min. Unbound transferrin was washed off, and cells were incubated in a water bath at 37°C. At indicated times, cells were placed in an ice-water bath (0°C) and digested with 10 μg/ml pronase for 60 min. Free and cell-associated transferrin were separated and radioactivity was counted in a gamma counter. Results are expressed as the mean ± S.D. of three determinations.

DISCUSSION AND CONCLUSION

I have summarized the results from our current studies on the effects of exogenous gangliosides as well as those of anti-glycolipid antibodies on cell growth. In both cases, the target molecules at the cell surface could be growth factor receptors, although our observations are limited and there may be other target molecules that we have not studied as yet. Interestingly, exogenous addition of GM3 to BHK cells appears to influence the receptor for FGF, while GM3 or GM1 added to Swiss 3T3 cells, KB cells, and A431 cells affects the function of the PDGF and EGF receptors. Since the function of these receptors has been shown clearly to be regulated by phosphorylation of the receptor, the effect of gangliosides has been investigated with focus on the phosphorylation of these receptors. Since our knowledge of the FGF receptor is incomplete, our major focus has been on the PDGF and EGF receptors. Clearly, tyrosine phosphorylation of both receptors was strongly inhibited in a dose-dependent manner when membrane preparations were examined *in vitro*. The inhibitory effect of GM3 on the EGF receptor was also reproduced in immunoprecipitates with anti-EGF receptor antibody. The inhibitory effect of GM3 on tyrosine phosphorylation of the EGF receptor was also demonstrated *in situ* in cells cultured in the presence of GM3. Those growth-arrested cells showed less tyrosine phosphorylation of their EGF receptors. The results of these studies clearly demonstrate the effects of gangliosides on receptor function, which are mediated by promotion or inhibition of phosphorylation of growth factor receptors. It has been shown incidentally that phosphatidylethanolamine, but not other phospholipids, promotes EGF receptor phosphorylation. This effect is due to inhibition of tyrosine phosphatase rather than enhancement of phosphorylation (Torres-Mendez. C.-R., Cooper, J., and Hakomori, S., unpublished observation). These results also indicate the important role of the membrane lipid environment in the function of growth factor receptors. An idealized view of the possible regulation of receptor function by gangliosides is illustrated in Figure 9.

The effect of polyclonal anti-glycolipid antibodies on cell growth was described previously (Lingwood and Hakomori, 1977). More recently, the effect of biotinyl monoclonal antibodies and avidin on cell growth has been studied (Okada et al., 1985). The mechanism of such antibody effects is highly complex. However, in L5178 lymphoma, which can grow in the absence of growth factors but requires transferrin, the target molecule could be the receptor for transferrin. Cells whose growth was inhibited in the presence of biotinyl antibody to Gg3Cer and avidin displayed a remarkable inhibition of transferrin internalization and separation of Gg3Cer from the transferrin receptor at the cell surface. It is assumed that internalization of the transferrin receptor with bound transferrin may require Gg3Cer, although this interpretation is still tentative. Obviously, ion transport and function of transporters will be important to explore.

Gangliosides have been detected in cell adhesion sites (Okada et al., 1984; Cheresh et al., 1984), and the exogenous addition of ganglioside at the initial stage of plating (i.e., when cells are detached and added on plates) inhibits cell adhesion on plates (Okada et al., 1984; Rauvala et al., 1981; Kleinman et al., 1979). It is possible, therefore, that membrane gangliosides may affect the function of adhesive proteins, or their receptors. As mentioned in the Introduction, I have only discussed in this arti-

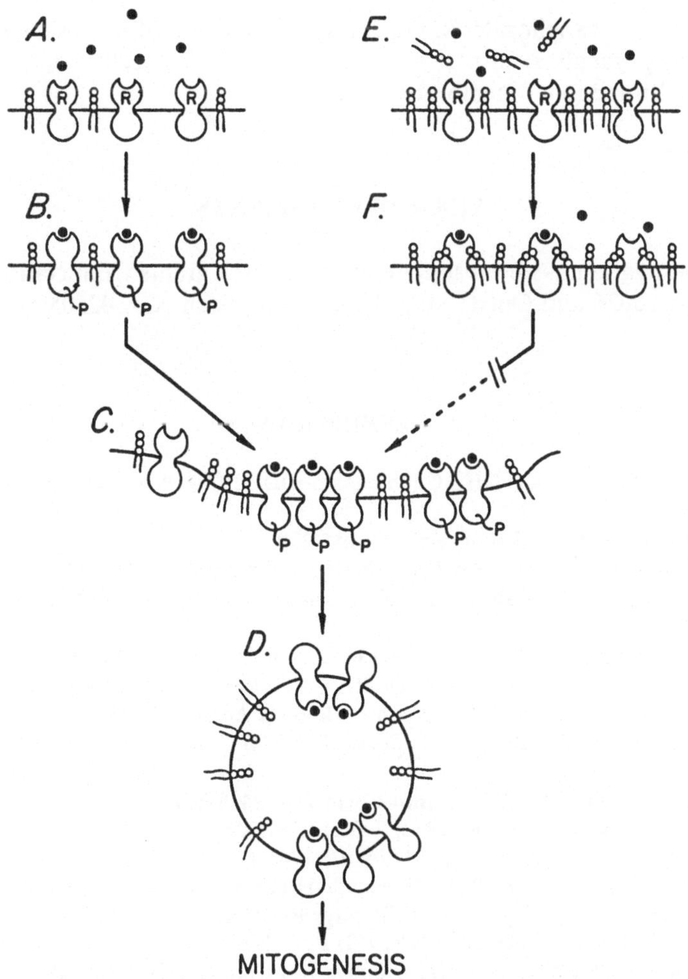

Figure 9. A possible mechanism of cell growth stimulation by growth factor, and inhibition of cell growth by gangliosides. Growth factor (•) added at the cell surface (A) is bound to the receptors (R). Subsequently, kinase activity of the cytoplasmic domain is activated, and receptors are phosphorylated (P). The phosphorylated receptors alter their conformation and aggregate together to form "coated pits" (C), which are in turn internalized (D). The internalized receptor-growth factor complex will be utilized for mitogenesis. GM3 and GM1 gangliosides are capable of interacting directly or indirectly with epidermal growth factor receptor or platelet-derived growth factor receptor to inhibit their kinase activity and autophosphorylation. The level of GM3 in normally-growing cells, however, is not sufficient to strongly inhibit phosphorylation of the receptor (A-B). Therefore, internalization of the receptor-growth factor complex can proceed (C-D). When cells were grown in medium with exogenous addition of GM3 ganglioside (E), GM3 ganglioside incorporated in the lipid bilayer interacts strongly with the growth factor receptor, and subsequently, tyrosine phosphorylation of the receptor is inhibited (F). Therefore, internalization of the growth factor-receptor complex does not proceed (F-G).

cle the basic role of gangliosides in regulating the function of intrinsic membrane proteins. Obviously, gangliosides play other major roles in cell-cell interactions; they are essentially multifunctional membrane modulators.

ACKNOWLEDGMENTS

This investigation has been supported by research grants from National Institutes of Health CA 20026 and Outstanding Investigator Grant CA 42505.

REFERENCES

Baker JB, Barsh GS, Carney DH, Cunningham DD (1978) Proc Natl Acad Sci USA 75: 1881-1886.

Barnes D, Sato G (1980) Anal Biochem 102: 255-270.

Bremer EG, Hakomori S (1982) Biochem Biophys Res Commun 106: 711-718.

Bremer EG, Hakomori S, Bowen-Pope DF, Raines E, Ross R (1984) J Biol Chem 259: 6818-6825.

Bremer EG, Schlessinger J, Hakomori S (1986) J Biol Chem 261: 2434-2440.

Cheresh DA, Harper JR, Schultz G, Reisfeld RA (1984) Proc Natl Acad Sci USA 81: 5767-5771.

Critchley DR, Macpherson IA (1973) Biochim Biophys Acta 296: 145-159.

Fishman PH, Simmons JL, Brady RO, Freese E (1974) Biochem Biophys Res Commun 59: 292-299.

Gahmberg CG, Hakomori S (1975) J Biol Chem 250: 2438-2446.

Glenn K, Bowen-Pope DF, Ross R (1982) J Biol Chem 257: 5172-5176.

Hakomori S (1970) Proc Natl Acad Sci USA 67: 1741-1747.

Keenan TW, Schmidt E, Franke WW, Wiegandt H (1975) Exp Cell Res 92: 259-270.

Kijimoto S, Hakomori S (1971) Biochem Biophys Res Commun 44: 557-563.

Kleinman HK, Martin GR, Fishman PH (1979) Proc Natl Acad Sci USA 76: 3367-3371.

Laine RA, Hakomori S (1973) Biochem Biophys Res Commun 54: 1039-1045.

Langenbach R, Kennedy S (1978) Exp Cell Res 112: 361-372.

Lingwood CA, Hakomori S (1977) Exp Cell Res 108: 385-391.

Maciag T, Kelley B, Cerundolo J, Ilsley S, Kelley PR, Gaudreau J, Forand R (1980) Cell Biol Int Report 4: 43-50.

Neufeld G, Gospodarowicz D (1985) J Biol Chem 260: 13860-13868.

Okada Y, Mugnai G, Bremer EG, Hakomori S (1984) Exp Cell Res 155: 448-456.

Okada Y, Matsuura H, Hakomori S (1985) Cancer Res 45: 2793-2801.

Patt LM, Itaya K, Hakomori S (1978) Nature 273: 379-381.

Pike LJ, Bowen-Pope DF, Ross R, Krebs EG (1983) J Biol Chem 258: 9383-9390.

Rauvala H, Carter WG, Hakomori S (1981) J Cell Biol 88: 127-137.

Rozengurt E, Brown KD, Pettican P (1981) J Biol Chem 256: 716-722.

Sakiyama H, Gross SK, Robbins PW (1972) Proc Natl Acad Sci USA 69: 872-876.

Shoyab M, Todaro GT (1981) Arch Biochem Biophys 206: 222-226.

Simmons JL, Fishman PH, Freese E, Brady RO (1975) J Cell Biol 66: 414-424.

Yogeeswaran G, Hakomori S (1975) Biochemistry 14: 2151-2156.

Young WW, Jr, MacDonald EMS, Nowinski RC, Hakomori S (1979) J Exp Med 150: 1008-1019.

Gangliosides and neuronal plasticity
G. Tettamanti, R.W. Ledeen, K. Sandhoff,
Y. Nagai, G. Toffano (eds.)
Fidia Research Series, vol. 6
Liviana Press, Padova, © 1986

NEURITIC RESPONSES TO GM1 GANGLIOSIDE IN SEVERAL IN VITRO SYSTEMS

Silvio Varon, Stephen D. Skaper, Ritsuko Katoh-Semba[1]

Department of Biology, School of Medicine, University of California, San Diego La Jolla, California 92093, USA and
[1]Department of Perinatology Institute for Developmental Research, Aichi Prefectural Colony, Kasugi, Aichi, Japan

INTRODUCTION

In explant or monolayer neural cultures, a neuron is subject to various influences from the medium, the substratum, and the other cells present. Medium components also alter the substratum properties, and cultured cells modify both medium and substratum besides themselves via direct contacts. This multiplicity of influence sources applies in vivo, and in pathological as well as normal circumstances. If the balance of such extrinsic influences is already optimal for the neural performance to be investigated, then experimental interventions cannot reveal agents with a promoting effect but only agents with an inhibiting influence. To seek and investigate a beneficial (promoting) agent one must rely on restrictive circumstances, either imposed experimentally or occuring naturally as may be the case in developmental or pathological situations.

The outgrowth of a neuronal process, or neurite, is controlled by a variety of extrinsic signals, both humoral and surface-anchored (Varon and Adler, 1980, 1981). Neuronotrophic agents are necessary for the general support of survival and growth capabilities of the neuron, but the control of neuritic growth is largely applied directly to the elongation machinery, represented by the neuritic growth cone and involving membrane and cytoskeletal dynamics. Most important appear to be the adhesive interactions between the surface of the growth cone membrane and the surfaces on which it will operate: cell surfaces (Adler and Varon, 1981; Noble et al., 1984), extracellular matrix (Sanes, 1983), or culture substratum and their coats (Letourneau, 1979).

Abbreviations: CNS, central nervous system; PNS, Peripheral nervous system; NGF, nerve growth factor; CNTF, ciliary neuronotrophic factor; DRG, dorsal root ganglion.

Adhesive interactions may be altered by substances modifying either the growth cone membrane or the substratum. Neurite promoting agents, therefore, may occur and act from the humoral environment or only after anchorage to the substratum. Several proteins have already been described to perform as neurite promoting factors (Campenot, 1982; Kligman, 1982; Henderson et al., 1981; Manthorpe et al., 1983; Gundersen, 1985; Davis et al., 1985b, c). Similarly, neurite-inhibiting agents must occur (also either as humoral or as substratum-anchored substances), and serum has been reported to include some such inhibitors (Skaper et al., 1983a, b; Davis et al., 1984).

To investigate new agents capable of promoting neuritic behaviors, one must design the experimental system in such a way that neuritic performance is less than optimal so as to allow recognition of a beneficial effect, and yet not so drastically restricted that the putative promoting agent cannot overcome the restriction. Such a fine tuning may require manipulation of any or all of the various neurite-modulating influences already listed: culture substratum and medium, and neuronotrophic, neurite-promoting and neurite-inhibiting agents. Time in culture must also be considered, since it is an important element both in the execution of a neuritic performance and in the changes occuring in medium, substratum and cell components of the culture. We shall describe here the application of these concepts to the investigations of GM1 ganglioside as a neurite modulating agent in vitro and show that GM1 promotes neuritic outgrowth from a variety of ganglionic explants, primary PNS and CNS neural monolayer cultures and PC12 cell cultures — provided that in each case the experimental conditions are suitably defined.

GANGLIONIC EXPLANT CULTURES

Neuritic outgrowth has been investigated for many years by use of explant cultures (Weiss and Hiscoe, 1948; Levi-Montalcini et al., 1954; Murray, 1965) and GM1 effects have been reported on the neuritic outgrowth from E8 (embryonic day 8) chick dorsal root ganglia (Roisen et al., 1981). Ganglionic explant cultures have been recently used to compare neuritic modulation by different substrata, the presence of absence of NGF, serum supplementation of the culture medium, and time in vitro (Skaper et al., 1985). Chick E8 dorsal root ganglia (DRG8), E11 sympathetic ganglia (SG11), and E8 ciliary ganglia (CG8) were explanted on polyornithine (PORN)-coated tissue culture plastic in a modified Eagle's Basal Medium (EBM) containing the serum-replacing, chemically defined N1 supplement (EBM/N1), and cultured for 48 hr. Experimental supplements were fetal calf serum, neuronotrophic factors and GM1, supplied at various concentrations either individually or in combinations. Figure 1 demonstrates the promoting effect of GM1 on the three types of ganglionic explants. Table 1 A lists the different sets of conditions needed in each case to optimize the display of such an effect.

DRG8 explants required omission of the trophic factor (Nerve Growth Factor, or NGF) and addition of serum (10%) — the latter imposing the need for a relatively high concentration of GM1 (Toffano et al., 1980). Inclusion of NGF would cause a massive outgrowth of neurites — the typical NGF "halo" — which masks any GM1 effect. Conversely, omission of the serum would allow enough neuritic outgrowth without

GM1 to make the GM1-induced differential less readily detected. A collagen (rather than PORN)-coated substratum would have been too restrictive on neuritic outgrowth (in the absence of NGF) for GM1 to overcome the restriction. Lastly, the GM1 differential was not detectable at 24 h and much less distinguishable after 72 hr — when untreated explants have almost caught up with their GM1-tested counterparts — illustrating the importance of the time element and raising the possibility that GM1 pro-

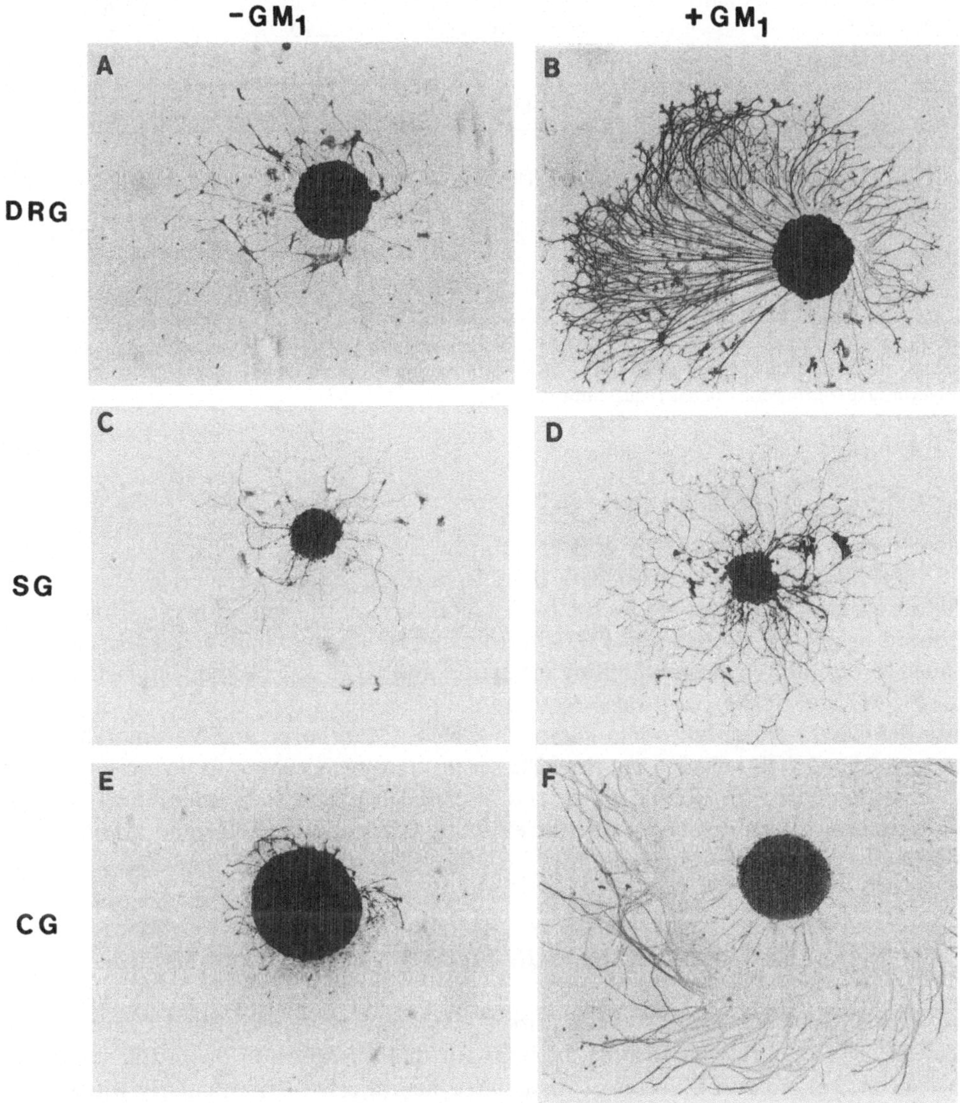

Figure 1. Effect of GM1 on neuritic outgrowth from 48 hr explant cultures of chick E8 dorsal root ganglia in 10% serum (A, B), E11 sympathetic ganglia in 10% serum and NGF (C, D), and E8 ciliary ganglia in 1% serum and CNTF (E, F). Control (A, C, E); GM1 (B, D, F).

Table 1. *Summary of Conditions for Optimal Neuritic Response to GM1 by Various Explant and Monolayer Cultures from PNS and CNS Tissues*

	DRG8	SG11	CG8
A. EXPLANTS			
Substratum	PORN	PORN	PORN
Time (hr)	48	48	48
Trophic agent (TU/ml)	0	0.15	10.0
Serum %	10	10	1
GM1 (M)	10^{-4}	10^{-4}	10^{-6}
	DRG8	CG8	CNS
B. MONOLAYERS			
Substratum	PORN	PORN	PORN/FN
Time (hr)	12-24	5-8	7-24
Trophic agent (TU/ml)	20-50	50	10^{-3} M (PYR)
Serum (%)	0.3-1	0	0
GM1 (M)	3×10^{-6}—3×10^{-5}	3×10^{-8}	10^{-7}

DRG8, chick embryo dorsal root ganglia 8; SG11; chick embryo symphopathetic ganglia 11; CG8, chick embryo ciliary ganglia 8; TU, trophic unit; PORN, polyornithine - coated tissue culture plastic; PORN/FN, polyornithine/fibronectin coated tissue culture plastic.

motes an earlier rather than a greater outgrowth of neurites.

SG11 explants, in contrast with the DRG8 ones, would not survive for 48 hr in the absence of their trophic factor NGF. NGF concentrations, however, had to be reduced below 0.5 trophic unit (TU)/ml to avoid its massive halo effect. Serum continued to be required as an inhibiting influence, and GM1 was consequently used again at 10^{-4} M. The CG8 explants also required the presence of their trophic factor, in this case the Ciliary Neuronotrophic Factor or CNTF (Manthorpe and Varon, 1985), for the survival of their neurons but the 10% serum treatment proved to be too inhibitory on neuritic outgrowth. Reducing the serum concentration to 1% maintained enough inhibition for a clear display of neuritic effect by GM1 (used at 100-fold lower concentrations because of the lower serum).

PRIMARY NEURONS IN MONOLAYER CULTURES

Explant cultures do not allow direct visualization of individual neurons and neurites, and require cumbersome methods for quantification of a neuritic response. In contrast, monolayer cultures of dissociated neural cells — particularly when non-neuronal elements have been kept at a minimum and the enriched neurons are seeded rather sparsely — permit the analysis, at any chosen time, of both the number of neurons that survive and the proportions of surviving neurons which display neuritic outgrowth. The latter determination defines the temporal recruitment of cells into a neuritic expression, an easy way to recognize and measure the impact of neurite-promoting agents. Neuritic responses, however, may involve several additional features of neuritic growth — number of neurites/cell, neuritic branching, neurite length, elongation rate, etc — the measurement of which are much more laborious. GM1 effects on neuritic growth have been reported in dissociated neuronal cultures from chick DRG (Roisen et al., 1981; Leon et al., 1984) and rat hippocampus (Seifert, 1981).

We have applied (Skaper et al., 1985) the concept of finely tuned restrictions to an analysis of GM1 neuritic effects in monolayer cultures of neurons from peripheral (DRG8, CG8) and central (E8 chick forebrain, E18 rat hippocampus) neural tissues. Cells dissociated from these sources were seeded into PORN-coated 6 mm microwells at 1000-1500 neurons/well, the seeding medium was replaced 4 hr later with Dulbecco's Modified Eagle's Medium plus N1 (DMEM/N1), containing the desired experimental supplements, and replicate cultures were fixed after various incubation times for numerical analyses. Table 1 B lists the optimal conditions under which a GM1 effect was readily measurable, and Figure 2 shows representative photographs of untreated and GM1-treated neuronal cultures.

DRG8 neurons required the presence of NGF for their survival (unlike their behavior in explant cultures) and the presence of serum (at 0.3% or higher concentrations) for the display of a GM1 effect on neuritic recruitment. Surprisingly, serum appeared to be required not to depress neuritic expression in control cultures but to enhance it under GM1 treatment. A 3-fold increase in neuritic recruitment by GM1 was already measurable at 12 hr. At subsequent times, recruitment in control cultures rose more rapidly than under GM1 treatment, thereby reducing the "relative" differential (none detectable any longer by 48 hr).

CG8 neurons also required their own trophic factor (CNTF) for survival and growth competence. Serum at 1% allowed only a slow and modest neuritic outgrowth, and no GM1-derived improvement. However, GM1 effects were measurable with 0.1% or lower serum and best determined in the complete absence of it — with a corresponding lowering of effective GM1 concentrations to the 10^{-8} range. In serum-free cultures, CNTF-supported CG8 neurons are fully recruited into a neuritic behavior by 24 hr. GM1 advanced considerably the recruitment in time, leading to a 3-fold increase over control by as early as 5-8 hr. As with the DRG neurons and the ganglionic explant studies, the GM1 benefits on neuritic recruitment were only temporary since control cultures rapidly caught up in this respect. Nevertheless, qualitative differences could still be detected at 24 hr in the overall extent of neuritic outgrowth, presumably reflecting the longer time spent on neuritic growth by the earlier-starting, GM1-treated neurons.

CNS neurons, whether from chick or rat embryonic tissues, require for their survival the availability in the medium of pyruvate — an ingredient of DMEM but not EBM (Selak et al., 1985). The PORN substratum, in the absence of serum, reduces CNS neuronal survival and even more so neuritic outgrowth, but pretreatment of the PORN with fibronectin (PORN/FN substratum) provided satisfactory conditions for both neuronal survival and a restrained neuritic extension. GM1 ganglioside (at 10^{-7}M, reflecting the absence of serum) increased neuritic recruitment 2-3 fold at any time between 7 and 24 hr culture. The same behaviors and GM1 effects were observed with other CNS neuronal cultures as well (data not shown).

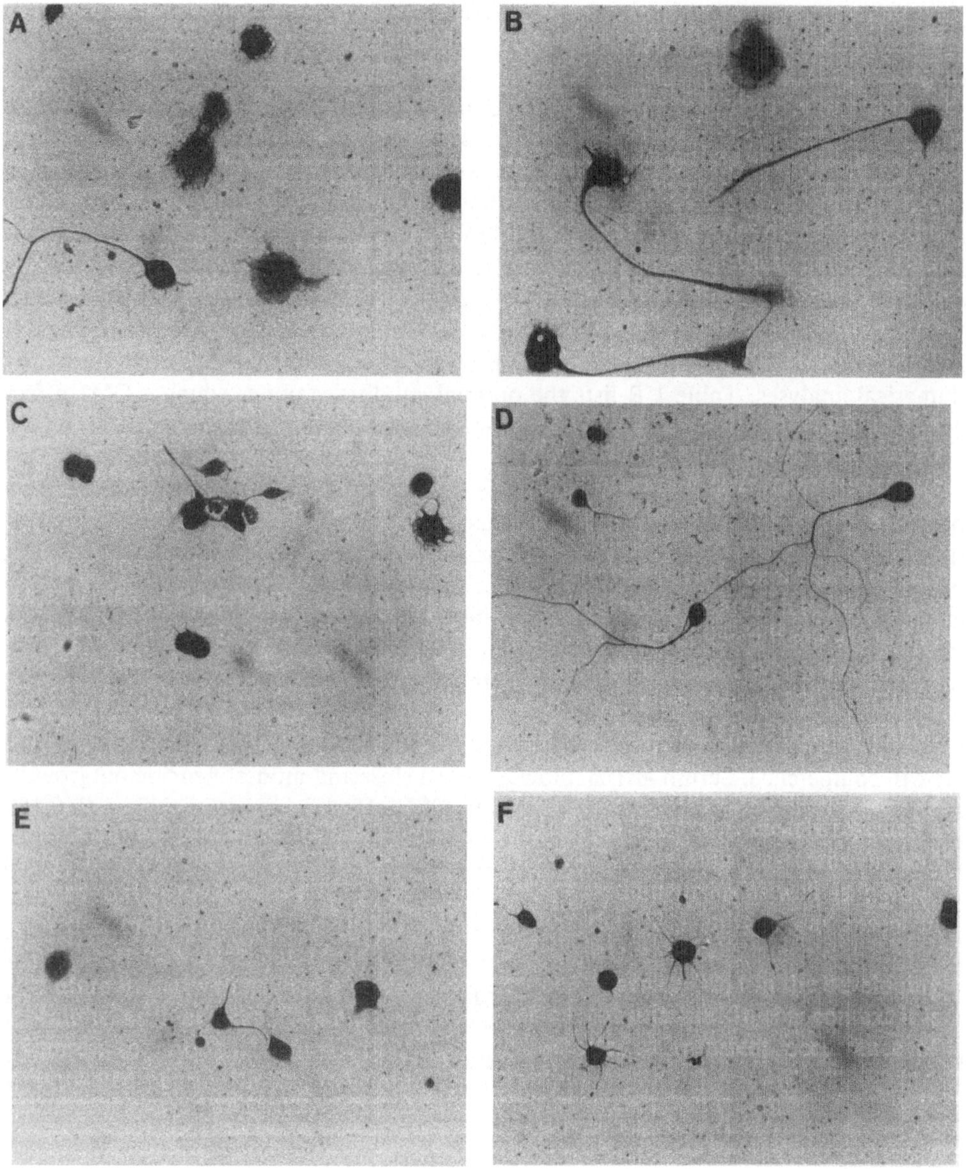

Figure 2. Microphotographs of chick E8 ciliary ganglionic neurons with CNTF at 8 hr (A, B) and dorsal root ganglionic neurons with NGF at 12 hr (C, D), and E18 rat hippocampal neurons with pyruvate at 14 hr (E, F). Control (A, C, E); GM1 (B, D, F).

Figure 3 shows a composite of GM1 titration curves obtained with DRG8, CG8 and CNS neuronal cultures each used under the optimal conditions and analyzed at several of the optimal times indicated in Table 1B. Optimal GM1 concentrations were the same at the different times for a given test neuronal population and a given set of conditions. CG8 neurons and CNS neurons, tested without serum, responded maximally to GM1 in the 10^{-8}—10^{-7}M range. DRG8 neurons, requiring 1% serum for a good response, reacted optimally to a correspondingly higher GM1 concentration range (around 10^{-5}M). Note that higher than optimal doses of GM1 lead to reduced effects.

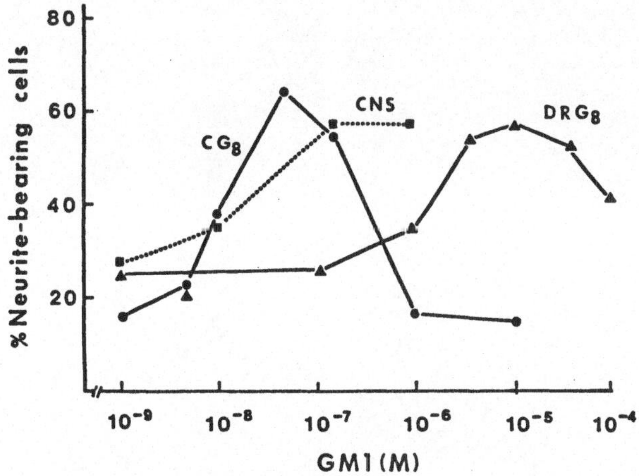

Figure 3. GM1 titration curves for chick E8 ciliary ganglion neurons (•) and dorsal root ganglion neurons (▲), and E18 rat hippocampal neurons (■), using the optimal conditions described in Table 1.

PC12 CELL CULTURES AND THEIR NGF-INDUCED NEUTRITE OUTGROWTH

PC12 cells, a clonal line from rat pheochromocytoma, respond to NGF treatment with an expression of neuronal properties, which include arrested proliferation and extension of neurites (Greene, 1978). GM1 has been reported to enhance the neuritic response of PC12 cells to NGF (Ferrari et al., 1983). Neuritic recruitment of PC12 cells by NGF, on the other hand, is considerably delayed by the presence of serum (Skaper et al., 1983b). We have, therefore, undertaken to analyze possible relationships among NGF, serum, GM1, and time effects on PC12 cell neuritic behaviors (Katoh-Semba et al., 1984; 1986). PC12 cells, grown in DMEM with 15% serum, were reseeded in serum free DMEM/N1 into PORN-coated 35 mm dishes at 20,000 cells/dish. The seeding medium was replaced at 2 hr (50% of the seeded cells were attached by then) with the experimental media (DMEM/N1 with the desired serum, NGF and/or GM1 supplementations), and the cultures periodically analyzed over 8 days without further medium changes (except for addition, every 3 days, of fresh NGF to NGF-containing cultures).

Figure 4 illustrates the morphology of 4 day PC12 cells cultured under the various conditions. NGF-untreated cells display no neurite-like processes (A). NGF treatment

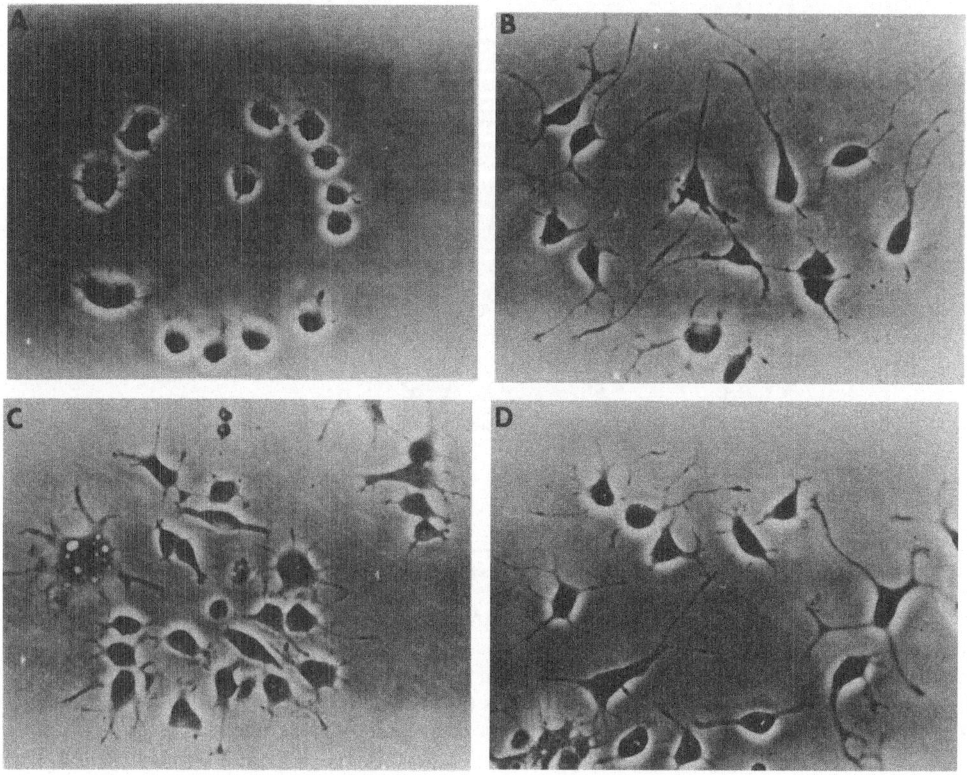

Figure 4. Photomicrographs of PC12 cells cultured for 4 days with: (A) 0.15% serum or 1% serum ± GM1; (B) 0.15% serum and NGF ± GM1; (C) 1% serum with NGF and without GM1; (D) 1% serum with NGF *and* GM1.

has recruited into a neuritic behavior most of the cells when supplied with low (0.15%) serum (B), but very few of them with 1% (or higher) serum levels (C). GM1 elicits no neurites in the absence of NGF (A), causes no recognizable enhancement when low serum allows an extensive response to NGF (B) but clearly improves the neuritic response to NGF in the presence of a higher serum level (D).

Quantitative analyses of these PC12 cell behaviors are shown in Figure 5. Proliferation (Fig. 5 A) proceeds exponentially in the absence of NGF, regardless of the serum concentration or the presence or absence of GM1, and also in the presence of NGF up to day 4 — after which time NGF stops numerical growth regardless of serum and/or GM1 treatments. Neuritic recruitment under NGF (Fig. 5 B; lack of neuritic expression in the absence of NGF is not shown) starts early and proceeds rapidly in low serum without being affected by GM1 — whereas in higher (1%) serum it displays both a 4 day lag and a slower rate. It is only under these serum-restricted circumstances that GM1 enhances neuritic recruitment of PC12 cells, largely by reducing the serum-imposed lag rather than minimizing the serum-induced rate decline. At any single time point, therefore, it would appear that GM1 increases the number of PC12 cells

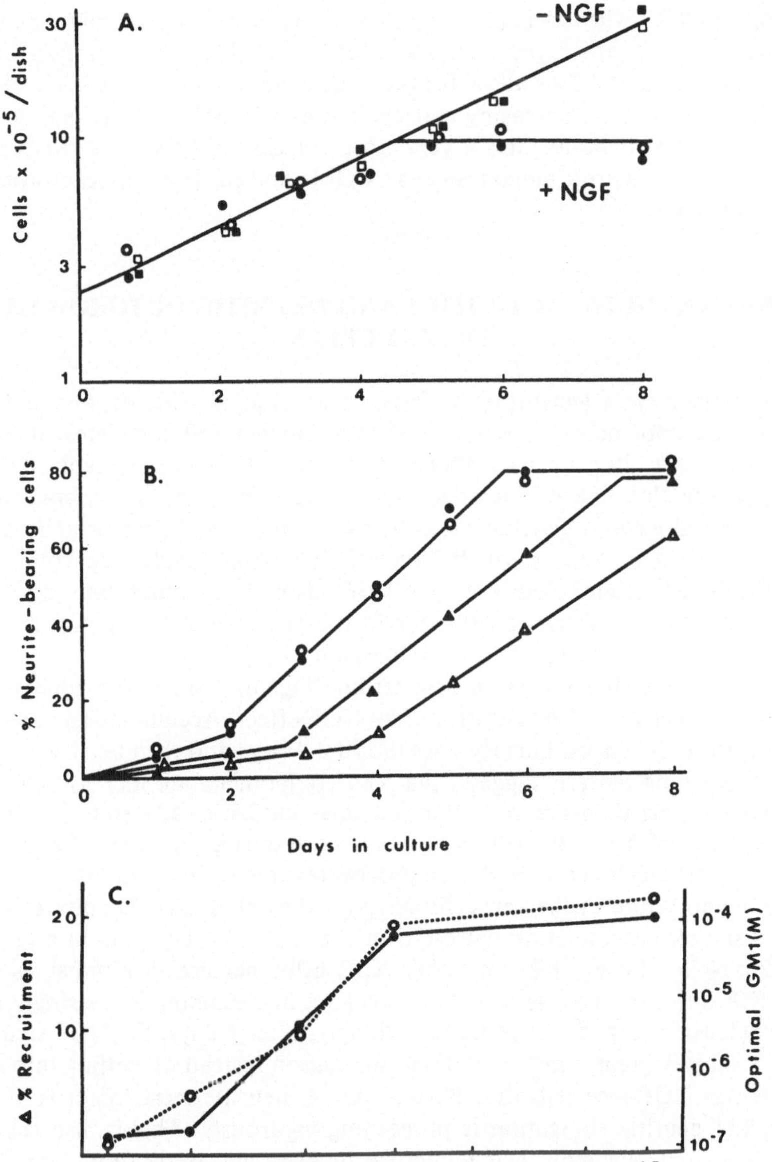

Figure 5. Proliferation (A), neuritic recruitment (B) and concentration-dependent effects of serum and GM1 on neuritic growth (C) of PC12 cells.
(A): 1% serum without (■) and with (□) GM1; with NGF (•); with NGF and GM1 (○). Identical results were obtained with O.15% serum. (B): 1% serum and NGF without (▵) or with (▴) GM1; 0.15% serum and NGF without (•) or with (○) GM1.
(C): ▵ % recruitment at 3 days (••) shows the difference in percentage of neurite-bearing cells between GM1-treated and control cultures; optimal GM1 concentration at different serum concentrations (○○).

responding to NGF, whereas in fact any given level of neuritic recruitment is merely achieved some 2 days earlier. Increasing concentrations of serum will impose increasing extents of restriction and thus allow for increasing amplitudes of the GM1 effect, but will also force the use of increasing concentrations of GM1 for maximal ganglioside benefits. Figure 5 C demonstrates a very close correlation between serum dosage influences on optimal neuritic enhancements by GM1 and the GM1 concentrations needed to achieve them.

BIOSYNTHETIC ACTIVITIES AND NEURITIC OUTGROWTH IN PC12 CELLS

The opportunity of generating different degrees of neuritic expression from the same PC12 cell pool makes it possible to seek biochemical correlates of either the neuritic performance itself or its modulation by the various agents involved. We have begun to pursue this line of investigation by examining the incorporation of ^{3}H-galactose into endogenous gangliosides, other glycolipids, and glycoproteins as well as the concurrent incorporation of ^{14}C-acetate into non-ganglioside total lipid of cultured PC12 cells (Katoh-Semba et al., 1986). Isotope accumulation (per 10^{6} cells) into these four classes of cell constituents in the presence of low serum is shown in Figure 6, and that in the presence of high serum in Figure 7.

The four temporal patterns in low serum (Fig. 6) were nearly identical, quite reproducible, and clearly demonstrative of an NGF effect. Accumulation per cell in the NGF-free cultures increased linearly over the first 4 days and then persisted at a constant level. This time pattern suggests that cell components are increasingly replaced with newly-synthesized, hence radio-labeled ones until a steady state is achieved (4 days): generation of new cells, which proceeds both during and after the first 4 days, will not affect the labeling accumulation process (expressed on a per cell basis) as long as size and composition of the new cells do not differ. NGF-treated cultures appeared to follow a similar accumulation pattern over the first four days, but at a more rapid rate. Under NGF, during this time, the PC12 cells increase in number as they do without NGF but also increase in volume because of becoming increasingly engaged in neurite extension (Fig. 5). Even more strikingly, after 4 days, the NGF-treated cells shifted to a sharply greater rate of radioaccumulation instead of settling into a steady state pattern as NGF-free cells did. Beyond day 4, neuritic recruitment is practically completed but neuritic elongation is proceeding vigorously. This is also the time at which NGF-treated cells stop their numerical increase and begin to undergo a somal enlargement. GM1 treatment under low serum, with or without NGF, had no effect at any time on radioaccumulation into any one of the four cell constituent classes — just as it revealed no impact on neuritic recruitment (Fig. 5).

Radioaccumulation patterns in the presence of high serum (Fig. 7) differed from those in low serum in two major respects:

(1) The first difference is the consistently higher radioaccumulation achieved in the presence than the absence of exogenous GM1. The GM1-induced increase concerns all three categories of lipids, but not the glycoprotein pool. The same 1.5-fold increase was imposed by GM1 on the labeling of lipid constituents in the absence and in the presence

of NGF, indicating independent and cumulative effects by the two agents. NGF itself (with or without GM1) displayed the same overall pattern (higher rate over the first 4 days, followed by a sharp further upturn) and nearly the same absolute levels under high serum as those seen with low serum. The GM1 stimulation was truly on biosynthesis of the lipid constituents, since neither the uptake of radioprecursors nor the

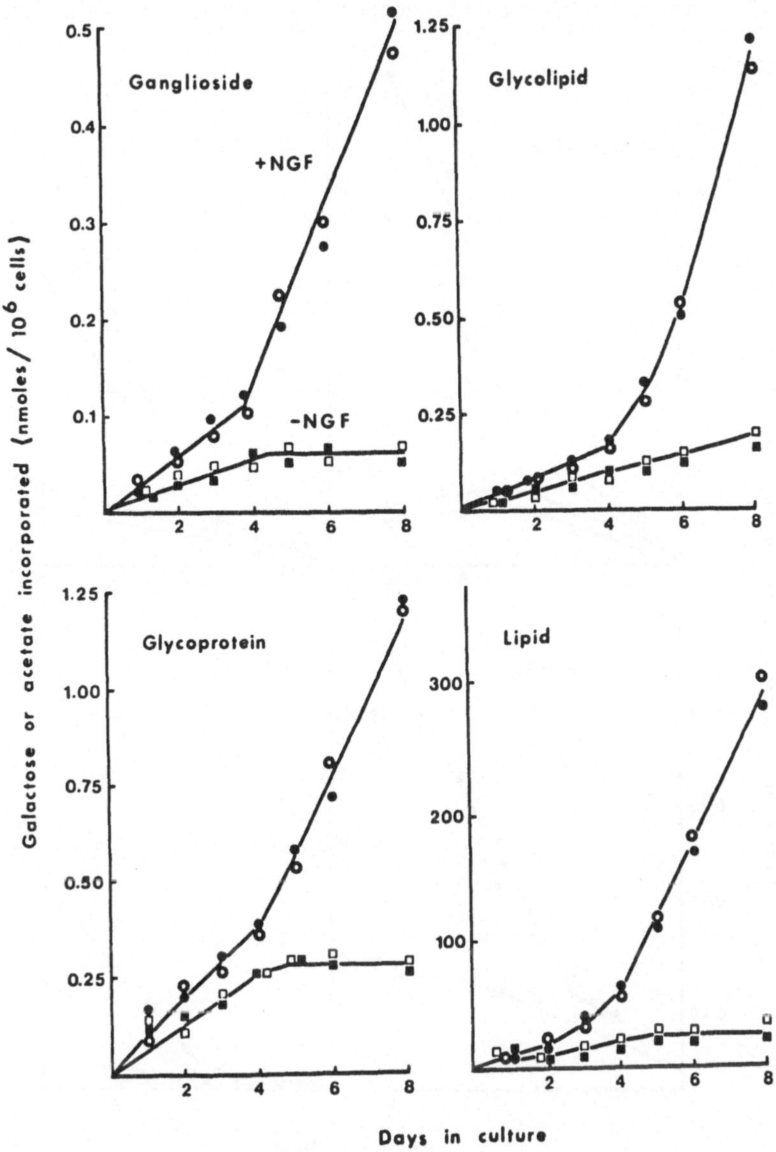

Figure 6. Temporal patterns of ^{3}H-galactose incorporation into gangliosides, non-ganglioside glycolipids and glycoproteins, and of ^{14}C-acetate into non-ganglioside total lipids, using 0.15% serum-supplemented medium. No additions (■); GM1 (□); NGF (•); GM1 + NGF (○).

degradation rates of prelabeled cell constituents were affected by the exogenous GM1 treatment. GM1 did not merely accelerate turnover of lipid components, a performance that might account for the earlier rate increase but not for the sustained higher levels in either the steady state period of NGF-free cultures or the ongoing radioaccumulation in NGF-treated ones. GM1 did not obviously increase the size of PC12 cells nor did it increase neurite extension in the absence of NGF. It appears from the available information, therefore, that exogenous GM1 may be altering the relative con-

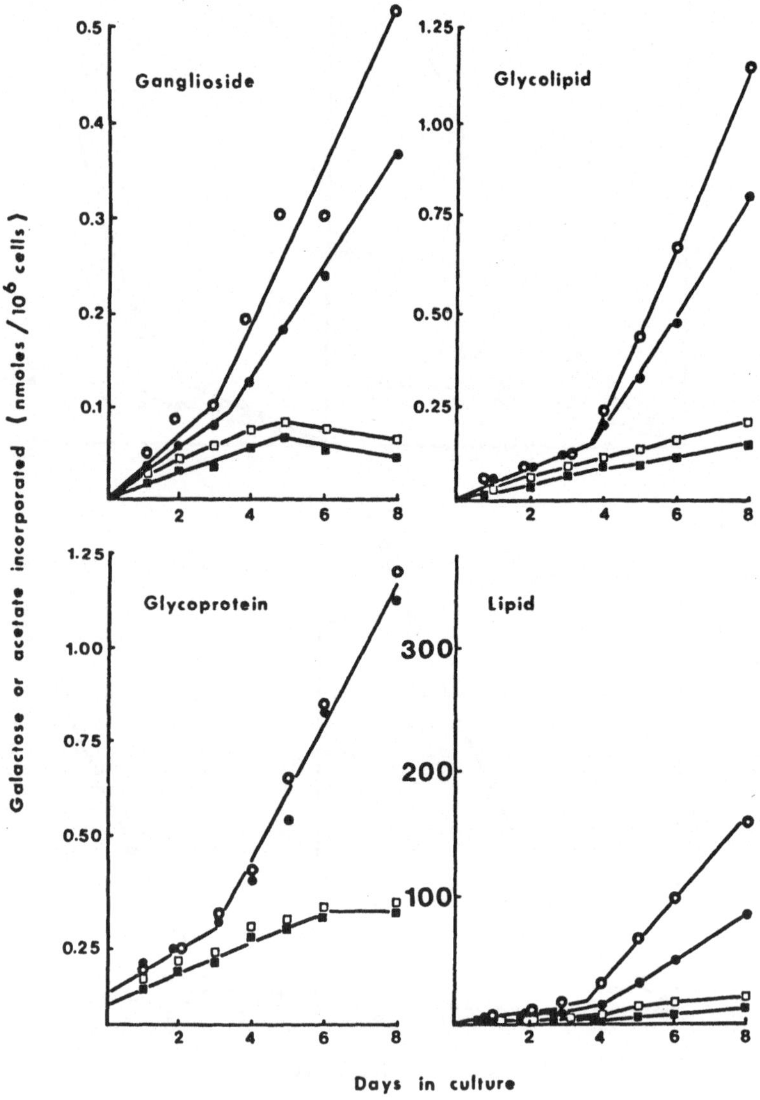

Figure 7. Temporal patterns of ^{3}H-galactose incorporation into gangliosides, non-ganglioside glycolipids and glycoproteins, and of ^{14}C-acetate into non-ganglioside total lipids, using 1% serum-supplemented medium. No additions (■); GM1 (□); NGF (•); GM1 + NGF (○).

tribution by lipid over protein syntheses to the biochemical composition of the PC12 cells cultured in high serum. Since this biochemical effect of GM1 could only be detected in the presence of high serum, it may either reflect the lifting of a serum-imposed restriction on lipid synthesis or represent an action of GM1 which requires concurrent support from some unidentified serum component.
(2) The second major difference concerned a considerable reduction in the overall levels at which non-ganglioside lipid incorporated ^{14}C-acetate in high versus low serum (compare the ordinate scales in Figs. 6 and 7). It appears that high serum inhibits the actual rate of lipid synthesis rather than reducing the specific activity of both precursor and product via substances which contribute to the intracellular acetate (though not galactose) pool used for lipid synthesis. Thus, serum-induced inhibitions of lipid synthesis and neuritic recruitment may relate to each other, as well as the ability of exogenous GM1 to relieve both.

This interpretation of the serum effect on total lipid radioaccumulation is strongly encouraged by the data shown in Figure 8. Neuritic recruitment (ordinate), measured in individual NGF-supported cultures regardless of serum and/or GM1 presence, is plotted against the radioactivity accumulated by the same individual cultures from either ^{3}H-galactose into ganglioside, glycolipid or glycoprotein (A) or ^{14}C-acetate into nonganglioside total lipid (B). The three glycoconjugate classes yelded identical patterns of which Figure 8, A is a composite. With the galactose incorporation, all points fall smoothly along two different curves, depending on whether low or high serum was involved. One underlying difference, of course, is time (Fig. 5). In low serum, any given level of neuritic recruitment is achieved sooner and is thus accompanied by less glyconjugate labeling. High serum delays neurite recruitment without similarly effecting galactose incorporation. Note, however, that the GM1-derived points fit the same curve as those from GM1-free cultures, despite their temporal differences — indicating that GM1 increases in parallel neuritic recruitment and galactose incorporation.

A dramatic contrast is offered by the ^{14}C-acetate-related plot (Fig. 8 B). Here, all points fit on a single curve regardless of high or low serum, and this single curve is identical to that obtained in low serum with the ^{3}H-galactose labeling. Thus, acetate label-

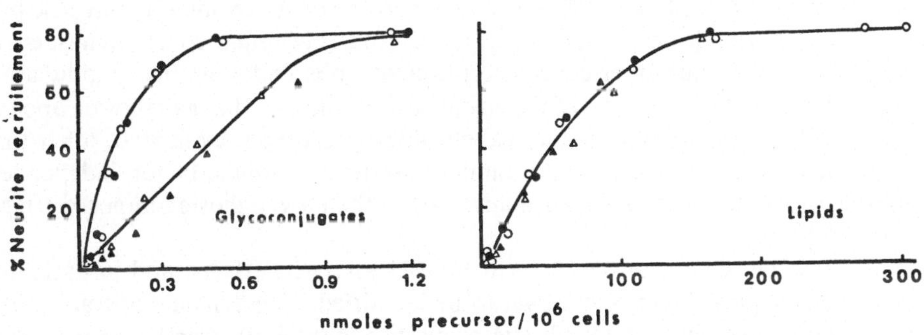

Figure 8. Relationships between neurite recruitment and ^{3}H-galactose incorporation into gangliosides, glycolipids or glycoproteins (A), or ^{14}C-acetate incorporation into non-ganglioside total lipid (B). 0.15% serum: NGF (•), NGF + GM1 (○); 1% serum: NGF (▲), NGF + GM1 (△).

ing of total lipid appears to represent faithfully the NGF-induced neuritic performance by PC12 cells, regardless of the modulations imposed on it by either or both serum and GM1. A clear question for the future, then, is whether lipid synthesis correlates so precisely with neuritic performance because lipid synthesis monitors neurite production or because it controls it.

DISCUSSION

We have shown that GM1 can promote neuritic outgrowth in vitro if appropriate restrictions are imposed on the optimal execution of a neuritic performance. Restrictions are appropriate when they are sufficient to reveal a stimulatory action by extrinsic agents, yet not extensive enough to overwhelm the latter. In the studies reviewed here, restrictions to neurite execution were imposed experimentally, by the selection of culture substrata and of graded concentrations of neuronotrophic agents and/or serum. Restrictions to neuritic, or other, neuronal performances are likely to occur also in vivo, during early development of the nervous system, in the process of neural aging, or in neuropathological and traumatic situations. Thus, success in styding in vivo effects of ganglioside administration may largely depend on an accurate recognition of the sets of circumstances where positive effects are possible — the therapy version of a "window of opportunity". Also worth noting is the observation that, in each in vitro situation, GM1 proved effective only in a narrow range of concentrations — higher as well as lower concentrations resulting in lesser on no benefits. Thus, variations in the local GM1 concentrations (which may result in vivo from variable distributions, absorption and/or degradation rates) may interfere with a successful utilization of an otherwise appropriate window of opportunity.

GM1 ganglioside was effective on a variety of primary neurons from both CNS and PNS, as well as on PC12 cells. The neuritic effect of ganglioside, therefore, is not mediated by interactions between GM1 and Nerve Growth Factor-associated mechanisms, since neuritic responses were elicited not only from neurons depending on NGF, but also on those depending on other neuronotrophic factors (e.g. CNTF) or on no known trophic factor (e.g. CNS neurons). Neuritic growth requires an intrinsic program (already available in primary neurons, presumably imparted or unmasked by NGF in PC12 cells), plus an effective execution which is under extrinsic modulation. It is on the latter — the execution of a neuritic performance — that ganglioside appears to exert its effects. It may also be that ganglioside effects reported in vivo, other than those on neuritic outgrowth, would similarly reflect a trophic-factor independent ganglioside modulation of the performance rather than a ganglioside-imposed trigger to a new task.

The biochemical mechanisms by which GM1 exerts its effects, and in particular those on neuritic performance, have yet to be identified. The dynamic operation of a growth cone involves surface membrane properties (adhesion), membrane plasticity, cytoskeletal modulation and, of course, accretion of membrane constituents and consumption of energy. Any substance modifying one or another of these features could exert an influence on the growth cone, and thus be seen as a "neurite-modulating" agent. Conversely, substances with apparently specific roles as neurite modulators may carry out their task via mechanisms involving one or another of the cell features mentioned above. Gangliosides have been reported to alter adenylate cyclase (Partington and Daly, 1979) cell membrane Na^+, K^+-ATPase (Mirzoyan et al., 1979), and phosphodiesterase activity (Davis and Daly, 1979) — all of which may have an impact on mechanisms of neurite elongation. The recent observation (Katoh-Semba et al., 1986) that GM1 stimulates or disinhibits lipid synthesis in PC12 cells may be relevant in this context. This possibility is strengthened by the finding that similar GM1 concentrations were effective for both neuritic and lipid synthesis promotions. On the other hand, lipid synthesis may have increased merely as a consequence of an earlier or greater neuritic outgrowth, promoted by GM1 through mechanisms unrelated to lipid synthesis itself. Thus, the studies reviewed here do not claim to shed light on GM1 ganglioside action, but only to set the stage for future investigations of it by defining choices and conditions for GM1-responsive in vitro experimental systems.

ACKNOWLEDGMENTS

This work was supported by a grant from FIDIA Research Laboratories.

REFERENCES

Adler R and Varon S (1981) Dev Biol 86: 69-80.
Campenot RB (1982) Dev Biol 93: 1-12.
Davis CW and Daly JW (1979) Mol Pharmacol 17: 206-211.
Davis G, Manthorpe M, Varon S (1985a) J Neurosci in press.
Davis G, Varon S, Manthorpe M (1985b) Trends Neurosci, 8: 528-532.
Davis G, Skaper SD, Manthorpe M, Moonen G, Varon S (1984) J Neurosci Res 12: 29-39.
Ferrari G, Fabris M, Gorio A (1983) Dev Brain Res 8: 215-222.
Greene LA (1978) Adv Pharmacol Ther 10: 197-206.
Gundersen RW (1985) J Neurosci Res 13: 199-212.
Henderson CE, Huchet M, Changeaux JP (1981) Proc Natl Acad Sci USA 78: 2625-2629.
Katoh-Semba R, Skaper SD, Varon S (1984) J Neurosci Res 12: 299-310.
Katoh-Semba R, Skaper SD, Varon S (1986) J Neurochem 46: 574-582.
Kligman R (1982) Brain Res 250: 93-100.
Letourneau P (1979) Exp Cell Res 124: 127-138.
Leon A, Benvegnu D, Dal Toso R, Presti D, Facci L, Giorgi O, Toffano G (1984) J Neurosci Res 12: 277-287.
Levi-Montalcini R, Meyer H, Hamburger V (1954) Cancer Res 14: 49-57.
Manthorpe M and Varon S (1985) Growth and Maturation Factors, Vol. 3 , J Wiley & Sons, New York, pp. 77-117.
Manthorpe M, Engvall E, Ruoslahti E, Longo FM, Davis GE, Varon S (1983) J Cell Biol 97: 1882-1890.
Mirzoyan SA, Mkheyan EE, Sekoyan ES, Sotskii OP, Akopov SE (1979) Bull Exp Biol Med 86: 1607-1610.
Murray MR (1965) in: Willmer EN (ed): Cells and Tissues in Culture, Academic Press, New York, pp. 373-455.
Noble M, Fok-Seang J, Cohen J (1984) J Neurosci 4: 1892-1904.
Partington CR and Daly JW (1979) Mol Pharmacol 15: 484-491.

Roisen FJ, Bartfield H, Nagele R, Yorke E (1981) Science 214: 577-578.
Sanes JR (1983) Ann Rev Physiol 45: 581-600.
Seifert W (1981) in: Rapport MM and Gorio A (eds): Gangliosides in Neurological and Neuromuscular Function, Development and Repair, Raven Press, New York, pp. 99-117.
Selak I, Skaper SD, Varon S (1985) J Neurosci 5: 23-28.
Skaper SD, Selak I, Varon S (1983a) J Neurosci Res 9: 359-369.
Skaper SD, Selak I, Varon S (1983b) J Neurosci Res 10: 303-315.
Skaper SD, Katoh-Semba R, Varon S (1985) Dev Brain Res 23: 19-26.
Toffano G, Benvegnu D, Bonetti AC, Facci L, Leon A, Orlando P, Ghidoni R, Tettamanti G (1980) J Neurochem 35: 861-866.
Varon S and Adler R (1980) Curr Topics Dev Biol 16: 207-252.
Varon S and Adler R (1981) Adv Cell Neurobiol 2: 115-163.
Weiss P and Hiscoe HB (1948) J Exp Zool 107: 315-396.

Gangliosides and neuronal plasticity
G. Tettamanti, R.W. Ledeen, K. Sandhoff,
Y. Nagai, G. Toffano (eds.)
Fidia Research Series, vol. 6
Liviana Press, Padova, © 1986

FUNCTIONAL ANALYSIS OF GANGLIOSIDE-RESPONDABLE HUMAN NEUROBLASTOMA CELL LINES

Y. Nagai[1,2], S. Tsuji[1], J. Nakajima[1], T. Sasaki[1]

[1]Department of Biochemistry, Faculty of Medicine, University of Tokyo, Hongo, Bunkyo-ku, Tokyo 113, Japan and
[2]Department of Neurobiology, Brain Research Institute, Niigata University, Niigata, Japan

BIOLOGICAL ACTIVITY OF GQlb, STRUCTURE AND ACTIVITY RELATIONSHIP

Previous reports from our laboratory (Tsuji et al., 1983) showed that ganglioside GQlb is capable of promoting cell proliferation, neurite number and neurite length in a highly specific manner in two human neuroblastoma cell lines, GOTO and NB-1. As little as a few nanomolar concentration of GQlb was sufficiently effective. Our subsequent analysis of the structure and activity relationship using various gangliosides revealed an interesting fact that two disialosyl residues of GQlb structure are absolutely necessary for the expression of the activity (Nakajima et al., 1986). As seen in Table 1, in which the results of the analysis are summarized schematically, any deletion of one sialic acid from either disialosyl residue results in complete loss of activity and the mere existence of four sialic acid residues does not assure the activity (e.g., compare with GQlc). Thus, the mode of action of GQlb is quite specific. The significance of this two disialosyl structure will be discussed again later. Furthermore, it was found that GQlb-oligosaccharide, which was prepared from GQlb by the ozonolysis method (Wiegandt and Bücking, 1970), could reproduce the activity but only at a 100 times the concentration of GQlb itself, and that its maximal activity to be attained remained half as much as that of GQlb (Nakajima et al., 1986). This fact implies an important role not only of the oligosaccharide portion of GQ1b but also of its ceramide portion, particularly in support of its possible involvement in the structure and function of the cell membrane.

Table 1. *Biological Activity of GQlb and Related Gangliosides: Analysis of the Structure and Activity Relationship*

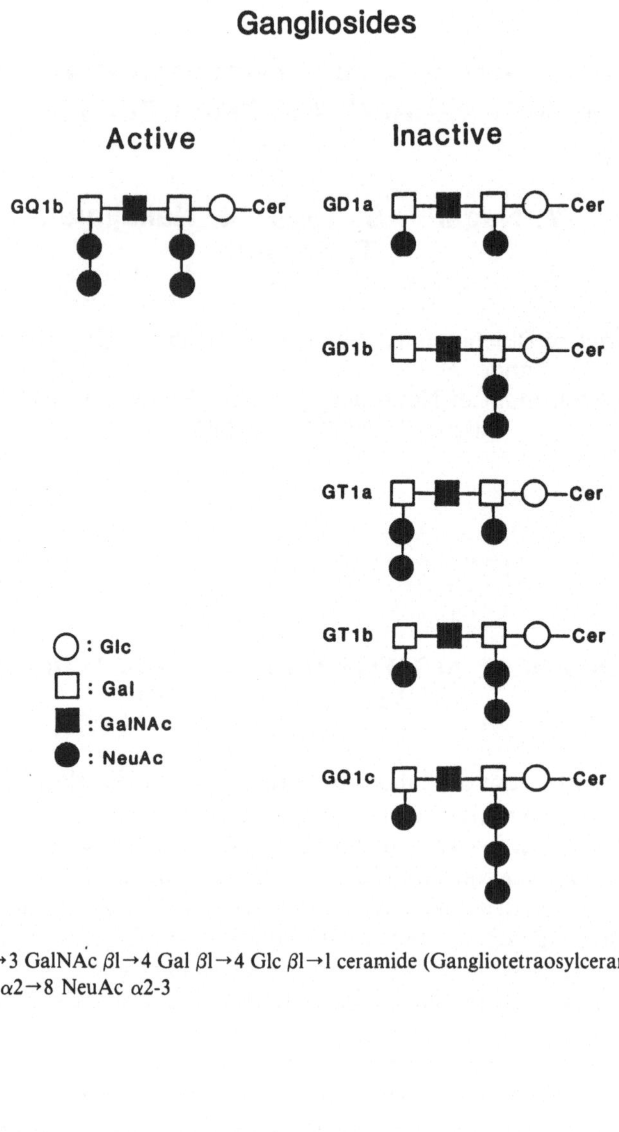

-Cer: Gal β1→3 GalNAc β1→4 Gal β1→4 Glc β1→1 ceramide (Gangliotetraosylceramide), : NeuAc α2→3, : NeuAc α2→8 NeuAc α2-3

THE MODE OF ACTION OF BIOACTIVE GANGLIOSIDE GQlb DIFFERS FROM THAT OF MOUSE 7S NERVE GROWTH FACTOR

The morphological response of the cells to GQlb seemingly did not differ greatly from that to 7S-nerve growth factor (NGF) of mouse salivary gland. But there were some differences between them with respect to biosynthetic response of protein, RNA and DNA and to other biochemical parameters which will be described below.

Behavior of GOTO and NB-1 Cell Lines after Exposure to Serum Deprivation

As we assayed the effects of GQlb using the cell culture deprived of serum, we have examined the effects of serum deprivation on the cells. As reported by Sekiguchi et al. (1979) and Miyake et al. (1973), both cell lines apparently started to differentiate morphologically on deprivation of serum. Our biochemical data also support this observation. Thus ^{3}H-thymidine incorporation was gradually decreased (i.e., T60 = 20 hrs in GOTO cells and 25 hrs in NB-1 cells), indicating the suppression of cell growth (Fig. 1). Meanwhile, RNA synthesis remained at almost the same level for 12 hrs as that in cultures with serum. On further incubation, its level was gradually lowered (T50 = 48 hrs in GOTO and 36 hrs in NB-1 cells, see Fig. 2). Protein synthesis also decreased gradually, and levelled off during 48 hrs to about 80% of the level in the presence of serum (Fig. 3). Intracellular concentration of cGMP was decreased in GOTO to 70% of the level in the serum-containing culture and 85% in NB-1 cells 24 hrs after deprivation of serum (data not shown), while cAMP was increased two-fold of that in the presence of serum during 24-48 hrs (Fig. 4).

Several enzyme activities also changed under serum deprivation as shown in Figure 5. Thus, the activity of acetylcholine esterase (AChE), the cholinergic key enzyme, was decreased to 33-35% during 24 hrs in both cell lines. The choline acetyltransferase (ChAT) activity was decreased more rapidly than AChE (Fig. 6). On the other hand, the activity of ornithine decarboxylase (ODC), the key enzyme for polyamine synthesis, increased about three-fold after 24 hrs incubation, and that of tyrosine hydroxylase synthesis increased 40% after 24 hrs in NB-1 cells. In GOTO cells, however, no activity of the latter enzyme was detectable under any conditions examined. The results thus indicate (i) decrease in the levels of DNA, RNA, protein synthesis and cGMP, (ii) elevation of cAMP level and (iii) activation of ODC and TH.

7S-Nerve Growth Factor in GOTO and NB-1 Cells Acts rather like One of the Survival and/or Differentiation Factors than like Growth Factor

NGF supplementation (100 ng/ml) to the medium without serum caused drastic changes in several biochemical parameters. DNA synthesis was abruptly reduced (T60 = 10-9.5 hrs in both cell lines) (Fig. 1), while RNA synthesis remained at almost the same level as that in cells cultured with serum during 12hrs (Fig. 2). Further incubation increased RNA synthesis up to 30-40% more than that in the presence of serum. Abrupt elevation of protein synthesis was observed on addition of NGF during the first 3 hrs, reaching 1.3-fold as high as that in the culture with serum after 12 hrs (Fig. 4). Protein synthesis in this case seemed to precede activation of RNA synthesis. The cGMP and cAMP levels attained by 24-48 hrs incubation were similar to those in the incubation without serum, though the response in cAMP elevation of the culture on addition of NGF was quicker than that occurring on exposure of the culture to serum deprivation. This quick response, however, does not seem to support a possibility that cAMP might be the first messenger of NGF-mediated signalling, because the cAMP elevation in this type of response should occur within a few minutes. Furthermore, a different morphological response was induced by exogenously added dibutylyl-cAMP when compared with that by NGF or GQlb. Dibutylyl-cAMP induced a so-called "hypertrophic state" in the two cell lines within 24-48 hrs.

With regard to the enzymatic response, elevation of ChAT and ODC was observed in GOTO and NB-1 and TH in NB-1, while AChE decreased in GOTO and NB-1 and

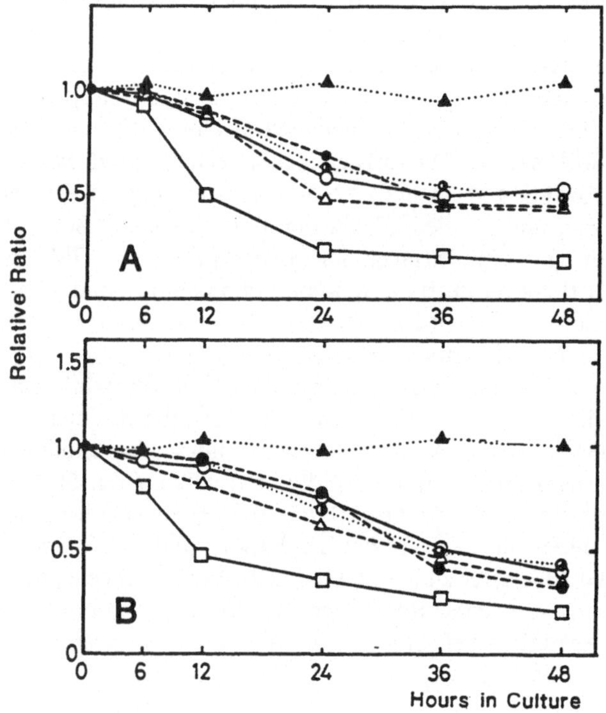

Figure 1. Effects of gangliosides or NGF on thymidine incorporation rate. Two cell lines, GOTO and NB-1, were usually cultured in medium containing 10% FCS (45% RPMI and 45% MEM, supplemented with sodium penicillin G (100 units/ml), and streptomycin sulfate (100 μg/ml). Cultures were maintained at 37°C in a humidified atmosphere of 5% CO_2 in air. For experiments in which the effects of serum were to be excluded, serum-free medium (50% RPMI and 50% MEM, supplemented with the above-mentioned antibiotics) was used. At each indicated time, 0.25 μCi/ml of ^{3}H-methyl thymidine (55 Ci/mmole) was added to the culture. Quadruplicate cultures were incubated at 37°C for 180 min, collected and washed three times with PBS by centrifugation (1500 rpm $\times$ 5 min). 10% of washed cells were dissolved with 1% SDS, and used for protein determination. The remaining washed cells were lyzed with 1% SDS-0.1% pronase (30°C for 30 min). In this step, ^{14}C-labelled-T7-DNA (1.0 $\times$ 10^4 cpm) was added for recovery compensation. After addition of 5 volumes of ethanol, cell lysate was stored at —20°C overnight. Precipitated nucleic acids were washed three times with cold ethanol, dissolved in 0.1 ml of NCS solubilizer, and counted in an Aloka scintilation counter. Incorporated thymidine counts were normalized with coprecipitated ^{14}C-T7-DNA counts, and devided by the amount of protein. The ordinate represents the relative ratio to the thymidine incorporation rate of zero hour cultured cells. ^{14}C-T7-DNA was prepared according to Saito and Miura (1963). A: GOTO, B: NB-l. ▲ 10% serum, • GQlb (5 ng/ml), △ without serum, ○ total gangliosides (50 ng/ml), ◑ GQ1lb (5 ng/ml) + GDla (20 ng/ml), □ NFG (100 ng/ml). The data with GD1a (20 ng/ml) was indistinguishable from that with the serum-free medium within experimental error.

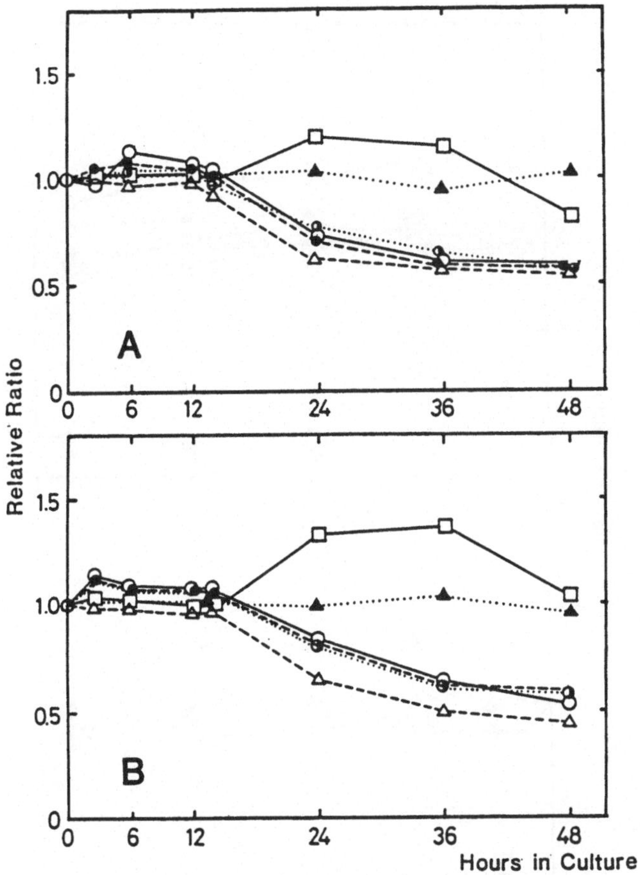

Figure 2. Effects of gangliosides or NGF on the rate of RNA synthesis. For estimation of transcriptional levels, 5,6-^{3}H-uridine (0.25 μCi/ml; specific activity: 42 Ci/mmole) was used in place of ^{3}H-thymidine as stated in the legends of Fig. 1. For recovery compensation, ^{14}C-labelled-poly(A-U) was used. ^{14}C-labelled-poly(A-U) was prepared according to Date et al. (1975). A: GOTO, B: NB-1. Further details and the explanation of symbols used are shown in the legends of Figure 1.

TH remained unchanged in GOTO. AChE and ODC responses were not appreciably different irrespective of the presence or absence of NGF, indicating that the changes of these two enzyme activities may be mainly ascribed to the effect of serum-deprivation *per se*. On the other hand, ChAT (Fig. 6) and TH activation might be related to the direct or indirect effect of NGF itself. The results indicate that NGF in the serum-free medium induces a different type of differentiation in the two cell lines from that seen in the serum- and NGF-free medium, particulary when the following facts are taken into consideration in the former case: (i) immediate cessation of DNA synthesis, (ii) increase in both RNA and protein synthesis and (iii) activation or induction in NB-1 of tyrosine hydroxylase. Thus, this type of NGF cell response might be

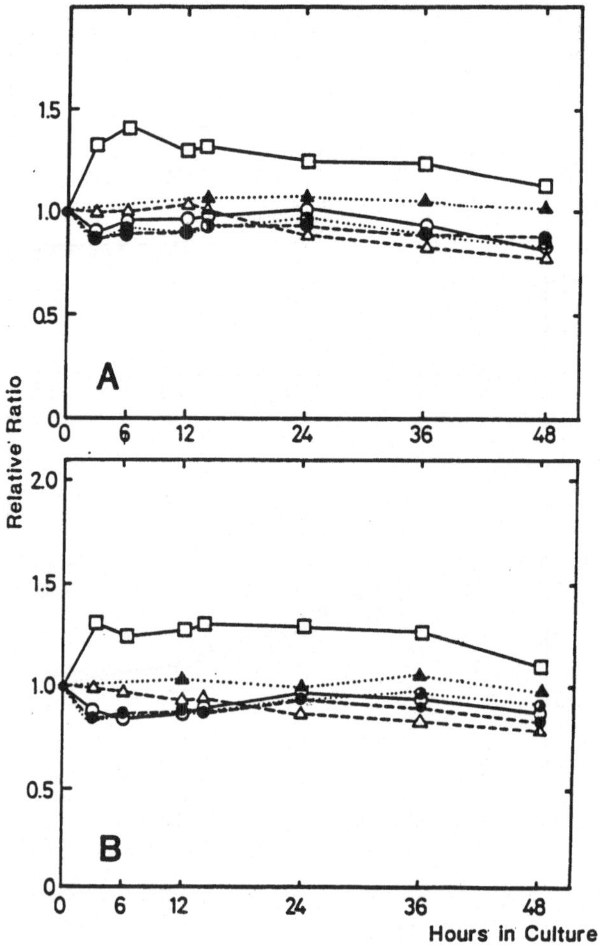

Figure 3. Effects of gangliosides or NGF on the rate of protein synthesis. For each indicated time point, the culture medium was replaced with 1 ml of leucine-free MEM containing 1 μCi/ml of L-3,4,5-^{3}H(N)-leucine (specific activity; 145 Ci/mmole). Quadruplicate cultures were incubated at 37°C for 60 min, and 1 ml of RPMI added. After further incubation at 37°C for 120 min the cultures were washed three times with PBS by centrifugation (1500 rpm × 5 min). Ten % of washed cells were dissolved in 1% SDS and then digested with 100 units/ml each of RNase T1 and DNase I prior to protein determination. The remaining washed cells were lyzed with 1% SDS and further digested with 100 units/ml each of RNase Tl and DNase I (30°C for 30 min). In this step, ^{14}C-methyl-BSA (1.0 × 10^4 cpm) was added for recovery compensation. Proteins were precipitated with 10% TCA (final concentration) and washed three times with 10% TCA by centrifugation. Resulting precipitates were dissolved in NCS solubilizer and toluene based scintilator, and counted. Incorporated ^{3}H counts were normalized with co-precipitated ^{14}C-methyl-BSA, and divided by the amounts of protein. ^{14}C-methyl-BSA was obtained by the reductive alkylation of BSA with ^{14}C-formaldehyde and sodium cyanoborohydride (Dottavio-Martin and Ravel 1978). A: GOTO, B: NB-1. The symbols used are the same as shown in the legends of Figure 1.

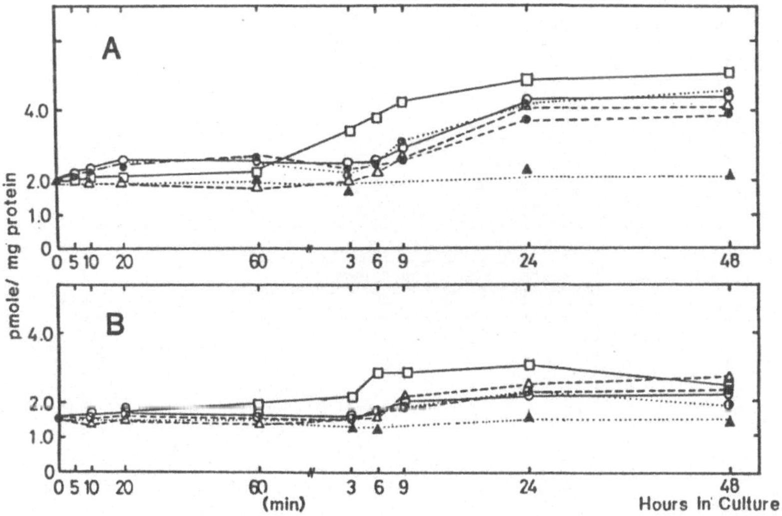

Figure 4. Effects of gangliosides or NGF on intracellular cAMP concentration. The cAMP concentration of each indicated time point was determined by radioimmunoassay according to Honma et al. (1977). A: GOTO, B: NB-1. The symbols used are the same as shown in the legends of Figure 1.

distinguished from the serum-deprivation (SD) type as mentioned in a preceding section.

The Mode of Action of Gangliosides

As reported previously, the effect of GQlb occurs rather transiently and is prolonged in coexistence with GDla (Arita et al., 1984), suggesting its synergistic action. Alteration of biochemical parameters, however, did not differ so much between both cases, and addition of GDla alone did not alter the parameters distinctly. ^{3}H-thymidine incorporation in the serum-free but GQlb-containing medium gradually fell, though its falling was slightly delayed compared with the case of serum-free medium without GQlb. GMl but neither GDla nor GTlb also suppressed ^{3}H-thymidine incorporation in both cell lines to the same degree as GQlb. From the results of ^{3}H-uridine incorporation experiments, the transcriptional level was raised slightly on addition of GQlb during 12 hrs and then gradually lowered. The time course of suppression of RNA synthesis was delayed compared with the case of serum-free medium without GQlb. The time course of protein synthesis (translational level) was indistinguishable from the case without serum. Enzymatic parameters were changed with some delay in time when either total brain gangliosides or GQlb alone were added, but finally settled to the same level as that in the serum-deprived cells. A typical example is given by ChAT activity (Fig. 6). The results suggest that biochemical response of the cells to added gangliosides should be "SD-type" rather than "NGF-type".

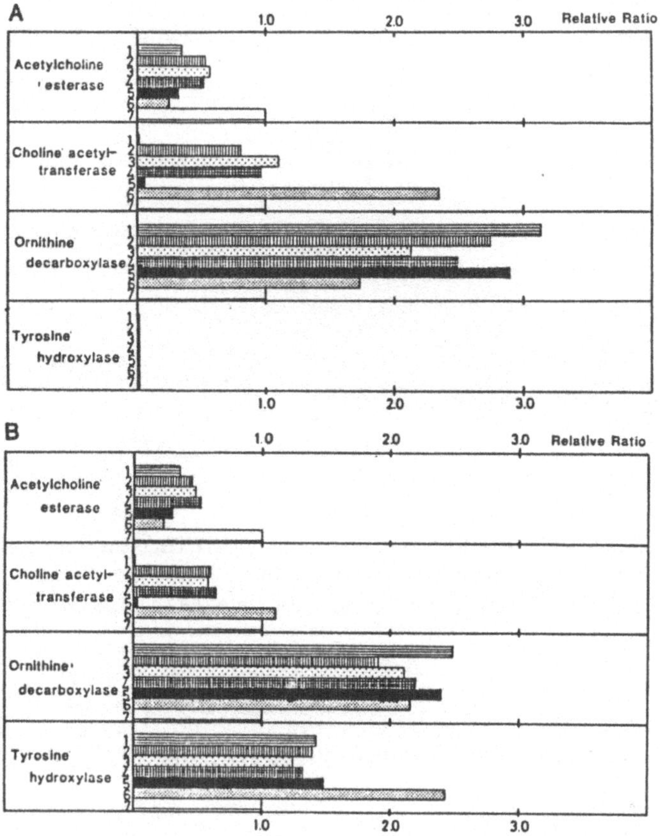

Figure 5. Effects of gangliosides or NGF on several enzyme activities. After 24 hour incubation, ganglioside(s)- or NGF-treated quadruplicate culture cells were collected and washed with PBS. Washed cells were suspended with an appropriate amount of 20 mM Hepes, (pH 7.4), containing 150 mM NaCl. The cell suspension was immediately homogenized by 30 strokes of the pestle. After centrifuging the homogenates, the supernatant was used for every enzyme assay and protein determination. The assay of tyrosine hydroxylase (TH) activity is based on the differing affinities of aromatic L-amino acids decarboxylase toward L-tyrosine and L-DOPA. Thus, starting with carboxyl labelled tyrosine as substrate, carboxyl labelled DOPA is formed as product. An excess amount of decarboxylase is then added to the tyrosine hydroxylase reaction mixture and the radioactive CO_2 preferentially liberated from the DOPA, which is formed during tyrosine hydroxylase-catalyzed reaction, is collected and measured. The aromatic L-amino acid decarboxylase was prepared from fresh hog kidney according to Christenson et al. (1970). Ornithine decarboxylase (ODC) activity was determined by measuring the release of $^{14}CO_2$ from L-l-^{14}C-ornithine according to the method of Russell and Snyder (1968). Acetylcholine esterase (AChE) was estimated essentially as described by Ellman et al. (1961). Choline acetyltransferase (ChAT) was assayed by the method of Fonnum (1975). Protein concentrations were determined by the method of Lowry et al. (1951) or Bradford (1976). The vertical axis shows the relative ratio of each enzyme activity to that of cells cultured with 10% serum. 1, without serum; 2, total gangliosides (50 ng/ml); 3, GQlb (5 ng/ml); 4, GQlb (5 ng/ml) + GDla (20 ng/ml); 5, GDla (20 ng/ml); 6, NGF (100 ng/ml); 7, 10% serum.

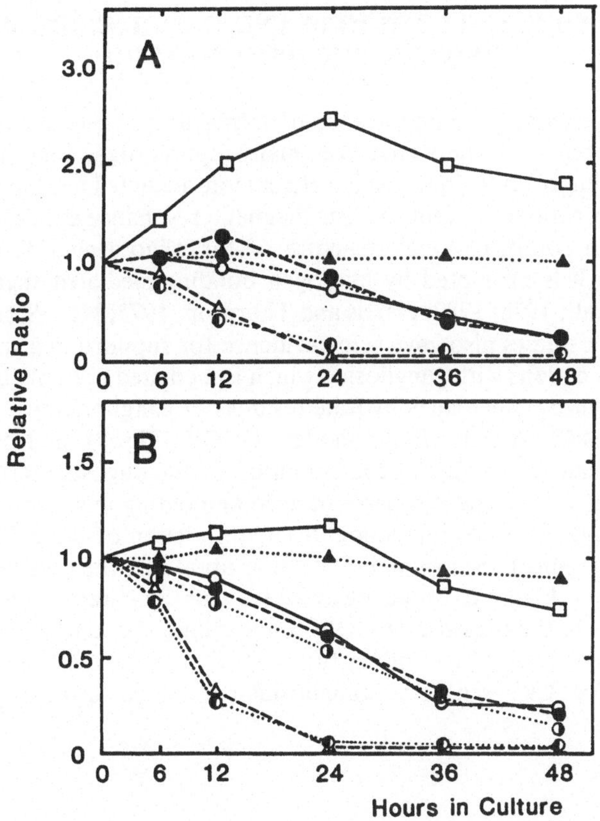

Figure 6. Effects of gangliosides or NGF on choline acetyltransferase activity. At indicated time point, ganglioside(s)- or NGF-treated quadruplicate culture cells were collected and washed with PBS. Choline acetyltransferase activity was measured as described in the legends of Fig. 5. A: GOTO, B: NB-1. ◐, GDla (20 ng/ml). Other symboles used are the same as shown in the legends of Figure 1.

Recently many biofactors that influence neuronal cells differently as well as characteristically have been reported. These factors may be classified into three categories, that is, growth factor, differentiation factor and survival factor. Growth factor will maintain or stimulate DNA synthesis and also RNA/protein synthesis depending upon conditions. Differentiation factor will stop cell growth (reduction of DNA synthesis) and will activate or induce some key metabolic enzymes (e.g., ODC, TH, AChE or ChAT). Survival factor, implication of which category is somewhat ambiguous, will maintain a certain degree of RNA and/or protein synthesis. From our observations, it might be concluded that neither NGF nor gangliosides are genuine growth factors, since, in both cases, DNA synthesis was neither enhanced nor maintained, but rather decreased and finally cell growth was stopped. Furthermore, morphologically the effects of NGF and GQlb are quite similar, but biochemically they are distinctly different.

GANGLIOSIDE, CALCIUM IONS AND GANGLIOSIDE-MEDIATED PROTEIN PHOSPHORYLATION

Above we discussed the importance of tetrasialosyl structure to the activity of GQlb. On the other hand, there have been many reports suggesting the possibility that gangliosides are importantly involved in calcium ion-mediated neuronal as well as other cell membrane functions. Recent nuclear magnetic resonance studies revealed formation of a unique conformational structure of the oligosaccharide moiety of brain ganglioside, which is promoted by hydrogen bonding-mediated sugar-sugar interactions (Sillerud et al., 1978; 1982; Harris and Thornton, 1978; Nerz-Stormes and Thornton, 1984). These studies also gave some evidence for supporting a special interaction of divalent metal cations with ganglioside which is mediated not only by sialic acid carboxylate anion but by other carbohydrate residues of gangliotetraose as well (Sillerud et al., 1978 on GMl). A space filling model of GQlb (Fig. 7) discloses an interesting fact: that GQlb can take a special conformation of the oligosaccharide in which two disialosyl residues form a space capable of accommodating metal cation, for example, Ca^{2+}ion. This suggests a possible ionophore-like function of the GQlb molecule. We then tried an experiment for examining Ca^{2+}ion uptake of the cells in the presence of GQlb. As shown in Figure 8, exogenously added Ca^{2+}ions were taken up progressively by GOTO cells in the presence of GQlb. Interestingly, protein phosphorylation was enhanced simultaneously. This finding further led us to the discovery of ganglioside-dependent and/or Ca^{2+}-sensitive, ganglioside-stimulated protein phosphorylation

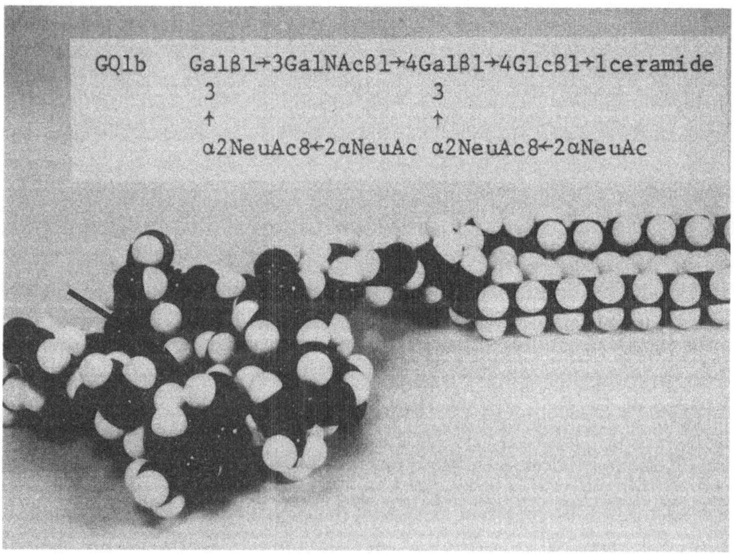

Figure 7. CPK space-filling model of ganglioside GQlb. Arrow indicates space to accommodate metal cation.

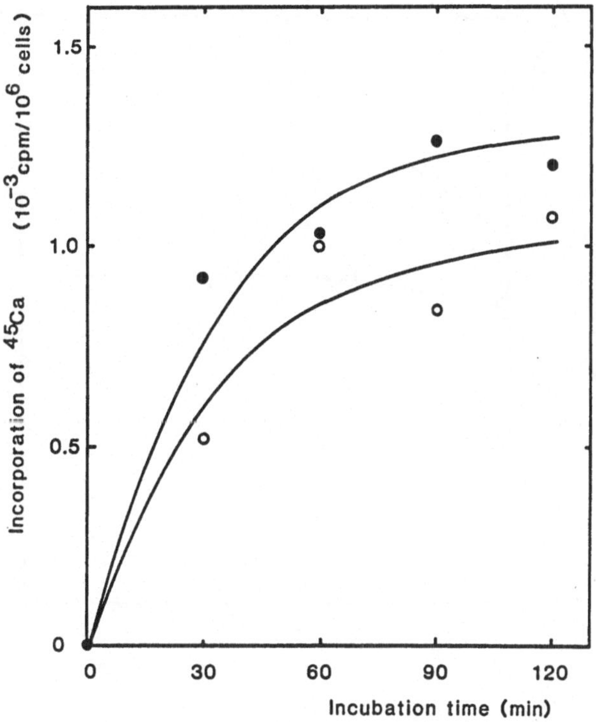

Figure 8. Stimulation by ganglioside GQlb of Ca^{2+}ion uptake. Cells (GOTO: 1×10^6) were seeded on 10cm polystyrene dish in culture medium containing 10% FCS. After 2 days, culture medium was changed to the serum-free medium containing both 2.5 μCi $^{45}Ca^{2+}$ions and 5 ng/ml GQlb. After indicated time periods, cells were collected on Millipore filter and washed with Ca^{2+}- and Mg^{2+}-free Hank's solution containing 5 mM EDTA, 10 mM HEPES and 0.6% glucose. Incorporated $^{45}Ca^{2+}$ions into the washed cells on filter were counted. •, G1lb (5 ng/ml); ○, control without addition of any exogenous effectors.

system in plasma membrane fraction of GOTO cells (Tsuji et al., 1985). A similar system was also reported independently by Goldenring et al. (1985) using rat brain membrane. Our membrane fraction exhibits autocatalytic protein phosphorylation in the presence of GQlb (5-10 ng/ml), Ca^{2+} (0.01-0.1 mM) and Mg^{2+} (1-10 mM) and can efficiently phosphorylate histone Hl in the presence of exogenous ganglioside GQlb; it can also phosphorylate tubulin but in rather broad preference with respect to the type of ganglioside added. Up to the present we have failed to demonstrate such phosphorylation with partially purified protein kinase fractions containing Ca^{2+}-dependent (C-kinase), Ca^{2+}- and calmodulin-dependent (Ca^{2+}/calmodulin kinase), cAMP-dependent (A-kinase) or cGMP-dependent (G-kinase) protein kinases. At present it is not clear whether or not such a new protein kinase system is involved in our GQlb-mediated biological action on human neuroblastoma cell lines. The investigation along this line is now in progress in our laboratory.

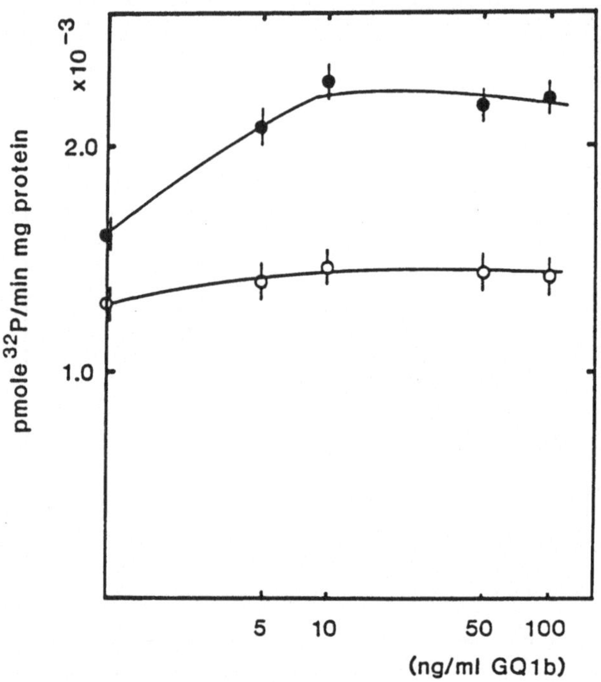

Figure 9. Stimulation of protein phosphorylation by ganglioside GQlb. Protein kinase assay was carried out in a final volume of 0.1 ml throughout the following procedures. Plasma membrane fractions (containing 50 μg protein) prepared from cultured cells (GOTO) according to Boone et al. (1969) were incubated with the indicated amounts of GQlb in 50 mM Tris-HCl, pH 7.2, and 10 mM $MgCl_2$, with or without 0.1 mM Ca^{2+} at 4°C for 15 min. The reaction was started by adding γ-^{32}p-ATP, 20 μg histone Hl and 10 μg tubulin at 37°C for 2 min, and stopped with an equivalent volume of 20% TCA. The precipitates were collected and washed with 10% TCA containing 5 mM Na-PPi, and counted. γ-^{32}p-ATP was prepared according to Schendel and Wells (1973). Purified histone Hl was prepared as described by Zeilig et al. (1981). Tubulin was purified from rat brain by the DEAE-cellulose pyrophosphate procedure of Eipper (1972). •, with Ca^{2+} (0.1 mM); ○, without Ca^{2+}. (See details in Tsuji et al., 1985).

ACKNOWLEDGMENTS

We acknowledge Y. Ohashi for her critical advice in studies of the CPK model of GQlb and M. Arita for his helpful assistance. This work was supported in part by a grant from the Ministry of Education, Science and Culture of Japan and the Special Coordination Funds for promoting science and technology from the Science and Technology Agency of Japan. Support by the Naito Foundation is also gratefully acknowledged.

REFERENCES

Arita M, Tsuji S, Omatsu M, Nagai Y (1984) Studies on bioactive gangliosides: II. Requirement of ganglioside GDla for prolonged GQlb-driven nerve growth promotion in neuroblastoma cell lines. J Neurosci Res 12: 289-297.

Boone CW, Ford LE, Bond HE, Stuart DC, Lorenz D (1969) Isolation of plasma membrane fragments from HeLa cells. J Cell Biol 41: 378-392.

Bradford MM (1976) A rapid and sensitive method for the quantitation of microgram quantities of protein utilizing the principle of protein-dye binding. Anal Biochem 72: 248-254.

Christenson JG, Dairman W, Udenfriend S (1970) Preparation and properties of a homogeneous aromatic L-amino acid decarboxylase from hog kidney. Arch Biochem Biophys 141: 356-367.

Date T, Suzuki K, Imahori K (1975) Studies on a thermophilic RNA polymerase which is active only on poly d(A-T) and poly dAdT. J Biochem 78: 955-967.

Dottavio-Martin D, Ravel JM (1978) Radiolabeling of proteins by reductive alkylation with ^{14}C-formaldehyde and sodium cyanborohydride. Anal Biochem 87: 562-565.

Eipper BA (1972) Rat brain microtubule protein: Purification and determination of covalently bound phosphate and carbohydrate. Proc Natl Acad Sci U S A 69: 2283-2287.

Ellman GL, Courtney KD, Andres V Jr, Featherstone RH (1961) A new and rapid colorimetric determination of acetylcholinesterase activity. Biochem Pharmacol 7: 88-95.

Fonnum F (1975) A rapid radiochemical method for the determination of choline acetyltransferase. J Neurochem 24: 407-409.

Goldenring JR, Otis LC, Yu RK, Delorenzo RJ (1985) Calcium/ganglioside-dependent proteine kinase activity in rat brain membrane. J Neurochem 44: 1229-1234.

Harris PL, Thornton ER (1978) Carbon-13 and proton nuclear magnetic resonance studies of gangliosides. J Am Chem Soc 100: 6738-6745.

Honma M, Satoh T, Takezawa J, Ui M (1977) An ultrasensitive method for the simultaneous determination of cyclic AMP and cyclic GMP in small-volume samples from blood and tissue. Biochemical Med 18: 257-273.

Lowry OH, Rosebrough NJ, Farr AL, Randall RJ (1951) Protein measurement with the Folin phenol reagent. J Biol Chem 193: 265-275.

Miyake S, Shimo Y, Kitamura T, Nojyo Y, Nakamura T, Imashuku S, Abe T (1973) Characteristics of continuous and functional cell line NB-1, derived from a human neuroblastoma. The Autonomic Nervous System 10: 115-120.

Nakajima J, Tsuji S, Nagai Y (1986) Bioactive gangliosides: V. Analysis of functional structures of neurites outgrowth promoting tetrasialoganglioside GQlb. Biochim Biophys Acta 876: 65-71.

Nerz-Stormes M, Thornton ER (1984) Carbon-13 spin-lattice relaxation studies of GDla micells. Limited segmental motion of head group saccharide units. J Am Chem Soc 106: 5240-5246.

Russell D, Snyder SH (1968) Amine synthesis in rapidly growing tissues: Ornithine decarboxylase activity in regenerating rat liver, chick embryo, and various tumors. Proc Natl Acad Sci U S A 60: 1420-1427.

Saito II, Miura K(1963) Preparation of transforming deoxyribonucleic acid by phenol treatment. Biochim Biophys Acta 72: 619-629.

Schendel PF, Wells RD (1973) The synthesis and purification of γ-^{32}p-adenosine triphosphate with high specific activity. J Biol Chem 248: 8319-8321.

Sekiguchi M, Oota T, Sakakibara K, Inui N, Fujii G (1979) Establishment and characterization of a human neuroblastoma cell line in tissue culture. Jpn J Exp Med 49: 67-83.

Sillerud LO, Prestegard JH, Yu RK, Schafer DE, Konigsberg WH (1978) Assigment of the ^{13}C nuclear magnetic resonance spectrum of aqueous ganglioside GMl micelles. J Am Chem Soc 17: 2619-2628.

Sillerud LO, Yu RK, Schafer DE (1982) Assignment of the carbon-13 nuclear magnetic resonance spectra of gangliosides GM4, GM3, GM2, GMl, GDla, GDlb and GTlb. Biochemistry 21: 1260-1271.

Tsuji S, Arita M, Nagai Y (1983) GQlb, a bioactive ganglioside that exhibits novel nerve growth factor (NGF)-like activities in the two neuroblastoma cell lines. J Biochem 94: 303-306.

Tsuji S, Nakajima J, Sasaki T, Nagai Y (1985) Bioactive gangliosides: IV. Ganglioside GQlb/Ca^{2+} dependent protein kinase activity exists in the plasma membrane fraction of neuroblastoma cell line, GOTO. J Biochem 97: 969-972.

Wiegandt H, Bücking HW (1970) Carbohydrate components of extraneuronal gangliosides from bovine and human spleen, and bovine kidney. Eur J Biochem 15: 287-292.

Zeilig CE, Langan TA, Glass DB (1981) Sites in histone Hl selectively phosphorylated by guanosine 3'5'-monophosphate-dependent protein kinase. J Biol Chem 256: 994-1001.

Gangliosides and neuronal plasticity
G. Tettamanti, R.W. Ledeen, K. Sandhoff,
Y. Nagai, G. Toffano (eds.)
Fidia Research Series, vol. 6
Liviana Press, Padova, © 1986

PRIMARY NEURAL CELL CULTURES AND GM1 MONOSIALOGANGLIOSIDE: A MODEL FOR COMPREHENSION OF THE MECHANISMS UNDERLYING GM1 EFFECTS IN CNS REPAIR PROCESS IN VIVO

R. Dal Toso, D. Presti, D. Benvegnù, G. Tettamanti[1], G. Toffano, A. Leon

Fidia Neurobiological Research Laboratories, Via Ponte della Fabbrica 3/A, 35031 Abano Terme, Italy
[1]Department of Medical Chemistry and Biochemistry, University of Milano, Via Saldini 50, 20133 Milano, Italy

INTRODUCTION

Contrary to long-held pessimistic notions, it is now recognized that not only the developing but also the adult mammalian central nervous system (C.N.S.) is capable of undergoing a series of repair processes following injury. Although the irreversibly damaged neurons cannot be replaced, repair in the adult can partly be accomplished by additional growth and reorganization of neuronal processes from undamaged axons in response to adjacent axonal and synaptic degeneration. The intensity of this process varies with the extent, localization and type of neuronal injury and is presumably determined by appropriate signalling, i.e. neuronotrophic factors, from humoral surroundings, cell-cell contacts and denervated target areas. In at least some situations, a close correlation between lesion-induced remodeling and behavioral recovery has been reported (Nieto-Sampedro et al., 1983). Agents or conditions capable of enhancing the availability or activity of neuronotrophic factors are currently being examined in the attempt to ameliorate functional recovery of damaged CNS neuronal tissue.

Studies from our own and other laboratories have recently documented the capability of monosialoganglioside GM1 to enhance functional recovery following specific lesions of adult mammalian brain (Toffano et al., 1983; Toffano et al., 1984a;

Abbreviations: CNS, central nervous system; BME, Eagle's basal medium; PBS, phosphate buffered saline; DA, dopamine; THF, tetrahydro furan; GABA, γ-aminobutyric acid; DABA, L-2,4-diaminobutyric acid; BZT, benztropine-mesylate; GFAP, glial fibrillary acidic protein; GIF, glyoxylic acid induced fluorescence; DRG, dorsal root ganglia; NGF, nerve growth factor; PDGF, platelet derived growth factor.

Toffano et al., 1984b; Agnati et al., 1983; Casamenti et al., 1985; Oderfeld-Nowak et al., 1984; Sabel et al., 1984; Kojima et al., 1984). In particular it has been reported that chronic administration of GM1 facilitates functional dopaminergic reinnervation of the striatum and maintains the number of dopaminergic cell bodies in the substantia nigra of rats following a partial mechanical lesion of the nigra-striatal pathway. Approaches for investigating such a pharmacological effect of the ganglioside essentially arise from evidence suggesting that gangliosides play a prominent role in regulating neurite outgrowth during development and in the adult state (Willinger and Schachner, 1980; Kasarskis et al., 1981). However, the cellular and molecular mechanisms determining the effects of the ganglioside administration in vivo are still obscure. Among several possibilities, it has been hypothesized that the GM1 effects occur via the modulation at the cell surface of neuronotrophic factors, the production of which is favoured by the lesion (Nieto-Sampedro et al., 1983).

In this paper we report our approach to understand GM1 effects on CNS repair processes in vivo by studying and analysing its effects on the development and survival of dopaminergic neurons in dissociated embryonic mesencephalic cells cultured in serum-free hormone-supplemented media.

MATERIALS AND METHODS

Cell Culture Preparation

Dissociated mesencephalic and striatal cell cultures were prepared as described in details elsewhere (Dal Toso et al., in preparation). Briefly, rostral mesencephalic tegmentum and corpus striatum, from brains of 13 and 15-day-old mouse embryos respectively, were dissected under microscope in sterile conditions and mechanically dissociated. The desired amount of cells was then plated on collagen-coated dishes in a serum-free medium consisting of an equal volume mixture of Eagle's Basal Medium (BME) and Ham's F12 supplemented with N_2 components (insulin, 25μg/ml; transferrin, 100 μg/ml; putrescine, 60μM; progesterone, 20 nM; sodium selenite, 30 nM) and triiodotyronine (30 nM). The cultures were incubated at 37°C in a water saturated (95% air and 5% CO_2) atmosphere.

GM1 Characterization and Preparation of GM1-Containing Media

GM1 ganglioside, extracted from calf brain according to Tettamanti et al. (1973), was purified and chemically characterized as described by Sonnino et al. (1978). Chromatographic analysis indicated that the purity of GM1 was over 99%. For preparation of GM1-containing media, desired amounts of purified GM1 were dissolved in chloroform-methanol (1:1 by volume). The residue, resuspended in chloroform-methanol (1:1 by volume), was dried again and redissolved in an appropriate volume of incubation medium in order to reach the final ganglioside concentration (from 10^{-8} M to 10^{-7} M). The solution was maintained at 37°C for at least 1 h with intermittent swirling. The recovery of GM1 by this procedure (measured by utilizing

^{3}H-GM1 labelled on the terminal galactose moiety, (Facci et al., 1984), exceeded 90%. The control medium was treated in the same way except for the omission of GM1. Typically, control or GM1-containing media were added to cultures at plating time and maintained for 4 or 8 days in vitro (see Results). The solvents were distilled before use.

Determination of GM1 Association to the Cells

Mesencephalic neurons on collagen-coated coverslips were incubated at 37°C for different times with ^{3}H-GM1 (10^{-8} M — 10^{-7} M) in serum-free medium. After incubation, coverslips were washed twice with PBS, once with the culture medium supplemented with 10% fetal calf serum and three more times with PBS. Radioactivity of each coverslip was extracted twice with 5 ml aliquots of tetrahydrofuran (THF) as described by Tettamanti et al. (1973). The THF extracts were added in glass vials, taken to dryness with a stream of N_2 and resuspended with 1 ml distilled water prior to addition of 10 ml of emulsifier (Instagel II). Samples were then counted by liquid scintillation with a Packard Tricarb (model 460 C) counter. Dried coverslips were then processed for DNA assay according to Erwin et al. (1981). THF treatment did neither interfere with DNA assay nor modify DNA content of coverslips.

^{3}H-Dopamine and ^{14}C-γ-Aminobutyric Acid Uptake

^{3}H-dopamine (DA) uptake and ^{14}C-γ-aminobutyric acid (GABA) uptake were assessed as previously described (Prochiantz et al., 1981). Typically benztropine-mesylate (BZT, 5 μM) and L-2, 4-diaminobutyric acid (DABA, 1 mM) were utilized so as to assess specific DA or GABA uptake respectively.

Immunocytochemistry and Histochemistry

Immunocytochemistry of cells and histochemical visualization of the dopaminergic cells in the culture system were performed as previously described (Berger et al., 1982). Assessment of the number of the surviving dopaminergic neurons was conducted as reported elsewhere (Dal Toso et al., in preparation).

RESULTS AND DISCUSSION

Immunocytochemical Characterization of Cell Types and Number of Dopaminergic Neurons in Culture

Figure 1 illustrates the typical appearance at phase contrast microscope of fetal mouse dissociated mesencephalic cells cultured for 4 days in serum-free hormone-supplemented medium. At all cell densities examined (0.5 to 2 $\times$ 10^6 plated cells), the immunocytochemical staining indicated that over 98% of the adhering cells could be classified as neuronal elements. This was evaluated utilizing a monoclonal antibody (RT97) against neurofilament protein and GFAP antiserum (Anderton et al., 1982; Raff et al., 1979).

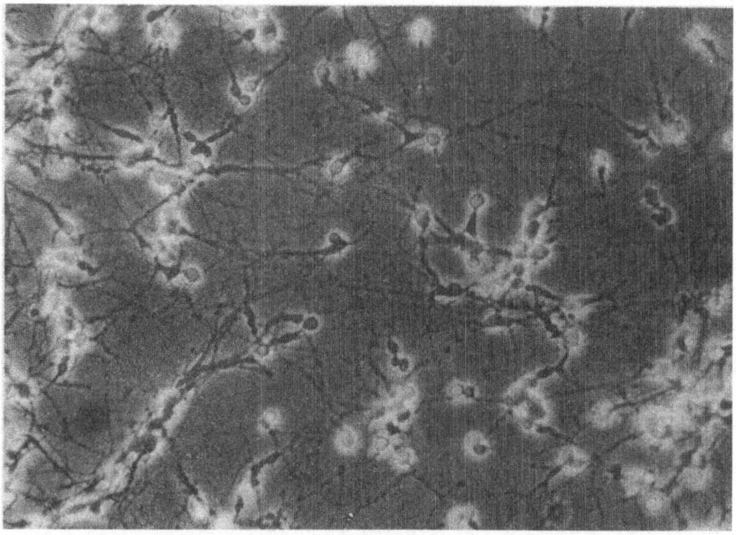

Figure 1. Phase contrast photomicrograph of dissociated mesencephalic cells cultured for 4 days in serum-free medium. Culture conditions are as described in Materials and Methods. 154 × magnification.

Evaluation of the number of dopaminergic neurons in vitro was conducted utilizing the catecholamine uptake procedure as previously reported by Berger et al. (1982) in association with the glyoxylic acid-induced fluorescence (GIF) technique. When utilizing this technique, approximately 0.15% of the total cultured mesencephalic cells were, irrespective of the cell density employed, positively labelled at day 4 in vitro. The GIF^+ neurons were completely absent after addition of BZT, suggesting that most, if not all, of the GIF^+ cells could be classified as dopaminergic. Most of these neurons displayed long and branched fluorescent neurites with regularly spaced varicosities.

GM1 Effects on Development and Survival of the Dopaminergic Cells in Culture

To assess possible GM1 effects on the DA neurons in the culture system, we monitored in diverse culture conditions the high affinity, BZT-sensitive ^{3}H-DA uptake together with the number of DA cells.

In control cell cultures, both the DA uptake and the number of DA cells varied as a function of time and cell density in vitro. Typically, with 1 or 2 x 10^6 cells initially plated per dish, the BZT-sensitive DA uptake reached maximal values at day 4 and then declined; furthermore, the relative number of DA neurons did not vary significantly up to day 4 but then also declined. Yet the DA uptake when referred per DA neuron was at all times more elevated in the more dense cultures (Fig. 2 C). In addition, the higher plating density resulted in a significantly smaller decline in the number of DA neurons (40% rather than 75%) between day 4 and 8 in vitro. These results indicate that development and survival of the DA cells, although scarse under the culture

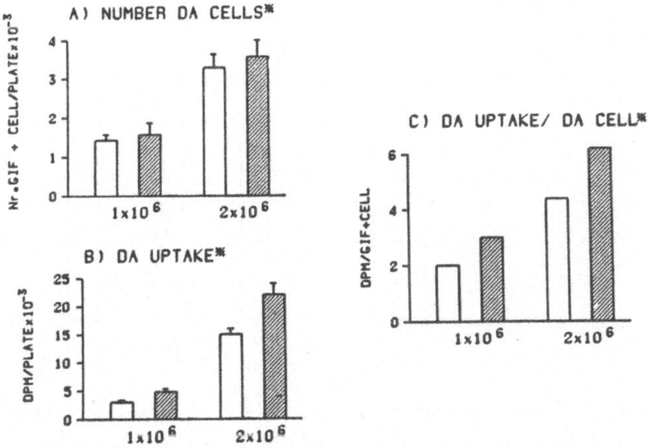

Figure 2. Effect of GM1 (10^{-7} M) on number of DA cells (A), BZT-sensitive DA uptake per plate (B), and per DA cell (C), in 4-day-old dissociated fetal mesencephalic cultures as a function of cell density. Culture conditions and DA uptake were conducted as reported in Materials and Methods. Control cultures (open histograms); GM1-treated cultures (hatched histograms). Values are the mean of triplicate analysis ± S.E.M.

conditions employed, are favoured by seeding at high cell density. These cell density effects most probably depend on the presence in vitro of self-supportive influences, i.e. neuronotrophic factors, produced by the cultured cells themselves.

In identical culture conditions as above, addition of GM1 (10^{-7} M) at plating time produced, when assessed at day 4, an increase in the BZT-sensitive ^{3}H-DA uptake per DA neuron (Fig. 2 C) without affecting BZT-insensitive DA uptake, number of DA cells per plate (Fig. 2 A), neuron to astroglial cell ratio and DNA content/plate. The GM1 effect was concentration-dependent (Fig. 3) and kinetically associated with a 3-fold reduction of the apparent K_m and 3-fold increase of the apparent V_{max} values. The latter effect may perhaps be viewed as indicative of increased neurite outgrowth. In addition, at day 8 in vitro there was a correlation between the GM1 effect on enhanced DA uptake per DA cell and enhanced survival of these cells (Fig. 4).

It is thus apparent from the above-mentioned results that GM1 addition to the mesencephalic cell culture system enhances not only early development but also survival of the DA cells in culture. This is in accord with previous reports concerning facilitation by GM1 of neuronal differentiation of a variety of both clonal and primary neuronal cell cultures (Morgan and Seifert, 1979; Dimpfel et al., 1981; Roisen et al., 1981; Leon et al., 1982; Ferrari et al., 1983; Facci et al., 1984; Byrne et al., 1983; Leon et al., 1984; Doherty et al., 1985; Hefti et al., 1985). Since, following such studies, GM1 has come to be viewed as a potential neuronotrophic or neuritogenic agent, a critical question is whether the exogenous GM1 behaves as a signal in its own right or operates through modification of membrane behavior. A related question is whether the effects are due to stable insertion of the ganglioside within the membrane or rather to simple association of the ganglioside to the membrane surface.

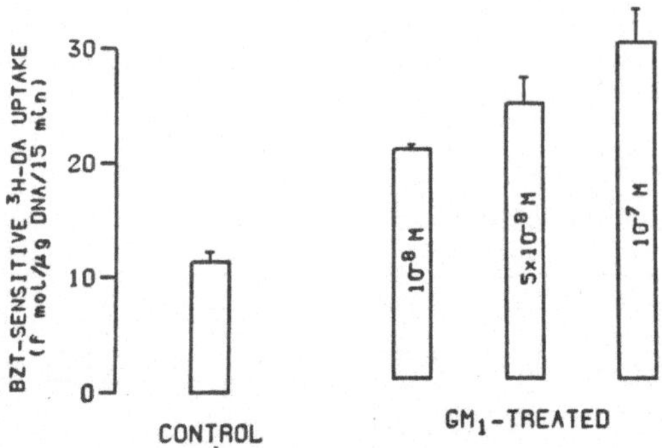

Figure 3. Concentration-dependent effect of GM1 on BZT-sensitive DA uptake in 4-day-old mesencephalic cell cultures (1-10^6 cells seeded per plate). Culture conditions and DA uptake were conducted as described in Materials and Methods. Values are the mean of triplicate analysis ± S.E.M.

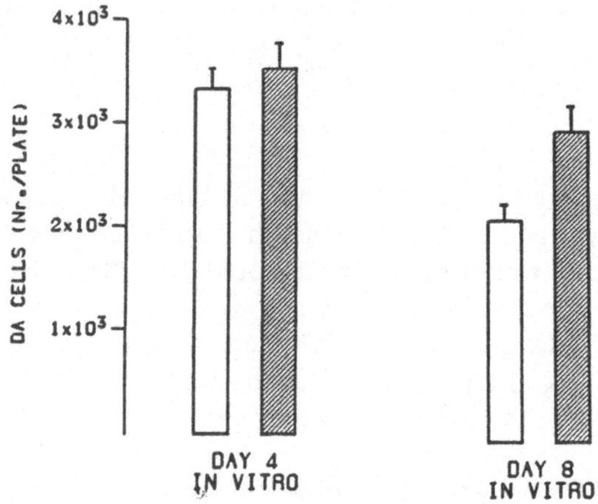

Figure 4. GM1 effect on long-term DA cell survival in vitro. Mesencephalic cells were seeded at 2.10^6 cells/plate and GM1 (10^{-7}M) was added at plating. Cells were visualized after 4 and 8 days in vitro as described in Materials and Methods. Control cultures (open histograms); GM1-treated cultures (hatched histograms). Values are the mean of triplicate analysis ± S.E.M.

GM1 is not a Neuronotrophic Agent but Modulates Cell Responsiveness to Neuronotrophic Agent(s)

As already reported, the mesencephalic cells produce in the culture system self-supportive neuronotrophic influences affecting both development and survival of the DA cells. The titers of these influences depend on the cell density employed. In fact, in control cultures of increasing cell density, the DA uptake when assessed at day 4 was linear up to an apparent cell density of 0.5×10^6 cells per plate. Above this cell density threshold value, the DA uptake increased proportionally to the square of the cell of the cell density. This has been taken as an indication of the presence of increasing cell density-dependent effects of self-supportive neuronotrophic influences produced in vitro by the cells themselves.

To assess whether GM1 was per se a neuronotrophic agent or not, we therefore investigated the effects of GM1 on BZT-sensitive DA uptake in cultures of increasing cell density. Interestingly, the addition of GM1 (10^{-7} M) was uneffective in enhancing DA uptake in low density cultures (i.e. cultures at or below threshold values) whereas it was at all times effective in higher density cultures (DA uptake approximately 1.5 to 2-fold greater). Thus, GM1 seems unable to substitute for the cell-derived influences but can either potentiate their action and/or exert independent influences to which neurons can only respond if appropriately supported. A similar relationship was observed between the effects of GM1 and NGF in pheochromocytoma (PC_{12}) and fetal chick dorsal root ganglionic (DRG) cells (Varon et al., 1986; Doherty et al., 1986). In this context it is noteworthy that NGF is uneffective on DA neurons in the mesencephalic culture system indicating that GM1 may affect the efficacy of different trophic agents acting on specific neuronal cells.

GM1 Effects are Related to its Association to the Cells

In the attempt to evaluate the relationship between association to cell membranes and functional effects in the mesencephalic system, we determined the extent of GM1 association to the cells (utilizing GM1 concentrations causing modifications of DA uptake). As already reported for neuroblastoma cells, serum stable association of GM1 increased with increasing GM1 concentrations (10^{-8} M - 10^{-7} M) and time of incubation (1 to 24 hours). Exposure of mesencephalic cells to 10^{-7} M GM1 for only the first 24 hours in culture followed by removal of culture medium, adequate washes and addition of new culture medium, was able to increase DA uptake when assessed at day 4. The effect was however smaller than in culture in which GM1 was present throughout the four days without changes of the culture medium (75% instead of 172% increase). Apparently, longer exposures to GM1 appear to be necessary for maximal GM1 effects.

The above results suggest that the portion of GM1 stably associated with the cells is most probably responsible for the effects observed on the DA neurons. In this context, incubation of various cells in the presence of gangliosides including GM1 has been reported to result in cell surface accumulation of the added glycolipid. Although some of the gangliosides can be released by treatment of the cells with serum or trypsin, considerable evidence suggests that a residual portion (trypsin-insensitive) is tightly associated with the outer leaflet of the plasma membrane (Facci et al., 1984). In addition, the stably associated ganglioside has been shown to be both functionally and metabolically active and to cause changes of membrane properties including enzyme activity (Leon et al., 1981), receptor binding (Berry-Kravis and Dawson, 1985) and ion permeability (Vyskočil et al., 1985). Noteworthy is that the stable insertion of very small amounts of GM1 modify membrane enzymatic activities of neuronal membranes (Leon et al., 1981). Analogously, very small amounts of inserted GM1 have also been reported to affect PDGF binding and PDGF-induced phosphorylation in cultured 3T3 fibroblasts (Bremer et al., 1984). This may indicate that the exogenous GM1 molecules become presumably inserted into the membrane at well-defined and preferential sites, the result of which is a change in functional membrane properties in defined microdomains of the plasma membrane bilayer.

Another possibility is that the stable association of exogenous gangliosides may be followed by ganglioside internalization and recycling into more complex gangliosides. These in turn may be responsible for the resulting modification in membrane properties. Which of these two possibilities plays a predominant role in the effects observed in this study is currently under investigation.

Neuronal and Molecular Specificity

To assess whether the GM1 effects were limited only to the DA cells in the culture system, parallel experiments were carried out on the GABAergic neurons present in mesencephalic or striatal cell cultures. Addition of GM1 (from 10^{-8} to 10^{-7} M) markedly enhanced, in a concentration-dependent manner, the DABA-sensitive ^{14}C-GABA uptake in 4-day-old cultured mesencephalic cells. Like the DA uptake, this effect was dependent on the cell density. Similarly, GM1 also increased the ^{14}C-GABA uptake in striatal dissociated cells. Again, no effect of GM1 was observed on cell viability and DNA content per plate. This indicates that the GM1 effects on CNS neurons are not limited to DA neurons present in the mesencephalic cell culture system utilized.

With respect to the molecular specificity, asialo GM1, sialic acid or the lipid-free oligosaccharide portion of GM1 were tested on specific DA and GABA uptake by the cultured mesencephalic cells. In contrast to GM1, asialo GM1 (10^{-7} M) had no activity. The addition of sialic acid (10^{-7} M) and the lipid-free oligosaccharide portion (10^{-7} M) also were uneffective.

CONCLUSION AND FUTURE PERSPECTIVES

The data reported show that exogenously supplied GM1, (but not asialo GM1, sialic acid or the oligosaccharide chain of GM1) enhances development and survival of the DA neurons present in mesencephalic cell cultures. The GM1 effects are not limited to the DA neurons and are most probably related to direct or indirect modifications of cell membrane properties. In addition, it appears that the attainment of these GM1 effects in vitro a) necessitates the occurrence of adequate extracellular signals, i.e. neuronotrophic agents, and b) involves GM1 potentiation of the acquisition and/or maintenance of the neuronotrophic-induced differentiative state. However, the molecular mechanisms underlying such effects are still obscure. An important step will be to identify the molecular reactions affected by GM1 and how such events are related to its effects in vitro.

Concluding, the present data are consistent with the hypothesis that the presence of exogenously supplied GM1 molecules on the neuronal cell surface affect the properties of the neuronal cell plasma membrane with respect to its capacity to respond, modulate and/or transfer signals from the extra to the intracellular spaces. In particular, several lines of evidence support the possibility that GM1 can modulate functional neuronal cell behaviors in response to neuronotrophic signals. Since injury-induced neuronotrophic molecules have been suggested to be essential for CNS repair processes in the adult (Nieto-Sampedro et al., 1983), these results may be relevant for the comprehension of GM1 facilitation of function recovery of injured CNS neurons in vivo. Verification of such a possibility inevitably necessitates identification, purification and study of the physiological relevance of neuronotrophic molecule(s) in damaged adult mammalian brain. Preliminary experiments along this line indicate that indeed trypsin-labile molecule(s) possessing neuronotrophic activity in vitro are present in the adult mammalian brain. Titers of the latter increase following certain types of CNS injury and, interestingly, GM1, at least in vitro, potentiates the extracted neuronotrophic activity.

REFERENCES

Agnati LF, Fuxe K, Calzà L, Benfenati F, Cavicchioli L, Toffano G, Goldstein M (1983) Gangliosides increase the survival of lesioned nigral dopamine neurons and favour the recovery of dopaminergic synaptic function in striatum of rats by collateral sprouting. Acta Physiol Scand 119: 347-363.

Anderton BH, Breinburg D, Downes MJ, Green PJ, Tomlinson BE, Ulrich J, Wood JN, Kahn J (1982) Monoclonal antibodies show that neurofibrillary tangles and neurofilaments share antigenic determinants. Nature 298: 84-86.

Berger B, Di Porzio U, Daguet MC, Gay M, Vignay A, Glowinsky J, Prochiantz A (1982) Long-term development of mesencephalic dopaminergic neurons of mouse embryos in dissociated primary cultures: morphological and histochemical characteristics. Neuroscience 7: 193-205.

Berry-Kravis E, Dawson G (1985) Possible role of gangliosides in regulating an adenylate cyclase-linked 5-hydroxytryptamine (5-HT_1) receptor. J Neurochem 45: 1739-1747.

Bremer EG, Hakomori SI, Bowen Pope D, Raines E, Ross R (1984) Ganglioside-mediated modulation of cell growth, growth factor binding, and receptor phosphorylation. J Biol Chem 259: 6818-6825.

Byrne MC, Ledeen RW, Roisen FJ, Yorke G, Scalafani JR (1983) Ganglioside-induced neuritogenesis: verification that gangliosides are the active agents and comparison of molecular species. J Neurochem 41: 1214-1222.

Casamenti F, Bracco L, Bartolini L, Pepeu G (1985) Effect of ganglioside treatment in rats with a lesion of the cholinergic forebrain nuclei. Brain Res 338: 45-52.

Dimpfel W, Moller W, Mengs U (1981) Ganglioside-induced formation in cultured neuroblastoma cells. In: Rapport MM, Gorio A (eds): Gangliosides in neurological and neuromuscular function, development and repair. Raven Press, New York, pp. 119-133.

Doherty P, Dickson JG, Flanigan TP, Walsh FS (1986) Molecular specificity of ganglioside action on neurite regeneration in cell cultures of sensory neurons, this volume.

Doherty P, Dickson JG, Flanigan TP, Walsh FS (1985) Ganglioside GM1 does not initiate, but enhances neurite regeneration of nerve growth factor-dependent sensory neurones. J Neurochem 44: 1259-1265.

Erwin BG, Stoscheck CM, Florini JR (1981) A rapid fluorimetric method for the estimation of DNA in cultures. Anal Biochem 110: 291-294.

Facci L, Leon A, Toffano G, Sonnino S, Ghidoni R, Tettamanti G (1984) Promotion of neuritogenesis in mouse neuroblastoma cells by exogenous ganglioside. Relationship between the effect and the cell association of ganglioside GM1. J Neurochem 42: 299-305.

Ferrari G, Fabris M, Gorio A (1983) Gangliosides enhance neurite outgrowth in PC12 cells. Dev Brain Res 8: 215-222.

Hefti F, Hartikka K, Frick W (1985) Gangliosides alter morphology and growth of astrocytes and increase the activity of choline acetyltransferase in cultures of dissociated septal cells. J Neurosci 5: 2086-2094.

Kasarskis EJ, Karpiak SE, Rapport MM, Yu RK, Bass NH (1981) Abnormal maturation of cerebral cortex and behavioral deficit in adult rats after neonatal administration of antibodies to ganglioside. Develop Bran Res 1: 25-35.

Kojima H, Gorio A, Janigro D, Jonsson G (1984) GM1 ganglioside enhances regrowth of noradrenaline nerve terminals in rat cerebral cortex lesioned by the neurotoxin 6-hydroxydopamine. Neuroscience 13: 1011-1022.

Leon A, Facci L, Toffano G, Sonnino S, Tettamanti G (1981) Activation of (Na^+, K^+)ATPase by nanomolar concentrations of GM1 ganglioside. J Neurochem 37: 350-357.

Leon A, Facci L, Benvegnù D, Toffano G (1982) Morphological and biochemical effects of gangliosides in neuroblastoma cells. Develop Neurosci 5: 108-144.

Leon A, Benvegnù D, Dal Toso R, Presti D, Facci L, Giorgi O, Toffano G (1984) Dorsal root ganglia and nerve growth factor: a model for understanding the mechanism of GM1 effects on neuronal repair. J Neurosci Res 12: 277-287.

Morgan JI, Seifert W (1979) Growth factors and gangliosides: a possible new perspective in neuronal growth control. J Supramol Struct 10: 111-124.

Nieto-Sampedro M, Manthorpe M, Barbin G, Varon S, Cotman CW (1983) Injury-induced neuronotrophic activity in adult rat brain: correlation with survival of delayed implants in the wound cavity. J Neurosci 3: 2219-2229.

Oderfeld-Nowak B, Skup M, Ulas J, Jezierska M, Gradkowska M, Zaremba M (1984) Effect of GM1 ganglioside treatment on postlesion responses of cholinergic enzymes in rat hippocampus after various partial deafferentations. J Neurosci Res 12: 409-420.

Prochiantz A, Daguet MC, Herbet A, Glowinski J (1981) Specific stimulation of in vitro maturation of mesencephalic dopaminergic neurons by striatal membranes. Nature 293: 570-572.

Raff MC, Fields LC, Hakomori S, Mirsky R, Pruss RM, Winter J (1979) Cell-type specific markers for distinguishing and studying neurons and the major classes of glial cells in culture. Brain Res 174: 283-308.

Roisen FJ, Bartfeld H, Nagele L, Yorke G (1981) Gangliosides stimulation of axonal sprouting in vitro. Science 214: 577-578.

Sabel BA, Dunbar GL, Stein DG (1984) Gangliosides minimize behavioral deficits and enhance structural repair after brain injury. J Neurosci Res 12: 429-443.

Sonnino G, Ghidoni R, Galli G, Tettamanti G (1978) On the structure of a new fucose containing ganglioside from pig cerebellum. J Neurochem 31: 947-956.

Tettamanti G, Bonali F, Marchesini S, Zambotti V (1973) A new procedure for the extraction and purification of brain gangliosides. Biochim Biophys Acta 296: 160-170.

Toffano G, Savoini G, Moroni F, Lombardi G, Calzà L, Agnati LF (1983) GM1 ganglioside stimulates the regeneration of dopaminergic neurons in the central nervous system. Brain Res 261: 163-166.

Toffano G, Savoini G, Moroni F, Lombardi G, Calzà L, Agnati LF (1984a) Chronic GM1 ganglioside treatment reduces dopamine cell body degeneration in the substantia nigra after unilateral hemitransection in rat. Brain Res 296: 233-239.

Toffano G, Savoini G, Aporti F, Calzolari S, Consolazione A, Maura G, Marchi M, Raiteri M, Agnati LF (1984b) The functional recovery of damaged brain: the effect of GM1 monosialoganglioside. J Neurosci Res 12: 397-408.

Varon S, Skaper DS, Katoh-Semba R (1986) Neuritic responses to GM1 ganglioside in several in vitro systems, this volume.

Vyskocil F, Di Gregorio F, Gorio A (1985) The facilitating effect of gangliosides on the electrogenic (Na^+/K^+) pump and on the resistance of the membrane potential to hypoxia in neuromuscular preparation. Pflügers Arch 403: 1-6.

Willinger M, Schachner M (1980): GM1 ganglioside as a marker for neuronal differentiation in mouse cerebellum. Develop Biol 74: 101-117.

Gangliosides and neuronal plasticity
G. Tettamanti, R.W. Ledeen, K. Sandhoff,
Y. Nagai, G. Toffano (eds.)
Fidia Research Series, vol. 6
Liviana Press, Padova, © 1986

MODULATION OF NEURITE OUTGROWTH AND GANGLIOSIDE ASSOCIATED CHANGES IN CEREBRAL NEURONS AND PC12 PHEOCHROMOCYTOMA CELLS

Ephraim Yavin

Dept of Neurobiology, Weizmann Institute of Science, Rehovot, Israel

STRUCTURE-FUNCTION RELATIONSHIP OF GANGLIOSIDES AT THE SURFACE OF THE NERVE CELL

The asymmetric position and molecular heterogeneity of gangliosides at the outer leaflet of the nerve plasma membrane has made this group of sialic acid containing glycosphingolipids a subject of expanding interest in neurobiology. Ever since the elucidation of the specific interaction between cholera toxin and GM1 (Moss and Vaughan, 1979; Bennet and Cuatrecasas, 1977) and the potential receptor function of gangliosides for thyrotropin hormone, a number of interactions between gangliosides and diverse ligands have been described (Kohn et al., 1981). Kohn and coworkers (1982) have suggested that gangliosides may serve as receptors for biologically active ligands to transduce signals across the bilayer. Basically, the model considers gangliosides as low affinity, high capacity binding sites which induce, upon interaction, conformational changes in the signal polypeptide. These changes evolve into additional short range hydrophobic interactions between the protein and the ganglioside which perturb the bilayer. An example for such perturbation is the generation of pores or channels in artificial bilayers which follows the interaction of cholera toxin (Tosteson and Tosteson, 1978) or tetanus toxin (Borochov-Neori et al., 1984) with the appropriate gangliosides.

The nervous system is particularly enriched with complex gangliosides that reside predominantly with the neuronal plasma membrane (Rösner, 1982; Sonnino et al., 1981; Ledeen, 1983). The mature brain contains more lipid-bound than protein-bound sialic acid (Tettamanti et al., 1973; Hilbig et al., 1983). It has been shown that at discrete periods in brain development, such as during axodendritic sprouting, the complexity of gangliosides and the net amount of lipid bound sialic acid increase substantially (Lund, 1978; Rösner, 1980). In tissue culture of nerve cells ganglioside ontogeny parallels neurite formation and synaptogenesis (Yavin and Yavin, 1979; Dreyfus et al.,

1980; Willinger, 1981). Artificially supplemented gangliosides have been considered as neuritogens although the molecular basis for this phenomenon is still unsolved. These results suggest that generation of gangliosides via synthesis is an essential step for neuronal cell differentiation (Haber and Gorio, 1984). Therefore, an understanding of the control mechanism of ganglioside diversity acquisition may shed light on the structure/function relation of the carbohydrate residue of these ubiquitous cell surface constituents.

MODEL SYSTEMS FOR NEURITOGENESIS AND GANGLIOSIDE ONTOGENESIS

In an attempt to establish whether or not there is a correlation between the initiation of neurite outgrowth and the metabolic regulation of gangliosides in this process, we have used three types of nerve cell cultures. The first system we have employed consists of dissociated cells from the fetal rat cerebral cortex. These preparations are highly enriched with neurons which have a built-in capacity to elaborate neurites and to mimic in vitro many aspects of mature nerve cells when seeded on appropriate substratum, even under the most stringent conditions (Yavin and Yavin, 1980). As early as 3-4 hours following attachment to the poly-L-lysine substratum, small protrusions are already apparent, and by 24 hours many cells have already elaborated long fibers (Fig. 1A). By 2-3 weeks the fiber outgrowth and intricacy is very extensive as visualized by the decoration of the microtubular elements by an antitubulin antibody (Fig. 1B). These cells extend neurites in serum-free medium without any supplements and have also the potential to synthesize polysialogangliosides (Yavin and Yavin, 1980; Yavin et al., 1984).

The second cell system we have used is a rat pheochromocytoma, PC12, a clone which was originally established by Greene and Tishler (1976) and which has become a valuable model for studying neuronal differentiation. These cells of adrenal medulla origin respond to nerve growth factor (NGF) by a decrease in cell division and the acquisition of a number of properties characteristic of mature sympathetic neurons (Greene and Shooter, 1980). After addition of nanogram quantities of NGF, the PC12 cells extend processes (Fig. 1C), and when maintained for long periods on collagen substratum, they form an intricate network of fibers (Fig. 1D). Some conflicting results concerning the composition of gangliosides in the PC12 cells have been reported (Seifert, 1981; Margolis et al., 1983). Nevertheless, there is agreement that in contrast to many transformed cell lines of neural origin (Yogeeswaran et al., 1973; Dawson, 1979; Dawson and Stoolmiller, 1976), PC12 cells contain practically all major ganglioside species including trisialo and tetrasialogangliosides and also some fucosyl derivatives (Margolis et al., 1984).

The third cell system we have employed, clone SB21B1, was obtained in the laboratory of M. Nirenberg (NIH, Bethesda) from a somatic cell fusion of mouse neuroblastoma NS20-6TG and mouse L cells. These cells cannot form synapses with striated muscle cells nor can they extend neurites following addition of agents which stimulate differentiation in the parent neuroblastoma line (Nirenberg et al., 1980). Also in contrast to the neuroblastoma parent, the hybrid synthesizes predominantly GM2

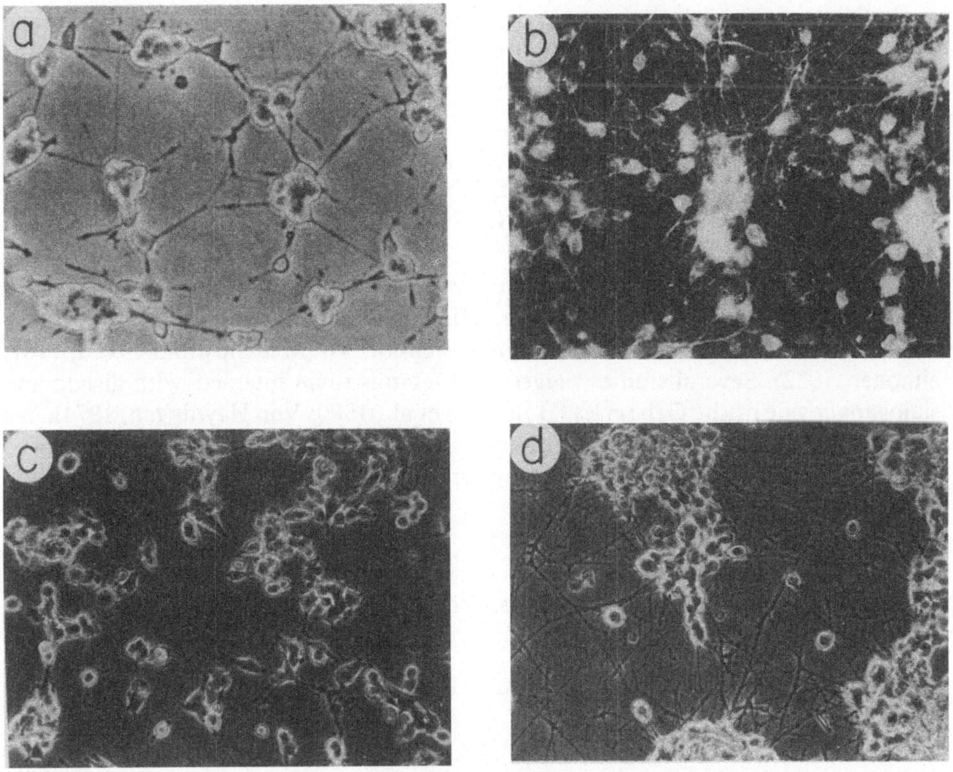

Figure 1. Phase contrast micrograph of dissociated rat cerebral neurons after 1 day (panel A) and immunofluorescence micrograph of the tubulin network of similar culture preparations after 18 days (panel B) in culture. Neurite outgrowth of PC12 rat pheochromocytoma cells visualized by phase microscopy after 1 day (panel C) and 8 days (panel D) in the presence of NGF (50 ng/ml).

and GM3 gangliosides (Yavin and Habig, 1984) similar to the parent fibroblast. When artificially supplemented with gangliosides in the growth medium, these cells undergo morphological differentiation expressed by neurite outgrowth (Ryback et al., 1983). Accompanied with this morphological change is a 9-fold increase in the expression of mRNA for tubulin. Removal of gangliosides reduces the mRNA level of tubulin to near normal levels, an effect suggesting a possible involvement of gangliosides in the regulation of tubulin gene expression.

A common denominator of the first two culture systems is the similarity of the ganglioside profile which resembles the normal developing nervous system. Additionally, both cell types exhibit a characteristic property of a developing nerve cell; i.e. fiber elongation. Perhaps on these common grounds, a concept of the role of gangliosides in neuritogenesis may emerge.

In this report I shall summarize the efforts in my laboratory to examine the morphological changes which take place during neuritogenesis in culture and the accompanying biochemical changes in glycoconjugate metabolism. The data suggest that although complex gangliosides are present at substantial levels in the 16 day old fetal rat brain or in the proliferating PC12 cells, there is no immediate dependency on further acquisition of these molecules to support early neurite outgrowth in both types of cells.

TETANUS TOXIN AS A TOOL FOR MEASURING G1b SERIES GANGLIOSIDE ONTOGENESIS

In recent years, tetanus toxin has become a widely used marker for neuronal cell identification (Mirsky et al., 1978). The affinity of tetanus toxin for nerve cells or plasma membranes of nervous tissue origin is well-known (Mellanby and Green, 1981; Wellhoner, 1982). Several studies suggest that tetanus toxin interacts with disialo and trisialoganglioside of the G1b series (Holmgren et al., 1980; Van Heyningen, 1974). We have used the affinity to G1b gangliosides to investigate the molecular mechanism underlying tetanus toxin internalization by nerve cells (Yavin et al., 1983) and to study the interaction with human erythrocytes artificially supplemented with these gangliosides (Lazarovici and Yavin, 1985).

The specificity of ^{125}I-labeled tetanus toxin toward the G1b gangliosides is well illustrated in Figure 2. Pure gangliosides applied by dot-spot technique on nitrocellulose

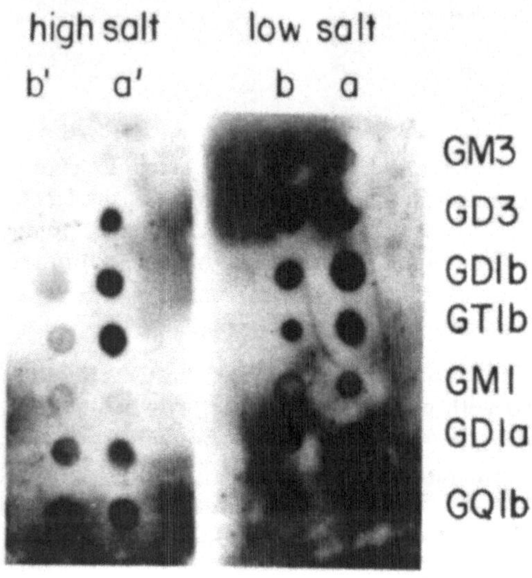

Figure 2. Fluorograph of binding of ^{125}I-labeled tetanus toxin to pure gangliosides on nitrocellulose paper. Pure gangliosides prepared and kindly provided by Dr B. Hauttecour, Institute Pasteur, Paris (0.5-1 μg) applied in small aliquots onto nitrocellulose filter paper were subjected for 30 seconds to a saturated solution of polyisobutyl metacrylate and thereafter soaked for 30 min in 0.2% gelatin in Pi-saline buffer. ^{125}I-labeled tetanus toxin (0.5×10^6 cpm/ml) in 25mM Tris chloride pH 7.4 buffer and 0.2% gelatin was added for 1h at room temperature. Excess toxin was removed by five rinses with 25 mM Tris-acetate in 0.2% gelatin (low ionic strength) or five rinses with Pi/saline (high ionic strength) buffer. After drying under I.R. lamp the paper was exposed to an x-ray film.

paper extensively bind ^{125}I-labeled tetanus toxin under low ionic strength conditions (25mM Tris-Cl in 0.2% gelatin). The toxin binds best to GD1b followed by GT1b and GD3 and also to GD1a and GQ1b ganglioside species. When 150mM NaCl is applied (high salt) to the toxin-ganglioside complex on the nitrocellulose paper, only GD1b, GT1b and GD3 retain the toxin. This suggests that electrostatic interactions dominate the initial binding between the toxin and the ganglioside. These electrostatic forces are also presumably responsible for the interaction of tetanus toxin with gangliosides such as GD1a and GQ1b. Binding of the toxin to GD3 is also notable. This ganglioside is abundant during early stages of neuronal development (Rösner, 1980; Hilbig et al., 1983; Seyfried et al., 1983) and is also present in cell cultures (Yavin and Yavin, 1979; Dreyfus et al., 1981). Since GD1b and GT1b gangliosides are present in both cerebral cell and PC12 cell cultures, it was of interest to compare binding of the toxin to these two types of cultures. As shown in Table 1, cerebral neurons exhibit a greater binding activity of tetanus toxin normalized per DNA than do PC12 cells. This could indicate that the ganglioside receptors for tetanus toxin may be different on the two cell types or even exhibit a different affinity. Another possibility is that binding of toxin requires additional surface components which are absent in PC12 cells.

Table 1. *Binding of ^{125}I-Tetanus Toxin to Cerebral Neurons and Pheochromocytoma (PC12)*

Experimental Conditions	Cerebral Cells	Pheochromocytoma PC12 Cells
	cpm/μg DNA	
Binding at:		
0-4°C	3479±515	575±42
37°C	6081±322	775±88

Cerebral neurons (2 weeks in culture) and proliferating PC12 cells were incubated with ^{125}I-labeled tetanus toxin (180.000 cpm/culture) in 20mM Tris-acetate, pH 7.4, 0.2M sucrose 5% serum and 2mM $CaCl_2$ for 90 min at 4°C or 37°C. Unbound toxin was removed by three rinses in Pi/saline supplemented with 0.25% bovine serum albumin. Cells were collected in 0.5% sodium dodecyl sulfate in 0.1M NaOH and diphenylamine reagent was added (Yavin et al., 1983). Cell associated radioactivity was determined in a gamma counter and DNA content read in a spectrophotometer.

Binding of tetanus toxin can be utilized to study the appearance of receptors on the cell surface. For instance, the binding of tetanus toxin by PC12 cells is more than doubled after 2 days of exposure to NGF (Fig. 3B). Binding is blocked by pretreatment of cells with sialidase prior to toxin addition. This is consistent with the observation that sialic acid residues are required for specific binding of the toxin. It is also in line with the finding that in the presence of NGF, gangliosides with specificity toward tetanus toxin are synthesized by the cells (Seifert, 1981). Figure 3 (panel A) also illustrates the age-in-culture-dependent association of ^{125}I-labeled tetanus toxin with the primary cerebral cells. A threefold increase of labeled toxin binding, normalized per cellular DNA, is observed after 2 weeks in culture. After this period, a gradual decrease in the amount of toxin bound is apparent. It is notable that these values are in good agreement with the changes in the lipid-bound sialic acid content of similar cultures (Yavin and Yavin, 1979) and that a marked reduction of the label is observed if the cells are pretreated with neuraminidase. The loss of binding is greater in cells at early times than it is in 2-week old cultures. Since the mature cultures exhibit an increase in complex sialoglycolipids of the G1b family, the possibility can be considered that if gangliosides play a role in this process, there is a selectivity which parallels the higher complexity. In this context it is worth mentioning that early neuritogenesis in cerebral cell cultures does not depend on the presence of disialo or higher gangliosides; sialidase added to freshly dissociated cells and left in the culture medium for several days does not change the course of neurite outgrowth in comparison to untreated cultures. Similarly, PC12 cells extend processes in the presence of sialidase with NGF (unpublished). By following the fate of ganglioside labeling by D-(^{14}C)- glucosamine or indirectly by tetanus toxin under these conditions we can show that most newly labeled polysialogangliosides turn into labeled GM1. Thus it would appear that the presence of GM1 at the cell surface is sufficient to permit the early steps in neuritogenesis, while gangliosides such as GD1a, GT1b and GQ1b presumably have a different role at some later phases of differentiation (Hilbig et al., 1983).

EFFECT OF TUNICAMYCIN ON NEURITOGENESIS AND GLYCOCONJUGATE METABOLISM

Studies on the structure-function relation of carbohydrate residues in glycopro-

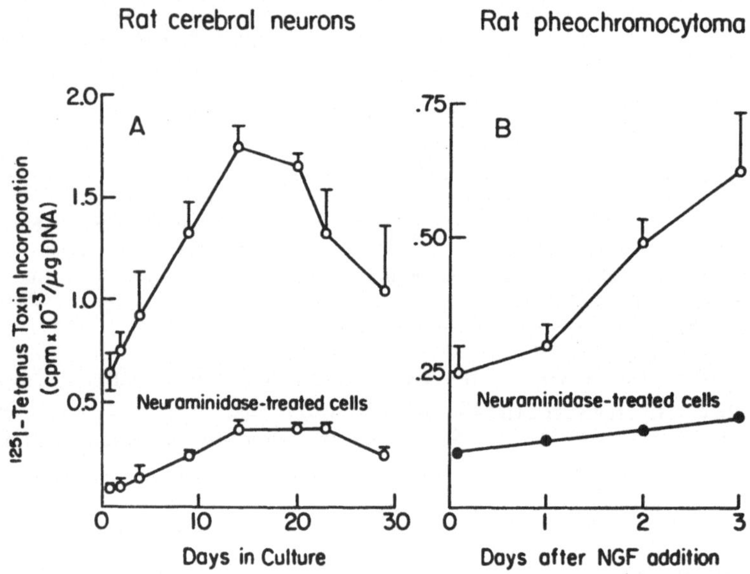

Figure 3. Acquisition of tetanus toxin binding activity by cerebral cells and PC12 cells after treatment with NGF. Cerebral cells (panel A) grown in culture for various time lengths and PC12 cells (panel B) after stimulation with 50 ng/ml NGF were incubated with ^{125}I-labeled tetanus toxin for 1h at 37°C. Binding conditions and determination of the cell associated toxin were essentially similar to those described in Table 1. Prior to toxin binding selected cultures were treated for 6h with Vibrio cholera neuraminidase (5 milliunits/culture, Calbiochem) as detailed elsewhere (Yavin *et al.*, 1983).

teins have been facilitated in cell culture by the use of specific glycosylation inhibitors. Tunicamycin, an amphipatic analog of UDP-GlcNAc is one such potent inhibitor which affects not only proteins (Olden et al., 1978) but also, as recently shown, lipid glycosylation in isolated Golgi vesicles (Yusuf et al., 1983). Tunicamycin blocked glucosamine labeling of both cellular gangliosides and, as expected, cellular glycoproteins when added to somatic neurohybrid NG108-C15 cells (Guarnaccia et al., 1983). It also affected cell adhesion and neurite outgrowth in N115 neuroblastoma cells (Richter-Landsberg and Duksin, 1983). We have extended these observations to examine the effect of tunicamycin on the initiation of neurite outgrowth and its possible correlation to glycoconjugate metabolism in normal rat cerebral neuron cultures (Yavin et al., 1984).

Al-tunicamycin added to the culture medium of cells maintained for 1 or 8 days in vitro caused 81.3% and 87.5% inhibition of D-[^{3}H]glucosamine uptake, respectively, in comparison to control values (Table 2). The data corroborate reports from other laboratories using cells in culture, which show selective inhibition of glycosylation after administration of tunicamycin (Yamada and Olden, 1982). About 78% of the cellular label from glucosamine in 1-day-old inhibited cultures resided in the delipidated cell fraction and consisted mainly of labeled proteins. In contrast, less than 45% of radioactivity was present in the protein fraction by day 8 in culture. In 8-day-old neuronal cultures, the majority of the radioactive label incorporated in the methanolic-

Table 2. *Inhibitory Effect of Tunicamycin on D-[^{3}H] Glucosamine and U-^{14}C Serine Uptake in Cerebral Cell Cultures*

Cell Treatment	U-[^{14}C] serine Total	D-[^{3}H] glucosamine uptake Total	Delipidated protein	Chloroform/Methanol pellet
	cpm × 10^{-4}/μg DNA	cpm × 10^{-3}/μg of DNA		
Day 1				
Control	5.2 ± 0.20	2.57 ± 0.18	2.01 ± 0.25	0.56
Tunicamycin	3.9	0.48 ± 0.20	0.09 ± 0.05	0.39
% inhibition	25	81.3	95.5	30.4
Day 8				
Control	7.4 ± 0.35	4.88 ± 0.32	2.16 ± 0.14	2.72
Tunicamycin	6.3	0.61 ± 0.04	0.23 ± 0.04	0.38
% inhibition	15	87.5	89.4	86.0

Cells were pulse-labeled for 24 hr with D-[^{3}H]glucosamine (5 μCi/0.5 ml per well) or with U-^{14}C-serine (5 μCi/0.5ml) in the absence or presence of A_1-tunicamycin (0.5μg/ml). Excess isotope was removed by rinsing twice with phosphate-buffered saline, and cells were transferred into Eppendorf conical tubes and precipitated with 5% trichloroacetic acid. The soluble radioactivity was discarded by centrifugation, and the insoluble pellet was extracted with chloroform/methanol, 1:2 (vol/vol), as described (Yavin et al., 1984). The pellet was hydrolyzed with 0.5 M NaOH, and aliquots were taken for DNA and radioactivity assays. Values represent averages of ± SEM of four culture wells.

water layer was found in glycolipids. This suggests that 8-day-old cerebral cells incorporate glucosamine into glycolipids more effectively than do the immature cells and also that the inhibitory effect of tunicamycin on ganglioside labeling in one-day old cells is less pronounced than in 8-day old cultures. Serine incorporation, on the other hand, was less affected than glucosamine uptake by tunicamycin. There was no inhibition of labeling of phosphatidyl serine nor inhibition of formation of phosphatidyl ethanolamine in the presence of tunicamycin (data not shown). In a similar manner, exposure of PC12 cells to the A1 analog of tunicamycin (Duksin and Mahoney, 1982) caused 74.6% and 71.8% inhibition of total ^{14}C-glucosamine cell-associated radioactivity of untreated and NGF treated cells, respectively (Table 3). These values are slightly smaller than those observed in nerve cells and may relate to the particular

Table 3. *Effect of A1 Tunicamycin on ^{14}C-Glucosamine Incorporation into PC12 Cells*

Cell Treatment	1-D-^{14}C-glucosamine uptake	
	Total	Chloroform/Methanol soluble
	cpm/μg protein	
Untreated	765 ± 82	180 ± 21
Tunicamycin (0.5 μg/ml)	194 ± 41	112 ± 25
% inhibition	74.6	37.8
NGF	772 ± 35	401 ± 23
Tunicamycin (0.5 μg/ml)	218 ± 48	100 ± 5
% inhibition	71.8	75.0

D-^{14}C-Glucosamine (3μCi/ml) and NGF (50ng/ml) were added to cells in 0.5% low serum containing medium for 48h. Harvesting of cells and determination of the radioactivity in the chloroform/methanol extract were done essentially as described in Table 2.

tunicamycin analog employed. A substantial fraction (23.5%) of glucosamine label is extracted by chloroform/methanol in actively dividing PC12 cells. This value is further increased after NGF treatment, a finding which suggests that the chloroform/methanol extractable glucosamine labeled products comprise the greater portion of the total cell-associated radioactivity. The data suggest that during differentiation of nerve cells there is an overall shift from a protein associated to a lipid associated glycosylation. The latter is blocked extensively (38 and 75% in untreated and NGF treated cells, respectively) after addition of tunicamycin. It should be noted that the chloroform/methanol soluble fraction contains, apart from glycolipids, sugar nucleotides and possibly other low molecular weight compounds (unpublished).

The patterns of [^{14}C]-glucosamine labeling of cellular gangliosides in control and tunicamycin (0.5 μg/ml) treated rat cerebral neurons (panel A) and PC12 cells (panel B) are illustrated in Figure 4. [^{14}C]-glucosamine is effectively incorporated into all major cellular gangliosides when added to cells in the absence of tunicamycin (panel A, track 1). The label is highest in GD1a (33.3%) followed by GT1b (26.9%), GD1b (14.8%) and GM1 (11.5%) as determined by radioactivity counting of the isolated

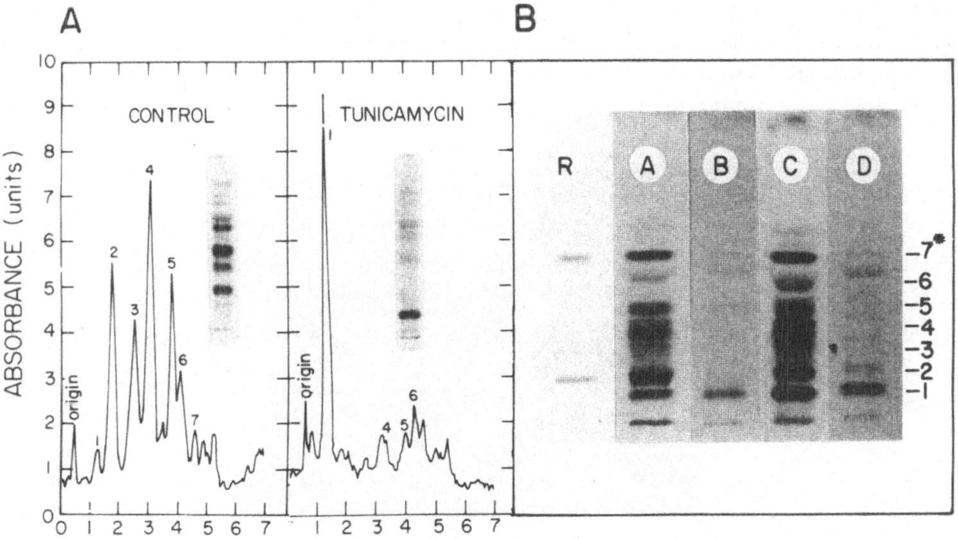

Figure 4. Comparison of D-[^{14}C]-glucosamine radioactivity incorporation into ganglioside species of cerebral neurons (panel A) and PC12 cells (panel B) before and after treatment with A1-tunicamycin. Experimental conditions for cell labeling were similar to those described in Tables 2 and 3. Lipid extracts were chromatographed on HPLC using a solvent system described elsewhere (Yavin et al., 1984). Panel B: Labeled gangliosides of untreated (track A) and tunicamycin treated (track B) proliferating PC12 cells is compared with untreated (track C) and tunicamycin treated (track D) NGF-primed (48 hrs, 50 ng/ml) cells. Resorcinol visualization of gangliosides of proliferating PC12 cells is depicted in track B. Note enhanced labeling of gangliosides after NGF treatment and the appearance of GQ1b in tunicamycin treated cultures. Abbreviations: (1) GQ1b; (2) GT1b; (3) GD1b; (4) GD1a; (5) GM1; (6) GM2; (7) GM3; (7) *GM3-tentative.

bands from the silica gel. A different picture emerges in the glycolipid extract separated on TLC from cells subjected to 0.5 μg tunicamycin. Of the total counts incorporated into the lipids in tunicamycin treated cells (1/5 of the normal values), nearly no radioactivity could be detected in the GT1b and GD1b ganglioside species (track 2). The major radioactive peak comigrated with tetrasialo (GQ1b) ganglioside (67.3%), the remaining counts being distributed between GD1a (10.9%), GM1 (3.7%) and GM2 (5.4%) ganglioside species. The major radioactive peak was nondialysable, pronase-insensitive, but exhibited limited sensitivity to sialidase treatment. Panel B in Figure 4 depicts the fluorograph of the ganglioside pattern of untreated pheochromocytoma cells (tracks A and B) and those treated with NGF (tracks C and D). Radioactivity from ^{14}C-glucosamine is effectively incorporated into all major gangliosides such as GQ1b, GT1b, GD1a, GM1 and GM3 (Panel B, track A) and it doubles in cells treated with NGF (track C). In the latter cells, a substantial increase in radioactivity is seen in the two bands which comigrate with GQ1b and GD1b gangliosides. The increase in ganglioside labeling following addition of NGF has been reported (Seifert, 1981; Margolis et al., 1983). When tunicamycin (0.5 μg/ml) is added to cells as illustrated in tracks B and D (Fig. 4), there is an almost complete inhibition of labeling of all major gangliosides. The only exception to this is GQ1b, which is still labeled both in the presence (track D) or in the absence (track B) of NGF. Thus, it would appear that PC12 cells and the primary neurons share common metabolic pathways for ganglioside

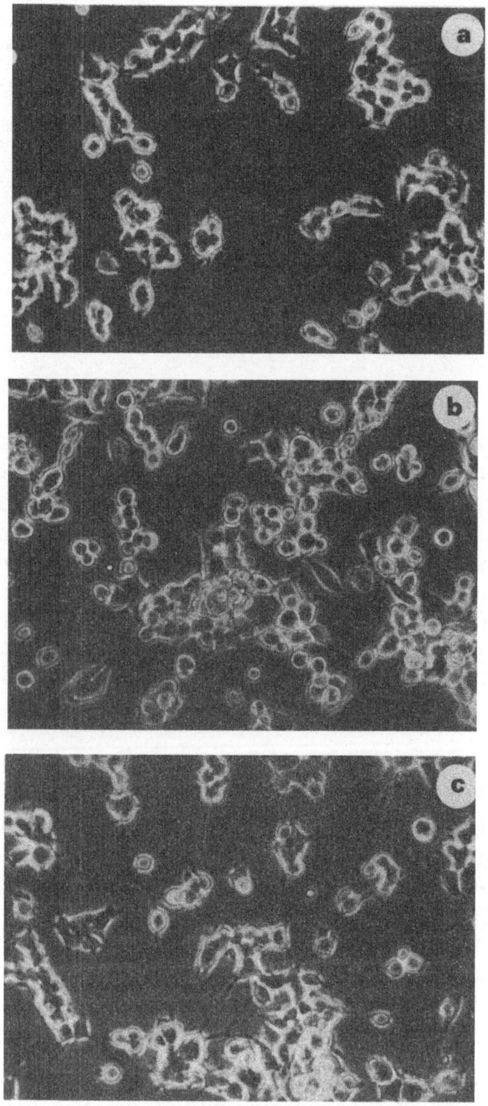

Figure 5. Effect of tunicamycin on NGF-induced fiber outgrowth. Phase contrast micrographs of PC12 cells primed for 48h with NGF in the absence (panel a) or in the presence of 0.2 μg tunicamycin (panel b). Note the substantial reduction of fiber outgrowth in panel b. Panel c depicts cells exposed for 3 days to NGF and subjected to 1 μg/ml tunicamycin at the end of the second day. The antibiotic does not affect outgrowth and cells are similar as compared to controls (micrograph not shown). Bar = 50μm.

biosynthesis. Furthermore, in both cell types tunicamycin does not block labeling, a finding which indicates the existence of a direct sialylation route independent of *de novo* synthesis (Maccioni et al., 1977; Yavin et al., 1984).

In addition to the effect on glycoconjugate metabolism, tunicamycin has a marked effect on neurite outgrowth. In a recent study, we reported that tunicamycin inhibits neuritogenesis in dissociated cerebral neurons at its onset following cell seeding (Yavin et al., 1984). When added to cells which already possess an established neurite network, the antibiotic is not as toxic and does not cause neurite retraction. Similary, PC12 cells also do not respond to NGF-stimulated neuritogenesis when 0.2μg of tunicamycin is added to the incubation medium several hours after cell plating (Fig. 5). However, NGF-primed cells (24-48h) continue to send out processes for at least 24h when treated with 1μg/ml tunicamycin. This suggests that the PC12 cells resemble the morphological behavior of normal nerve cells rather than that of transformed N115 neuroblastoma cells following tunicamycin addition. The latter retract the fibers induced by serum removal in the presence of tunicamycin (Richter-Landsberg and Duksin, 1983). Thus, it would appear that the neuritic outgrowth following NGF addition and that following serum removal may be two entirely different processes.

ACKNOWLEDGMENTS

The author whishes to thank Drs. L. Kohn, W. Habig, C. Richter-Landsberg, D. Duksin and Z. Yavin for interest and participation in some aspects of these studies. Mrs. S. Gil is acknowledged for technical assistance.

REFERENCES

Bennett V, Cuatrecasas P (1977) Membrane gangliosides and activation of adenylate cyclase. In: Cuatrecasas P and Greaves MF (eds): The Specificity and Action of Animal, Bacterial and Plant Toxins, Chapman and Hall, London, p. 3-66.

Borochov-Neori H, Yavin E, Montal M (1984) Tetanus toxin forms channels in planar lipid bilayers containing gangliosides. Biophys J 45: 83-85.

Dawson G (1979) Complex carbohydrates of cultured neuronal and glial cell lines. In: Margolis R U and Margolis RK (eds): Complex carbohydrates of Nervous Tissue. Plenum Press, New York, pp. 291-325.

Dawson G, Stoolmiller A (1976) Comparison of the ganglioside composition of established mouse neuroblastoma cell strains grown *in vivo* and in tissue culture. J Neurochem 26: 225-226.

Dreyfus H, Harth S, Massarelli R, Louis JC (1981) Mechanism of differentiation in cultured neurons: Involvement of gangliosides. In: Rapport MM and Gorio A (eds): Gangliosides in Neurological and Neuromuscular Function, Development and Repair, Raven Press, New York, pp. 151-170.

Dreyfus H, Louis JC, Harth S, Mandel P (1980) Gangliosides in cultured neurons. Neurosci 5: 1647-1655.

Duksin D, Mahoney WC (1982) Relationship of the structure and biological activity of the natural homologues of tunicamycin. J Biol Chem 257: 3105-3109.

Greene LA, Shooter EM (1980) The nerve growth factor: Biochemistry, synthesis and mechanism of action. Annu Rev Neurosci 3: 353-402.

Greene LA, Tischler AS (1976) Establishment of a noradrenergic clonal line of rat adrenal pheochromocytoma cells which respond to nerve growth factor. Proc Natl Acad Sci USA 73: 2424-2428.

Guarnaccia SP, Shaper JH, Schnaar RL (1983) Tunicamycin inhibits ganglioside biosynthesis in neuronal cells. Proc Natl Acad Sci USA 80: 1551-1555.

Haber B, Gorio A (eds) (1984) Neurobiology of gangliosides J Neurosci Res 12: 147-509.

Hilbig R, Lauke G, Rahmann H (1983) Brain gangliosides during the lifespan (embryogenesis to scenescence) of the rat. Dev Neurosci 6: 260-270.

Holmgren J, Elwing H, Fredman P, Svennerholm L (1980) Polystyrene absorbed ganglioside for investigation of the structure of the tetanus toxin receptor. Eur J Biochem 106: 371-379.

Kohn LD. Aloj SM, Beguinot F, Vitti P, Yavin E, Yavin Z, Laccetti P, Grollman EF, Valente WA (1982) Molecular interactions at the cell surface: Role of glycoconjugates and membrane lipids in receptor recognition processes. In: Shepard J, Anderson VE, Eaton J (eds): Membranes and Genetic Diseases, Alan R Liss Inc, NY, pp. 55-83.

Kohn LD, Consiglio E, Aloj SM, Beguinot F, De Wolf MJS, Yavin E, Yavin Z, Meldolesi MF, Shifrin S, Gill DI, Vitti P, Lee G, Valente WA, Grollman EF (1981) The structure of the thyrotropin (TSH) receptor: potential role of gangliosides and relationship to receptors for interferon and bacterial toxins. In: Schweiger HG (ed): Springer-Verlag, Berlin, p. 696-706.

Lazarovici P, Yavin E (1985) Tetanus toxin interaction with human erythrocytes I. Properties of polysialoganglioside association with the cell surface. Biochim Biophys Acta 812: 523-531.

Ledeen RW (1983) Gangliosides. In: Lajtha A (ed): Handbook of Neurochemistry, Vol. 3, Plenum Press, New York, pp. 41-90.

Lund RD (1978) Development and Plasticity of the Brain, Oxford University Press, New York, pp. 1-370.

Maccioni HF, Landa CA, Arce A, Capputto R (1977) The biosynthesis of brain gangliosides. Evidence for a transient pool and an end product pool of gangliosides. Adv Exp Med Biol 83: 267-281.

Margolis RK, Salton SRJ, Margolis RV (1983) Complex carbohydrates of cultured PC12 pheochromocytoma cells. Effects of nerve growth factor and comparison with neonatal and mature rat brain. J Biol Chem 258: 4110-4117.

Margolis RU, Mazzulla M, Greene LA, Margolis RK (1984) Fucosyl gangliosides of PC12 pheochromocytoma cells. FEBS Lett 172: 339-342.

Mellanby J, Green J (1981) How does tetanus toxin act? Neurosci 6: 281-300.

Mirsky R, Wendon LMB, Black P, Stolkin C, Bray D (1978) Tetanus toxin: a cell surface marker for neurons in culture. Brain Res 148: 251-259.

Moss J, Vaughan M (1979) Activation of adenylate cyclase by choleragen. Annu Rev Biochem 48: 581-600.

Nirenberg M, Wilson S, Higashida H, Thompson J, Eisenbarth G, Walsh F, Rotter A, Kenimer J, Sabol S (1980) Synapse plasticity. Pont Acad Sci Scri Varia 45: 123-127.

Olden K, Pratt RM, Yamada KM (1978) Role of carbohydrates in protein secretion and turnover: Effect of tunicamycin on the major cell surface glycoprotein of chick embryo fibroblasts. Cell 13: 461-473.

Richter-Landsberg C, Duksin D (1983) Role of glycoproteins in neuronal differentiation. Exptl Cell Res 149: 335-345.

Rösner H (1980) Ganglioside changes in the chicken optic lobes and cerebrum during embryonic development: Transient occurrence of novel multisialogangliosides. Wilhelm Roux's Arch Dev Biol 188: 205-213.

Rösner H (1982) Ganglioside changes in the chicken optic lobes as biochemical indicators of brain development and maturation. Brain Res 236: 46-61.

Rybak S, Ginzburg I, Yavin E (1983) Gangliosides stimulate neurite outgrowth and induce tubulin mRNA accumulation in neural cells. Biochim Biophys Res Commun 116: 974-980.

Seifert W (1981) Gangliosides in nerve cell cultures. In: Rapport MM, Gorio A (eds): Gangliosides in Neurological and Neuromuscular Function, Development and Repair, Raven Press, New York, pp. 99-117.

Seyfreid TN, Miyazawa N, Yu RK (1983) Cellular localization of gangliosides in the developing mouse cerebellum: Analysis using the Weaver mutant. J Neurochem 41: 491-505.

Sonnino S, Ghidoni R, Masserini M, Aporti F, Tettamanti G (1981) Changes in rabbit brain cytosolic and membrane bound gangliosides during prenatal life. J Neurochem 36: 227-232.

Tettamanti G, Bonali F, Marchesini S, Zambotti V (1973) A new procedure for the extraction, purification and fractionation of brain gangliosides. Biochim Biophys Acta 296: 160-170.

Tosteson MT, Tosteson DC (1978) Bilayers containing gangliosides develop channels when exposed to cholera toxin. Nature 275: 142-144.

Van Heyningen WE (1974) Gangliosides as membrane receptors for tetanus toxin, cholera toxin and serotonin. Nature 249: 415-417.

Wellhoner HH (1982) Tetanus neurotoxin. Rev. Physiol Biochem Pharmacol 93: 1-68.

Willinger M (1981) The expression of GM1 ganglioside during neuronal differentiation. In: Rapport MM, Gorio A (eds): Gangliosides in Neurological and Neuromuscular Function, Development and Repair, Raven Press, New York, pp. 17-27.

Yamada KM, Olden K (1982) In: Famura G (ed): Biology of tunicamycin, Japan Scientific Societies Press, Tokyo, pp. 119-144.

Yavin E, Richter-Landsberg C, Duksin D, Yavin Z (1984) Tunicamycin blocks neuritogenesis and glucosamine labeling of gangliosides in developing cerebral neuron cultures. Proc Natl Acad Sci USA 81: 5638-5642.

Yavin E, Yavin Z (1979) Ganglioside profiles during neural tissue development: acquisition in the prenatal rat brain and cerebral cell cultures. Dev Neurosci 2: 25-37.

Yavin E, Yavin Z, Kohn LD (1983) Temperature mediated interaction of tetanus toxin with cerebral neuron cultures: characterization of a neuraminidase-insensitive toxin receptor complex. J Neurochem 40: 1212-1219.

Yavin E, Habig WH (1984) Ganglioside-mediated tetanus toxin interaction with neural hybrid clones: Binding to synaptic competent and incompetent cells. J Neurochem 42: 1313-1320.

Yavin Z, Yavin E (1980) Survival and maturation of cerebral neurons on poly-L-lysine surfaces in the absence of serum. Dev Biol 75: 454-459.

Yogeeswaran G, Murray RK, Pearson ML, Sanwal BD, McMorris FA, Ruddle FH (1973) Glycophingolipids of clonal lines of mouse neuroblastoma and neuroblastoma XL cell hybrids. J Biol Chem 248: 1231-1239.

Yusuf HKM, Pohlentz G, Sandhoff K (1983) Tunicamycin inhibits ganglioside biosynthesis in rat liver Golgi apparatus by blocking sugar nucleotide transport across the membrane vesicles. Proc Natl Acad Sci USA 80: 7075-7079.

Gangliosides and neuronal plasticity
G. Tettamanti, R.W. Ledeen, K. Sandhoff,
Y. Nagai, G. Toffano (eds.)
Fidia Research Series, vol. 6
Liviana Press, Padova, © 1986

GANGLIOSIDE EFFECTS ON ASTROGLIAL CELLS IN VITRO

Stephen D. Skaper, Ritsuko Katoh-Semba[1], Laura Facci[2], Silvio Varon

Department of Biology, School of Medicine, University of California, San Diego, La Jolla, California 92093 USA,
[1]Department of Perinatology, Institute for Developmental Research, Aichi Prefectural Colony, Kasugi, Aichi, Japan,
[2]Laboratory of Biochemistry, Fidia Research Laboratories, Via Ponte della Fabbrica, 3a, Abano Terme, Padova, Italy

INTRODUCTION

It is increasingly recognized that neuronal maintenance, function and repair capabilities may depend in vivo — as they do in vitro — on the availability to the neuron of appropriate extrinsic agents. Three classes of such agents have already been recognized (Varon and Adler, 1980; 1981; Varon et al., 1984; Varon, 1985). Neuronotrophic agents are presumed to control neuronal survival and general capabilities for growth and/or function. Neurite promoting agents are specifically involved in modulating the execution of neuritic programs, i.e. the extension of axons and dendrites. Transmitter modulating agents are concerned with the choice of transmitter modalities expressed by a nerve cell. As research in this field proceeds, it will become possible to specify additional categories of neuronoactive, extrinsic agents. In vivo, as in vitro, extrinsic neuronoactive agents may occur i) free in the humoral environment (extracellular fluid, culture media), ii) anchored to interstitial structures (extracellular matrices, culture substrata), and/or iii) on the surfaces of other cells with which neurons come into contact.

In the course of early development, adult performance and reactions to pathological insults in the central nervous system (CNS), astroglial cells are in a position to present CNS neurons with critical supplies of neuroactive extrinsic agents. In turn, astroglial cells are subject to regulation by agents supplied by neurons, other glial cells, or blood and cerebrospinal fluids. Investigation of glia-modulating agents is best approached by use of in vitro astroglial cell cultures (Schoffeniels et al., 1978; Varon

Abbreviations: CNS, central nervous system; PNS, peripheral nervous system; PORN, polyornithine; FN, fibronectin; cyclic AMP, cyclic adenylic acid; cyclic GMP, cyclic guanilic acid.

and Somjen, 1979; Manthorpe et al., 1985). We have recently developed a procedure by which thousands of replicate microcultures of rat cerebral astroglial cells can be set up and maintaned in a minimal, chemically defined medium on a defined substratum (Rudge et al., 1985). These microcultures release into their medium at least three types of neuronoactive agents: protein trophic factors addressing PNS neurons, low molecular weight trophic agents addressing CNS neurons (and also PNS neurons in concurrence with their protein trophic factors), and large laminin-like neurite promoting factors.

Gangliosides are normal membrane constituents, largely localized to the outer leaflet of the plasma membrane and thus presenting their carbohydrate moieties on the surface of the cells (Yamakawa and Nagai, 1978). They are abundant in neural tissue (particularly neurons), and the ganglioside compositions of neuronal and glial cell membranes appear to be different (Ledeen, 1978). Exogenously supplied gangliosides have been shown to promote neuritic outgrowth in vitro (Varon et al., 1985) and in vivo (Gorio et al., 1980; 1983; Sparrow and Grafstein, 1982), and also to affect a number of other neuronal behaviors in the CNS when administered in vivo (Toffano et al., 1983; Karpiak et al., 1984; Sabel et al., 1984). There have been, however, no studies addressing possible interactions of exogenous gangliosides with glial cells, which in turn could alter glial regulations of neuronal performances and thus, ultimately, neuronal behaviors themselves.

In this presentation, we review two recent studies carried out in our laboratory by use of the new astroglial microculture system, which demonstrate that: (1) exogenously presented GM1 ganglioside (as well as other gangliosides) interact directly with astroglial cells to modify their behaviors; (2) GM1 blocks the ability of astroglial cells to assume in vitro the stellate morphology typical of a fibrous astrocyte; and, (3) astroglial proliferation, which can be re-triggered in those cultures by the introduction of serum, can be equally re-triggered by presentation of ganglioside in the absence of serum.

ASTROGLIAL CELL CONVERSION FROM FLAT TO STELLATE MORPHOLOGY

Neonatal rat cerebra are dissociated and the cells grown on tissue culture plastic in serum-supplemented medium for 10 days before being harvested. The procedure provides for a reproducible collection of 5×10^6 cells/brain, of which 95% or more are astroglial (AG) cells by their content of Glial Fibrillary Acidic Protein (Rudge et al., 1985). For subsequent studies, the harvested AG cells are re-seeded with modified Eagle's basal medium (HEBM) into 6 mm microwells sequentially precoated with polyornithine and fibronectin (PORN/FN). The seeding medium is replaced with fresh HEBM at 2 hr (at which time 12,000 of the 16,000 seeded cells/well are already attached), and again at 24 hr so as to remove undefined substances carried over into the secondary microcultures from the primary culture or the harvesting medium. The 24 hr medium change is also used to supply the cultures with the desired agents under investigation. The 24 hr re-fed cultures are designated as D1 (day one) cultures and used in all subsequent analyses.

The D1 cultures, from a 16,000 cells/well seeding, constitute a nearly confluent monolayer of flat AG cells, as shown in Figure 1 A. It is well known, in several other

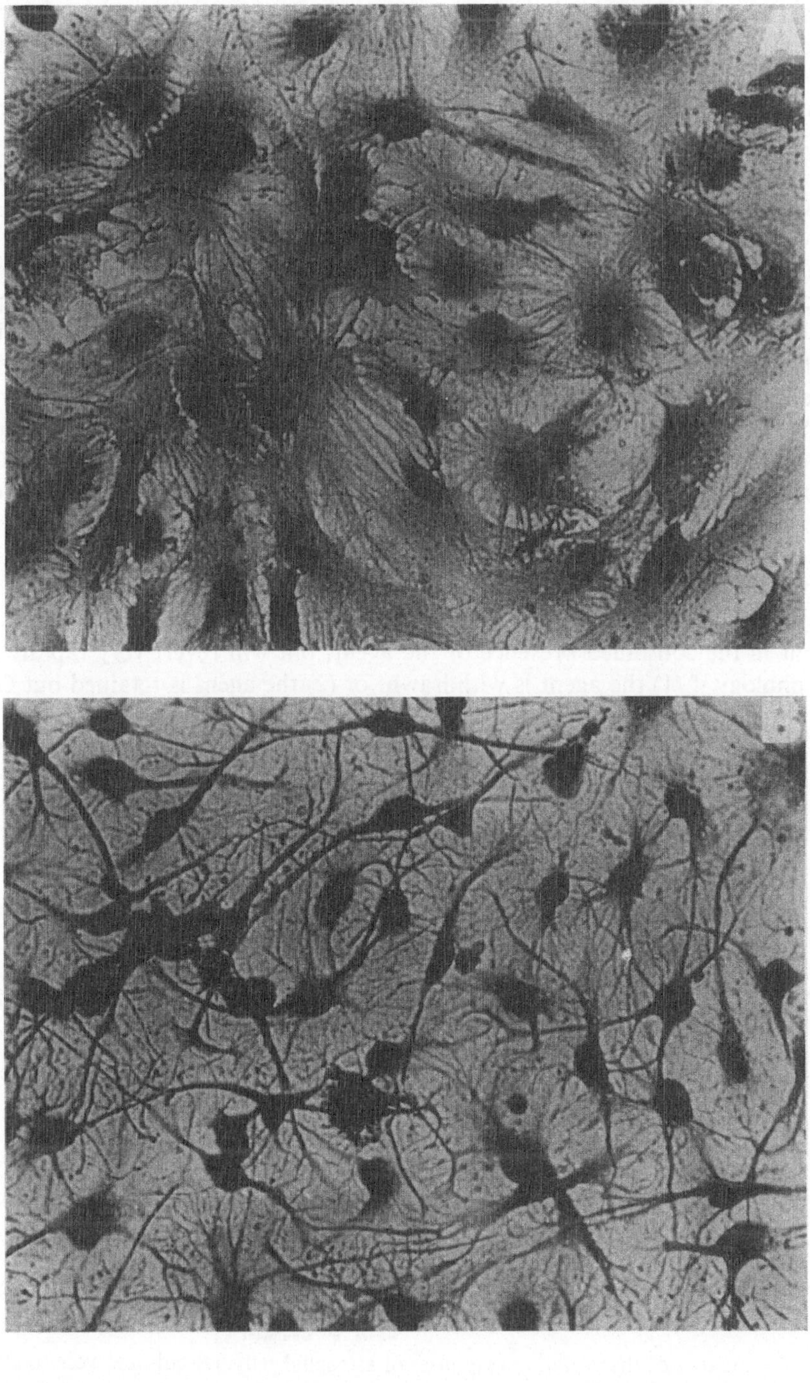

Figure 1. Morphological conversion of astroglial cells from flat (A) to stellate (B) shapes. D1 cultures were treated with 1 mM dibutyryl cyclic AMP or none for 4 hr.

astroglial cell culture systems, that certain treatments of these flat cells in the absence of serum cause them to assume a stellate morphology (Moonen et al., 1976; Manthorpe et al., 1979; Magistretti et al., 1983). In the new, chemically defined system of AG microcultures the flat AG cells also can be made to undergo this "stellation" process, illustrated in Figure 1 B, by presentation to the D1 cultures of dibutyryl cyclic AMP (DBcA), forskolin, or cholera toxin — three agents known to elevate cyclic AMP in several cell types.

A detailed study of the stellation response has been carried out in the AG microcultures (Skaper et al., 1986). This study has revealed that GM1 ganglioside is not soliciting the AG conversion from flat to stellate morphologies, despite its reported ability to stimulate adenylate cyclase in rat cerebral cortical membranes (Partington and Daly, 1979). Rather, exogenous GM1 both prevents and reverses the stellation response by AG cells to the already identified stellation-inducing agents. Figure 2 illustrates the GM1 action, as well as several temporal aspects of the stellation response. On presentation of the inducing agent (e.g. DBcA) alone, the proportion of AG cells displaying a stellate morphology rises from a baseline of 1-5% to 95-100% of the total population. The conversion starts almost immediately after presentation of the agent and is completed in less than 2 hr. If, however, the agent is presented together with GM1 (bottom broken line) no conversion takes place even at the earliest times. Dibutyryl cyclic AMP-treated cultures will maintain their stellate morphology for at least 24 hr in the continued presence of the agent, but will revert very rapidly to the flat morphology if (1) the agent is withdrawn, or (2) the agent is retained but GM1 is added.

An interesting feature of the AG microculture system (unrelated to the GM1 effect) was revealed in the course of these studies, namely that i) the stellate morphology

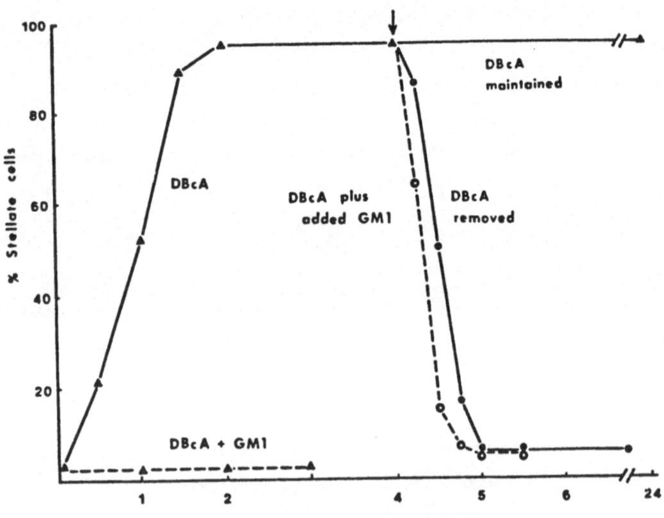

Figure 2. Time course of the stellation responses of astroglial cells. D1 cultures were treated with 1 mM dBcA with (▲—▲) or without (▲. . . ▲) 6 × 10^{-5}M GM1 and the percent stellate cells evaluated at the times indicated. Replicate cultures, after 4 hr of dBcA treatment (arrow), had their media replaced with dBcA (▲—▲), no dBcA (●—●) or dBcA plus GM1 (○---○).

could not be held for 48 hr, despite persistence or renewal of the inducing agent, and ii) the ability to respond to a first presentation of the agent was itself declining with time in culture, so that D3 cultures responded only to higher concentrations of the agent and D5 cultures failed to respond altogether. These temporal declines in astroglial stellation performances, while not fully defined, were not attributable to peculiarities of the medium or the substratum used in this culture system (Skaper et al., 1986).

Figure 3 displays titration curves of the ability by GM1 to block or reverse the astroglial stellation response to DBcA, forskolin, or cholera toxin. The ganglioside ID50 (inhibitory dose for 50% effect), about 2×10^{-5}M, was found to be identical against all three agents (and was also unaffected by 10-fold excesses of the agents), making it unlikely that the interference by GM1 results from the formation of inactive complexes between GM1 and three such different molecular agents. The titration curve also remained unchanged by a 24 hr pre-exposure of GM1-containing medium to the PORN/FN coated microwells, revealing no loss of GM1 potency by time-related inactivation or by possible adsorption or binding to the culture substratum itself. Other experiments showed that neither the stellation response of astroglial cells nor its blockade by GM1 reflected specific properties of the PORN/FN substratum, since both performances occurred equally on culture plastic appropriately precoated with collagen, PORN only, fibronectin only, or with laminin alone or in conjunction with PORN. These and other data strongly support the view that a direct association of GM1 with the cultured astroglial cells is responsible for the interference by the ganglioside with the astroglial stellation behavior.

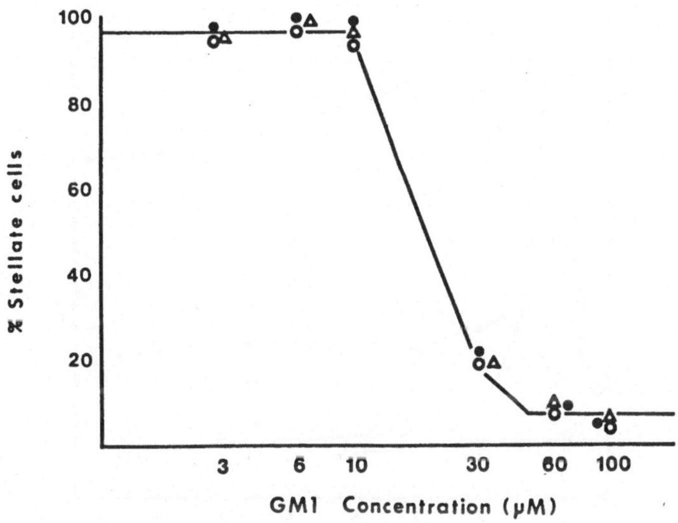

Figure 3. Titration curves of the stellation-blocking effect of GM1. 1 mM dBcA (•), 10^{-5}M forskolin (○, FSK), 6×10^{-8}M cholera toxin (△, CTX). Serial dilutions of GM1 were prepared in the presence of the indicated drugs, and presented to D1 cultures for 4 hr. Identical results were obtained with GM1 freshly prepared, or stored for 24 hr in polyethylene tubes or in polyornithine/fibronectin coated microwells.

ASTROGLIAL CELL PROLIFERATION IN THE MICROCULTURE SYSTEM

The standard astroglial microcultures maintain a nearly constant cell number over at least an 8-day period. Nevertheless, a 24 hr pulse of ^{3}H-thymidine labels a modest proportion of cell nuclei even when applied at the end of 1 week (Rudge et al., 1985), suggesting the occurrence of some moderate turnover (i.e. proliferation matched by cell losses) — which might underlie the slow change in culture properties already revealed by the stellation studies. A more detailed study of the proliferative properties of the astroglial microcultures has been carried out recently (Katoh-Semba et al., 1986). Besides providing further evidence for a temporal evolution of culture properties, this study has demonstrated that exogenous ganglioside triggers proliferative activities in astroglial cells to about the same extent as does serum.

Figure 4 illustrates the proliferative behaviors of a standard astroglial microculture seeded at 16,000 cell/well and presented, at D1, with either control or GM1-containing medium. Replicate cultures were pulsed with ^{3}H-thymidine for 2 hr before each of the culture times indicated, and both cell counts and TCA-precipitable radioactivity measured in each case. Control cultures (broken lines) showed a practically constant cell number, and a low DNA labeling capability which declined even further with culture time. Presentation of GM1 ganglioside to D1 cultures caused no change in DNA labeling or cell number for the first 12 hr, after which both rose very substantially. DNA labeling increased about 5-fold over the next 12 hr, but then declined to nearly the starting level in the following day. Cell numbers increased more progressively between 12 and 48 hr to 2.5 times the starting number, and then stabilized at about that

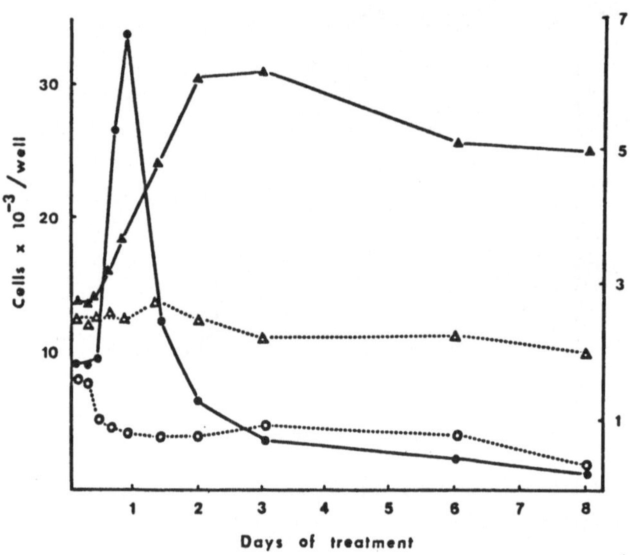

Figure 4. Proliferative behavior of astroglial cells in response to exogenous GM1. Astroglial cells were seeded at 16,000/microwell and treated with 6×10^{-5}M GM1 or none at D1. ^{3}H-Thymidine incorporation into DNA (•, ○) and cell numbers (▲, △) for GM1-treated (•, ▲) and control cultures (○, △).

level. These data i) demonstrate an unequivocal and substantial stimulation by GM1 ganglioside on astroglial cell proliferation, and ii) point to a restriction in either the proliferative response or the proliferative behavior of the cultured astroglial cells, since DNA synthesis was only stimulated for 1 day and population increase ceased within the next day.

The "mitogenic" effect of GM1 on astroglial cells is specific to intact ganglioside molecules, but not restricted to the GM1 species of ganglioside. Figure 5 shows titration curves of various ganglioside-related compounds with regard to their 24 hr effect on DNA labeling (identical dose-response curves can be obtained by measuring cell number increases at 24 or 48 hr). All ganglioside species tested (GM1, GD1a, GD1b and GT1b) proved competent, whereas molecules lacking one or another moiety of the ganglioside structure elicited no proliferative response. Of the three ganglioside classes, GM1 was in fact the least potent and GT1b the most potent, suggesting an interesting relationship between sialic acid content and "mitogenic" potency of the gangliosides. Figure 5 also illustrates two features we find to characterize all the in vitro studies carried out thus far on ganglioside potency toward ganglioside-responsive cells (Varon et

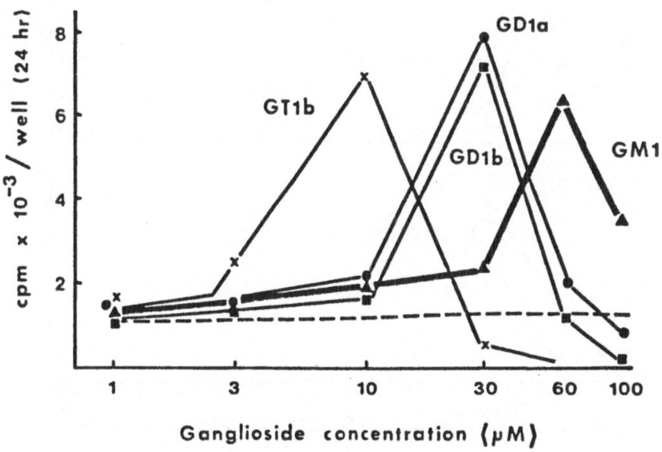

Figure 5. Concentration dependence of the ganglioside-triggered proliferative response in astroglial cells. DNA labeling was measured at 24 hr of treatment of D1 cultures. GM1 (▲), GD1a (•), GD1b (■), GT1b (x), asialo GM1 or sialic acid (---).

al. 1986). One feature is the limited concentration range in which a ganglioside can be effective, since too little or too much ganglioside is equally ineffective: the fact that patterns identical to those in Figure 5 are obtained by counting the number of astroglial cells after 48 hr of ganglioside treatment suggests that excess concentrations of ganglioside are actually toxic to the astroglial cells, at least in the present culture system. The second feature, which derives from the first one, is that dose-response curves are essential before one can recognize or rule out the competence of a putative agent — since a test at only 6×10^{-5}M (at which GM1 was optimally active) would have shown no response to any of the other three gangliosides tested.

The transient proliferative response of astroglial cells in the microculture reflects a restrictive feature of the culture system and not a limitation of the ganglioside "mitogenic" capability. The most explicit evidence for such a conclusion is that identical proliferative responses are observed under serum as under GM1 treatments. Figure 6 shows (top panel) that both DNA labeling and cell numbers change coincidentally in D1 cultures (from a 16,000 cells/well seed) presented with 6×10^{-5}M GM1 or 10% fetal calf serum (the optimal serum concentration) — with the expected transient rise of DNA labelling in the first day of treatment and plateau in cell numbers after two days. With both GM1 and serum, moreover, the timing as well as the amplitude of the responses varied coincidentally with the cell seeding density used. Seeding 6,000 and 2,000 (instead of 16,000) cells/well led to: i) declining plating efficiencies, increas-

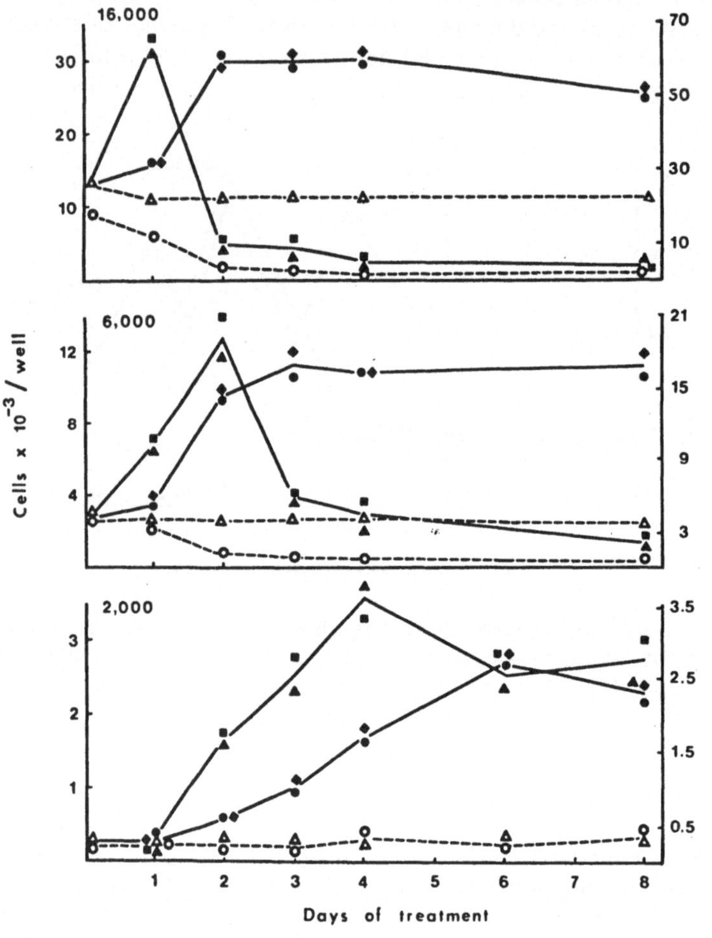

Figure 6. Effects of astroglial cell seeding density on cell proliferative response to GM1 (6×10^{-5}M) and fetal calf serum (10%). The experimental protocol of Fig. 1 was used here. ^{3}H-Thymidine labeling (▲, ■, ○) and cell numbers (◆, •, △) for GM1-treated (▲, ◆), serum-treated (■, •) and control (○, △) cultures.

ingly delayed peak times for both DNA labeling and cell increases, and decreasing absolute amplitudes but markedly increasing relative amplitudes of both responses.

These new data indicate that the restrictions observed with either GM1- or serum-induced proliferation cannot be attributed to: i) cell age in vitro (proliferation ceases at 2, 3 or 6 days with decreasing seeding densities), ii) contact inhibition (cell numbers/well level off at 30,000, 11,000 or 2,500, respectively), or iii) a preimposed, fixed number of divisions (low seeded cells must undergo 3 times more divisions than high-seeded ones). In addition, as with the stellation response and its decline in time, both the proliferative responses to GM1 or serum and their transient nature did not reflect special features of the substratum or medium, since they continued to be observed with other substrata and media — even though different substrata affected both absolute and relative dimensions of the astroglial proliferative responses.

Additional information on the temporal decline of the proliferative responses is shown in Figure 7. In one set of experiments (Fig. 7 A), the test (or control) medium applied to D1 cultures was replaced daily (instead of not at all) with medium containing the same agent (GM1 or serum) at the same optimal concentration, and responses measured at the end of each treatment day. No additional or sustained stimulation of DNA labeling was achieved, nor did the cell population increase beyond its expected 48 hr plateau (data not shown; but see Fig. 4) — excluding a depletion of the mitogenic signal as a potential cause of the transient response. In another set of experiments (Fig. 7 B), GM1 or serum were supplied only once but to increasingly "older" cultures. Astroglial cells responded identically in D1, D2 or D3 cultures — ruling out an early loss of astroglial susceptibility to the mitogenic signals. However, responses to GM1 declined in D4 cultures to level off at less than 10% of the initial response by D5, and responses to serum dropped one day later (D5) to a 50% final level in D6 and older cultures. The loss of the proliferative response to GM1 occurs over the same time

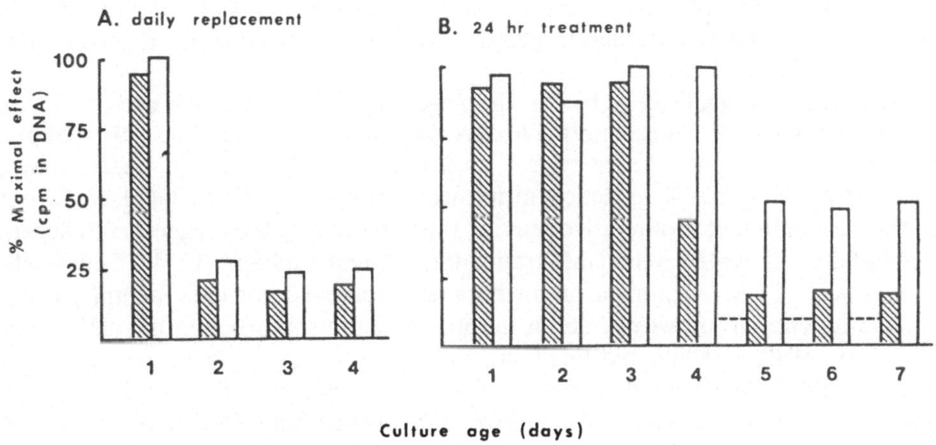

Figure 7. Effect of daily replacement of GM1 or serum-containing medium (A) or 24 hr treatment with GM1 or serum at progressively advanced culture ages (B) on astroglial cell proliferative behaviors, as measured by ^{3}H-thymidine labeling. GM1 (6×10^{-5}M, ▨); serum (10%, ▭); control cultures (---).

period as that of the stellation response to the same agent. The more delayed and less drastic loss of proliferative responses to serum suggests differences in the mode of action of the two "mitogens". Lastly, media collected from untreated D3 or D5 cultures and supplied to D1 cultures together with GM1 or serum elicited normal responses, disproving the possibility of an accumulation with time of mitotic inhibitors.

DISCUSSION

We have reviewed two recently discovered responses to GM1 and other gangliosides by cultured rat astroglial cells. Both responses are: i) highly reproducible, ii) massive, iii) early, and iv) readily quantifiable — and, therefore, offer excellent opportunities for future investigations of the biochemical events underlying either the mode of action of the ganglioside molecule or the execution machinery of the target astroglial cell.

One ganglioside effect is prevention or reversal of the assumption by flat astroglial cells of a stellate morphology under the influence of certain identified agents — which include neuron-released molecules such as neurotransmitters and neuropeptides (Magistretti et al., 1983). The stellation response presumably involves the encounter between agent and cell, a transduction mechanism leading to a potential second message, and the operation (in response to the latter) of various cell machineries required for the execution of the morphological conversion. GM1 could interfere in principle at any one of these levels. It has been generally assumed, but not yet fully established, that stellation occurs in response to an intracellular increase of cyclic AMP, which so many of the stellation-inducing agents are known to elicit in several cells. Future experiments can determine, if that is the case, whether GM1 interferes with the rise in cyclic AMP or with subsequent events concerning membrane, cytoskeleton or, possibly, ionic pumps. What is particularly thought-provoking is the plausibility that similar molecular targets are involved in the action of exogenous GM1 on neuronal cells as well.

The second ganglioside effect is the triggering of proliferative activity in the cultured astroglial cells. It is tempting to speculate, once again, that the astroglial proliferative response and the astroglial stellation-blocking response share similar molecular mechanisms. A reduction in intracellular cyclic AMP by GM1 could well underlie resumption of proliferative activity, the latter having been suggested to depend on reduced cyclic AMP/cyclic GMP ratios (Seifert and Rudland, 1974). Conversely, consequences of GM1 treatment on membrane, cytoskeleton or ionic pump(s) would fit well with current knowledge about mitotic regulations (Farmer et al., 1978; Ben-Ze'ev et al., 1980; McClain and Edelman, 1980; Friedkin et al., 1979; Rozengurt and Mendoza, 1980). Interestingly, serum, which GM1 mimics with regard to mitotic stimulation of astroglial cells, is also known to favor their flat rather than stellate morphology (Manthorpe et al., 1979) as does GM1 — another point of commonality between the two responses.

One can only speculate, at present, on the potential relevance in vivo of an interaction between "extrinsic" gangliosides and astroglial cells. Two directions are worth particular attention:

(1) The proposition that astroglial cells can produce and release neuronoactive agents, directly demonstrated in vitro, is also receiving increasing support from in vivo data. For example, localized lesions in the CNS lead to accumulation of neuronotrophic factors in the injured tissue and its fluid exudate (Manthorpe et al., 1983; Nieto-Sampedro et al., 1983; Longo et al., 1984), and chemical lesions by kainic acid injections lead to similar local accumulation of neuronotrophic factors (Nieto-Sampedro et al., 1983). In both cases the lesions are accompanied by substantial gliosis, suggesting that astroglial cells may indeed be a major source of these and other factors and that astroglial output of neuronoactive substances may itself be a regulated behavior (e.g. triggered by local injury).

(2) The in vitro conversion of flat to stellate astroglial cells does not yet have an authenticated counterpart in physiological or pathological in vivo behaviors (Varon and Somjen, 1979). Nevertheless, it has been proposed (Fedoroff et al., 1984) that converted cells resemble "reactive" astroglial elements and that, therefore, the conversion process may provide an in vitro model for at least some aspects of the gliosis observed in vivo under CNS injury. Gliosis often involves an increase in "fibrillary" astrocytes as well as a spurt of astroglial proliferation (Latov et al., 1979; Varon and Somjen, 1979). Exogenous GM1 and other gangliosides appear involved in the in vitro counterparts of both features, albeit in opposite directions to each other.

An important new task, along either one of the above two directions, is to identify molecular modulators of astroglial behaviors, which may modify the nature or amplitude of the support provided by glial cells to neurons and thus ultimately alter the neuronal behavior itself. The findings described here provide the first evidence that gangliosides are to be included among such glial modulators.

ACKNOWLEDGMENTS

This work was supported by a grant from Fidia Research Laboratories. We wish to thank Dr. John S. Rudge for helpful advice and discussion in preparing this manuscript.

REFERENCES

Asou H and Brunngraber EG (1984) Neurosci Let 46: 115-118.

Ben-Ze'ev A, Farmer SR, Penman S (1980) Cell 21: 365-372.

Farmer SR, Ben-Ze'ev A, Benecke B-J, Penman S (1978) Cell 15: 627-637.

Fedoroff S, Neal J, Opas M, Kalnins VI (1984) J Neurocytol 13: 1-12.

Friedkin M, Legg A, Rozengurt E (1979) Proc Natl Acad Sci USA 76: 3909-3912.

Gorio A, Carmignoto G, Facci L, Finesso M (1980) Brain Res 197: 236-241.

Gorio A, Marini P, Zanoni R (1983) Neuroscience 8: 417-429.

Karpiak SE, Vilim F, Mahadik SP (1984) Dev Neurosci 6: 127-135.

Katoh-Semba R, Facci L, Skaper SD, Varon S (1986) J Cell Physiol 126: 147-153.

Latov N, Nilaver G, Zimmermann EA, Johnson WG, Silverman AJ, Defendini R, Cote L (1979) Dev Biol 72: 381-384.

Ledeen RW (1978) J Supramol Struct 8: 1-17.

Longo FM, Selak I, Zovickian J, Manthorpe M, Varon S, U H-S (1984) Exp Neurol 84: 207-218.
Magistretti PJ, Manthorpe M, Bloom FE, Varon S (1983) Regulatory Peptides 6: 71-80.
Manthorpe M, Adler R, Varon S (1979) J Neurocytol 8: 605-621.
Manthorpe M, Nieto-Sampedro M, Skaper SD, Lewis ER, Barbin G, Longo FM, Cotman CW, Varon S (1983) Brain Res 267: 47-56.
Manthorpe M, Rudge J, Varon S (1986) In: Fedoroff S (ed): Astiocytes, Vol 2. Academic Press, New York, in press.
McClain DA and Edelman GM (1980) Proc Natl Acad Sci USA 77: 2748-2752.
Moonen G, Heinen E, Goessens G (1976) Cell Tissue Res 167: 221-227.
Nieto-Sampedro M, Manthorpe M, Barbin G, Varon S, Cotman CW (1983) J Neurosci 3: 2219-2229.
Partington CR and Daly JW (1979) Molecular Pharmacol 15: 484-491.
Rozengurt E and Mendoza S (1980) Ann NY Acad Sci 339: 175-190.
Rudge JS, Manthorpe M, Varon S (1985) Dev Brain Res 19: 161-172.
Sabel BA, Slovin MD, Stein DG (1984) Science 225: 340-342.
Schoffeniels E, Franck G, Hertz L, Lower D (eds) (1978) Dynamic Properties of Glial Cells, Pergamon Press, New York.
Seifert W and Rudland PS (1974) Proc Natl Acad Sci USA 71: 4920-4924.
Skaper SD, Facci L, Rudge JS, Katoh-Semba R, Manthorpe M, Varon S (1986) Dev Brain Res 25: 21-31.
Sparrow JR and Grafstein B (1982) Exp Neurol 77: 230-235.
Toffano G, Savoini G, Moroni F, Lombardi G, Calza L, Agnati LF (1983) Brain Res 261: 163-166.
Varon S (1985) Discussions in Neurosciences, Vol. II, 3, Foundation FESN, Geneva.
Varon S and Adler R (1980) Curr Topics Dev Biol 16: 207-252.
Varon S and Adler R (1981) Adv Cell Neurobiol 2: 115-163.
Varon S and Somjen G (1979) Neurosci Res Prog Bull 17: 1-239.
Varon S, Manthorpe M, Williams LR (1984) Dev Neuroscience 6 (2): 73-100.
Varon S, Skaper SD, Katoh-Semba R (1986), this volume.
Yamakawa T and Nagai Y (1978) Trends Biochem Sci 3: 128-131.

Gangliosides and neuronal plasticity
G. Tettamanti, R.W. Ledeen, K. Sandhoff,
Y. Nagai, G. Toffano (eds.)
Fidia Research Series, vol. 6
Liviana Press, Padova, © 1986

THE ROLE OF GANGLIOSIDES IN NEUROTROPHIC INTERACTION IN VITRO

F.J. Roisen*, S.G. Matta*, G. Yorke*, M.M. Rapport[1]

Dept. of Anatomy, University of Medicine and Dentistry of New Jersey,
Rutgers Medical School, Piscataway, New Jersey, USA;
[1]Div. of Neuroscience, New York State Psychiatric Institute, New York and
Dept. of Biochemistry and Molecular Biophysics, Columbia University,
College of Physicians and Surgeons, New York, New York, USA

INTRODUCTION

Gangliosides are membrane-associated acidic glycolipids that have been implicated in the regulation of cell processes such as differentiation (Moskol et al., 1974; Roisen et al., 1981a, b; Dimpfel et al., 1981; Leon et al., 1982), growth (Fishman et al., 1977; Langenback and Kennedy, 1978; Morgan and Seifert, 1979) and axonal regeneration (Obata et al., 1977; Caccia et al., 1979; Norido et al., 1982; Spirman et al., 1982). Their concentration in the brain is 14 times higher than in liver and nearly 60 times higher than in muscle (Seyfried et al., 1978; Leeden and Yu, 1982). Patterns of ganglioside synthesis have been reported to change during intense periods of neuritogenesis and synaptogenesis (Rösner, 1980; Sonnino et al., 1981). Their abundance, distribution and temporal occurrence in neuronal membranes suggest that gangliosides play a key role in the development of the nervous system.

Our laboratory (Roisen et al., 1981a; b; 1984) and others (Dimpfel et al., 1981; Leon et al., 1982) have shown that addition of bovine brain gangliosides (BBG) to primary and established neuronal cultures enhanced their neuritogenic responses. These responses were characterized by a series of temporally-related events. Within one minute of exposure to gangliosides, dramatic changes in surface topography were observed which included the formation of microvilli and ridge-like ruffled membranes. Accompanying this burst of surface activity and perhaps providing its cytoskeletal base, there was a reorganization of the subcortical microfilamentous meshwork into bundles of microfilaments which formed core elements of the newly developed surface

Abbreviations: BBG, bovine brain gangliosides; DRG, dorsal root ganglia; NGF, nerve growth factor; ODC, ornithine decarboxylase; SM, standard medium; SFM, serum-free medium.

* Present address: Dept of Anatomy, Health Sciences Center, University of Louisville, Louisville, Kentucky 40292, USA.

projections (Spero and Roisen, 1984). Increased levels of intracellular cAMP and protein phosphorylation (Spero et al., 1985) were observed after twenty minutes. One of the phosphorylated proteins, identified as the intermediate filament protein vimentin, may play a structural role in the initial stages of axonal sprouting. Metabolism, as measured by ornithine decarboxylase (ODC) activity, the rate-limiting enzyme in polyamine biosynthesis, was also increased by ganglioside treatment and reached maximal levels after 6 hr exposure. During this period, the initial stages of neurite development were evident. After 24 to 48 hr, the number and length of processes of Neuro-2a neuroblastoma and the number of neurites in organized cultures of chick embryonic dorsal root ganglia (DRG) were increased. Although the Neuro-2a processes formed during ganglioside exposure appeared equivalent to those formed in response to other neuritogenic conditions, i.e., serum deprivation, ultrastructural studies revealed complex cytoskeletal cores composed of microfilament bundles which were present only in the ganglioside-induced neurites (Spero and Roisen, 1984).

In our present study we have examined the neuronotrophic interactions of the four major brain gangliosides in the BBG mixture on primary and established *in vitro* models. Gangliosides, especially GM1, have been reported to potentiate the actions of the well-known trophic agent Nerve Growth Factor (NGF) (Spirman et al., 1982; Schwartz and Spirman, 1982; Matta et al., 1986). Therefore, we investigated the capacity of individual gangliosides to potentiate the neuritogenesis of three different models: a. sensory ganglia (DRG) which are NGF-dependent, b. a clonal pheochromocytoma line (PC12) which is NGF-responsive but nonobligatory and c. a neuroblastomal line (Neuro-2a) which is NGF-insensitive. To probe NGF's neuritogenic activity, six antibodies were used: five mouse monoclonal antibodies prepared against ganglioside GM1 and a rabbit polyclonal antibody prepared against BBG and purified by affinity absorbtion to GM1-containing liposomes.

METHODS

Cell Culture

Neuro-2a murine neuroblastoma (ATCC-CCL 131) stock cultures were grown in 90% Minimum Essential Medium with Hanks' balanced salt solution (GIBCO, Grand Island, NY), 10% heat-inactivated fetal bovine serum (Irvine Scientific, Irvine, CA), 10 mg% gentamicin (Schering, Kenilworth, NJ), 75 mg% additional $NaHCO_3$ and 0.1 mM non-essential amino acids at 35°C in an atmosphere of 5% CO_2. The stock cultures were passed via trypsin-EDTA (0.05% and 1 mM, respectively) approximately every 5 days at a cell density of 2×10^5/ml. Test cultures were plated in medium containing the material under evaluation and maintained for 22 hr before examination. NGF was prepared as described by Cohen (1960) and added to the medium at 5 μg/ml. Gangliosides, supplied generously by Fidia Research Laboratories, Abano Terme, Italy, were dissolved in balanced salt solution and added to the culture medium to a final concentration of 200 μg/ml.

PC12 rat pheochromocytoma (a gift from Dr. L. Greene) were maintained in medium RPMI 1640 supplemented with 10% heat-inactivated horse serum (GIBCO) and 5% fetal bovine serum (Irvine) as stock cultures in a humidified atmosphere of 5%

CO_2 at 35°C. The cells were detached by mechanical shock and passed every 5 days. For morphological study the cells were plated at a density of 2×10^5/ml onto collagen-coated coverslips and maintained as described previously (Matta et al., 1986). NGF was incorporated into the medium at a concentration (20 μg/ml) which insured sufficient but submaximal development. For the study of the effect of GM1 "priming" on ODC activity, a lower concentration of NGF (5μg/ml) was used to prevent the NGF-mediated elevation of ODC from masking the ganglioside potentiation. Gangliosides were added at a concentration of 200 μg/ml.

Dorsal root ganglia were excised aseptically from 8.5 day old White Leghorn chick embryos and explanted onto collagen-coated coverslips. The collagen was prepared by photo-reconstitution after the method of Masurovsky and Peterson (1973) and equilibrated in medium containing serum for 24 hr. After explantation the ganglia were fed with a standard medium (SM) which consisted of Medium 199 (GIBCO) supplemented with 10% dialyzed heat-inactivated fetal bovine serum, in the presence of NGF (5μg/ml) or its absence. Following the addition of gangliosides (200 μg/ml) to the SM, the cultures were incubated as lying drop preparations in Maximow chambers at 35°C. At 46 to 48 hr, they were removed from incubation and their development was evaluated microscopically as described previously (Roisen et al., 1972).

Ornithine Decarboxylase Assay: Index of Metabolic Activity

The activity of ornithine decarboxylase (ODC) was assayed by the method of Chen et al. (1976) as modified by Roisen et al. (1981b). Since serum stimulates cellular metabolism, a serum-free medium (SFM) was used for the DRG cultures and a serum-depleted (spent) medium was used for the cell lines. Test agents were added to either the SFM or the spent media and applied to the cultures for 6 hr prior to ODC assay. Protein content was determined by dye staining (BioRad method, 1977). A minimum of 100 ganglia were assayed per treatment. All cell line experiments were performed in triplicate.

Morphological Evaluation

To quantify subjective evaluation of neurite initiation (number) and neurite elongation (length), a computer-assisted image analysis system was employed as previously described (Matta et al., 1986). The system included a Dage Newvicon camera with video overlay coupled with an Apple IIe computer and Optomax software which allowed determination of cell area, perimeter, shape and length.

RESULTS

Neuro-2a Neuroblastoma

Exposure of Neuro-2a cells to BBG or to each of the four major gangliosides in the mixture (GM1, GD1a, GD1b or GT1b) increased number and length of neuronal processes (Fig. 1). The responses elicited by each ganglioside were not equivalent;

GT1b had the most neuritogenic activity as evidenced by the large number of long neurites that developed in its presence (Table 1). GD1a and GT1b increased neurite number maximally, while GD1b and GT1b produced the longest neurites. GM1 had moderate neuritogenic activity by comparison. Neuro-2a neuritogenesis was unaltered by NGF either in the presence or absence of ganglioside. On the other hand, ODC levels, unaltered by NGF in the absence of ganglioside, were elevated when NGF was administered in the presence of either GM1 or GD1a.

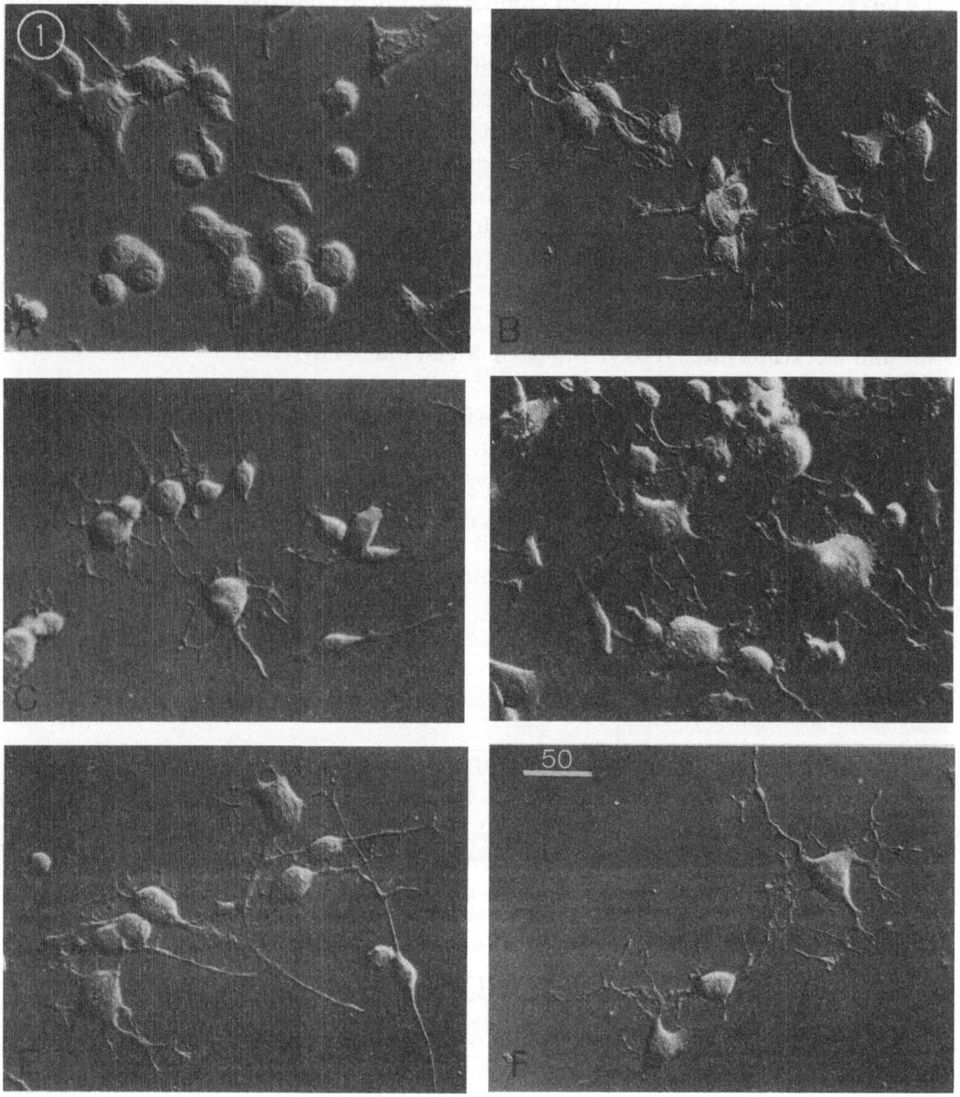

Figure 1. Neuro-2a neuroblastoma 24 hr in vitro in medium containing 200 μg/ml of the specified ganglioside. A. Control, B. Bovine brain ganglioside, C. GM1, D. GD1a, E. GD1b, F. GT1b. Nomarski optics, bar = 50 μm.

Table 1. *Effects of Individual Gangliosides on Neuro-2a Neuritogenesis*

	GMl	GDla	GDlb	GTlb
Number of neurites	1.18*	2.14*	1.48*	1.98*
Neurite length	1.67*	1.92*	2.43*	2.03*
ODC specific activity (—NGF)	1.16	1.38	1.64	2.07*
(+NGF)	1.62*	1.46*	1.20	1.17

Data for the 3 parameters are expressed as elevation above control levels (1.00). * indicates significance at $p<0.05$. The presence of NGF (5 μg/ml) did not affect neuritic outgrowth.

PC12 Pheochromocytoma

PC12 cells did not respond morphologically either to BBG or individual gangliosides in the absence of NGF. All of the gangliosides increased some aspect of NGF-mediated neuritogenesis after 4 days exposure (Fig. 2). GM1 was more effective than other gangliosides in potentiating NGF's neuritogenic activity. However, it was no more effective than GD1a or GD1b in potentiating NGF's induction of ODC activity which was assayed at 6 hr, the time the maximal stimulation was reached (Table 2). To examine the temporal basis of ganglioside-NGF interactions, PC12 cells were incubated for periods of up to 22.5 hr in GM1 and washed prior to NGF exposure. The results of these "priming" experiments are summarized in Table 3. Treatment of PC12

Table 2. *Effects of Gangliosides on NGF-Mediated PC12 Neuritogenesis*

	GMl	GDla	GDlb	GTlb
Number of neurites	1.11*	1.00	0.93	1.00
Neurite length	2.67*	1.49*	1.66*	2.16*
ODC specific activity	1.16*	1.32*	1.22*	1.07

Data for the 3 parameters are normalized to the control value of 1.00. Cells were treated with 20 μg/ml NGF and 200 μg/ml ganglioside. * indicates significance at $p<0.05$.

Table 3. *Effects of GMl Priming on ODC Induction by NGF in PC12 Cells*

Pretreatment (22 hr)	Priming (30 min)	Treatment (6 hr)	Specific Activity
SM	SM	SM	1.00
SM	GMl	SM	1.84
SM	SM	GMl	1.62
SM	SM	NGF	24.3
SM	GM1	NGF	27.2
SM	SM	NGF+GM1	34.8
GMl	GMl	NGF	32.5
NGF	NGF	GMl	8.1

The effects of sequential treatments with GMl and NGF on ODC activity were evaluated in three separate studies; a representative one is shown. During treatment, the test agents were incorporated into spent medium; NGF was applied at 5 μg/ml and GMl at 200 μg/ml.

cultures with GM1 for 30 min prior to 6 hr incubation in SM resulted in an approximate two-fold increase in ODC activity. A substantial increase was produced by 30 min GM1 exposure prior to 6 hr of NGF treatment compared to levels obtained by NGF alone. Pretreatment of PC12 cells with GM1 for 22.5 hr, followed by exposure to NGF for

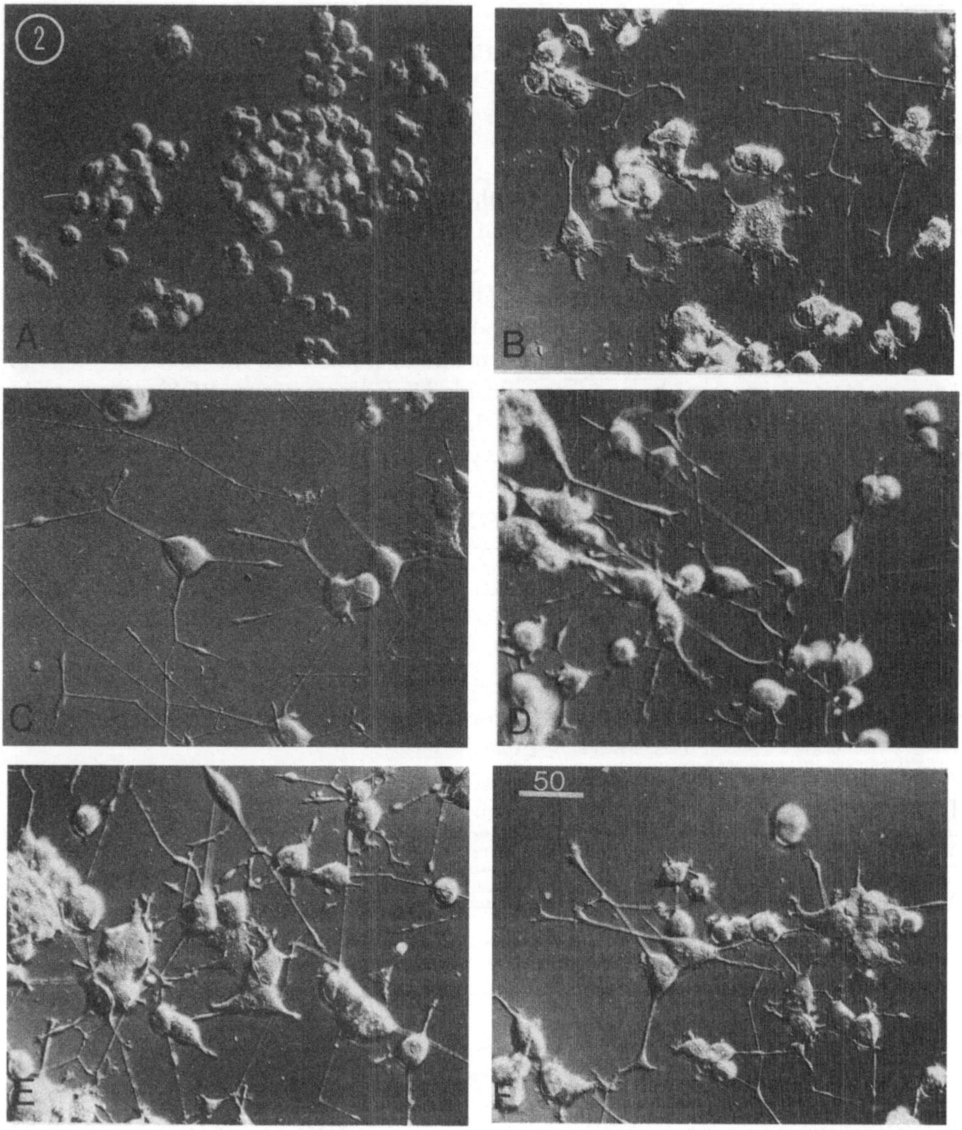

Figure 2. PC12 pheochromocytoma 96 hr in vitro in medium containing 20 μg/ml NGF plus the indicated ganglioside. A. Control, B. NGF, C. GM1, D. GD1a, E. GD1b, F. GT1b. Nomarski optics, bar = 50 μm.

6 hr, produced the same level of ODC activity as 6 hr of simultaneous exposure to NGF and GM1. In contrast, treatment with NGF for 22.5 hr prior to exposure to GM1 for 6 hr was less stimulatory.

Chick Sensory Ganglia

Embryonic chick DRG treated with SM containing BBG or GT1b for 48 hr have an increased number of processes over controls. The presence of NGF in the medium enhanced substantially neurite length as well as number. The effect of simultaneous application of NGF and individual gangliosides is summarized in Table 4. GM1 and GD1b potentiated NGF-mediated neuritogenesis, but only GM1 elevated ODC levels (data not shown). Affinity purified polyclonal antibodies against GM1 were incorporated into SM and applied to DRG for 48 hr. The effect of these antibodies was to diminish specifically NGF's capacity to increase neurite length and number (Table 5), but they had only limited effect on the radial migration of accompanying nonneuronal cells (Fig. 3). Five different monoclonal antibodies (designated B6, C3, C4h2, D1 and D3) against GM1 were added individually to SM supplemented with NGF and applied to DRG for 48 hr. All of the monoclonal antibodies (mAbs) were incorporated into the SM at a concentration of 20%. The mAbs did not produce equivalent responses: C4h2 and D1 inhibited NGF-mediated neurite initiation (number) and elongation (length); B6 and C3 inhibited only initiation; whereas, D3 only inhibited elongation (Table 5). Three mAbs (B6, C3 and C4h2) blocked NGF's induction of ODC activity. Incorporation of ascites fluid control into the SM had no effect on neuritogenesis.

Table 4. *Effects of Gangliosides on NGF-Mediated DRG Neuritogenesis*

	GMl	GDla	GDlb	GTlb
Number of neurites	1.22*	0.94	1.27*	1.10
Neurite length	1.10*	1.00	1.00	1.00

Data are expressed in relation to the normalized control value of 1.00. * indicates elevation over control levels ($p<0.05$).

Table 5. *Effects of Antibodies to GM1 on NGF-Mediated DRG Neuritogenesis*

	Polyclonal anti-GMl			Monoclonal anti-GMl				
	1 μg/ml	100 μg/ml	200 μg/ml	B6	C3	C4h2	D1	D3
Neurite number	0.65*	0.45*	NGr	0.43*	0.38*	0.50*	0.38*	0.75
Neurite length	1.08	0.54*	NGr	0.94	0.85	0.73*	0.57*	0.60*
ODC specific activity	—	—	0.27*	0.65*	0.55*	0.56*	0.84	0.92

Data are expressed in relation to the normalized control value of 1.00. NGF was present at a final concentration of 5 μg/ml.
* indicates significant reduction below control level ($p<0.05$). NGr = No Growth

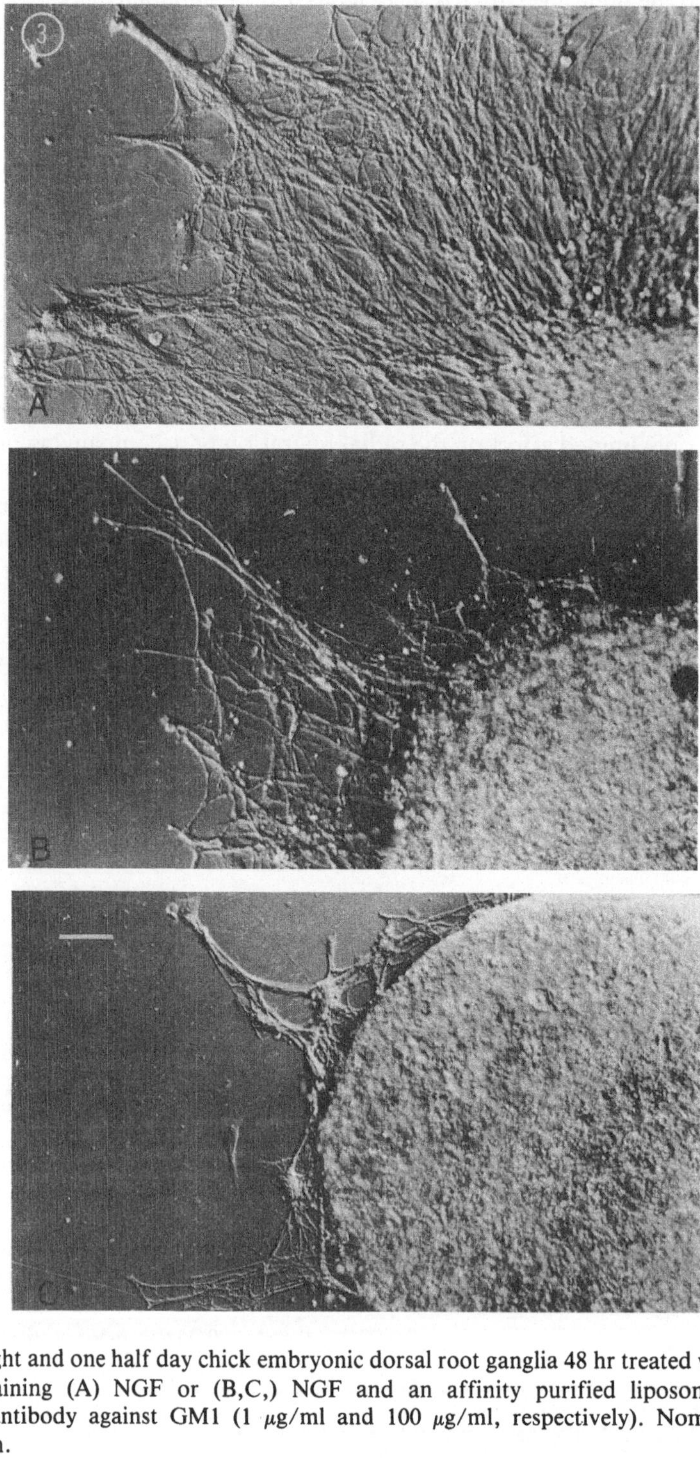

Figure 3. Eight and one half day chick embryonic dorsal root ganglia 48 hr treated with standard media containing (A) NGF or (B,C,) NGF and an affinity purified liposomally derived polyclonal antibody against GM1 (1 μg/ml and 100 μg/ml, respectively). Nomarski optics, bar = 100 μm.

DISCUSSION

Each of the four major gangliosides contained in bovine brain enhanced the neuritic sprouting and elongation of Neuro-2a cells. The neurite-promoting capacities of these gangliosides were not equivalent. Furthermore, only one ganglioside, GT1b, increased ODC activity. Although Neuro-2a cells did not respond morphologically or metabolically to NGF alone, exposure of these neurons simultaneously to GM1 or GD1a and NGF increased ODC activity. Therefore, under these conditions, treatment of NGF-unresponsive cells with GM1 and NGF produced a response, but did not promote the growth of neurites more than GM1 alone.

To examine the relationship between gangliosides and NGF, we employed the PC12 cell model that has been well-characterized as being sensitive to, but not dependent on, NGF. None of the four gangliosides increased PC12 neurite formation or ODC activity. In contrast, all were able to potentiate NGF-mediated neuritogenic activity. GM1 was the most effective in potentiating NGF's actions, a finding consistent with other experiments indicating that GM1 is the ganglioside most closely associated with NGF function. Temporally, induction of ODC activity in PC12 cells by GM1 followed by NGF was as effective as the simultaneous exposure to NGF and GM1. These experiments demonstrated that the incorporation of GM1 into PC12 membranes facilitated later NGF-receptor mediated interactions. In the opposite direction, priming of PC12 cells with NGF prior to GM1 was less stimulatory. It has been suggested that GM1 may be the NGF receptor (Spirman et al., 1982; Schwartz and Spirman, 1982), but this possibility seems unlikely, since binding studies did not reveal increased NGF binding by GM1 treated cells (Ferrari et al., 1983).

Further evidence in support of a role for GM1 in NGF-mediated neuronal development comes from experiments utilizing organized cultures of embryonic chick sensory ganglia (DRG). Previously, we reported that exposure of DRG to BBG increased neurite formation (Roisen et al., 1981a). Investigators (Schwartz and Spirman, 1982; Spirman et al., 1982) have shown GM1, in comparison to the other gangliosides, to be most effective in potentiating neuritogenesis of this NGF-obligatory culture system. We used an affinity purified polyclonal antibody against GM1 to probe NGF's ability to increase DRG neuritogenesis. The antibody blocked both NGF-mediated development and ODC induction. To probe further the mechanism of this inhibition, the effects of five different mAbs directed against GM1 were evaluated. The individual mAbs produced different responses: some blocked both neurite initiation and elongation, some blocked only initiation, others only elongation. The effects of the mAbs on NGF-mediated ODC induction were not uniform. Since each mAb probably recognizes a different epitope of GM1, these experiments suggest that specific regions of the ganglioside molecule may provide important signals for different aspects of the neuritogenic response. Different regions of the GM1 molecule may: a. amplify or potentiate the neuritogenic responses produced by NGF and b. regulate and select the specific character of the response.

It is unlikely that the ganglioside-mediated neuritogenesis or potentiation of NGF were due to nonspecific membrane insertion or ionic interaction, since other agents including cerebrosides, cardiolipin, inositol, asialo-GM1 and sialic acid did not produce similar responses. The ganglioside may act by enhancing the entry of NGF into the cell.

This might occur through a change in membrane properties due to: a. the unmasking of partially masked NGF receptors, b. a clustering or change in receptor distribution, c. a change in receptor microenvironment, or d. an alteration of receptor turnover. Gangliosides may also regulate surface receptors for NGF by altering the cytoskeletal anchorage of membranes. We demonstrated previously that ganglioside exposure resulted in a rapid redistribution of cytochalasin-D-sensitive subcortical microfilaments (Spero and Roisen, 1984). Shooter et al. (1981) reported that the disruption of actin-containing microfilaments inhibited the degradation of NGF receptors. In addition, these microfilaments have been shown to regulate the shuttling of NGF receptors between the lysosomal and nuclear compartments (Yanker and Shooter, 1982). Furthermore, Schechter and Bothwell (1981) suggested a direct association between NGF high affinity receptors and a cytoskeletal component.

The role of gangliosides in the broad cascade of biochemical, morphological and behavioral events triggered by NGF in vivo is uncertain. In vitro experiments have suggested that these molecules may have neuritogenic properties of their own. No reports have appeared which demonstrate clearly the ability of gangliosides to potentiate trophic agents other than NGF. Whether this activity is independent of other trophic agents or is in concert with a variety of trophic agents remains to be elucidated. From the present studies, it seems clear that gangliosides have the potential to define, regulate and amplify neuronotrophic interactions. The possibility that in vivo gangliosides interact with non-neuronal cells, e.g., glia (Hefti et al., 1985) and muscle, in addition to neurons should not be overlooked. Indeed, the in vitro studies suggest that ganglioside treatment in the absence of outside trophic influence results in neurite promoting activity and in the presence of external trophic factors stimulates neuronotrophic as well as neurite promoting activity.

ACKNOWLEDGMENTS

This work was supported by NIH grants NS-11299 and NS-11605.

REFERENCES

Bio-Rad Protein Assay, BioRad Laboratories Technical Bulletin 1051, April 1977.

Caccia MR, Meola G, Cerri C, Frattola L, Scarlato G, Aporti F (1979) Muscle & Nerve 2: 381-389.

Chen KY, Heller J, Canellakis ES (1976) Biochem Biophys Res Commun 68: 401-408.

Cohen S (1960) Proc Nat Acad Sci 46: 302-311.

Dimpfel W, Moller W, Mengs U (1981) in: Rapport MM and Gorio A (eds): Gangliosides in Neurological and Neuromuscular Function, Development, and Repair. Raven Press, New York, pp. 119-134.

Ferrari G, Fabris M, Gorio A (1983) Dev Brain Res 8: 215-221.

Fishman PH, Moss J, Manganiello VC (1977) Biochem 16: 1871-1875.

Hefti F, Hartikka J, Frick W (1985) J Neurosci 5: 2086-2094.

Langenback R and Kennedy S (1978) Exp Cell Res 112: 361-372.

Ledeen RW and Yu RK (1982) in: Ginsburg V (ed): Methods in Enzymology. Academic Press, New York, pp. 139-191.

Leon A, Facci L, Benvegnu D, Toffano G (1982) Dev Neurosci 5: 108-114.
Masurovsky EB and Peterson ER (1973) Exp Cell Res 76: 447-448.
Matta SG, Yorke G, Roisen FJ (1986) Dev Brain Res 27: 243-252.
Morgan JI and Seifert W (1979) J Supramol Struct 10: 111-124.
Moskol JR, Gardner DA, Basu S (1974) Biochem Biophys Res Commun 61: 751-758.
Norido F, Cannella R, Gorio A (1982) Muscle & Nerve 5: 107-110.
Obata K, Momoko M, Handa S (1977) Nature 266: 369-371.
Roisen FJ, Murphy RA, Braden WG (1972) J Neurobiol 4: 347-368.
Roisen FJ, Bartfeld H, Nagele R, Yorke G (1981a) Science 214: 577-578.
Roisen FJ, Bartfeld H, Rapport MM (1981b) in: Rapport MM and Gorio A (eds): Gangliosides in Neurological and Neuromuscular Function, Development, and Repair. Raven Press, New York, pp. 135-150.
Roisen FJ, Spero DA, Held SJ, Yorke G, Bartfeld H (1984) in: Ledeen RW, Yu RK, Rapport MM, Suzuki K (eds): Ganglioside Structure, Function, and Biomedical Potential. Plenum Press, New York, pp. 499-511.
Rösner H (1980) Wilhelm-Roux's Arch 188: 205-213.
Schechter AL and Bothwell MA (1981) Cell 24: 807-874.
Schwartz M and Spirman N (1982) Proc Nat Acad Sci 79: 6080-6083.
Seyfried TN, Ando S, Yu RK (1978) J Lipid Res 19: 538-543.
Shooter EM, Yanker BA, Landreth GE, Sutter S (1981) Rec Prog Horm Res 37: 417-446.
Sonnino S, Ghidoni R, Masserini M, Aporti F, Tettamanti G (1981) J Neurochem 36: 227-232.
Spero DA and Roisen FJ (1984) Dev Brain Res 13: 37-48.
Spero DA and Roisen FJ (1986) Int J Dev Neurosci 3: 631-642.
Spero DA, Browning ET, Roisen FJ (1986) NY Acad Sci 715: 143-145.
Spirman N, Sela BA, Schwartz M (1982) J Neurochem 39: 874-877.
Yanker BA and Shooter EM (1982) Ann Rev Biochem 51: 845-868.

Gangliosides and neuronal plasticity
G. Tettamanti, R.W. Ledeen, K. Sandhoff,
Y. Nagai, G. Toffano (eds.)
Fidia Research Series, vol. 6
Liviana Press, Padova, © 1986

EVIDENCE FOR THE EFFECTS OF GANGLIOSIDES ON THE DEVELOPMENT OF NEURONS IN PRIMARY CULTURES

Michèle Durand, Bernard Guérold, Dominique Lombard-Golly, Henri Dreyfus

Unité 44 de l'INSERM and Centre de Neurochimie du CNRS,
5, rue Blaise Pascal, 67084 Strasbourg Cedex, France

INTRODUCTION

A useful tool to investigate the involvement of gangliosides in the molecular mechanisms leading to cellular differentiation and maturation is to culture cells in the presence of the glycolipids (Byrne et al., 1983; Fishman et al., 1983; Facci et al., 1984; Dreyfus et al., 1984a; Dreyfus et al., 1984b; Massarelli et al., 1985b) which are known to incorporate into the plasma membranes (Schwarzmann et al., 1985). In neurons, much experimental evidence has shown a neuronotrophic and a neuritogenic effect due to gangliosides (Byrne et al., 1983; Facci et al., 1984; Dreyfus et al., 1984b; Massarelli et al., 1985b; Ledeen, 1984). Previous experiments in our laboratory have shown that the addition of a mixture of gangliosides to the growth medium in primary cultures of chick brain neurons during the period of synaptogenesis led to stimulation of sprouting of the secondary neuronal processes (Dreyfus et al., 1984b; Massarelli et al., 1985a, 1985b). Furthermore, we were interested in the study of the effect of exogenous gangliosides added at seeding in neuronal cultures grown without serum. Some preliminary results have already been reported (Dreyfus et al., 1984c).

In this manuscript we mainly present some aspects of cellular metabolism in chick neurons cultured in the presence of exogenously added gangliosides. DNA and protein synthesis were examined through the incorporation of ^{3}H[thymidine] and ^{3}H[leucine] into cultured neuronal cells. Changes in the amounts of sialoglycoproteins and gangliosides were studied and the metabolism of the gangliosides was followed by using tritiated sialoglycolipids labelled on their sphingosine moiety. Modifications of the

Abbreviations: DMEM, Dulbecco's modified Eagle's medium; DMEMS, Dulbecco's modified Eagle's medium with 20% fetal calf serum; BSM, modified Bottenstein-Sato defined medium; 5-HT, 5-hydroxytryptamine; GL, total neutral glycolipids; GP, total glycoprotein; GG, total ganglioside.

aminergic properties occuring in the neuronal cultures grown without serum were also investigated. Most of the techniques used in this study have been described elsewhere (Dreyfus et al., 1980; 1984a; 1984c).

RESULTS

Effect of Exogenous GM1 and GT1b on the Metabolism of Primary Cultures of Neurons

Pure neuronal cell cultures from 8-day-old chick embryo hemispheres were obtained as described earlier (Pettmann et al., 1979). In our studies some "control" cultured neurons were grown continuously in Dulbecco's modified Eagle's medium (DMEM) in the presence of 20% fetal calf serum (DMEMS). Other "control" cultures were seeded and grown 1 day in DMEMS, then in a modified Bottenstein-Sato defined medium (BSM) using DMEM supplemented with insulin, transferrin, progesterone, putrescine, selenium (Bottenstein and Sato, 1979) and also by 10^{-12} M estradiol (DMEMS/BSM). Cultures were also seeded and grown in the presence of 10^{-6}M GM1, 10^{-6} MGT1b or 2×10^{-6}M GT1b dispersed in BSM. Cultures grown in DMEMS/BSM and in GM1/GT1b-BSM were analysed from 0 to 6 days in culture (d.i.c.) after washing the cells in cold saline solution (NaCl 0.9%) containing 5% fetal calf serum and then in saline alone in order to remove the adsorbed gangliosides.

Morphology and Amine Content of Neurons

Figure 1A and B shows an electron microscope scan photograph of neuronal cells seeded and grown in 10^{-6}M GM1 and 2×10^{-6} MGT1b in BSM, respectively. Neurons seeded and grown in BSM alone degenerated rapidly. However the addition of gangliosides to this medium prevented the degeneration (Dreyfus et al., 1984c) and the cells reached a high level of development and maturation (Dreyfus et al., 1984c). The shape of the 6-day-old neurons closely resembled that obtained when cells were seeded in DMEMS and afterwards in BSM (Dreyfus et al., 1984c). The degree of neuronal purity of these different cultures was analysed; the activity of acetylcholinesterase, which is a specific marker for chick brain neurons, was determined by cytochemical methods. At 6 d.i.c. the neuronal purity was about 90% and 95% for GM1- and GT1b-treated cells, respectively.

Observation of the neurons seeded and grown in DMEMS and DMEMS/BSM has shown that the first symmetrical electron dense membrane thickenings appear at 3 d.i.c.. The number of thickenings was larger in GT1b- than in GM1-treated cells. After 6 d.i.c. synaptic-like junctions were observed, some of them containing dense core vesicles (Dreyfus et al., 1984c). Therefore we analyzed by HPLC with electrochemical detection (Felice et al., 1978) the amine content of these cultures (Fig. 2). For cells grown in DMEMS, serotonin (5-HT), dopamine (DA) and noradrenaline (NA) were detected. When cells were seeded and grown for 1 day in DMEMS and then in BSM, and also for cells grown in ganglioside/BSM, the 5-HT capacity disappeared whereas catecholaminergic properties were expressed. Moreover, in GM1/BSM and

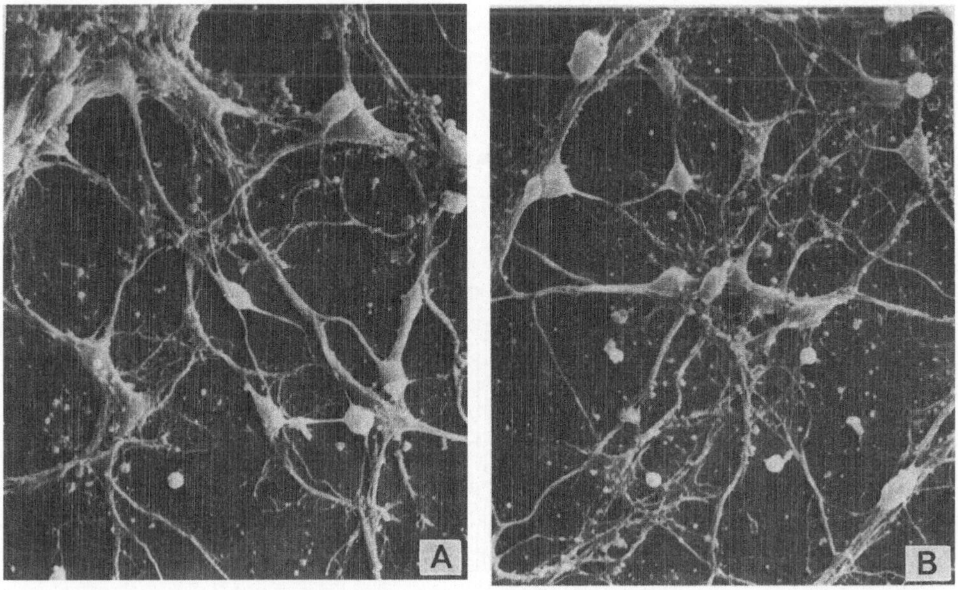

Figure 1. Scan electron micrographs (×630) of 6-day-old cultured neurons grown from 0 to 6 days in culture in BSM medium containing 10^{-6}M GM1 (A) or 2.10^{-6}M GT1b (B).

GT1b/BSM treated cells, NA concentrations were higher than those of DA when compared to control cells and the ratio NA/DA was the highest in GT1b treated cells. For the neurons grown in each particular condition, precursors of neurotransmitter (tyrosine- and tryptophan) and catabolic relevant compounds (3-methoxytyramine, homovanillic acid and 5-hydroxyindolacetic acid) were also detected.

The results indicate that variations in the components of the growing media and in the serum may influence the morphological and catecholaminergic properties of the neurons.

Incorporation of [^{3}H] Thymidine and [^{3}H]Leucine into Neurons

These experiments were performed to characterize the periods of cell division, cell differentiation and cell maturation.

To study the period of cell division we have measured the DNA content and the incorporation of [methyl-^{3}H]thymidine into DNA during 24 h at each day of culture. To determine the periods of differentiation and maturation of neurons in culture, besides the observations by electron microscopy and the determination of the amines, we have measured the protein/DNA ratio and the incorporation of [^{3}H]leucine during 1 h at each day of culture.

The DNA content of cultures treated or not treated with gangliosides increased up to 2 d.i.c. and decreased thereafter, mostly because of the change of the growth medium at days 3 and 5. We already reported the lower protein to DNA ratios of cultures seeded in the presence of DMEMS and then grown in BSM, as compared to

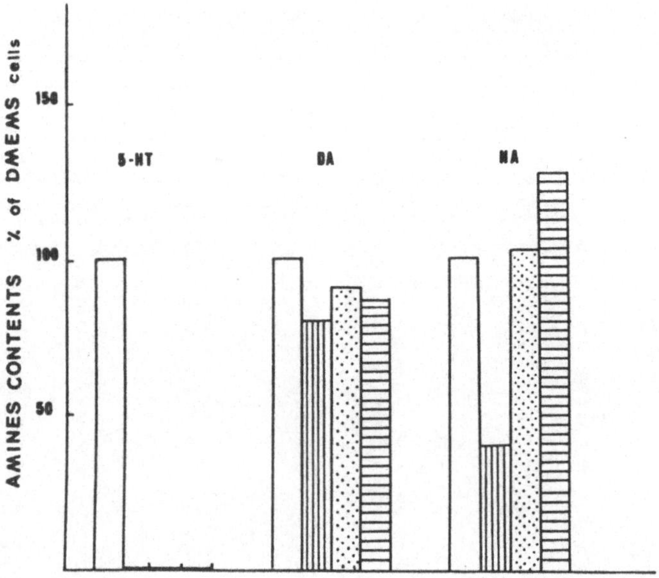

Figure 2. Amine contents (DA = dopamine; 5-HT: serotonine; NA: noradrenaline) analyzed by HPLC in 6-day-old cultured neurons grown:

- (open bar) from 0 to 6 days in DMEMS;
- (vertical hatching) from 0 to 1 day in DMEMS and from 1 to 6 days in BSM;
- (dotted) from 0 to 6 days in BSM containing 10^{-6}M GM1;
- (horizontal hatching) from 0 to 6 days in BSM containing 2.10^{-6}M GT1b.

Results are expressed as percentages of the amine contents measured in 6-day-old control cells grown in DMEMS.

cultures grown continuously in DMEMS, However, the delay in protein synthesis was considerably diminished in GT1b treated neurons (Dreyfus et al., 1984c) when compared to cells grown in the presence of DMEMS/BSM.

During the cell division period, the incorporation of [methyl-^{3}H]thymidine (Fig. 3) increased during the first two days in culture in parallel to the division of neuroblasts. Maximum incorporation was observed for cultures grown in the presence of GM1, which had the best effect on cell proliferation. Thus serum substitution with GM1 appears to facilitate the multiplication of neuroblasts. Moreover, when neurons were seeded in the synthetic medium (BSM) alone in the absence of gangliosides, thymidine incorporation was reduced during the period of cell division. This indicated that the effect produced by gangliosides cannot be simply due to the presence of the synthetic medium (BSM) at seeding.

After 48 h of an intense proliferation period, neuroblasts differentiated and the level of [^{3}H]thymidine incorporation was reduced from 3 to 6 d.i.c. This remaining incorporation may be related to some mechanisms of DNA restoration and repair in the cells and also to the appearance of some glioblasts in cultures grown in the presence of GM1.

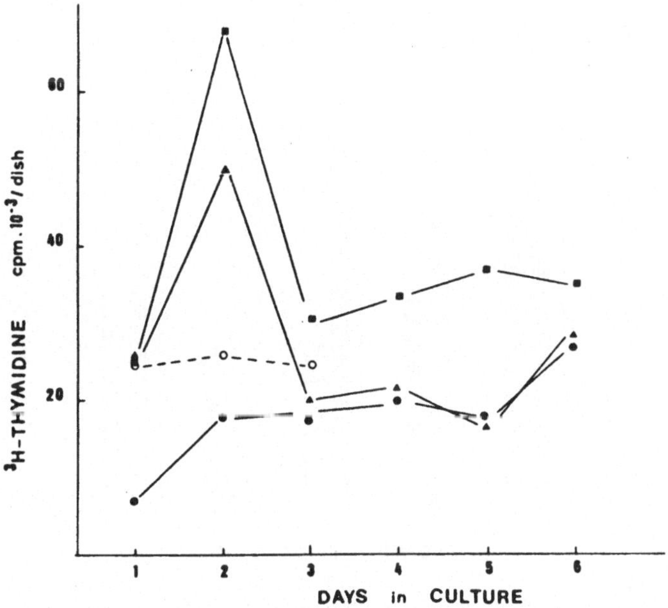

Figure 3. Incorporation of [methyl-^{3}H]thymidine (86 Ci/mmol, Amersham) during the maturation of primary cultures of neurons after 24h of incubation in:
• DMEMS;
○ BSM;
■ BSM + 10^{-6}M GM1;
▲ BSM + 2.10^{-6}M GT1b.

The incorporation of [^{3}H]leucine into the protein was determined at each day for one hour. Two maxima of incorporation were observed (Fig. 4) at days 2 and 5 and at days 3 and 6 in cells grown in DMEMS and in GT1b/BSM respectively. For cultures grown in the presence of GM1 the second peak was smaller. The first peak is concomitant with cell proliferation and the second one with cell maturation. This second peak did not appear in BSM cultured cells. Both maxima indicated an intense protein synthesis and the second peak could therefore explain the marked increase in protein/DNA ratio observed at these periods (Dreyfus et al., 1984c).

Hence we may conclude that gangliosides could act to facilitate the development and maturation of chick neurons in cultures grown in a synthetic medium.

Metabolism of Sialoglycoconjugates

Sialic Acid Content during Neuron Maturation

Figure 5 shows the developmental profiles of ganglioside- and glycoprotein-NeuAc in the neurons seeded and grown in different media. In general ganglioside (GA)-NeuAc increased as a function of time of culture whereas the glycoprotein (GL)-NeuAc decreased. For GA-NeuAc, the development was similar for cells grown in

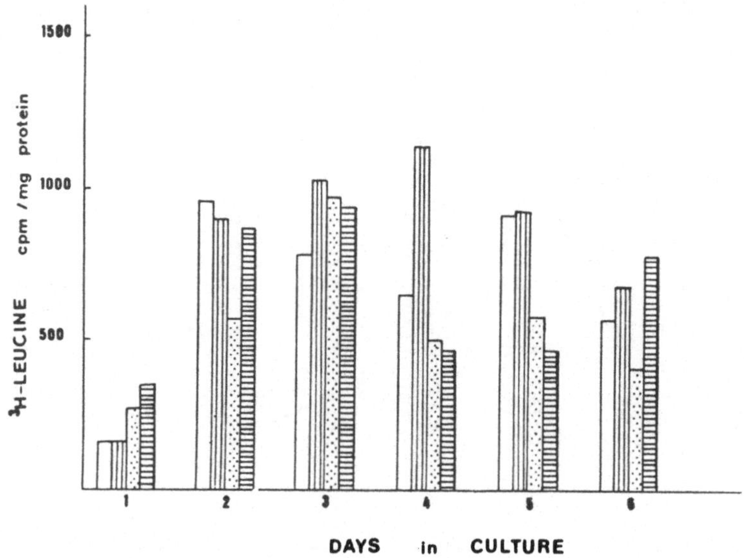

Figure 4. Incorporation of [^{3}H]leucine (120 Ci/mmol, Amersham) during the maturation of primary cultures of neurons after 1 h of incubation in:
▥ DMEMS (1 day) and BSM;
□ DMEMS;
▦ BSM + 10^{-6}M GM1;
▤ BSM + 2.10^{-6}M GT1b.

DMEMS and DMEMS/BSM whereas lower amounts were observed when cultures were performed in BSM alone. However, cultures seeded and grown in the presence of GM1 and GT1b in BSM showed an increase in their NeuAc content which was maximal at 6 d.i.c. where it reached a two-fold value. These results showed that the culture medium influenced the content of glycoconjugate-NeuAc and that the addition of gangliosides led to increase in both GA- and GL-NeuAc. One can also assume, on the basis of the GL-NeuAc changes, that the incorporation of gangliosides could modify activities of enzymes involved in the glycoconjugate-NeuAc metabolism.

Developmental Profiles of Gangliosides

Figure 6A,B,C shows the amount of ganglioside-NeuAc/mg protein in neurons cultured in DMEM/BSM, GM1/BSM and GT1b/BSM respectively. In control cells (Fig. 6A), the content of GD1a and GT1b increased after 3 d.i.c., those of GD3 reached a maximum at 5 d.i.c. and dropped thereafter. Treatment of the neurons by GM1 (Fig. 6B) and GT1b (Fig. 6C) led to a dramatic increase of the added ganglioside but concomitant changes in other ganglioside species could also be detected. The addition of GM1 produced an enhancement in the amount of this ganglioside and also of GD3 after the 5 d.i.c., and to a lesser extent of GD1b. By using GT1b the quantities of GT1b increased as well as those of GD1b and GD3, when compared to control cells. This in-

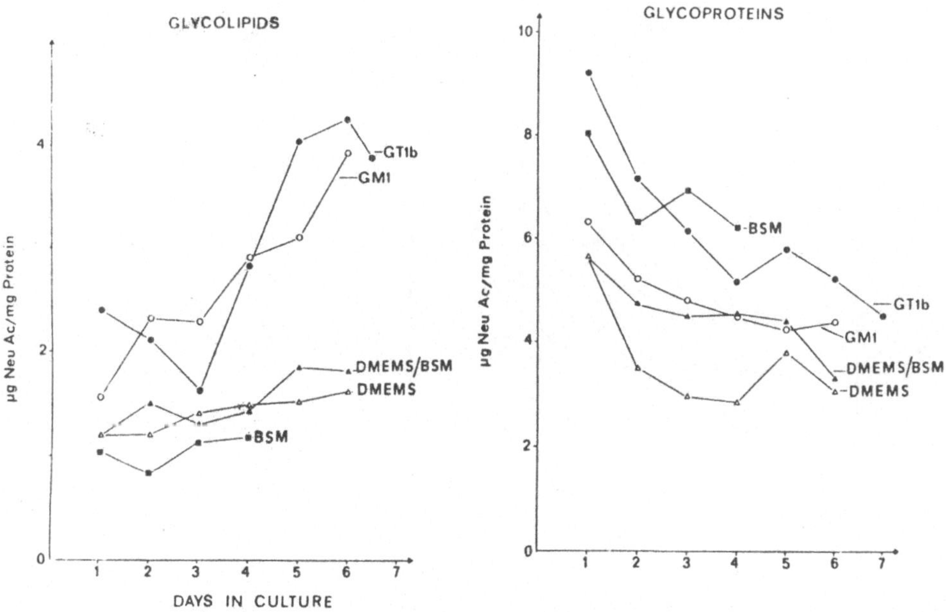

Figure 5 Variations of ganglioside-NeuAc and glycoprotein-NeuAc during the maturation of primary cultures of neuronal cells. NeuAc was determined by the method of Denny et al. (1981; 1983).

■ Seeding and culture in BSM;
△ Seeding and culture in DMEMS;
▲ Seeding and 1 day culture in DMEMS, then in BSM;
○ Seeding and culture in BSM containing 10^{-6}M GM1;
● Seeding and culture in BSM containing 2.10^{-6}M GT1b.

dicated that the addition of exogenous gangliosides to the culture medium led to changes in the amount of the added ganglioside(s) but also in the amount of other gangliosides which were metabolically derived from the added one(s).

Studies with [^{3}H] Labelled Gangliosides

In order to ascertain the results obtained with exogenous gangliosides, the study was performed with GM1, GD1a, GD1b and GT1b tritiated in their sphingosine moiety by the method of Ghidoni et al. (1981). These gangliosides (sp. act. about 450 Ci/mole) were added at seeding and at each medium renewal (1,3,5 d.i.c.). Figures 7 and 8 show for 6 d.i.c. the distribution of the radioactivity in each glycoconjugate fraction (total neutral glycolipids: GL; total glycoprotein: GP; total gangliosides: GG and individual gangliosides) after culturing the cells in the presence of 10^{-6}M [^{3}H]GM1 (Fig. 7 A-B), 10^{-6}M [^{3}H]GT1b (Fig. 7 C-D), 10^{-6}M [^{3}H]GD1b (Fig. 8 A-B) and 10^{-6}M [^{3}H] GD1a (Fig. 8 C-D). The amount of [^{3}H] labelling found in the ganglioside fractions

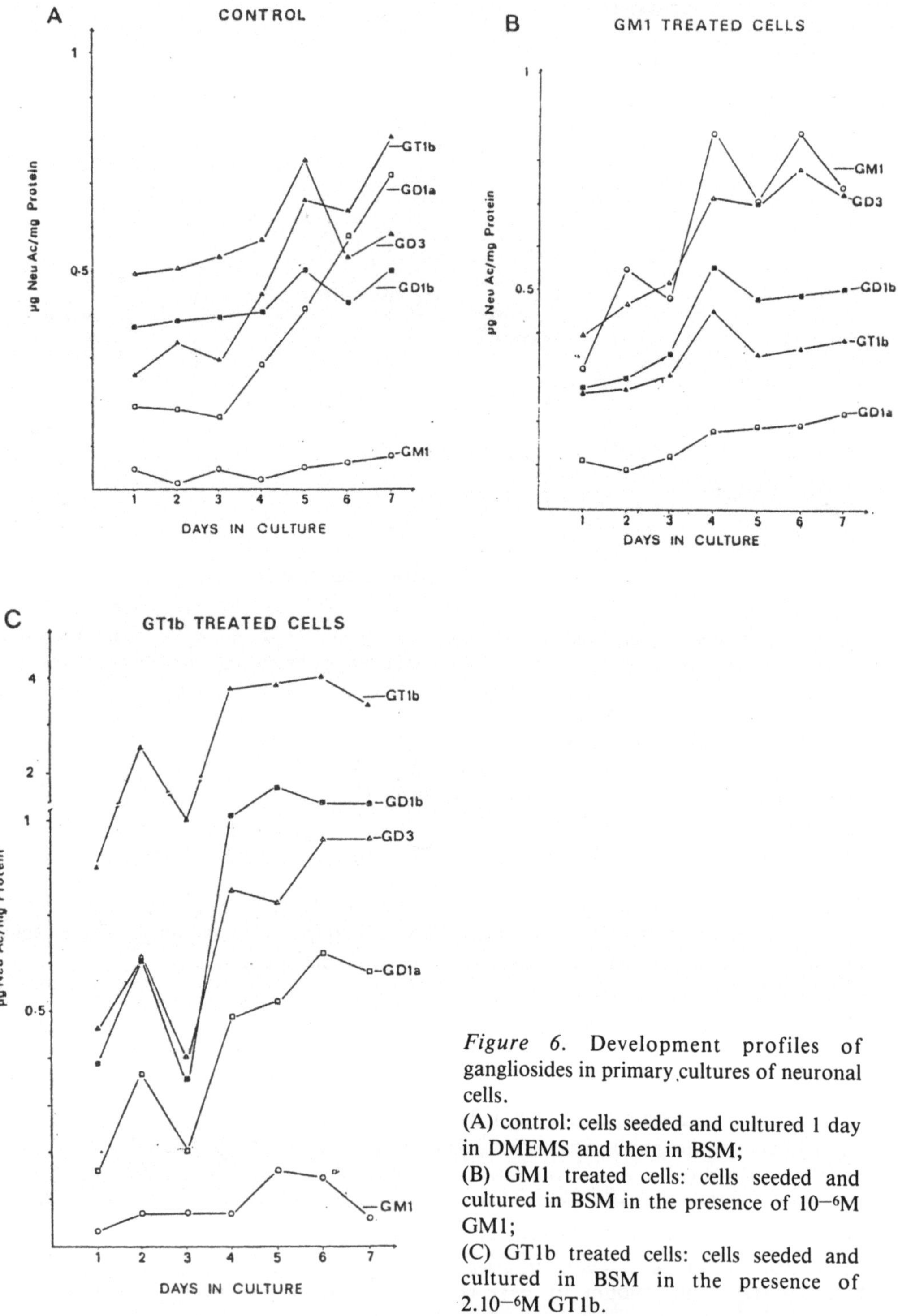

Figure 6. Development profiles of gangliosides in primary cultures of neuronal cells.

(A) control: cells seeded and cultured 1 day in DMEMS and then in BSM;

(B) GM1 treated cells: cells seeded and cultured in BSM in the presence of 10^{-6}M GM1;

(C) GT1b treated cells: cells seeded and cultured in BSM in the presence of 2.10^{-6}M GT1b.

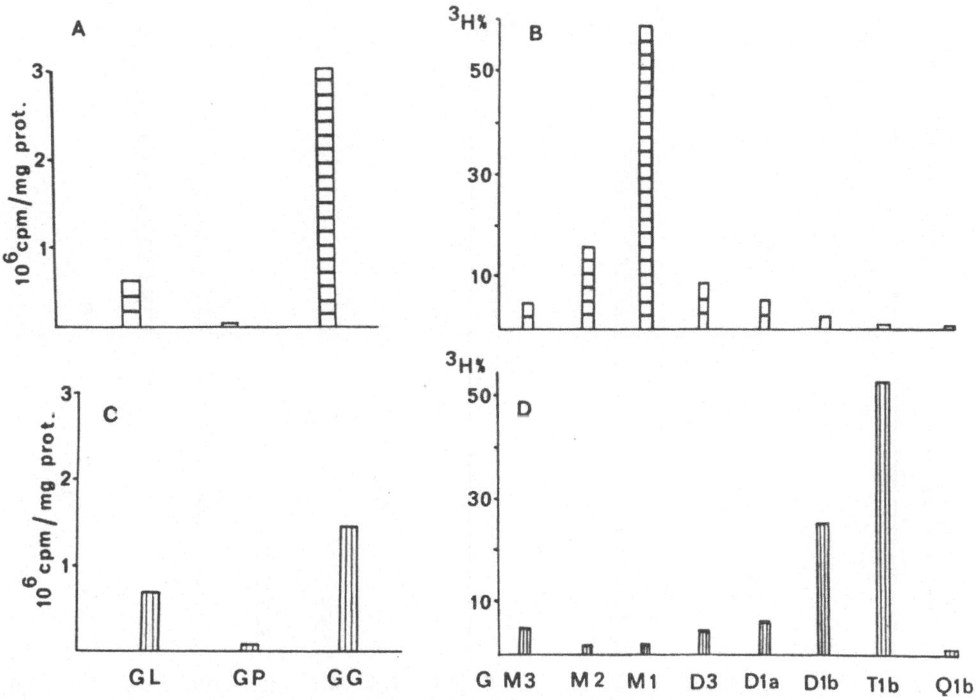

Figures 7. Distributions of radioactivity in the different glycoconjugate fractions in 6-day-old neurons cultured at days 0,1,3 and 5 in BSM containing 10^{-6}M [^{3}H]GM1 (450 mCi/mmol, A-B), and 10^{-6}M [^{3}H]GT1b (460 mCi/mmol, C-D) labelled on the sphingosine moiety.
GL: neutral glycolipids;
GP: glycoproteins;
GG: gangliosides.

represented 91.2%, 87.5%, 84.6% and 68.5% of total radioactivity recovered in the glycolipids after the treatment with GD1b, GD1a, GM1 and GT1b respectively. This result indicated that the exogenously incorporated gangliosides were metabolized and that following degradation the [^{3}H]sphingosine moiety was reincorporated into neutral glycolipids.

The distribution of the radioactivity in the different gangliosides was as follows: 1) with [^{3}H]GM1, 59% in GM1, 16% in GM2, 9% in GD3 and 6% in GD1a and GM3, the remainder in GD1b and GT1b; 2) with [^{3}H]GT1b, 54% in GT1b, 26% in GD1b, 6% in GD1a and GM3, the remainder in GD3, GM1 and GM2; 3) with [^{3}H]GD1b, 59% in GD1b, 29% in GD1a, 6% in GT1b and the remainder in GM3, GD3, GM2, GM1 and GQ1b; 4) with [^{3}H]GD1a, 48% in GD1a, 25% in GD3, 10% in GM1, 7% in GM3, 6% in GD1b and traces in GM2, GT1b and GQ1b.

This study showed that exogenously added gangliosides entered the routes of ganglioside metabolism in vitro. The extent of the labelling of the different gangliosides

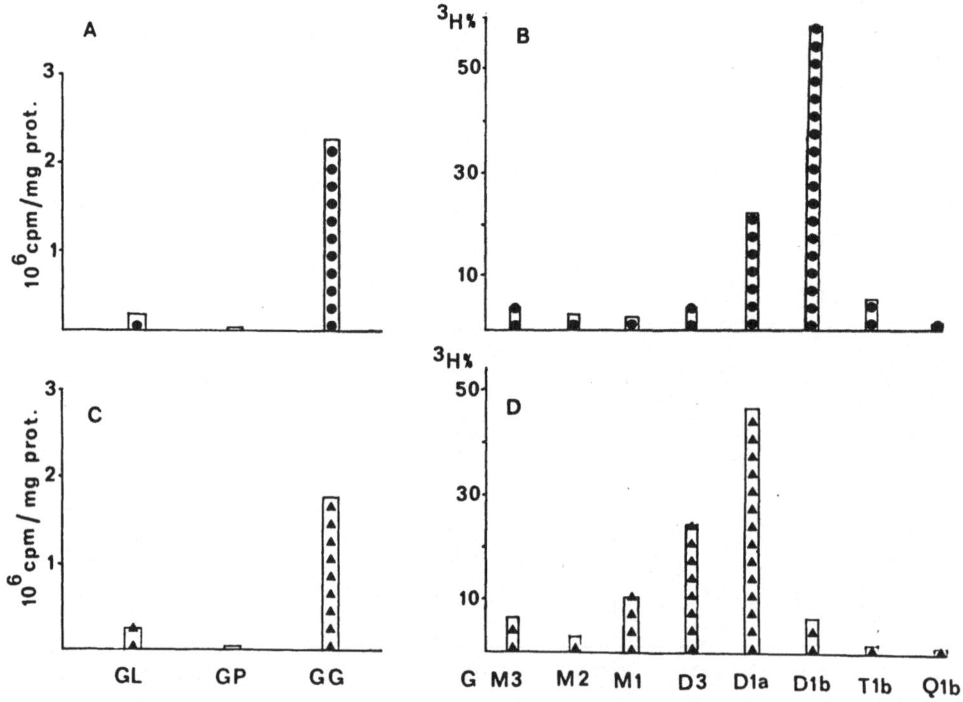

Figure 8. Distributions of radioactivity in the different glycoconjugate fractions in 6-day-old neurons cultured at days 0,1,3 and 5 with 10^{-6}M [^{3}H]GD1b (460 mCi/mmol, A-B) and 10–^{6}MGD1a (460 mCi/mmol) labelled on the sphingosine moiety.
GL: neutral glycolipids; GP: glycoproteins; GG: gangliosides.

could indicate some preferential pathways and could also influence the metabolism of the other gangliosides species. Therefore, in neurons GM1 is preferentially used as substrate for GM2. In vivo studies (Tettamanti et al., 1985) as well as in vitro studies (Fishman et al., 1983) have shown that [^{3}H]GM2 was detected by using GM1 labelled in the lipid portion of the molecule. Our results showed also that GT1b was desialylated to GD1b; GD1b could be metabolized to GD1a via GT1b and GD3 could be obtained from GD1a via GM1 and GD1b. All these results implicated the activities of enzymes present in the multiglycosyltransferase and -hydrolase complexes. Another interesting result concerned the fact that the percentage of radioactivity obtained in each ganglioside following the treatment by single ganglioside species [^{3}H]GD1b, [^{3}H]GD1a and [^{3}H]GT1b was similar when a mixture of equal amounts of these gangliosides was used. In Figure 9 are reported the values obtained during the maturation of the neuronal cultures. It can be seen that the final radioactivity values were rapidly reached. From this study it could also be assumed that each individual incorporated ganglioside determined its own pathway which was not influenced by the additional presence of other ganglioside species.

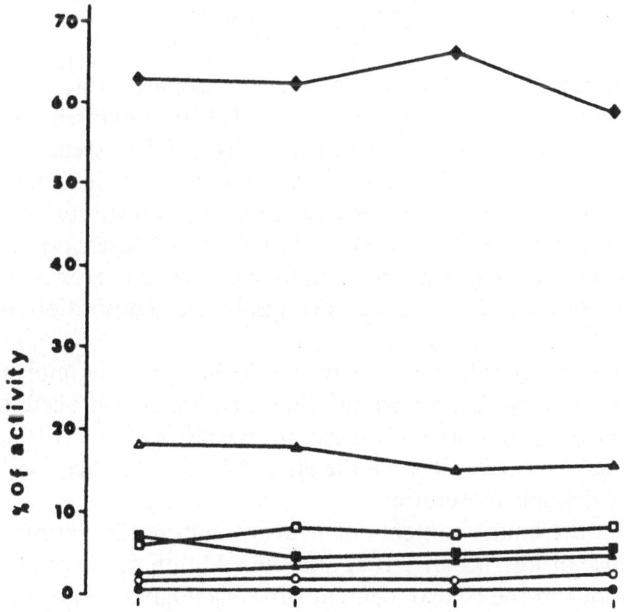

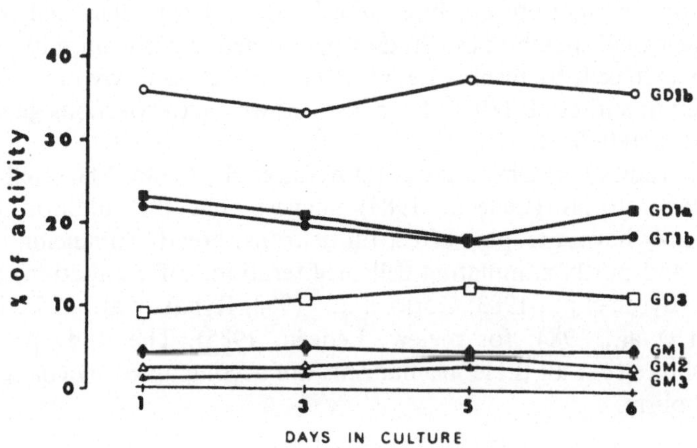

Figure 9. Distributions of radioactivity in gangliosides during maturation of neurons cultured at days 0,1,3 and 5 in BSM containing 10^{-6}M [^{3}H]GM1 (450 mCi/mmol, A) and 10^{-6}M [^{3}H]GT1b-GD1b-GD1a in equal amounts (460 mCi/mmol, B) labelled on the sphingosine moiety.

CONCLUSIONS

The studies reported here showed that when neuronal cultures were grown in the presence of exogenous gangliosides GM1 and GT1b, modifications in the morphological and biochemical development of the cells could be seen. This could be due to the strong adhesion and/or insertion of the sialoglycolipids into the neuronal membranes. In our culture conditions, gangliosides are able to substitute for serum: neurons seeded and grown in a synthetic medium (without serum) degenerate rapidly but present a normal development in the presence of gangliosides. However, the neuronal maturation was accompanied by various changes in the composition and metabolism of cell components:

1) Stimulation of thymidine incorporation during some definite steps of the cell growth depending on the considered ganglioside. Treatment of the cells by gangliosides led also to modifications in protein synthesis as revealed by the [^{3}H]leucine incorporation. GM1 seems to be more involved in the phase of cell multiplication, whereas GT1b seems to facilitate cellular maturation.

2) Changes in the amine contents of neurons cultured in various conditions including the presence or absence of serum and the addition of exogenous gangliosides.

3) Modifications in the NeuAc content of the sialoglycoconjugates (gangliosides and sialoglycoproteins) during the maturation of the neuronal cells.

4) Variations in ganglioside pattern and metabolism after ganglioside treatment. The present study has shown that exogenously incorporated GM1 and GT1b were rapidly metabolized in neuronal cell cultures. The different pathways included mechanisms of degradation and biosynthesis which led to the final composition of ganglioside neuronal membranes. Studies performed in vitro on murine NCTC 2071 and rat glioma C6 cells by Fishman et al. (1983) as well as in vivo by injection in mice or rats by Tettamanti et al. (1985) have also shown that exogenous gangliosides were taken up and metabolized.

Previous studies performed in our (Dreyfus et al., 1984a; Massarelli et al., 1985b) and other laboratories (Ledeen, 1984) suggest that the addition of exogenous gangliosides may have a trophic effect on neurons, greatly enhancing the number of cell contacts and partly stimulating cell proliferation, differentiation, sprouting and regeneration (Byrne et al., 1983; Gorio et al., 1983; Rybak et al., 1983; Dreyfus et al., 1984b; Facci et al., 1984; for review, Ledeen, 1985). The study presented in this manuscript shows that all these mechanisms are accompanied by changes at the level of cell metabolism.

ACKNOWLEDGMENTS

The secretarial assistance of Ms. C. Thomassin-Orphanides is gratefully acknowledged. The unlabelled gangliosides GM1 and GT1b were kindly provided by Fidia Research Laboratories.

REFERENCES

Bottenstein JE and Sato G (1979) Proc Natl Acad Sci USA 76: 514-517.

Byrne MC, Ledeen RW, Roisen FJ, Yorke G, Sclafani JR (1983) J Neurochem 41: 1214-1222.

Denny FC, Denny PA, Allerton SE (1981) Clinica Chim Acta 116: 409-415.

Denny FC, Denny PA, Allerton SE (1983) Clinica Chim Acta 131: 333-336.

Dreyfus H, Louis JC, Harth S, Mandel P (1980) Neuroscience 5: 1647-1655.

Dreyfus H, Ferret B, Harth S, Gorio A, Durand M, Freysz L, Massarelli R (1984a) J Neurosci Res 12: 311-322.

Dreyfus H, Ferret B, Harth S, Gorio A, Freysz L, Massarelli R (1984b) in: Ledeen RW, Yu RK, Rapport MM, Suzuki K (eds): Ganglioside Structure, Function and Biomedical Potential. Plenum Publishing Corporation, New York, pp. 513-524.

Dreyfus H, Massarelli R, Lombard D, Gorio A, Freysz L, Durand M (1984c) in: Caciagli F, Giacobini E, Paoletti R (eds): Developmental Neuroscience: Physiological, Pharmacological and Clinical Aspects. Elsevier Science Publishers BV, Amsterdam, pp. 27-32.

Facci L, Leon A, Toffano G, Sonnino S, Ghidoni R, Tettamanti G (1984) J Neurochem 42: 299-305.

Felice LJ, Felice JD, Kissinger PT (1978) J Neurochem 31: 1461-1466.

Fishman PH, Bradley RM, Hom BE, Moss Y (1983) J Lipid Res 24: 1002-1011.

Ghidoni R, Sonnino S, Masserini M, Orlando P, Tettamanti G (1981) J Lipid Res 22: 1286-1295.

Gorio A, Marini P, Zanoni R (1983) Neuroscience 8: 417-429.

Ledeen RW (1984) J Neurosci Res 12: 147-159.

Ledeen RW (1985) Trends in Neurosciences 8: 169-174.

Massarelli R, Durand M, Guerold B, Ferret B, Lombard D, Freysz L, Dreyfus H (1985a) in: Dreyfus H, Massarelli R, Freysz L, Rebel G (eds): Cellular and Pathological Aspects of Glycoconjugate Metabolism. Les Editions INSERM, Paris, pp. 309-334.

Massarelli R, Ferret B, Gorio A, Durand M, Dreyfus H (1985b) Int J Devl Neuroscience 3: 341-348.

Pettmann B, Louis JC, Sensenbrenner M (1979) Nature 281: 378-380.

Rybak S, Ginsburg I, Yavin E (1983) Biochem Biophys Res Comm 116: 974-980.

Schwarzmann G, Sonderfeld S, Conzelmann F, Marsh D, Sandhoff K (1985) in: Dreyfus H, Massarelli R, Freysz L, Rebel G (eds): Cellular and Pathological Aspects of Glycoconjugate Metabolism. Les Editions INSERM, Paris, pp. 195-210.

Tettamanti G, Ghidoni R, Sonnino S, Chigorno V, Venerando B, Giuliani A, Fiorilli A (1984) in: Ledeen RW, Yu RK, Rapport MM, Suzuki K (eds): Ganglioside Structure, Function and Biomedical Potential. Plenum Publishing Corporation, New York, pp. 273-284.

Gangliosides and neuronal plasticity
G. Tettamanti, R.W. Ledeen, K. Sandhoff,
Y. Nagai, G. Toffano (eds.)
Fidia Research Series, vol. 6
Liviana Press, Padova, © 1986

FACILITATED-ESTABLISHMENT OF CONTACTS AND SYNAPSES IN NEURONAL CULTURES: GANGLIOSIDE-MEDIATED NEURITE SPROUTING AND OUTGROWTH

P.E. Spoerri

Department of Neurosurgery, University of Göttingen, Robert-Koch-Str.40, D-3400 Göttingen, Federal Republic of Germany

INTRODUCTION

Gangliosides are natural components of the neuronal membranes, and their addition to cultures of cells of neural origin, or administration to animal with neurolesions was shown to modulate cell growth and differentiation, nerve regeneration and reinnervation (see the review by Ledeen, 1984).
The molecular mechanism by which exogenously applied gangliosides produce the various neuroplastic cellular modifications is not known. However, the neuritogenic effect influencing the growth, length and branching of neurites in cultures, accompanied by synaptogenesis has been well documented by light and electron microscopy. Moreover, the neuritogenic and neurotrophic effect facilitating further the action of growth factors in cultures of young and adult neurons is portrayed.

MATERIALS AND METHODS

Preparation of Cultures

Stock cultures of C1300 mouse neuroblastoma, clone N-2A, purchased from the American Type Culture Collection (Rockville, Md.), were maintained at 37°C in an atmosphere of 5% CO_2 and 95% air. The cultures were grown as monolayers in plastic petri-dishes (35 mm, Nunc) or in plastic flasks (Corning) and were fed twice to three times a week with Eagle's minimum essential medium or Ham-F-10 medium supplemented with 10% fetal calf serum (FCS) and 1% of 200 mM L-glutamine.

Abbreviations: CNS, central nervous system; NGF, nerve grouth factor; FCS, fetal calf serum; GABA, γ-aminobutyric acid; BME, eagle's basal medium; SCG, superior cervical ganglion; cAMP, adenosine 3',5'-cyclic monophosphate.

Addition of γ-Aminobutyric Acid

After 5 days in culture the cells were fed with normal medium or with the same medium containing 1-2% instead of 10% FCS to which GABA (γ-aminobutyric acid) or NaBr was added (Spoerri, 1983). These substances were dissolved in double distilled water and added to the normal medium producing concentrations of 10^{-4} - 10^{-5}M. GABA or NaBr was applied to the cultures for a maximum of 2 days.

Primary Cultures

Cultures were prepared from the cerebra of 8-day-old chick embryos as suggested by Pettmann et al. (1979). The cerebral hemispheres were dissected out and the meningeal membranes removed. The brain tissue was then placed in a pool of Hank's balanced salt solution and pieces approximately 1-2 mm were cut out. The tissue was then gently dissociated by passage through a 22-gauge needle into and out of a 5 ml syringe (five times). After mechanical dissociation, Eagle's basal medium (BME), containing 26.4 mM $NaHCO_3$, 333 mM D-glucose, 2 mM L-glutamine and 10% FCS, was added to poly-D-lysine treated petri-dishes (35 mm, Nunc), containing 0.2 ml of cell suspension (5 × 10^4 cell density). The cultures were incubated at 37°C in a humidified atmosphere consisting of 5% CO_2 and 95% air.

Addition of Gangliosides

Purified bovine brain gangliosides (Seromed, Munich, presently Serolab, Aidenbach), were purchased singly and where then added together to make a mixture consisting of 27% GMl, 40%GDla, 16% GDlb and 19% GT. The gangliosides were dissolved in medium and introduced into the culture medium in concentrations ranging from 200 to 250 μg/ml. They were added to 6-7 day-old N-2A cultures. The effect of added ganglioside mixtures or/and single species (GMl, GDla) on preconfluent cells (24-48 hr after seeding) was determined as follows. The cells were washed with medium containing 1% FCS. After washing, 2.5 ml of fresh medium containing 1% FCS plus 200-250 μg/ml of the appropriate ganglioside(s) was added to the cells.

Primary cultures from embryonic chick hemispheres received a mixture of individual gangliosides (GMl, GDla, GDlb and GTlb, gifts from Fidia Research Lab.). These were put together to form a mixture of 21.6% GMl, 40% GDla, 14.7% GDlb and 18.5% GTlb. They were dissolved in Eagle's basal medium and introduced into the culture medium at a concentration of 150 μg/ml.

Electron Microscopy

After draining of the medium the cells were fixed in situ at 36°C for 30 min in 2.5% glutaraldehyde in a phospate buffer (pH 6.8). The cells were post-fixed in 1% osmium tetroxide in 0.2M phosphate-buffered saccharose for 1 h. After dehydration in ethanol, the cultures were embedded in Epon 812. Selected cells were mounted and cut as described (Spoerri et al., 1980a). The ultrathin sections needed for electron microscopy were stained on grids with a saturated aqueous solution of uranyl acetate and lead citrate. For light microscopy, 0.5-1.0 μm sections were stained with toluidine blue.

Culture of Young and Adult Superior Cervical Ganglion Explants

The superior cervical ganglion (SCG) was removed aseptically from male mice ageing 3, 6 and 18 months respectively. The ganglia were washed in oxygenated Ringer's solution (Ames and Hastings, 1956; Piper et al., 1982). The Ringer's solution contained in mM: NaCl, 125.4; KCl, 3.6; $MgCl_2$, 1.2; $NaHCO_3$, 22.6; $NaHPO_4$, 0.1; Na_2HPO_4, 0.4; Na_2SO_4, 1.2; $CaCl_2$, 1.15; and glucose 10. In the same solution the capsel was removed and the ganglion was divided into 4 equal pieces for explants in culture. They were placed in 0.1 μg/ml poly-L-lysine coated petri dishes and cultured as previously described (Spoerri et al., 1984). The culture medium was made up of L-15(Leibowitz)medium supplemented with 15% FCS, 1% of 200 mM L-glutamine, 2.4% bicarbonate, 6 mg/ml glucose and 10 ng/ml NGF (2.5S, Sigma).

Addition of GMl

Experimental cultures received additionally 10^{-6}M GMl (Fidia Research Lab.). The GMl was dissolved in an appropriate volume of incubation medium in order to reach the final ganglioside concentration. The solution was maintained at 37°C for at least 1 hr with intermittent swirling as described by Leon et al. (1984).

At intervals of over 10 days, the number of explants with processes (axons) and the length of processes was determined by measuring the distance between the end of the longest axons and the surface of the ganglion under a Leitz tissue culture microscope at a magnification of 100x, using an ocular micrometer. The statistical evaluation was made by the Student t test.

RESULTS

Neuronal Cells in Culture. Effects of Ganglioside Addition

N-2A cells grown in monolayer cultures in the presence of serum possess few neurites and display a phenotype characterized by slow differentiation and rapid proliferation. This is in contrast to the type of growth observed when serum is deleted or low concentrations are used. Upon addition of GABA or NaBr to the non-synchronous cell population, extensive neurite formation with elaborate contacts is observed within 20 - 24 h. This effect has been extensively described elsewhere (Spoerri et al., 1980b; Spoerri and Wolff, 1981; 1982; Eins et al., 1983; Spoerri, 1984).

Similarly, the addition of gangliosides to the N-2A cells enhanced the rate and degree of neurite formation (Fig. 1a, b, c). Addition of the GMl fraction alone to these cells resulted in a widely spread neuritic outgrowth (Fig. 2). By contrast, the type of growth seen when GDla is added to the culture medium is characterized by simple and relatively short processes (Fig. 3).

Electron microscopy revealed that when various concentrations of a purified mix-

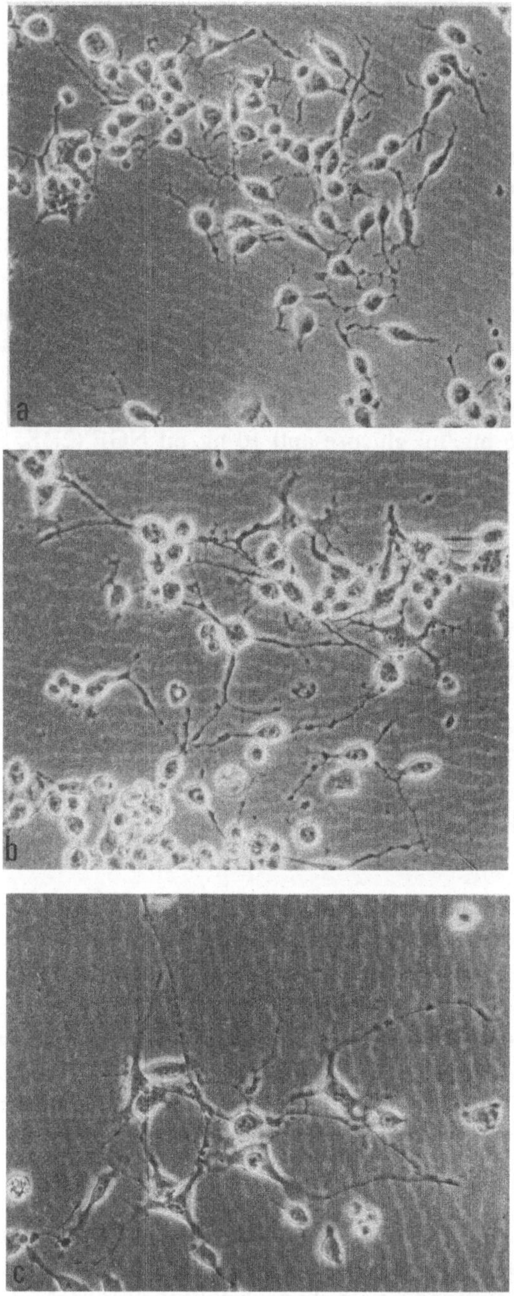

Figure 1. a. N-2A cells, living preparation after 24h in medium containing 1% fetal calf serum (FCS) and gangliosides. b, c. Same cells after 48 h in medium containing 1% FCS and 200 μg/ml gangliosides (GMl, GDla, GDlb, GT), depict extensive neurite formation with contacts. Phase contrast, × 314.

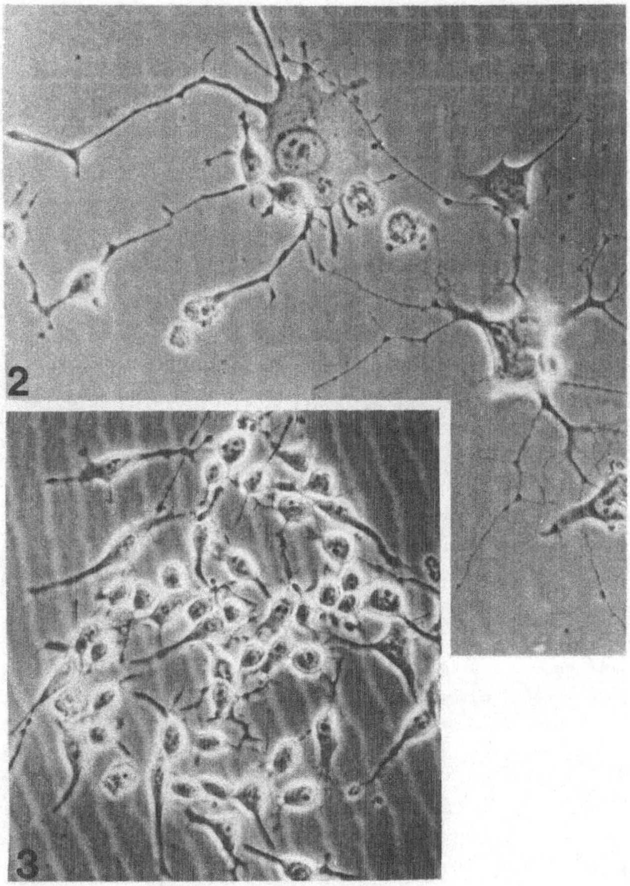

Figure 2. N-2A cells after treatment with GMl alone. Note the extensively long, fine, branching neurites with numerous fine projections. Phase contrast, x 464.

Figure 3. GDla treatment alone (48 h). The number of processes is reduced and so is their length. Phase contrast, × 379.

ture of single species of ganglioside (GMl, GDla, GDlb, GT) were added to GABA-differentiated cells, mature synapse-like contacts were encountered (Figs. 4a, b, 5, 6, 7, 8, 9).

Ultrastructural Morphology of Synapses

All synapses have the presynaptic element with vesicle aggregation, opposite the postsynaptic element. Mitochondria were present and the synaptic vesicles were spherical to slightly ovoid in shape, and their size ranging from 40 - 200 nm in diameter (Figs. 4b, 5, 7, 8).

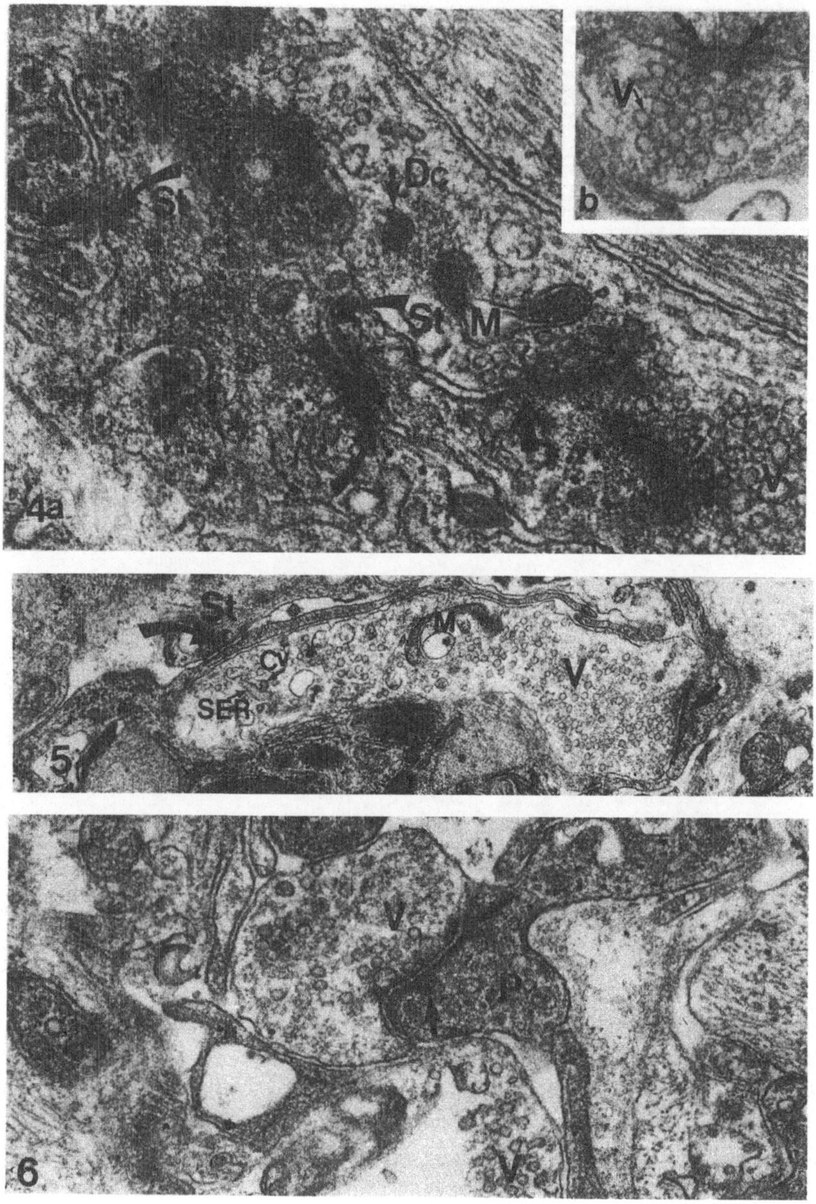

Figures 4a, b, 5, 6. Electron micrographs displaying varicosities which establish synaptic contacts with neurites (P) or somata (S). Note the membrane thickenings (straight arrows), the presynaptic dense projection with vesicle aggregation (V) and the electron-dense postsynaptic thickening with mostly one dense projection (two in Fig. 6, arrows). These terminal swellings display numerous round-to-oval vesicles (V) of varying diameter (40 - 200 nm). Additionally, a dense-core vesicle (Dc), mitochondria (M) and smooth endoplasmic reticulum (SER) with coated vesicles (Cv, small arrows). Numerous symmetrical thickenings (St) or desmosome-like junctions are present (curved arrows). From 9-day-old cultures treated with 10^{-4}M GABA for 2 days followed by treatment with a mixture of bovine brain gangliosides for another 2 days. (After Spoerri, 1983) (4a) × 77 343; (4b) × 38 919; (5) × 22 344; (6) × 40 390. Reproduced by permission of the International Society of Developmental Neuroscience.

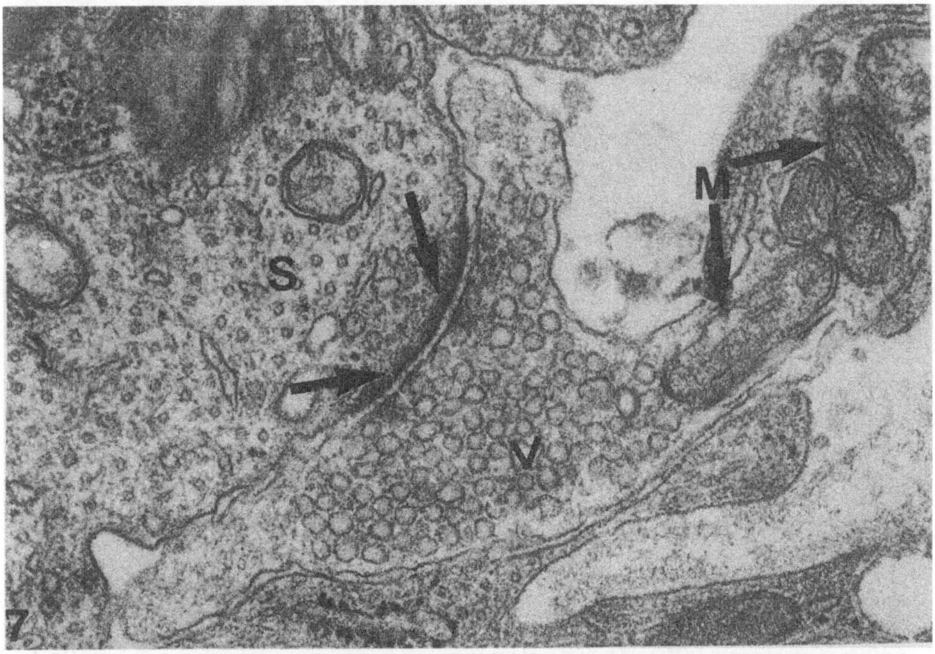

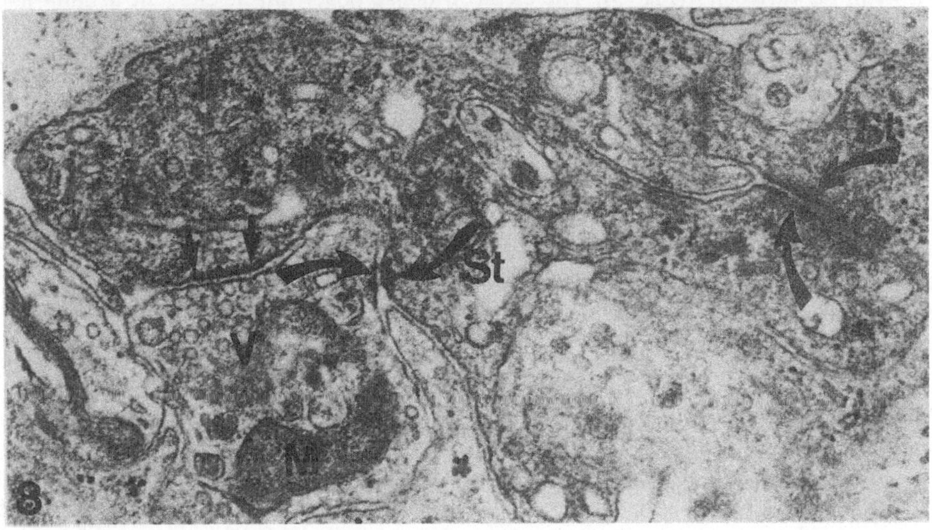

Figures 7 and 8. Some other varicosities showing membrane thickenings (straight arrows), aggregation of vesicles (V) on the presynaptic side and a mitochondrion (M). Note once more the presence of symmetrical thickenings (St) between the short neurites (P curved arrows). From a 9 day-old pretreated culture followed by addition of gangliosides for another 2 days. (Fig. 8, after Spoerri, 1983). × 47 000. Reproduced by permission of the International Society of Developmental Neuroscience.

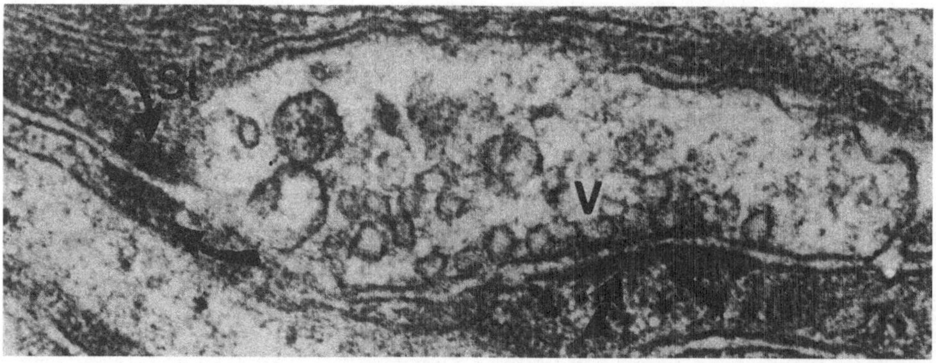

Figure 9. A magnified view of a synapse. Note the short dense postsynaptic thickening (straight arrows) and the presynaptic dense projection with a small number of large-vesicle (V) aggregations. A symmetrical thickening (St, curved arrows) is also present. (After Spoerri, 1983). × 90 000. Reproduced by permission of the International Society of Developmental Neuroscience.

Dense-core vesicles were scarce in the mature boutons as previously noted (Spoerri and Wolff, 1981; 1982). Smooth endoplasmic reticulum with coated vesicles could be seen within a varicosity (Fig. 5). The electron-dense postsynaptic element usually had one, but could have two dense projections (Fig. 6). There was usually one presynaptic dense projection per presynaptic element. This projection was short and narrow while the postsynaptic density was long with a relatively dark profile, the synapses being mostly asymmetrical and resembling those found in primary cultures.

Removal of the GMl fraction from the ganglioside mixture affects the formation of mature synapses. Apparent presynaptic elements which may be interpreted as primitive stages of synaptogenesis were numerous. The varicosities or boutons were filled with clear and dense-core vesicles; membrane thickenings could only be seen on the postynaptic side or were non-existent (Fig. 10a, b). Symmetrical thickenings or desmosome-like junctions containing no presynaptic dense-projections with vesicle aggregation were numerous (Fig. 11).

Ganglioside Addition to Primary Cultures

The development of primary neuronal cell cultures from embryonic chick brain was characterized by the early aggregation of cells into clusters and by the consequent emmission of bundles of neurites interconnecting these clusters. This occurred 2-3 days after cell seeding. Addition of mixtures of purified single gangliosides (GMl, GDla, GDlb, GTlb) to the culture medium at 150 μg/ml concentration, caused an earlier formation of clusters, as a consequence an earlier outgrowth of neurites. In cultures supplemented with gangliosides the network was already evident 1$^{1}/_{2}$ - 2 days after cell seeding, while in the control some kind of similar pattern became evident 3 days later (Fig. 12a, b, c).

Electron microscopy of the young, 4-5 day-old ganglioside-treated and untreated

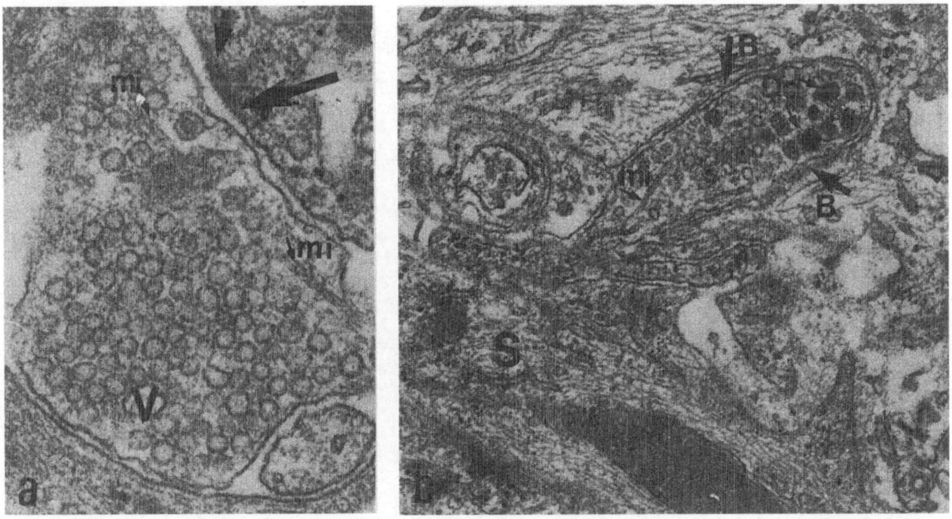

Figure 10a, b. Immature boutons (B) or terminal swellings (straight arrows). Without membrane thickenings, with the exception of an electron-dense projection on the postsynaptic side in (a) (straight arrows). The boutons (B) are filled with numerous clear-core vesicles (V) and plenty of dense-core vesicles (Dc) and microtubules (mi). The apparent presynaptic elements are from 9-day-old cultures treated with 10^{-4}M GABA for 2 days followed by ganglioside treatment without the GM1 fraction for an additional 2 days. (a) × 45 000; (b) × 38 000 (After Spoerri, 1983). Reproduced by permission of the International Society of Developmental Neuroscience.

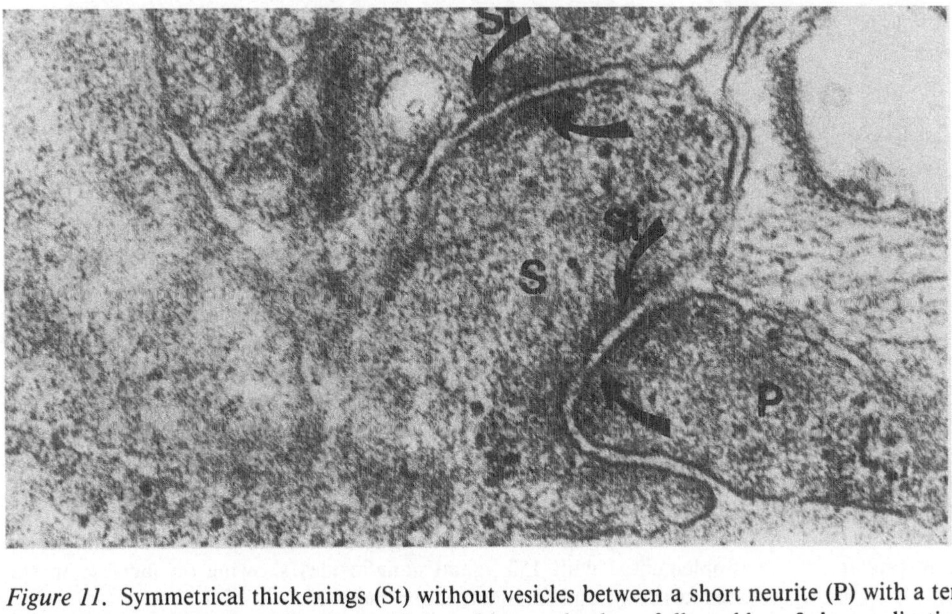

Figure 11. Symmetrical thickenings (St) without vesicles between a short neurite (P) with a terminal swelling and somata (S). From a 9-day-old treated culture followed by a 2-day application of the ganglioside mixture without the GMl fraction. (After Spoerri, 1983) × 90 000. Reproduced by permission of the International Society of Developmental Neuroscience.

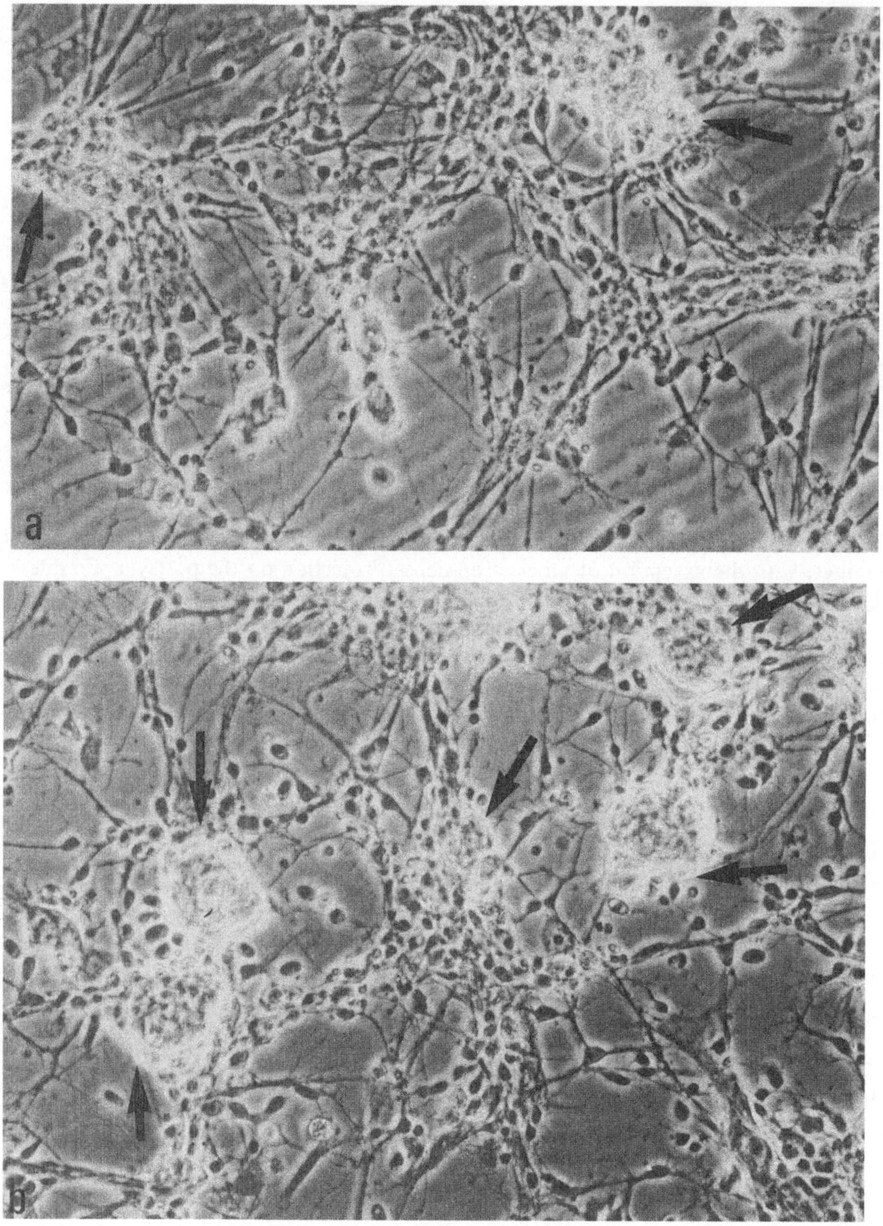

Figure 12a-c. a. Photomicrographs taken 2 days after plating CNS neurons from 8-day chick embryos in medium supplemented with serum. Note the early aggregation of clusters of neurons (arrows) and the emmission of single or bundles of long neurites.

b. A similar culture supplemented with 150 μg/ml gangliosides showing an increase in the number of cell aggregates (arrows), and interconnecting neurites.

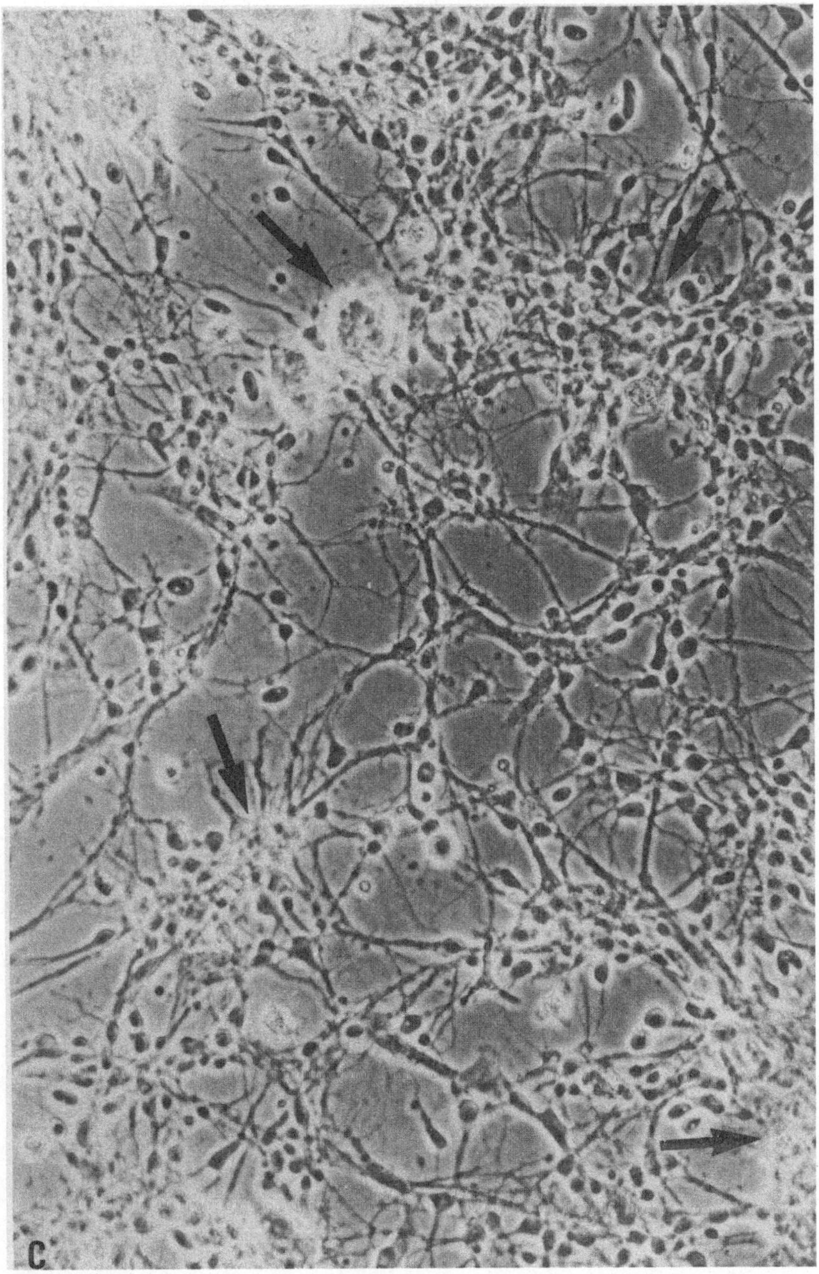

c. In ganglioside supplemented medium, network formation due to interocommunicating neurites with fine side branches is already evident 2 days after seedings. (Arrows indicate clusters) Phase contrast, × 350.

cultures, revealed mature synapses in the cultures to which gangliosides were added (Figs. 13, 14, 15). The synapses resembled Gray's type 1, which have a somewhat wider

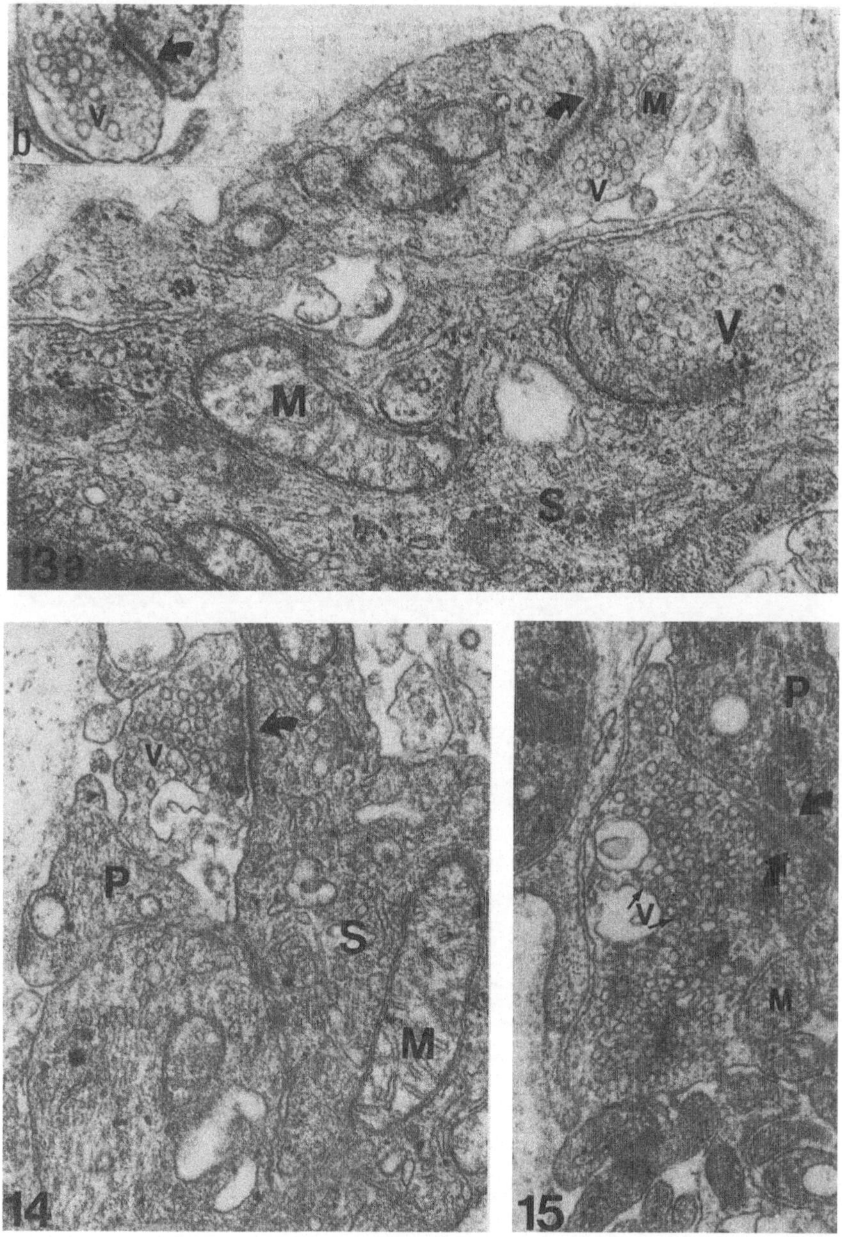

Figures 13a, b, 14, 15. Electron micrographs depicting mature synapses (thick curved arrows), of neurons from 4-5-day-old chick embryo cultures. The vesicles (V) are round to oval, also seen within the soma (S). Mitochondria (M) are also present. P process. (13a, b; 14) × 38 675; (15) × 35 002.

cleft and more pronounced synaptic density. There was usually one presynaptic dense projection per presynaptic element and it was short and narrow, typical of a young synapse (Burry and Lasher, 1978). The size and shape of the synaptic vesicles was the same as in other cultures. Symmetrical thickenings or specialized contacts, which may represent early synapses (Spoerri et al., 1980b), were also numerous in the untreated cultures.

Addition of GMl to NGF Supplemented Adult Mouse SCG Explants

Upon addition of GMl to NGF-supplemented cultures of superior cervical ganglion of explants from young and relatively adult mice, the number of explants showing processes increased and so did the outgrowth of processes, which might have approached the level of responsiveness of young neurons. The speed of outgrowth of processes in the 18 month old explants increased to at nearly the same rate as that of the younger animals. Additionally, the length of the axons in young and adult neurons did not show pronounced differences (Fig. 16).

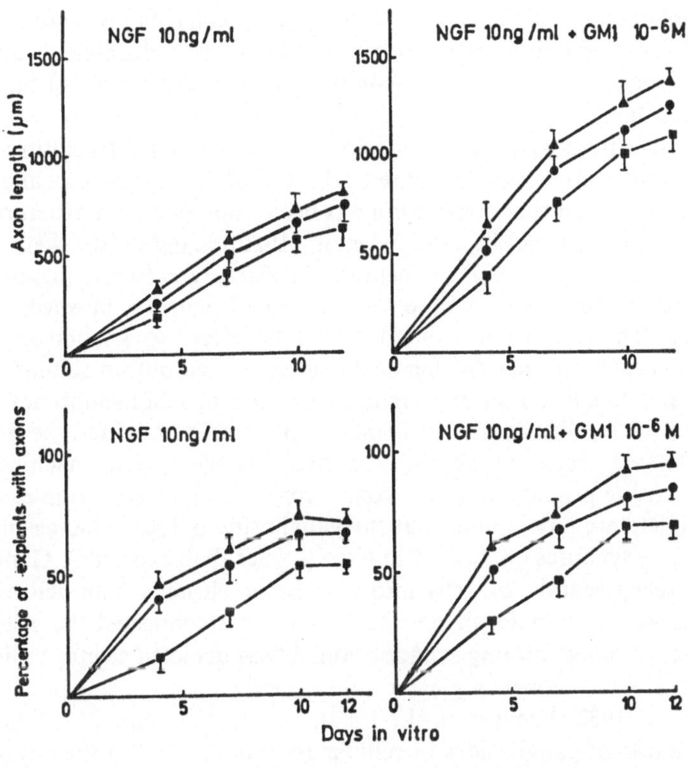

Figure 16. The length of axons or neurites coming out of the explants (upper) and the percentage of SCG explants with process or axon growth (lower), at different times in culture and under different conditions, in the presence of NGF with or without GMl. SCG explants were obtained from 3— (▲), 6— (•), 18 (■) month old animals. Each point is the mean ± SE of 4 observations from cultures from 4 donors.

DISCUSSION

The results reported here show that supplementing the culture medium of neuronal cells (cell-lines or primary cultures) with gangliosides leads to changes in cell morphology, increasing the number and length of neurites and forming numerous network-like contacts. It is of considerable interest that in addition to enhancing neurite growth, exogenous gangliosides also induce the formation of mature synapse-like contacts in N-2A cells pretreated with γ-aminobutyric acid or NaBr (Spoerri, 1983) and as presently described. The synapses resemble Gray's type 1, which have a somewhat wider cleft and more pronounced synaptic density than those of type 2 when they reach maturity (Kojima, 1975). Gangliosides potentiated the neurotrophic or hyperpolarizing effect of GABA or NaBr.

Previously, the possible role played by gangliosides in the process leading an undifferentiated neural cell to its ultimate expression, the establishment of synapse, was approached differently. Massarelli et al. (1982) treated neuronal membranes with purified neuraminidase and pointed out that the negative charges of sialic acid may have an effect on the fluxes of ions, hence on their chemical gradients, and consequently upon the polarization of the neuronal membrane, which can promote the establishment of contacts. The release of neurotransmitter is Ca^{++} dependent, and a correlation has been suggested between sialocompounds and Ca^{++} (Probst and Rahmann, 1980; Svennerholm, 1980).

In the present investigation, gangliosides have been added to cultures of neurons at a culture age that corresponds to the beginning of synaptogenesis and their effect during the stages of neuronal maturation has been examined ultrastructurally. Similar studies have been carried out by Dreyfus et al. (1980). Gangliosides were found to induce changes in the morphology of neurons in primary cultures, increasing the cell number, the size of the cell body and the number of neurites; however, no synapses were observed. The authors attributed a trophic effect to gangliosides. Recently, Dreyfus et al. (1984) reported biochemical studies carried out on similar cultures and suggested the possible insertion of gangliosides into neuronal membranes. This would lead to structural modifications, influencing enzymatic activities, neurotransmitter transport and some nerve cell mechanisms involving the nuclear machinery.

Presently, using primary cultures from embryonic chick cerebrum as did Dreyfus et al. (1980; 1984), we have found that the ganglioside-induced changes involved formation of mature synapses on day 4-5 in vitro (earlier than expected). Gangliosides accelerated the reaggregation of cells into clumps or clusters. The neurons gave out elaborate processes or bundles of neurites, which interconnected the cells or cell aggregates with each other forming synaptic contacts as depicted electron microscopically. Thus, the present morphological results, combined with the biochemical ones (Massarelli et al., 1982; Dreyfus et al., 1984), add further support to the hypothesis that suggests a role of gangliosides in cellular recognition and in the establishment of membrane contacts. It is not surprising that exogenous gangliosides produce the effects described above as these substances play an important role in the developing nervous system, where their abundance coincides in time and space with the outgrowth of neurites and with the establishment of neuronal connections (Vanier et al., 1971; Willinger and Schachner, 1980; Bass, 1981). Additionally, gangliosides ensure the linking between pre- and postsynaptic membranes as suggested by Dreyfus et al. (1984). The linking role in maintaining the contact between two neuronal membranes can easily be seen using time-lapse cinematography (Spoerri, 1984). It takes more than one ganglioside to attain this effect. The interaction of 3 to 4 different gangliosides are definitely necessary for the events of recognition and linking that lead to synaptogenesis, followed by maturation of the contacts into functional communicating synapses. Moreover, these substances seem to play a role as mediator molecules in neuritogenic-neurotrophic interactions regulating neurite outgrowth and converting the cells from an unstable intermediate state into a stably-differentiated resting state, with previously and presently acquired synapses (Morgan and Seifert, 1979; Spoerri, 1983).

Recently, particular interest has developed with respect to the concept of neuroplasticity (Cotman, 1978) which can be elicited by gangliosides (Rapport and Gorio, 1981). GMl administered in vivo promoted the recovery of CNS functions after lesion and caused regenerative neuronal sprouting in response to injury (Toffano et al., 1983a, b; 1984a, b). According to Leon et al. (1984), the presence of GMl molecules in the neuronal cell surface, either as endogenous or as stably inserted exogenous molecules, are of importance in determining the properties of the plasma membrane, particularly with respect to the capacity of modulating and responding to signals from the extra- to the intracellular spaces. Deletion of the GMl fraction from the culture medium containing GDla, GDlb, GT, led to lack of synaptic elements as presently described. This correlates with other findings (Leon et al., 1982) indicating that GMl induced differentiation is accompanied by an increase in the intracellular cAMP. This is in accord with previous observations that agents inducing morphological differentiation elevate cAMP during the Gl phase of the cell cycle (Waymire et al., 1978). The addition of dibutyryl cAMP to N-2A cultures produces the same morphologic effect as gangliosides (Spero and Roisen, 1984) but no synapses have been mentioned. GMl potentiated the NGF-induced effects in primary cultures of dorsal root ganglia (Levi-Montalcini and Hamburger, 1951; Leon et al., 1984). However, without NGF, GMl was totally ineffective. In cells with NGF requirement GMl cannot substitute for the neuronotrophic agent but can only facilitate the neuronotrophic effect. GMl or gangliosides may also enhance the efficacy of various neurotrophic agents (Spoerri and Roisen, in preparation).

It has been believed that NGF has no effect on nervous tissue after the neonatal period, except for the sympathetic neurons during embryonic and neonatal periods. There are only a few reports on the action of NGF upon the growth or regeneration of adult peripheral sympathetic ganglion cells (Argiro and Johnson, 1982; Uchida and Tomonaga, 1985). We report here that GMl potentiated the NGF-induced effects in young and adult superior cervical ganglia of mice, by augmenting the number of axons per explant and increasing the axon length, when compared to the NGF-effect alone.

In conclusion, the findings considered in the current in vitro investigation provide further information concerning the ganglioside facilitation of neuronal development and differentiation leading to synapse formation and regeneration.

ACKNOWLEDGMENTS

Thanks are due to Drs. Gino Toffano and Alberta Leon for encouragement and advice. I am grateful to Fidia Research Laboratories for providing the gangliosides, and to the German Science Foundation for support.
I also thank Mrs. Sigrid Peilert for meticulous typing of the manuscript.

REFERENCES

Ames A and Hastings AB (1956) J Neurophysiol 19: 201-212.
Argiro V and Johnson MI (1982) J Neurosci 2: 503-512.
Bass NH (1981) in: Rapport MM and Gorio A (eds): Gangliosides in Neurological and Neuromuscular Function, Development and Repair. Raven Press, New York, pp. 29-43.
Burry RW and Lasher SR (1978) Brain Res 147: 1-15.
Cotman CW (1978) Neuronal Plasticity, Raven Press, New York.
Dreyfus H, Louis JC, Harth S, Mandel P (1980) Neurosci 5: 1647-1655.
Dreyfus H, Ferret B, Harth S, Gorio A, Durand M, Freysz L, Massarelli R (1984) J Neurosci Res 12: 311-322.
Eins S, Spoerri PE, Heyder E (1983) Cell Tissue Res 229: 457-460.
Kojima T, Saito K, Kakimi S (1975) An Electron Microscopic Atlas of Neurons. University of Tokyo Press, Tokyo.
Leeden RW (1984) J Neurosci Res 12: 147-159.
Leon A, Facci L, Benvegnu D, Toffano G (1982) Dev Neurosci 5: 108-114.
Leon A, Benvegnu D, Dal Toso R, Presti D, Facci L, Giorgi O, Toffano G (1984) J Neurosci Res 12: 277-287.
Levi-Montalcini R and Hamburger V (1951) J Exp Zool 116: 321-362.
Massarelli R, Wong TH, Harth S, Louis JC, Freysz L, Dreyfus H (1982) Neuroschem Res 7: 301-316.
Morgan JI, Seifert W (1979) J Supramol Struct 10: 111-124.
Pettmann B, Louis JC, Sensenbrenner M (1979) Nature 281: 378-380.
Piper HM, Probst J, Schwartz P, Hunter FJ, Spieckermann PG (1982) J Mol Cell Card 14: 397-412.
Probst W and Rahmann H (1980) J Therm Biol 5: 243-247.
Rapport MM and Gorio A (1981) Gangliosides in Neurological and Neuromuscular Function, Development and Repair. Raven Press, New York.
Spero DA and Roisen FJ (1984) Dev Brain Res 13: 37-48.
Spoerri PE, Dresp W, Heyder E (1980a) Acta anat 107: 221-223.
Spoerri PE, Glees P, Dresp W (1980b) Cell Tissue Res 205: 411-421.
Spoerri PE and Wolff JR (1981) Cell Tissue Res 218: 567-579.
Spoerri PE and Wolff JR (1982) Cell Tissue Res 222: 379-388.
Spoerri PE (1983) Int J Devl Neurosci 6: 383-391.
Spoerri PE (1984) Sektion Medizin, Serie 6, Nummer 15, Film B 1544, Institut für den Wissenschaftlichen Film, Göttingen.
Spoerri PE, Kelley K, Armstrong D, Ellis A (1984) Opthalmic Res 16: 307-314.
Spoerri PE, Ludwig HC, Ogawa Y (1985) Acta anat 123: 64-66.
Svennerholm L (1980) in: Svennerholm L, Mandel P, Dreyfus H, Urban PF (eds): Structures and Functions of Gangliosides. Plenum Press, New York, pp. 533-544.
Toffano G, Savoini G, Moroni F, Lombardi G, Calza L, Agnati LF (1983a) Brain Research 261: 163-166.

Toffano G, Savoini G, Aldino C, Valenti G, Dal Toso R, Leon A, Calza L, Zini I, Agnati LF, Fuxe K (1983b) in: Proceedings of International Society for Neurochemistry. Vancouver, BC, Canada, Plenum Press, New York, pp. 475-488.

Toffano G, Savoini G, Moroni F, Lombardi G, Calza L, Agnati LF (1984a) Brain Res 296: 233-239.

Toffano G, Savoini F, Aporti S, Calzolari S, Consolazione A, Maura G, Marchi M, Raiteri M, Agnati LF (1984b) Neurosci Res 12: 397-408.

Uchida Y and Tomonaga M (1985) Age 8: 19-20.

Vanier MT, Holm M, Ohman R, Svennerholm L (1971) J Neurochem 18: 581-592.

Waymire JC, Gilmer-Waymire K, Kaycock JW (1978) Nature 276; 194-195.

Willinger M and Schachner M (1980) Develop Biol 74: 101-107.

Gangliosides and neuronal plasticity
G. Tettamanti, R.W. Ledeen, K. Sandhoff,
Y. Nagai, G. Toffano (eds.)
Fidia Research Series, vol. 6
Liviana Press, Padova, © 1986

EFFECT OF ADDED GANGLIOSIDES ON THE DEVELOPMENT OF FETAL MOUSE SPINAL CORD-DORSAL ROOT GANGLION EXPLANTS CULTURED IN A CHEMICALLY DEFINED MEDIUM

Robert E. Baker

Netherlands Institute for Brain Research, Meibergdreef 33,
1105AZ Amsterdam, The Netherlands

THE COMPOSITION OF CULTURE MEDIA IN PRIMARY CULTURES OF NEURAL CELLS

The Effect of the Presence of Fetal Calf Serum

The use of organotypic fetal mouse spinal cord-dorsal root ganglion explants (SC-DRG) as a model system for studying the development of selective interneuronal connections under in vitro conditions was pioneered in the early 1960's (Crain, 1973). The medium used was serum-supplemented, however, and often contained embryo extracts. The location of the DRG afferent fiber terminals within the cord explants was determined both electrophysiologically and histologically using horseradish peroxidase (HRP) and was shown to be preferentially located within the dorsal regions of the cord (Crain et al., 1968; Crain and Peterson, 1981; Smalheiser et al., 1981). Moreover, SC-DRG explants grown under these conditions extensively flatten and become virtually a monolayer of cells (Fig. 1A), leading to a general loss of cord cytoarchitecture which frequently makes it difficult to identify the different regions of the explant.

The Use of Chemically Defined Culture Media

Drawbacks in the use of serum-supplemented media have been noted for many years: in short, the composition of the sera vary from batch to batch, making it impossible to know the exact levels of hormones, proteins, lipids, sugars etc. bathing the explants. Recent improvements in the preparation of serum-free media (CDM) (Romijn et al., 1984) led us to compare the selectivity of DRG afferent ingrowth and fiber distribution in cord explants grown in a variety of CDM. Such explants exhibit a greatly enhanced preservation of their cytoarchitecture throughout the culturing

Abbreviations: SC-DRG, spinal cord-dorsal root ganglion; HRP, horseradish peroxidase; CDM, chemically-defined medium; SBA, spontaneous bioelectric activity; HRP,

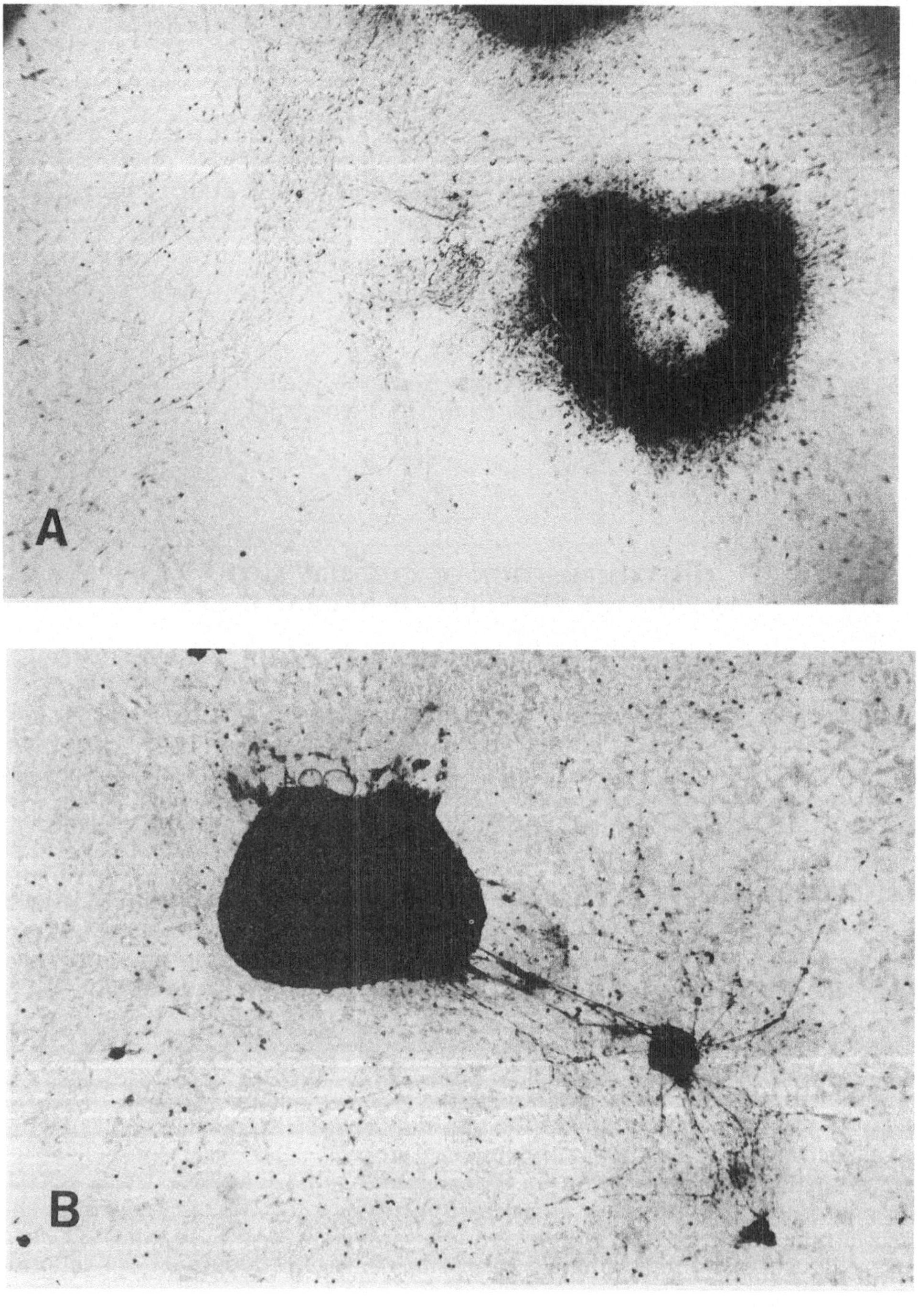

Figure 1. A. SC-DRG explant, 12 days in vitro, grown in a serum-supplemented medium. B. SC-DRG explant, 28 days in vitro, grown in a D(+)galactose-supplemented serum-free medium.

period (up to six weeks, Fig. 1B); see Habets et al., 1981; Baker et al., 1982). Flattening of the explants is minimal and the cords retain their typical cross-sectional appearance. In addition, there appears to be an enhancement of spontaneous bioelectric activity (SBA). In the basal CDM, however, the distribution of DRG afferents is no longer restricted to dorsal cord regions, as had been seen to occur in serum-supplemented medium, but is equally divided between dorsal and ventral halves of the explant (Fig. 2). Sensory fiber pathways entered the cord explant differently in the two media: there was an equal dorsal-ventral distribution of entrances in CDM-grown cords compared to predominantly dorsal entry in serum-grown explants.

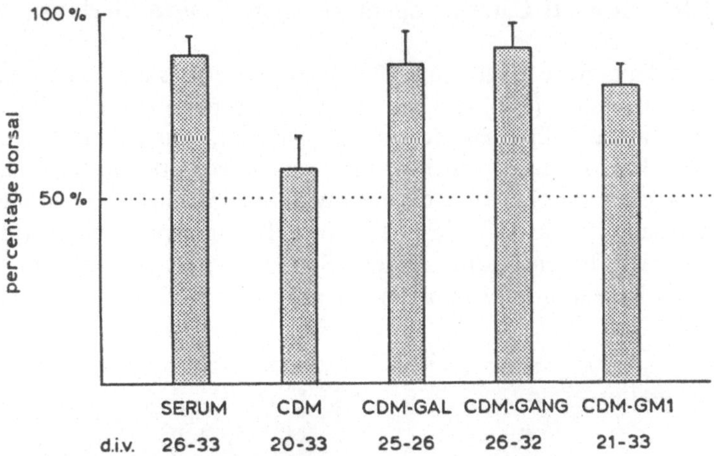

Figure 2. Bar graph representing the dorsal percentages of DRG afferent terminals and/or fibres present in SC explants chronically exposed to various growth media. Serum-grown (n = 8) and all galactose (galactose-l-phosphate, n = 8; D(+) galactose, n = 6) and ganglioside (mixture, n = 18, GM1, n = 18) grown explants evinced dorsal afferent projection patterns which significantly differed both from the 50% level and from control, CDM-grown explants (n = 20) ($p < 05$, Student's t-test, one-tailed).

Effect of the Addition of Membrane Costituents, particularly Gangliosides

It was therefore assumed that some serum-borne factor(s) must be responsible for the development of selective DRG connections within dorsal cord. A variety of membrane components are thought to be responsible for the development of cell-cell adhesion and/or recognition. Many of these components have oligosaccharide chains extending into the extracellular spaces, which are thought to be the active sites for intercellular recognition and adhesion. Based on these observations it was decided to examine several membrane constituents which could be involved in cell-cell recognition/adhesion. One group of glycolipids in particular, the gangliosides, have been implicated in several cell-cell selectivity studies (Marchase, 1977; Obata et al., 1977), and have also been shown to exert considerable growth-promoting effects on such tissues as DRG cells (Roisen et al., 1981; Hauw et al., 1981; Gorio et al., 1983). Since these compounds contain considerable amounts of galactose in their oligosaccharide chains, and inasmuch as galactose is not present in our CDM, the effects of exogenously added galactose on both sensory afferent projection patterns and on central cord excitability were also studied.

FUNCTIONAL BEHAVIOUR OF SC-DRG EXPLANTS IN CHEMICALLY DEFINED MEDIA CONTAINING GALACTOSE AND GALACTOSIDES

Selection of DRG Explants Carrying Spontaneous or Evoked Bioelectric Activity

SC-DRG explants were taken from 13 day mouse fetuses and plated on collagen-coated dishes or coverslips (Fig. 3). Care was taken to plate and examine only those explants which retained their cross-sectional appearance, thus eliminating any doubts about which was the dorsal and which was the ventral cord region (Fig. 1B). The explants were cultured in a serum-free, chemically defined medium containing either (a) 10 μg/ml galactose (either as D(+) galactose or galactose-l-phosphate), (b) 50 μg/ml mixture of purified bovine gangliosides (Sigma) or (c) 10^{-5}M purified GM1 ganglioside (Fidia Research Laboratories). The explants were grown and maintained

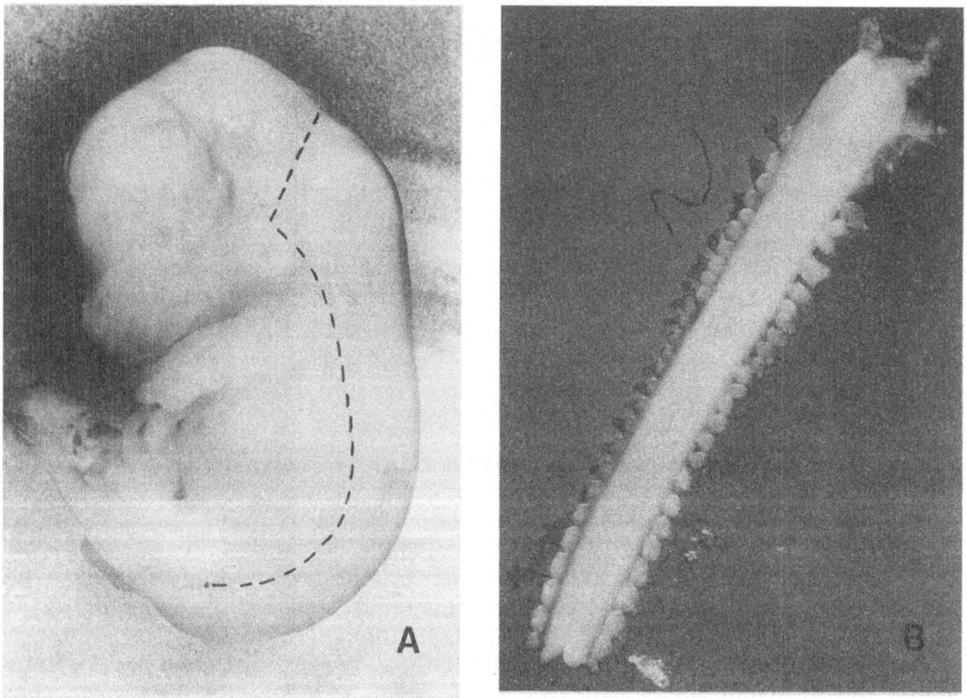

Figure 3. A. 13 days mouse fetus. A block of tissue containing the developing spinal cord and dorsal root ganglia was cut free from the fetus (dashed lines) and the cord with attached DRG removed (B). The SC-DRG tissues were then cut into cross-sections of approximately 0.5 to 1 mm thicknesses and plated.

in their respective media for 21-34 days at 36°C with constant gassing by 95% air and 5% CO_2. Medium changes were carried out once a week.

Cultures were selected for examination on the basis of their having a DRG which had migrated some distance away from the cord explant and which was connected to the cord by a distinct fiber bundle bridge. Explants were transferred to a temperature-controlled, vibration-damped table and perfused with Eagle's minimum essential medium under constant gassing with 95% air and 5% CO_2. Spontaneous and DRG-evoked bioelectric activity within the cord was recorded from 16 evenly spaced points across the surface of the explant (Habets et al., 1981; Baker et al., 1982). For each culture the number of active points and the percentage of these points located dorsally (D%) were calculated for both spontaneous discharges and DRG-evoked activities. The points of entry of the DRG fiber bundles into the cord were scored as being predominantly dorsal, ventral or indeterminate. Statistical evaluation of the data was carried out using Student's t-test.

Cultures were examined histologically using a HRP staining technique (Udin and Fisher, 1983; Baker, 1985). The selected DRG was iontophoresed with the HRP mixture for 30 min. Following this the explants were maintained under standard culturing conditions for 1-4 hours before fixation and staining.

Effect of Galactose, Galactose-1-P and Gangliosides on Functional DRG Terminals in Vitro

The mean number of spontaneously active points did not differ in control, galactose-l-phosphate and purified bovine ganglioside cultures (Table 1). There was, however, a slight but significant reduction in the number of SBA points measured in

Table 1. *Number and Distribution of Spontaneously Active and DRG-Evoked Points in SC-DRG Explants*

Group	n	Spontaneous Potentials	% D	Evoked Potentials	% D (+)
Control	20	12.2±0.8	47±3	10.9±1.1	46±4
D (+) gal.	6	9.0±1.7**	57±6	9.3±1.7	58±6
Gal-1-P.	8	14.4±0.5	52±2	10.6±1.3	51±2
Gang. Mix.	18	10.8±0.9	51±3	8.3±0.7	49±3
GM1	18	9.7±0.6**	55±2	5.3±0.8*	47±9

** $p<.0025$; * $p<0.01$, Student's t-test, one-tailed.
(+) % D, percentage of active points located dorsally. See also text.

D (+)galactose-grown explants as compared with those cords cultured in galactose-l-phosphate. Spontaneously active points recorded in GM1-grown explants showed no differences from control, D(+) galactose, or mixed ganglioside-grown cord values, but were also significantly different from the galactose-l-phosphate values. In other words, galactose-l-phosphate seems to enhance SBA. The distribution of the SBA points in all groups was evenly divided between dorsal and ventral halves of the cord explant.

The mean number of recorded DRG-evoked responses per explant showed little variation in all culture groups except for the GM1-grown explants, which had significantly fewer areas of the cord from which evoked activity could be elicited following DRG stimulation (Table 1; evoked activity refers to the spike barrages occurring within the cord explant following stimulation of the DRG). If the spikes showed a latency variation of less than 0.5 msec they were regarded as coming from an afferent fiber termination site in the vicinity of the recording electrode (Baker et al., 1982). The distribution of all these points, as in the other culture groups, consistently was divided between the dorsal and ventral halves of the cord explant.

The distribution of functional DRG terminals (as determined electrophysiologically) and/or visible fibers (as determined histologically) within the cord reflected the type of medium in which the explants were grown. Terminals and fibers in control cultures (i.e., CDM with any additives) were never seen to deviate significantly from a 1:1 dorso-ventral distribution pattern (Fig. 2). Explants exposed to the two galactose sugars and the two ganglioside-supplemented CDM, in contrast, showed a clearcut increase in dorsal innervation preferences on the part of the DRG afferents. The dorsal preponderance of fibers observed in these experimental cultures was equal to the values measured in serum-grown explants, and differed significantly both from a 50:50 distribution and from control values.

CONCLUSIONS

The results of the various experiments reported above show conclusively that an in vitro model system, i.e., the fetal mouse organotypic SC-DRG explant, can be utilized effectively in the study of factors underlying the development of selective interneuronal connections. Explants grown in a serum-free medium devoid of galactose or gangliosides show no DRG sensory afferent preference for dorsal cord regions. The addition of galactose sugars or various gangliosides to the serum-free medium restores these dorsal cord preferences.

It is not known how either galactose alone or gangliosides, when added to CDM, exert their actions on or within SC-DRG explants so as to restore the dorsal cord preferences to the level observed in serum-grown cultures. One possibility would be that the restoration of selectivity patterns by the sensory afferents is a membrane-associated phenomenon, conceivably associated with the demonstrated effects of gangliosides on enzymatic and ionic processes. These, in turn, might effect the complexity of macromolecular membrane constituents which have been implicated in the maturation of neural tissues. Alternatively, such additives could be internalized and enter into metabolic pathways associated with production, transportation and insertion into the membranes of the various substances that underly this selectivity. Whatever the underlying basis, the end result of exposing the explants to additives is consistent: a restoration of dorsal cord preferences by the DRG afferents.

REFERENCES

Baker RE (1985) Horseradish peroxidase tracing of dorsal root ganglion afferents within fetal mouse spinal cord explants chronically exposed to tetrodotoxin. Brain Res 334: 357-360.

Baker RE, Habets AMMC, Brenner E, Corner MA (1982) Influence of growth medium, age in vitro and spontaneous bioelectric activity on the distribution of sensory ganglion-evoked activity in spinal cord explants. Dev Brain Res 5: 329-341.

Crain SM (1973) Microelectrode recording in brain tissue cultures. In: RF Thompson and MM Patterson. (eds): Academic Press, New York, pp. 39-75.

Crain SM, Bornstein MB, Peterson ER (1968) Maturation of cultured embryonic CNS tissues during chronic exposure to agents which prevent bioelectric activity. Brain Res 8: 363-372.

Crain SM, Peterson ER (1981) Selective innervation of target regions within fetal mouse spinal cord and medulla explants by isolated dorsal root ganglia in organotypic co-cultures. Dev Brain Res 254: 341-362.

Gorio A, Marini P, Zanoni R (1983) Muscle innervation. III. Motoneuron sprouting capacity, enhancement by exogenous gangliosides. Neurosci 8: 417-429.

Habets AMMC, Baker RE, Brenner E, Romijn HJ (1981) Chemically defined medium enhances bioelectric activity in mouse spinal cord dorsal root ganglion cultures. Neurosci Let 22: 51-56.

Hauw JJ, Fenelon S, Boutry JM, Escourolle R (1981) Effects des gangliosides sur la croissance de ganglions spinaux de cobaye en culture in vitro. Resultats preliminaires concernant une preparation de gangliosides de cortex cerebral de boeuf. CR Acad Sc Paris 292: 569-571.

Marchase RB (1977) Biochemical investigation of retinotectal adhesive specificity. J Cell Biol 75: 237-257.

Obata K, Oide M, Handa S (1977) Effects of glycolipids on in vitro development of neuromuscular junction. Nature 266: 369-371.

Roisen FJ, Barfeld H, Nagele R, Yorke G (1981) Ganglioside stimulation of axonal sprouting in vitro. Science 214: 577-578.

Romijn HF, Huizen F, Wolters PS (1984) Towards an improved serum-free, chemically defined medium for long-term culturing of cerebral cortex tissue, Neurosci Biobehav Rev 8: 301-334.

Smalheiser NR, Peterson ER, Crain SM (1981) Specific neuritic pathways and arborizations formed by fetal mouse dorsal root ganglion cells within organized spinal cord explants in culture: a peroxidase-labelling study. Dev Brain Res 1: 383-396.

Udin SB, Fisher MD (1983) Visualization of HRP-filled axons in unsectioned, flattened optic tecta of frogs. J Neurosci Meth 9: 283-285.

Gangliosides and neuronal plasticity
G. Tettamanti, R.W. Ledeen, K. Sandhoff,
Y. Nagai, G. Toffano (eds.)
Fidia Research Series, vol. 6
Liviana Press, Padova, © 1986

MOLECULAR SPECIFICITY OF GANGLIOSIDE ACTION ON NEURITE REGENERATION IN CELL CULTURES OF SENSORY NEURONS

Patrick Doherty, John G. Dickson, Thomas P. Flanigan, Frank S. Walsh

Institute of Neurology, Queen Square, London, United Kingdom

INTRODUCTION

Studies of the antigenic sites recognised by the monoclonal antibody (McAb) A2B5 (Eisenbarth et al., 1979) and on the binding receptors for tetanus toxin (Van Heyningen, 1963; Mirsky et al., 1978) have demonstrated that neurons can express a relatively unique family of higher order gangliosides at their cell surface. The high abundance of gangliosides in nervous tissue (Wiegandt, 1971; Ledeen, 1978) considered together with the qualitative and quantitative changes apparent in the total ganglioside pool during the development of several brain areas (Kasai and Yu, 1983; Hilbig et al., 1984) lends support to the concept that endogenous gangliosides may somehow function in the process of neuritogenesis in the embryo and/or neurite regeneration in the adult. Supportive evidence for such a role comes from the observation that a complex mixture of bovine brain gangliosides can stimulate motor neurone sprouting in vivo (Gorio et al., 1983).

Despite the widespread observation that both a mixture of bovine brain gangliosides as well as the purified monosialoganglioside GM1 can stimulate neuronal sprouting in vitro (Roisen et al., 1981; Toffano et al., 1983; Rybak et al., 1983; Facci et al., 1984) the molecular specificity and mechanisms by which gangliosides exert their effects on neurite regeneration remain unknown.

To date a considerable number of in vitro studies have been concerned with the effect of exogenous gangliosides on the process of neuroblastoma cell differentiation. A difficulty in interpreting the mode of action of gangliosides in such studies is that

---

Abbreviations: NGF, nerve growth factor; DRG, dorsal root ganglia; McAb, monoclonal antibody; HRP-RAM, horse radish peroxidase-conjugated rabbit antimouse Ig; DMEM, Dulbecco's modified Eagle's medium; FCS, fetal calf serum; PBS, phosphate-buffered saline; HS, horse serum; OPD, orthophenylenediamine.

there are several distinct mechanisms, including inhibition of cell division, that can promote both the biochemical and morphological differentiation of neuroblastoma cells. Primary cell cultures of nervous tissue remain the most relevant in vitro model system for the study of the effect of gangliosides on neuronal cell survival and growth. However, in general, several factors including the unequivocal identification of specific cell types together with the density and complexity of neuritic outgrowth have tended to limit such studies to low density enriched neuronal cultures grown over relatively short periods of time.

With this in mind, we have recently developed a novel bioassay that provides an arbitrary measurement of the degree of neuronal cell survival and neuritic outgrowth in primary cell cultures. Significant increases in the level of neurofilament protein induced by NGF in primary cultures of both human and chick dorsal root ganglia (DRG) (Doherty et al., 1984a, b) or alternatively by a human muscle derived trophic factor in primary cultures of spinal cord (Doherty et al., 1985) have been quantified by an enzyme-linked immunoadsorbent assay (ELISA) utilising McAb reagents specifically reactive with neurofilament protein. The specific nature of this index has allowed us to embark on a series of studies on the effects of exogenous gangliosides on neurofilament protein expression in vitro. In the present study as well as reporting on the interaction of the monosialoganglioside GM1 with NGF, we present evidence that higher order gangliosides are highly effective stimulators of neurite regeneration.

MATERIALS AND METHODS

Antibody Reagents

The neurofilament McAb RT97 was obtained as an ascites fluid (Anderton et al., 1982) and was used at 1:500 - 1:750 dilution. Horseradish peroxidase-conjugated rabbit anti-mouse Ig (HRP-RAM) was obtained from Miles Chemical Company. Anti-serum to the 2.5S active component of NGF (Collaborative Res Inc, MA, USA) was purchased as a lyophilised whole sera preparation and was reconstituted in Dulbecco's modified Eagle's medium (DMEM) supplemented with 10% (v/v) heat inactivated foetal calf serum (FCS). All antibody dilutions for the enzyme-linked immunoadsorbent assay (ELISA) were into phosphate-buffered saline (PBS) supplemented with 10% foetal calf serum (FCS).

Cell Culture of Chick DRG

DRG obtained from 7 or 9-day chick embryos were dissected free of meningeal tissue and dissociated as previously described (Doherty et al., 1984b, 1985). Cell cultures were initiated by seeding 100 μl of the dissociated cell suspension $(0.5 \times 10^5$ cells/ml) into individual wells of a 96-well collagen coated microtitre plate. All cultures were maintained at 37°C under a water-saturated atmosphere of 90% air - 10% CO_2. Cytosine arabinoside (10^{-5}M) was routinely added on day two of culture to suppress the growth of non-neuronal cells. Culture media was DMEM supplemented with 10% FCS or 10% horse-serum (HS).

Ganglioside Treatment of DRG Cell Cultures

Highly purified preparations of individual gangliosides (prepared according to Tettamanti et al., 1973) were supplied by Dr A. Leon and Dr G. Toffano (Fidia Res. Labs. Abano Terme, Italy). Medium containing gangliosides was prepared as follows: individual gangliosides were taken to dryness under a stream of nitrogen from a chloroform-methanol (2:1, v/v) solution and directly dissolved in DMEM. This solution was sterilized by passage through a 0.2 μm filter and either added directly to the microtitre wells prior to cell seeding, or alternatively introduced to established neuronal cultures that had been grown for two days in the presence of 5 ng/ml NGF. Prior to the introduction of gangliosides, these latter cultures were washed two times with DMEM supplemented with 10% FCS.

Quantitication of Neurofilament Protein in Cell Cultures

The ELISA assay for neurofilament protein has been described in detail elsewhere (Doherty et al., 1984a, b). Briefly following fixation in 4% (w/v) paraformaldehyde at 4°C, cultures were permeabilized in methanol (—20°C, 15 min) washed with PBS and incubated sequentially (60 min at room temperature) with (1) PBS containing 10% FCS (2) RT97 ascites diluted 1:500 or 1:750 and (3) a 1:2000 dilution of horseradish peroxidase-conjugated rabbit anti-mouse Ig. Cultures were then washed three times with PBS, twice with distilled H_2O and finally incubated with 50μl of 0.2% (w/v) o-phenylenediamine (OPD), 0.02% (v/v) H_2O_2 in 50 mM citrate buffer. After 30 min the conversion of OPD to its oxidised product was stopped by the addition of 50μl of 4.5 M H_2SO_4. Aliquots of the reaction product were then transferred to a fresh microtitre plate and optical density at 492 nm determined using a Flow Titretek Multiscan apparatus. Appropriate control incubation mixtures were used as reference samples.

Indirect Immunofluorescence

Indirect immunofluorescence staining was carried out as previously described (Dickson et al., 1982; Doherty et al., 1984a, b) using McAb RT97 to identify and localise neurofilaments in cultures. Cultures grown in 35-mm tissue culture dishes were processed identically as for neurofilament assay up to and including incubation with a 1:200 dilution of McAb RT97 ascitic fluid. Cultures were then washed three times with PBS containing 10% FCS and incubated with rhodamine-conjugated sheep anti-mouse Ig for 60 min at room temperature. After further washing, coverslips were mounted on the cultures in 50% (vol/vol) glycerol in PBS and cells viewed on a Leitz Dialux microscope.

RESULTS

Ganglioside GM1 Has no Intrinsic Trophic Activity, but Can Act synergistically with NGF

Primary cultures of chick DRG cells developed over 5 days in vitro to yield a sparse layer of non-neuronal cells over which, depending on the presence of NGF, individual or small groups of sensory neurons with rounded, phase-bright perikarya were also present. As shown in Figure 1, there was no obvious difference in the non-neuronal cell population in cultures treated with NGF (5ng/ml), GM1 (100 μg/ml) or NGF plus GM1. There was very little neuronal cell survival in both control and ganglioside treated cultures. However, there was a marked increase in neuronal cell survival and differentiation in the NGF treated cultures with a further increase in the density and complexity of the neuritic network apparent in cultures simultaneously treated with NGF and GM1.

Using the ELISA assay for neurofilament protein, significant increases in RT97 binding were measured at day five in vitro to cultures grown in the continuous presence of 0.2 ng/ml NGF, with half-maximal and maximal responses apparent at about 1 and 5 ng/ml (Fig. 2a). In contrast, as predicted from a morphological examination of treated cultures, GM1 (100 μg/ml) had no significant effect on RT97 binding to cultures grown in the absence of NGF. However, there was a significantly greater response to NGF (5ng/ml) in cultures simultaneously treated with GM1 (100 μg/ml) as compared to untreated controls (Fig. 2b). In all cultures, McAb RT97 binding can be localised to the neuritic network (Fig. 1 a-c), supporting our previous conclusion that changes in the level of its binding reflect changes in the relative expression of neurofilament protein intimately related to the degree of neuronal cell survival and/or neurite regeneration in the cultures.

Separation of GM1 and NGF Stimulation of Neurofilament Protein Expression

The addition of GM1 (0-200 μg/ml) at 48 hr. in vitro to cultures exposed to NGF only for this initial period (see Materials and Methods) resulted in a marked dose-dependant increase in the level of RT97 binding to the cultures at day five in vitro (Fig. 3a). Significant effects were found at 2 μg/ml, the lowest GM1 concentration used and reached a peak at 100 μg/ml GM1. The increased binding of McAb RT97 was again associated with an increase in the density and complexity of neurite outgrowth in the ganglioside treated cultures (data not shown). In contrast, the re-addition of NGF (5ng/ml) or an anti NGF serum (1:1000) over the day two-five period of culture did not significantly influence neurofilament protein expression (Fig. 3b). Thus, over the day two-five period of culture, GM1 can modulate the survival and/or neurite regenerative capacity of the population of NGF-dependant neurons in the apparent absence of the trophic factor.

The GM1 Response is not Inhibited by an Anti-NGF Serum

The data outlined above suggests that neuronal survival and growth over the day 2 - day 5 period of culture is independent of both endogenous and exogenous NGF. It would therefore appear unlikely that GM1, when added to cultures at day 2, stimulates neurofilament protein expression by directly interacting with NGF. To further test this contention sensory neuronal cultures, again grown at a maximal NGF concentration for two days, were washed and GM1 (100 μg/ml) added in the presence and absence of the anti-NGF serum (1:1000 dilution). Figure 4 shows that the GM1-induced

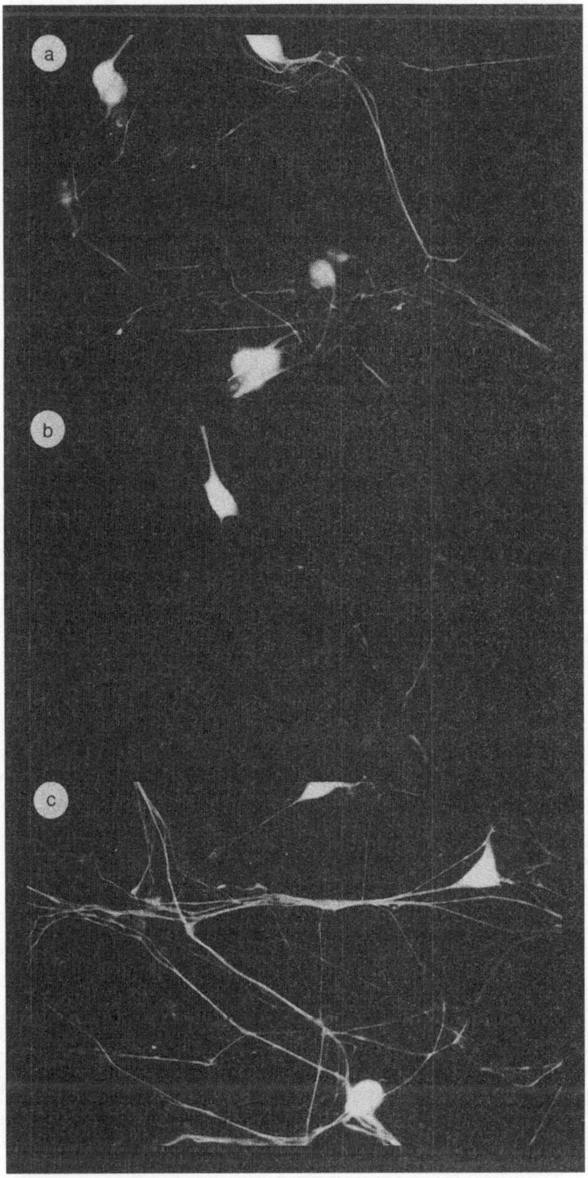

Figure 1. Immunofluorescence analysis of RT97 binding in NGF treated (a) GM1 treated (b) and NGF/GM1 treated (c) cultures of 9-day embryonic chick DRG cells. Cultures were fixed at day 5 in vitro, permeabilised, and incubated sequentially with RT97 (1:200 dilution) followed by rhodamine-conjugated sheep anti-mouse Ig. Cells were viewed with epifluorescence optics. Scale bars represent 50 μm.

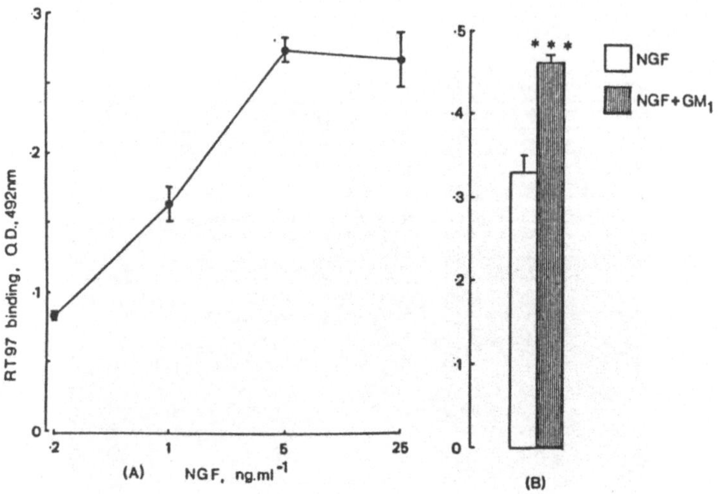

Figure 2. A Binding of RT97 to DRG cultures grown for 5 days in the presence of varying concentrations (0-25 ng/ml) of NGF. The results are expressed as the absolute increase in O.D. measured in cultures grown in the continuous presence of NGF as compared with untreated controls, with each value representing the mean ± S.E. of 6 independent determinations with the conjugate control value subtracted. Conjugate control values were determined by omitting RT97 from the reaction sequence and were taken as the mean of two determinations for each point. The basal binding of RT97 to cultures grown in the absence of NGF was 0.18 ± 0.004 O.D. units (mean ± S.E., n=6).
B. Effect of NGF on the binding of RT97 to DRG cultures grown in the presence and absence of GM1 (100μg/ml). The results show the absolute increase in O.D. over the basal value measured for control cultures grown in the absence of NGF. The basal binding of RT97 to control cultures was 0.35 ± 0.005 and 0.30 ± 0.14 O.D. units for cultures grown in the absence and presence of GM1 (100 μg/ml) respectively. Each value is the mean + S.E. of six independent determinations. *** $P < 0.01$.

increase in neurofilament protein expression is not inhibited by the anti-serum to NGF. We have previously shown that the anti-NGF serum can fully block the biological activity of a maximally active concentration of NGF (Doherty et al., 1984b).

The Molecular Specificity of Ganglioside Action over the Day Two - Five Period of Culture

Figure 5 shows that when added at a level of 100 μg/ml on day two of culture, GM1 is not unique in its ability to promote increases in neurofilament protein expression over the subsequent three days of culture. A similar effect was found in cultures treated with GM2 and GM3. Small, but consistent effects were also apparent with GD_3, GDla and GDlb. However, the most striking effects were observed with the higher order gangliosides GQlb and GTlb. A morphological examination of the cultures was consistent with the higher order gangliosides having a considerably greater effect on the capacity of the NGF-dependant population of sensory neurons to survive and regenerate in vitro over the day two - five period of culture. An increased number of surviving neurons, a marked hyperplasia of neuronal cell bodies and a more extensive and dense network of neuritic outgrowth were clearly apparent in the GQlb and GTlb treated cultures (Fig. 6). As with GM1, the above responses were not inhibited by the anti-NGF sera (data not shown).

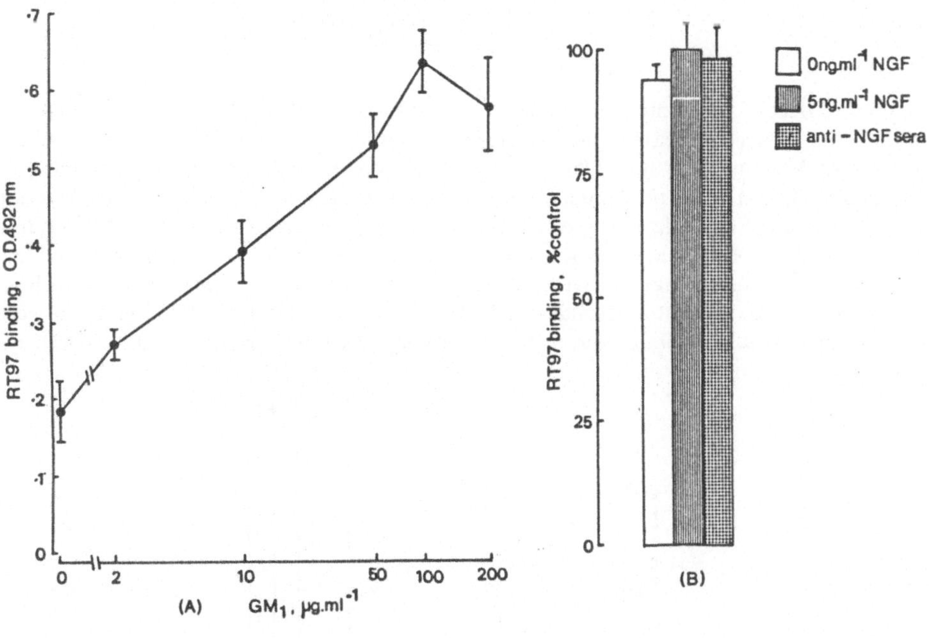

Figure 3. A Effect of GM1, added at day 2 of culture on the binding of RT97 to cultures fixed at day five. Cultures established in the presence of 5ng/ml NGF for 48 hrs in vitro, were washed with DMEM and fresh media containing 0-200 μg/ml GM1 added. The results show the absolute increase in RT97 binding measured over the day 2 - day 5 period of culture. Each value is the mean ± S.E. of six independent determinations. The binding of RT97 at day 2 of culture was measured as 0.06 ± 0.003 O.D. units (mean ± S.E. n=6).

B. Effect of NGF and an anti-NGF sera on the binding of RT97 over the day 2 - day 5 period of culture. Cultures established in the presence of 5ng/ml NGF for 48 hrs in vitro were washed and fresh media containing Ong/ml NGF (□), 5ng/ml NGF (▩) or an anti-NGF sera (▦) added. Cultures were fixed at day 5 and RT97 binding determined. The results show the binding of RT97 as a percentage of that measured in cultures grown in the continuous presence of 5ng/ml NGF. This latter value was 0.51 ± 0.07 (mean ± S.D. n=6). Results show the mean + 1 S.E. of six independent determinations with conjugate control value subtracted.

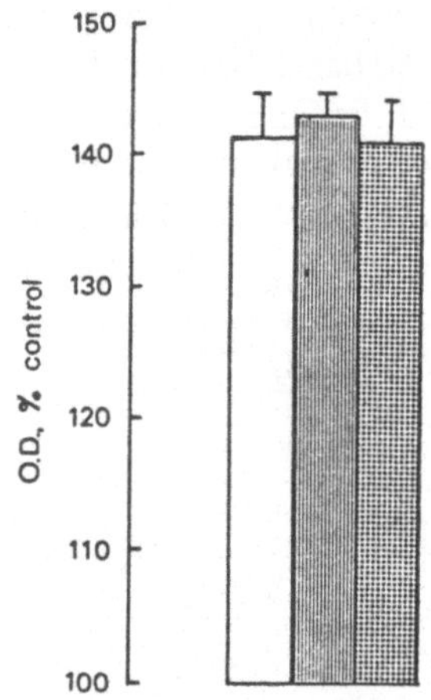

Figure 4. Effect of pre-immune rabbit sera (1:100 dilution) and anti-NGF sera (1:1000 dilution) on the GM1 induced increase in neurofilament protein expression. Cultures, established in the presence of 5 ng/ml NGF for 48 hrs in vitro, were washed and fresh media containing GM1 (100μg/ml, □), GM1 plus pre-immune sera (▩), or alternatively GM1 plus anti-NGF sera (▩) added. After a further 3 days of growth, cultures were fixed and assayed for RT97 binding as described in Methods. The results show the relative increase in RT97 binding over the day 2 - day 5 period as a percentage of that measured in cultures grown in control media. Each value is the mean + S.E. of eight independent determinations. The binding of RT97 at day of culture was measured as 0.24 ± 0.013 and at day 5 for control cultures as 1.09 ± 0.017 O.D. units.

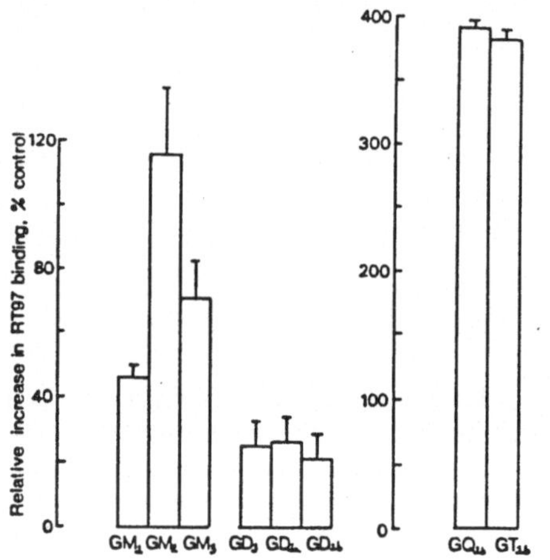

Figure 5. Effect of various gangliosides, added at day 2 of culture, on RT97 binding measured at day 5 of culture. Cultures, established in the presence of 5ng/ml NGF for 48 hrs in vitro, were washed and fresh media supplemented with 100μg/ml ganglioside added. After a further 3 days of growth cultures were fixed and assayed for RT97 binding. The results show the relative increase in RT97 binding over the day 2 - day 5 period as the percentage increase over that measured in cultures grown in control media. Each value is the mean + S.E. of six independent determinations. The binding of RT97 at day 2 was measured at 0.36 ± 0.004 (8) O.D. units and at day 5 to control cultures as 0.60 ± 0.01 (16) O.D. units.

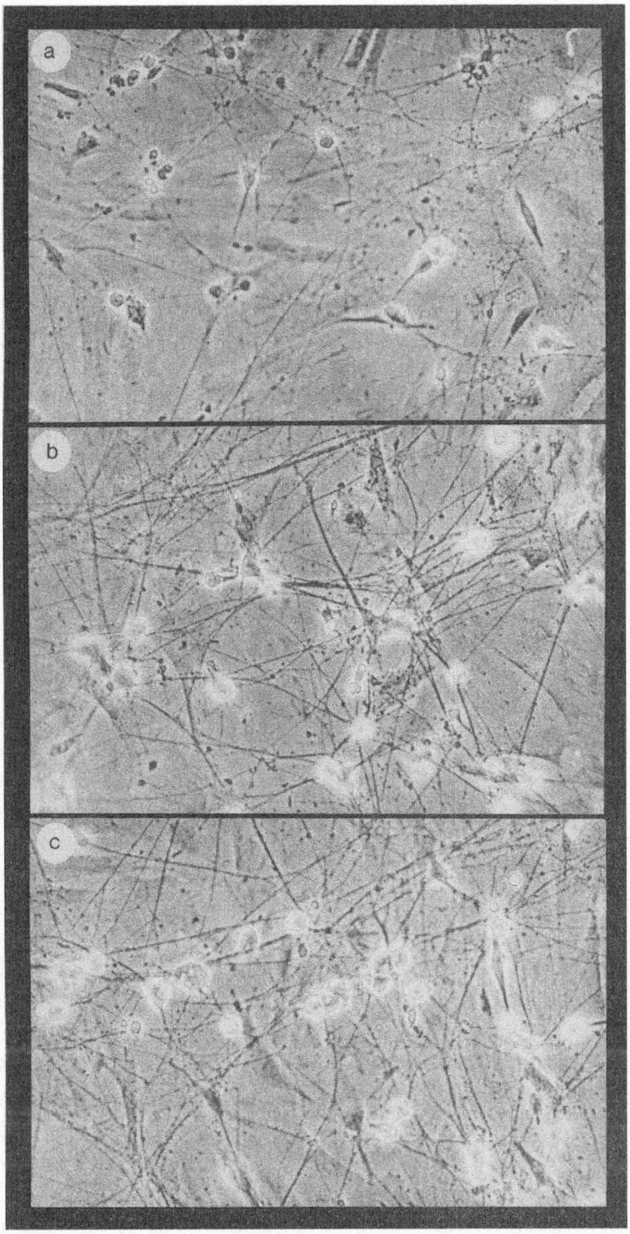

Figure 6. Morphological examination of neuronal survival and neurite regeneration in control (a), GTlb (b) and GQlb (c) treated cultures. Cultures grown as described in Figure 5 were photographed using phase contrast optics following the ELISA assay procedure.

DISCUSSION

Considerable evidence, obtained using a wide variety of experimental and analytical techniques, supports the general hypothesis that gangliosides may play a functional role in the development of the nervous system. The specific nature of the role for gangliosides, whether acting as receptor/recognition molecules or alternatively acting as modulators of ligand/receptor activity, lacks consensus. Direct evidence that even the monosialoganglioside GM1 can function in both roles comes from studies on the binding properties of cholera toxin (Van Heyningen et al., 1971) and from growth factor induced mitosis of fibroblasts (Bremer et al., 1984). The observation that exogenous gangliosides can become stably associated with neuronal membranes (Wiegandt, 1976; Toffano et al., 1980) and that it is cell associated ganglioside that promotes neuritogenesis in mouse neuroblastoma cells (Facci et al., 1984) is consistent with the hypothesis that they may facilitate differentiation by acting as receptors and/or modulating receptor ligand activity for differentiation factors present in the serum (Roisen et al., 1981; Facci et al., 1984). Alternatively it has been suggested that GM1 may function as an acceptor molecule for NGF (Schwartz and Spirman, 1981) or may directly provide a local stimulus that determines the site of formation of new neuronal branches (Ferrari et al., 1983).

In the present study we have adopted a direct experimental approach to determine if a variety of highly purified gangliosides can modulate the survival and neurite regeneration of embryonic chick DRG neurons. Barde et al., (1980) have documented the changing requirement of this population of neurons for survival factors when grown in primary culture over a 48 hr period in vitro. At its maximum efficacy, and under standard conditions of tissue culture, NGF will support the survival of approximately 40% of the neuronal population. In the absence of NGF essentially no neurons survive a two day period of culture. In the present study we have shown that over a five day period in vitro, 5ng/ml NGF will provide a maximal survival and differentiation of sensory neurons as assessed by the ELISA assay and confirmed by a morphological examination of treated cultures. In contrast, our results have shown that over the same time period GM1 does not in itself possess intrinsic trophic activity. However, the addition of GM1 together with NGF was associated with a ganglioside-induced increase in the expression of neurofilament protein.

We have also shown that providing NGF is available for the initial 48 hrs of culture, it can be withdrawn with no detrimental effect on neurofilament protein expression at up to day five of culture. Furthermore, the addition of an anti-NGF sera, or alternatively the re-addition of NGF were without effect. In contrast, when added at day 2 of culture, GM1 greatly increased the expression of neurofilament protein. Thus, over the day two-five period of culture, GM1 can stimulate the survival and/or regenerative capacity of an NGF-dependant population of sensory neurons independently of the simultaneous presence of NGF in the culture media. The possibility that GM1 interacts with either residual or endogenous NGF is unlikely as the response was not affected by an anti-NGF sera. These data would seem to exclude the possibility that the GM1 response is mediated via an increased binding of NGF to the neuronal cells, a model for ganglioside action that has been proposed by Schwartz and Spirman (1981). In this context it is noteworthy that Ferrari et al. (1983) have also reported that

GM1 enhancement of NGF-induced differentiation of PC12 pheochromocytoma cells does not involve changes in the level of specific binding of NGF to the monolayer cultures.

A perhaps not surprising observation, in view of their increased abundance and regulated expression pattern in developing nervous tissue, is that the higher order gangliosides GQlb and GTlb were found to exhibit much more pronounced effects on the survival and regenerative capacity of the sensory neurons over the day 2 - 5 period of culture. The possibility that the activity in lower order ganglioside preparations reflects minor contamination with, for example, GQlb is unlikely as the effective molar concentration range is similar for a variety of gangliosides with only the magnitude of the cellular response being different (Doherty et al., unpublished observation). Alternatively, the greater efficacy of the higher order gangliosides may reflect a functional role for their endogenous counterparts, a role that can perhaps only be poorly mimicked by lower order gangliosides.

The exact nature of exogenous ganglioside action remains to be determined. However, the results of this study clearly suggest that their effects on DRG neurons do not reflect a direct interaction with the physiological trophic factor for that neuronal population. The possibility that gangliosides act as receptor molecules or modulators of receptor activity for other differentiation factors present in serum, or alternatively act by sequestering inhibitory factors from serum (Skaper and Varon, 1985) remains to be determined. However, based on the observation that GM1 does not in itself possess intrinsic trophic activity, but can increase the capacity of the axotomised neurone to regenerate in response to a well-characterized neuronotrophic signal, we would suggest that its most likely site of action is at the level of signal execution rather than signal reception.

REFERENCES

Anderton BH, Breinburg D, Downes MJ, Green PJ, Tomlison BE, Ulrich J, Wood JN, Kahn J (1982) Nature 298: 84-86.

Barde Y-A, Edgar D, Thoenen H (1980) Proc Natl Acad Sci USA 77: 1199-1203.

Bremer EG, Hakomori S, Bowen-Pope DF, Raines E, Ross R (1984) J Biol Chem 259: 6818-6825.

Dickson JB, Flanigan TP, Walsh FS (1982) in: Rowland LP (ed): Human Motor Neuron Diseases. Raven Press, New York, pp. 435-451.

Doherty P, Dickson JG, Flanigan TP, Walsh FS (1984a) J Neurochem 42: 1116-1122.

Doherty P, Dickson JG, Flanigan TP, Walsh FS (1984b) Neurosci Lett 51: 55-60.

Doherty P, Dickson JG, Flanigan TP, Walsh FS (1985) J Physiol 360: 43.P.

Eisenbarth GS, Walsh FS, Nirenberg M (1979) Proc Natl Acad Sci USA 76: 4913-4917.

Facci L, Leon A, Toffano G, Sonnino S, Ghidoni R, Tettamanti G (1984) J Neurochem 42: 299-305.

Ferrari G, Fabris M, Gorio A (1983) Dev Brain Res 8: 215-221.

Gorio A, Marini P, Zanoni R (1983) Neuroscience 8: 417-429.

Van Heynigen WE (1963) J Gen Microbiol 31: 375-387.

Van Heyningen WE, Carpenter CCJ, Pierce NF, Greenough WB (1971) J Infect Dis 124: 415-418.

Hilbig, Lanke G, Rahman H (1984) Dev Neurosci 6: 260-270.
Kasai N and Yu RK (1983) Brain Res 277: 155-158.
Ledeen RW (1978) J Supramol Struct 8: 1-17.
Mirsky R, Wendon LMB, Black P, Stolkin C, Bray D (1978) Brain Res 148: 251-259.
Roisen FJ, Bartfeld H, Nagele L, Yorke G (1981) Science 214: 577-578.
Rybak S, Ginzburg I, Yavin E (1983) Biochem Biophys Res Commun 116: 974-980.
Schwartz M and Spirman N (1982) Proc Natl Acad Sci USA 79: 6080-6083.
Skaper, SD and Varon S (1985) Int J Dev Neurosci 3: 187-198.
Tettamanti G, Bonali F, Marchesini S, Zambotti V (1973) Biochim Biophys Acta 296: 160-170.
Toffano G, Benvegnù D, Bonetti AC, Facci L, Leon A, Orlando P, Ghidoni R, Tettamanti G (1980) J Neurochem 35: 861-866.
Toffano G, Savoini G, Moroni F, Lombardi G, Calza L, Agnati LF (1983) Brain Res 261: 163-168.
Wiegandt H (1971) Adv Lipid Res 9: 249-289.
Wiegandt H (1976) Adv Exp Med Biol 71: 3-14.

Gangliosides and neuronal plasticity
G. Tettamanti, R.W. Ledeen, K. Sandhoff,
Y. Nagai, G. Toffano (eds.)
Fidia Research Series, vol. 6
Liviana Press, Padova, © 1986

NEW EVIDENCE FOR THE MORPHOFUNCTIONAL RECOVERY OF STRIATAL FUNCTION BY GANGLIOSIDE GM1 TREATMENT FOLLOWING A PARTIAL HEMITRANSECTION OF RATS. STUDIES ON DOPAMINE NEURONS AND PROTEIN PHOSPHORYLATION

Kjell Fuxe, Luigi F. Agnati[1], Fabio Benfenati[1], Isabella Zini[1], G. Gavioli[1], G. Toffano[2]

Department of Histology, Karolinska Institutet, Stockholm, Sweden;
[1]Department of Human Physiology, University of Modena, Modena, Italy;
[2]Fidia Research Laboratories, Abano Terme, Padova, Italy

INTRODUCTION

In previous studies it has been shown that chronic treatment with GM1 can increase the survival of dopamine (DA) cell bodies with their dendrites in substantia nigra following a partial unilateral hemitransection of rats (Agnati et al., 1983a; 1984; Fuxe and Agnati, 1984; Toffano et al., 1983; 1984a). This in turn may be responsible for the enhancement of collateral sprouting observed from remaining striatal DA nerve terminals, leading to recovery of dopaminergic synaptic function in the striatum. This action has been found to be specific and not related to antiinflammatory effects of ganglioside GM1, since neither treatment with dexamethazone nor treatment with acetylsalicylic acid has been able to produce any trophic action on the nigral DA nerve cell bodies following a partial hemitransection in rats (Agnati et al., 1984). Recently we have also analyzed the effects of ganglioside GM1 treatment in unilateral partially hemitransected rats on striatal energy metabolism using the radioactive deoxyglycose technique and on striatal blood flow using radiolabelled iodoantipyrine as tracer (Agnati et al., 1985a). The results showed that GM1 can counteract the imbalance in striatal energy metabolism and in striatal blood flow found between the striatum of the lesioned and unlesioned side. This action may be related to excitatory effects of GM1 on the lesioned side and to inhibitory effects of GM1 on the unlesioned side. These results underline the evidence obtained in previous functional studies indicating that

Abbreviations: DA, dopamine; α-MT, αmethyl-p-d1-tyrosine methyl ester; cyclic AMP, cyclic adenylic acid; SDS, sodium dodecyl sulphate; CCK, the metabolism of a striatum undergoing degenerative and regenerative changes can be at least partially restored following chronic GM1 treatment (Agnati et al., 1985a).

In the present paper we will report new evidence, using quantitative microfluorimetry, that chronic GM1 treatment can prevent the DA cell bodies in the substantia nigra and the ventral tegmental area and the DA nerve terminal networks in the striatum and nuc. accumbens from degenerating following a partial unilateral hemitransection in the rat. Effects of acute and chronic GM1 treatment have also been analyzed on DA utilization in the substantia nigra and in the basal ganglia in the partially hemitransected rat. Finally, the effects of ganglioside GM1 administration on cyclic AMP and calcium induced protein phosphorylation in striatal membranes (P2 fraction) from unilaterally partially hemitransected rats are summarized (see Agnati et al., 1985a). Studies on protein phosphorylation emphasize the ability of chronic GM1 treatment to restore striatal function following a partial hemitransection.

STUDIES ON MESOSTRIATAL DA NEURONS FOLLOWING A PARTIAL UNILATERAL HEMITRANSECTION IN THE RAT. EFFECTS OF CHRONIC GANGLIOSIDE GM1 TREATMENT

Male specific pathogen free Sprague-Dawley rats were used and partial unilateral hemitransection of the ascending mesostriatal DA pathway was performed as previously described by Agnati et al. (1983a). The DA levels were determined in DA cell bodies of the ventral midbrain and in various DA nerve terminal rich areas of the striatum, nuc. accumbens and tuberculum olfactorium by means of quantitative histofluorimetrical measurements in sections in which the catecholamines had been converted into strongly fluorescent compounds by means of the formaldehyde fluorescence method of Falck-Hillarp (Agnati et al., 1979; Andersson et al., 1985). Studies on DA utilization were performed by means of the tyrosine hydroxylase inhibition method using the α-methyl-p-dl-tyrosine methylester (α-MT, 250 mg/kg, i.p., 2 h before killing) (Andersson et al., 1985). The GM1 treatment started immediately following the stereotaxic operation and lasted for 14 days. The dose given i.p. was 10 mg/kg (once daily). The last dose was administered 2 h before decapitation.

Studies in the Ventral Midbrain

The results obtained in the substantia nigra and in the ventral tegmental area are illustrated in Figures 1 and 2 and in Table 1. It is seen that on the lesioned side there is a clearcut reduction in the number and fluorescence intensity of the DA nerve cell bodies of the medial sustantia nigra (A9 group, Fig. 1) and of the ventral tegmental area (A10 group, Fig. 2). By treating for 14 days with GM1 it is possible to at least partly counteract the disappearance of fluorescent cell body profiles and to restore in part the fluorescence intensity in the DA nerve cell bodies. In Table 1 some quantitative histofluorimetrical results are illustrated from the substantia nigra. DA fluorescence intensity has been measured in 40 DA cell bodies in each animal. In other rats the tyrosine hydroxylase inhibitor αMT was also administered (2 h before killing) in order to evaluate possible changes in DA utilization in relation to treatment with GM1. The

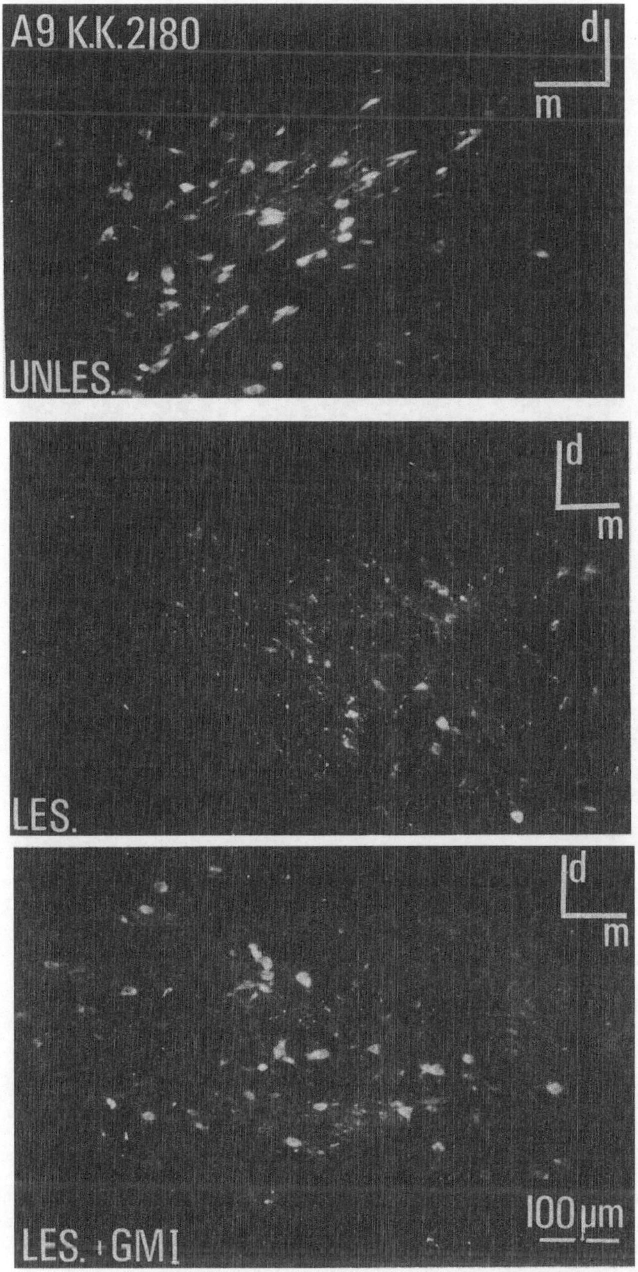

Figure 1. Fluorescence microphotograph from a coronal section showing dopamine cell bodies in the medial part of the substantia nigra of the intact side (upper panel) and of the lesioned (LED) side of a partially hemitransected rat after saline or GM1-treatment as described in text (daily i.p. injections of 10 mg/kg of GM1 for two weeks). The hemitransection was performed immediately before the onset of GM1 treatment. Formaldehyde fluorescence histochemistry was used to demonstrate the dopamine stores in the nigral cells. König-Klippel level was 2180 μm. d = dorsal direction; m = medial direction.

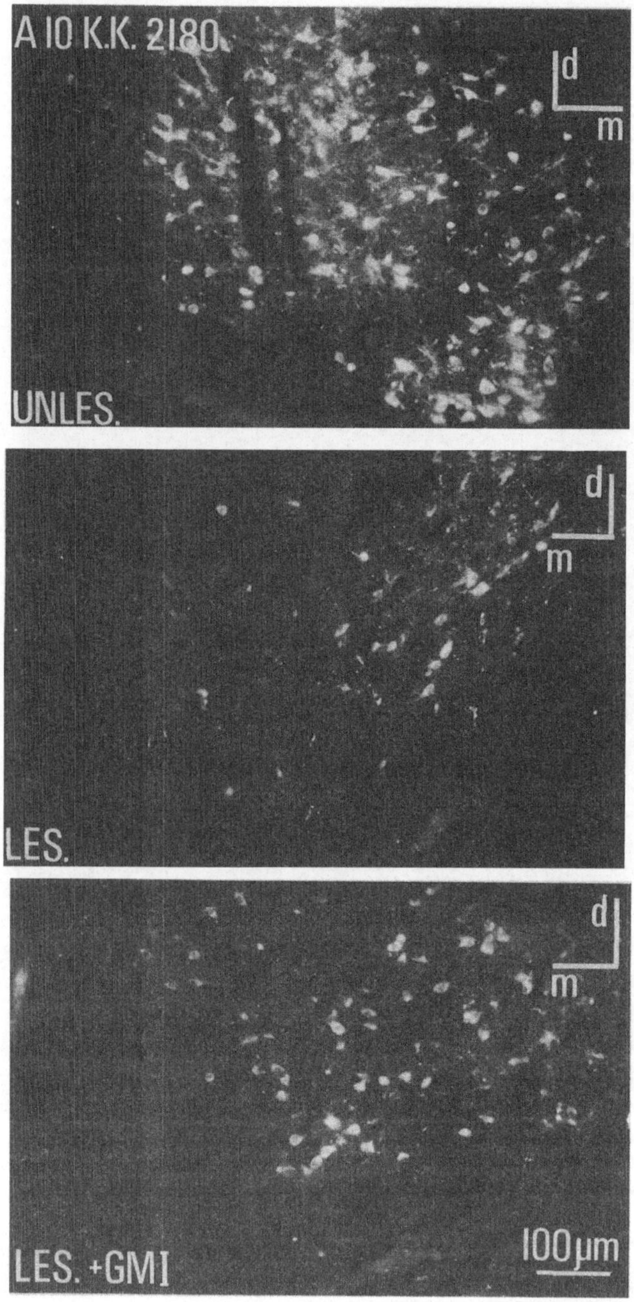

Figure 2. Fluorescence microphotograph of a coronal section showing dopamine cell bodies in the ventral tegmental area (A10) of the intact side (upper panel) and of the lesioned side of a partially hemistransected rat after saline or GM1 treatment. For further details, see text and legend to Figure 1.

Table 1. *Effects GM1 Treatment on the DA Levels and Utilisation in the Substantia Nigra (A9) of the Lesioned and Unlesioned Side of partially Hemitransected Rats at the Tel-Diencephalic Level*

Treatment	Dose mg/kg	A9 DA fluorescence unlesioned side (nmol/g)	lesioned side (nmol/g)
unoperated		115 ± 0.5	115 ± 0.5
hemitransection + saline		88 ± 2	48 ± 2
hemitransection + GM1, 14 days, once daily	10	96 ± 1	89 ± 1.5
		DA fluorescence % depletion after αMT unlesioned	lesioned
unoperated		33 ± 2.1	33 ± 2.1
hemitransection + saline		42 ± 1.7	46 ± 2
hemitransection + GM1, 14 days, once daily	10	21 ± 0.5	28 ± 1.1

Means ± s.e.m. Number of observations = 40.

results support the qualitative findings illustrated in Figure 1. Thus, there is a reduction in the DA levels in the DA cell bodies on the lesioned side and this reduction in DA levels is partly counteracted by chronic treatment with GM1. A borderline statistical significance has been obtained. Furthermore, the results indicate that GM1 has no substantial effect on the DA levels in the DA cell bodies on the unlesioned side. Furthermore, as also seen in Table 1 (αMT experiments), the ability of αMT to produce a disappearance of the DA stores after tyrosine hydroxylase inhibition is maintained and if anything increased following chronic GM1 treatment. In this case, however, this trend for an enhancement of DA utilization by the DA cell bodies following ganglioside GM1 treatment may also be true for the unlesioned side. Taken together, the results indicate an increased survival of DA cell bodies in the substantia nigra of the lesioned side following chronic GM1 treatment and that these DA cell bodies may have increased DA levels and turnover rates compared with those present on the lesioned side in animals treated with saline.

Studies in the Forebrain

In Figure 3 it is shown that the partial hemitransection produces a marked disappearance of DA fluorescence in all parts of the striatum and a substantial disappearance of DA fluorescence in the diffuse types (CCK-negative) of DA nerve terminals in the nuc. accumbens and tuberculum olfactorium. However, the dotted type (CCK positive) of DA nerve terminals in the nuc. accumbens and tuberculum olfactorium appear to be unaffected by the partial hemitransection, giving evidence for a separate origin. It is shown that chronic GM1 treatment in part restores DA fluorescence in all parts of the striatum as well as in the diffuse types of DA terminals of the nuc. accumbens. These results are compatible with the view that chronic ganglioside GM1 treatment, probably via its action in the substantia nigra and in the ventral tegmental area, can enhance the collateral sprouting from remaining DA nerve terminal networks in the striatum and, as demonstrated in this study, also in the nuc. accumbens. As seen in Figures 1 and 2 chronic GM1 treatment has no action on the DA levels in the various forebrain DA nerve terminal systems analyzed on the unlesioned side.

In previous papers studying the DA nerve terminal networks in the striatum by means of tyrosine hydroxylase immunoreactivity in combination with image analysis, chronic GM1 treatment was found to induce a much higher degree of DA reinnervation

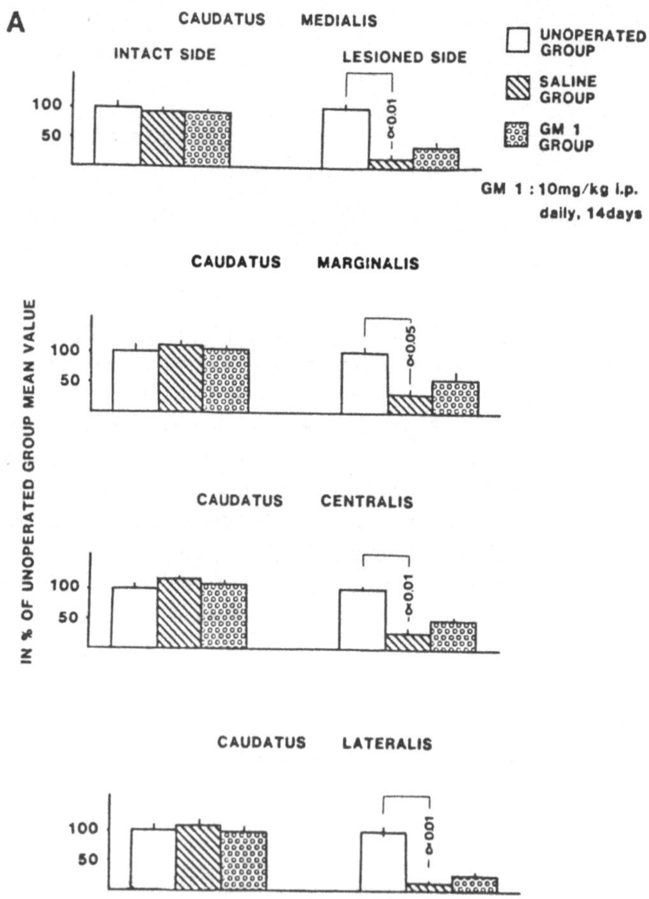

Figure 3. Continued next page.

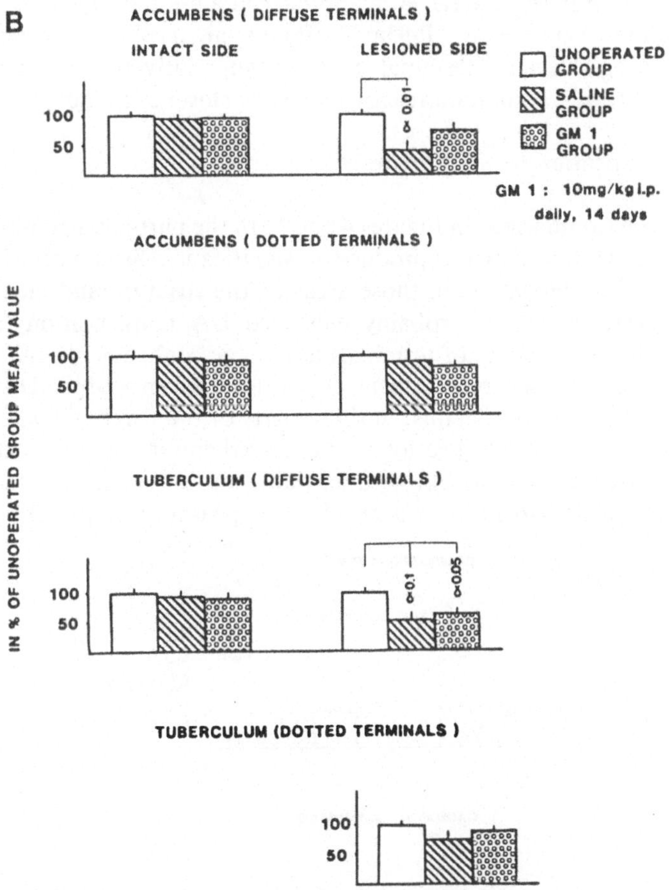

Figure 3. Dopamine levels in different parts of the nucleus caudatus, nucleus accumbens and tuberculum olfatorium of unoperated rats and on the intact and lesioned side of partially hemitransected rats treated with saline or GM1 as described in text and in text to previous figures. Quantitative histofluorimetrical determinations have been performed (Andersson et al., 1985). Means ± s.e.m. are shown (n = 6). The values are expressed as % of unoperated group mean value. Statistical analysis was performed according to Dunn-test comparing all possible pairs of treatment at an experimental wise error rate of $\alpha = 0.05$ or $\alpha = 0.01$. Caudatus medialis = medial part of the nucleus caudatus 100% = 140 nmol/g of tissue; Caudatus marginalis = marginal zone of the nucleus caudatus 100% = 161; Caudatus centralis = central part of the nucleus caudatus, 100% = 156; Caudatus lateralis = lateral part of the nucleus caudatus 100% = 134; Accumbens (diffuse terminals) = diffuse type (CCK negative) of dopamine terminals of the anterior nucleus accumbens 100% = 134; Accumbens (dotted terminals) = dotted type (CCK immunoreactive) of dopamine terminals of the posterior and dorsal part of nucleus accumbens, 100% = 332; Tuberculum (diffuse terminals) = diffuse type (CCK negative) of dopamine nerve terminals of the lateral and posterior part of the tuberculum olfatorium 100% = 152. Tuberculum (dotted terminals) = dotted type (CCK immunoreactive) of dopamine nerve terminals in the medial and posterior part of tuberculum olfatorium 100% = 228. All areas were analyzed at König-Klippel level A8620 μm except accumbens (diffuse terminals) which was analyzed at König-Klippel level A9410 μm.

of the striatum, which in some areas even approached a normal degree of DA innervation of the striatum (Agnati et al., 1983a; Toffano et al., 1983). The explanation may be that the sprouting DA nerve terminals may contain relatively low DA levels (Sachs et al., 1970), while their TH immunoreactivity may be closer to that of intact terminals.

Studies on DA Utilization in the Forebrain

The results are summarized in Figures 4 and 5. In the partially hemitransected rats chronic GM1 treatment is shown to produce no significant effects on the lesioned side nor on the intact side. However, in those areas of the striatum and nuc. accumbens where chronic GM1 treatment probably enhanced DA reinnervation (all parts of striatum analyzed, diffuse types of terminals in nuc. accumbens), chronic GM1 treatment produced a trend for an enhancement of the disappearance of the DA stores after tyrosine hydroxylase inhibition. Thus, in those parts of the forebrain where DA reinnervation takes place, the DA utilization is maintained and if anything increased compared with saline treated animals having a partial hemitransection. On the intact side chronic GM1 treatment had no action on the disappearance of the DA stores after

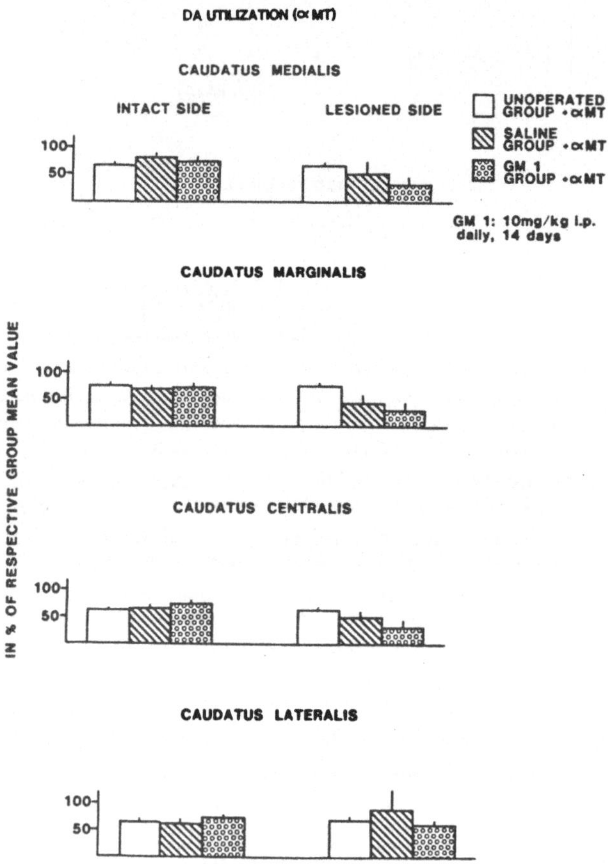

Figure 4. Continued next page.

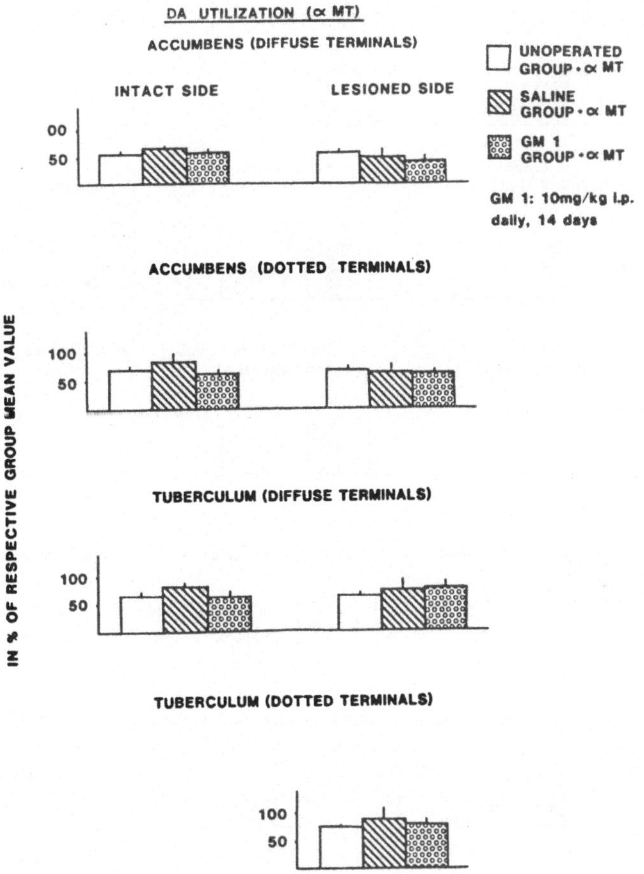

Figure 4. Dopamine utilization in various parts of the nucleus caudatus, nucleus accumbens and tuberculum olfatorium in unoperated rats and on the intact and lesioned side of partially hemitransected rats following saline or chronic GM1 treatment as described in previous figure legend and in text. Dopamine utilization was studied by measuring the decline of the dopamine stores after treatment with the tyrosine hydroxylase inhibitor, α-methyl tyrosine methylester (α-MT, H44/68, 250 mg/kg, i.p., 2h before killing). The formaldehyde fluorescence method in combination with quantitative histofluorimetry was used to measure the dopamine stores. Means ± s.e.m. are shown (n = 4) and expressed as % of respective group mean value at the time of the injection of the α-MT in the respective group and area analyzed. Statistical analysis was performed by means of Dunn test. No significances were found. For abbreviations see text to Figure 3.

tyrosine hydroxylase inhibition, indicating no effects on DA utilization on the intact side.

For comparison, the effects of acute and chronic GM1 treatment in the same dose as performed in the experiments on lesioned animals have been studied on DA levels and utilization in intact male rats. The results are summarized in Figure 5. Acute GM1 treatment (10 mg/kg, 2 h) had no action on DA levels and utilization in any of the

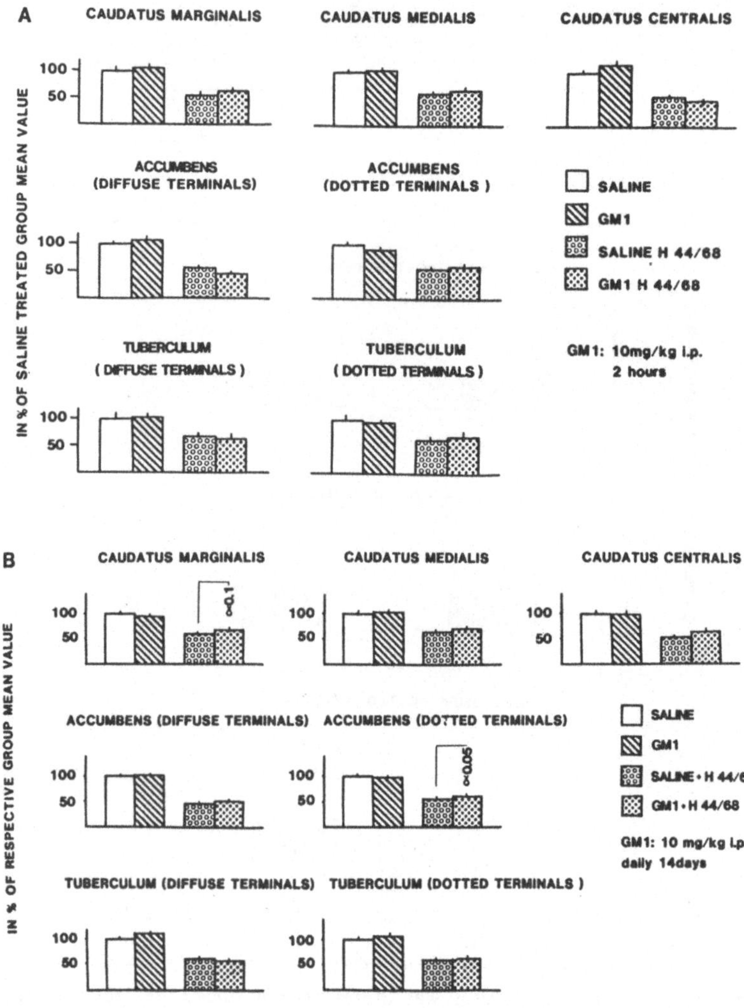

Figure 5. Effects of acute (A) and chronic (B) GM1 treatment on dopamine stores and dopamine utilization in various parts of the nucleus caudatus, nucleus accumbens and tuberculum olfactorium of the unoperated male rat. All rats were killed by decapitation 2h after an injection of 10 mg/kg of GM1. This was true both in the acute and chronic experiments. Means ± s.e.m. are shown (n = 6). In the acute experiments (A) the values obtained following treatment with the tyrosine hydroxylase inhibitor α-MT (H44/68, 250 mg/kg, i.p. 2h before killing) were expressed as % of the saline treated group mean value. In the chronic (B) experiments the values of the dopamine stores in the various areas observed after saline or GM1 treatment were expressed as % of the respective group mean value in the saline and GM1 treated group. Mann-Whitney U-test was used in the statistical analysis. Caudatus marginalis: 100% = 159 (acute) and 136 nmol/g (chronic); caudatus medialis: 100% = 138 (acute) and 163 nmol/g (chronic); caudatus centralis: 100% = 142 (acute) and 117 nmol/g (chronic); accumbens (diffuse): 100% = 112 (acute) and 114 nmol/g (chronic); accumbens (dotted): 100% = 351 (acute) and 388 nmol/g (chronic); tuberculum (dotted): 100% = 254 (acute) and 284 nmol/g (chronic); tuberculum (diffuse): 100% = 154 (acute) and 131 nmol/g (chronic).

various forebrain DA nerve terminal systems studied in the nucleus caudatus putamen, nucleus accumbens and tuberculum olfactorium. Also, chronic GM1 treatment had little effect on regional DA levels and utilization in the intact rat. The only significant change was a reduction in the disappearance of the DA stores after tyrosine hydroxylase inhibition in the dotted type of the DA nerve terminals of the nucleus accumbens and a trend for a reduced depletion of DA stores after tyrosine hydroxylase inhibition within the DA nerve terminals of the marginal zone (CAUD marg.). Thus, only in those areas on the lesioned side where signs of an enhanced DA reinnervation has taken place after GM1 treatment can a trend for an enhancement for DA utilization and turnover be observed following chronic GM1 treatment. These results also give evidence that ganglioside GM1, administered acutely or chronically, does not directly interfere with the pre- and postsynaptic dopaminergic mechanisms but mainly exerts trophic actions on lesioned DA neurons (Agnati et al., 1983a; 1984; Toffano et al., 1983). In agreement with our hypothesis, it has also been possible to show that gangliosides can counteract the shrinkage of choline acetylase immunoreactive nerve cells in the nucleus basalis following unilateral cortical lesions (Cuello et al., 1986).

Hypothesis on the Trophic Activity of Ganglioside GM1 on the Mesostriatal DA Neurons

The hypothesis is summarized in Figure 6. It is suggested that the systemically administered GM1 can to a significant degree penetrate the blood brain barrier and enter the brain (Ghidoni et al., 1986). The GM1 prepartion is also very pure (99%).

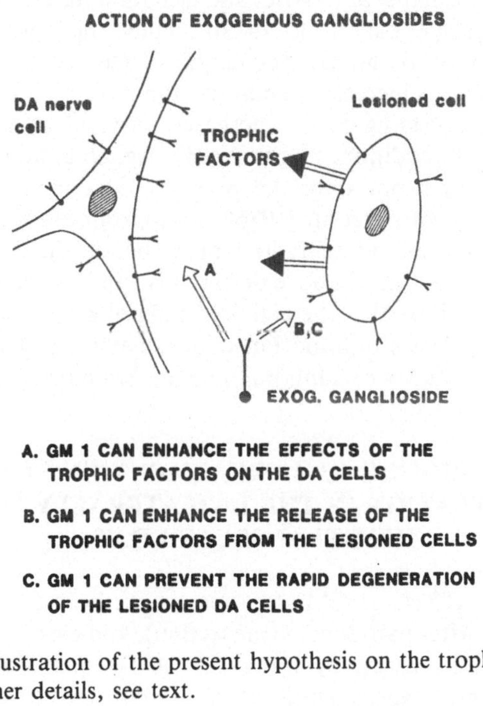
ACTION OF EXOGENOUS GANGLIOSIDES
DA nerve cell
Lesioned cell
TROPHIC FACTORS
A
B,C
EXOG. GANGLIOSIDE
A. GM 1 CAN ENHANCE THE EFFECTS OF THE TROPHIC FACTORS ON THE DA CELLS
B. GM 1 CAN ENHANCE THE RELEASE OF THE TROPHIC FACTORS FROM THE LESIONED CELLS
C. GM 1 CAN PREVENT THE RAPID DEGENERATION OF THE LESIONED DA CELLS

Figure 6. Schematic illustration of the present hypothesis on the trophic actions of exogenous gangliosides. For further details, see text.

Therefore, the trophic action of GM1 is most probably produced by GM1 itself. In view of the fact that the trophic activity of GM1 is associated with neuronal membranes (Leon et al., 1984; Facci et al., 1984), it is suggested that GM1 exerts its action by being incorporated in the neuronal membranes of the lesioned and unlesioned DA nerve cell bodies of the substantia nigra and ventral tegmental area. The incorporation of GM1 into the DA neuronal membranes may then cause an activation of Na,K-ATPase leading to a maintenance of ion-pump processes, which are of importance for the stabilization of membrane ion conductance (Karpiak et al., 1986; Gorio et al., 1986). Due to such a membrane action it may be envisaged that GM1 treatment can increase the survival of lesioned DA nerve cell bodies and also delay the death of many lesioned DA nerve cells. It is also assumed that the lesion has triggered a marked increase in the synthesis of trophic factors in the lesioned DA nerve cell bodies. By increasing the survival of the lesioned DA cell bodies, increased amounts of trophic factors can therefore be released over time from the lesioned cells, which may then reach the types of DA cell bodies of the substantia nigra that are unlesioned. As indicated in Figure 6, it is also possible that GM1 by itself can additionally enhance the release of the trophic factors from the lesioned DA cells. It is also assumed that the ganglioside GM1 is incorporated in the membrane of the unlesioned DA nerve cell bodies, where it may enhance the action of the trophic factors released from the lesioned nigral cells via *e.g.* increasing the coupling of the growth factor receptors to their biological effector (Hakomori, 1984). In line with this hypothesis we have previously shown that gangliosides are also modulators of transmitter receptors, since chronic ganglioside treatment counteracts the biochemical signs of DA supersensitivity induced by chronic haloperidol treatment (Agnati et al., 1983c) and also modulates 5-HT-2 receptors in the cerebral cortex of the rat (Agnati et al., 1983b). Thus, the key to the trophic activity of GM1 may be its ability to enhance the action of trophic factors in the nerve cell membrane and its ability to increase membrane function by a putative activation on Na^+,K-ATPase activity, leading to an increased survival of the lesioned DA and other nigral nerve cells. In addition, it is shown by Agnati et al. (1986) that the action of the ganglioside requires not only the presence of trophic factors but also of polyamines. As stated by Varon et al. (1986), an appropriate window of opportunity is required for the ganglioside to exert its trophic action on the nerve cells and the biochemical mechanisms described above define this window. Also the results of Consolazione et al. (1985) indicate that the activity in nigral afferents as well as in striatal afferents may be another factor of importance for the ability of GM1 to facilitate functional recovery following injury of adult mammalian brain (see Toffano et al., 1984b).

EFFECTS OF GANGLIOSIDE GM1 TREATMENT ON STRIATAL PROTEIN PHOSPHORYLATION IN THE UNILATERALLY PARTIALLY HEMITRANSECTED RAT

In this analysis the effects of GM1 treatment have been analyzed in unoperated, acutely lesioned (2 days after partial hemitransection), and chronically lesioned rats (15 days after partial hemitransection) on cyclic AMP and Ca^{2+} induced protein phosphorylation. The three experimental groups were treated either acutely with GM1

(10 mg/kg, i.p., 6 h before killing) or chronically with GM1 as described above (Agnati et al., 1985a).

Protein phosphorylation was analyzed in the striatal P2 fraction. For details on the protein phosphorylation procedures, see Agnati et al. (1985a). Synaptosomal phosphorylation processes were studied either in basal conditions or after stimulation with cyclic AMP (5mM) and $CaCl_2$ (1.1 mM). After the phosphorylation assay the solubilized proteins were subjected to slab gel electrophoresis, protein staining and autoradiography. The proteins were separated by one dimensional SDS polyacrylamide electrophoresis according to the method of Laemmli (1970). The darkness of the bands on the film was measured quantitatively by a scanning microdensitometer and the optical density was expressed in arbitrary units. Seven peaks were identified and analyzed on the densitograms and numbered according to their migration from the origin of the gel. In Figure 7 we have summarized the effects of acute and chronic GM1 treatment on cyclic AMP and calcium induced phosphorylation in striatal P2 membranes of intact and acutely and chronically hemitransected rats. We have only studied the seven peaks mentioned above. The overall action on striatal P2 protein phosphorylation was evaluated by considering the sum of the seven peak heights on the lesioned side and on the intact side. Furthermore, the optical density measurements made on each of the seven identified peaks was related to the same peak in the saline treated unoperated rats (control rats) by subtracting the corresponding values obtained in the control rats.

In Figure 7 it is shown that cyclic AMP and Ca^{2+} induced protein phosphorylation is reduced preferentially in the striatum of the lesioned side. The only exception was calcium induced protein phosphorylation studied 2 days after hemitransection, at which time interval the opposite was true. It must also be noted that both acute and chronic GM1 treatment tended to depress cyclic AMP and CA^{2+} induced protein phosphorylation in unoperated rats (controls) (Agnati et al., 1985a).

As seen in Figure 7A, acute GM1 treatment both in acutely hemitransected and chronically hemitransected rats favor the cyclic AMP induced protein phosphorylation on the lesioned side. Similar results were also obtained following chronic GM1 treatment. In this way both acute and chronic GM1 treatment could restore the balance between the two sides with regard to cyclic AMP induced phosphorylation processes.

In Figure 7B it is shown that acute and chronic GM1 treatment induce similar but more clearcut changes in calcium induced protein phosphorylation in the striatal membrane fraction than on cyclic AMP induced protein phosphorylation. Thus, after both acute and chronic partial hemitransection, acute and chronic GM1 treatment favored calcium induced protein phosphorylation and restored the balance between the two sides with regard to calcium induced phosphorylation processes. The action of GM1 was in this case especially pronounced for certain striatal proteins (Fig. 8), restoring *e.g.* the balance in 64K dalton protein phosphorylation on the two sides of 2 and 16 day old hemitransected rats, respectively.

The present results demonstrate that the ganglioside GM1 can exert both inhibitory and stimulatory effects on cyclic AMP and calcium induced protein phosphorylation processes in the striatal P2 fraction of rats. The present results also show that chronic treatment with GM1 is not essential, since effects are observed also following 6 h of treatment. Also it is not essential that an ingrowth of new terminals takes place in the striatum, since the effects can be observed also two days after

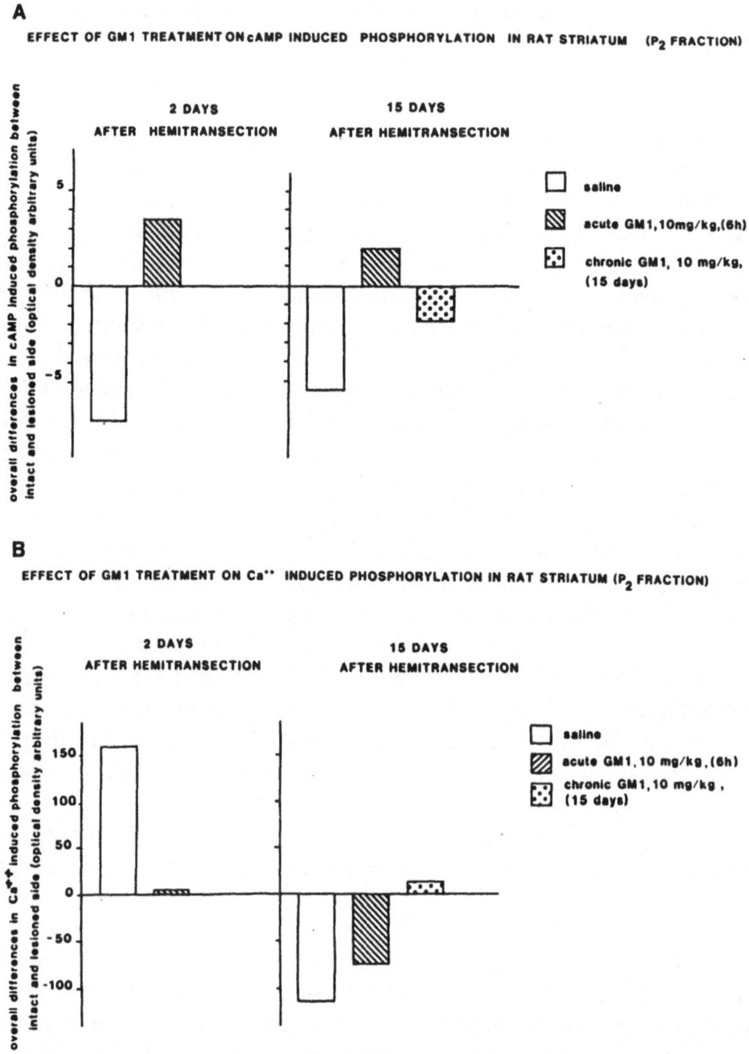

Figure 7. Overall summary of the effects of acute (10 mg/kg, i.p., 6h before killing) and chronic (10 mg/kg, i.p., once daily for 15 days) GM1 treatment on cyclic AMP (A) and Ca^{++} (B) induced protein phosphorylation on the intact and transected side (2 or a 15 day hemistransection) of rats. The seven major protein bands were identified in the densitograms (Agnati et al., 1985a), and all of them were considered. In the analysis the overall difference in phosphorylation of the intact and lesioned side has been obtained by plotting the overall difference in optical density. This value was obtained by subtracting the optical density for each of the 7 proteins on the intact side (always related to that in the unoperated rat by subtraction of the corresponding mean optical density value in the unoperated group) with the corresponding optical density value on the lesioned side (always related to that in the unoperated rat, see above). The sum of the differences was then plotted. Negative values indicate preferential decrease in protein phosphorylation on the lesioned side. No difference between the lesioned and unlesioned side means the same degree of overall phosphorylation as in the saline treated unoperated rat. (see Agnati et al., 1985a).

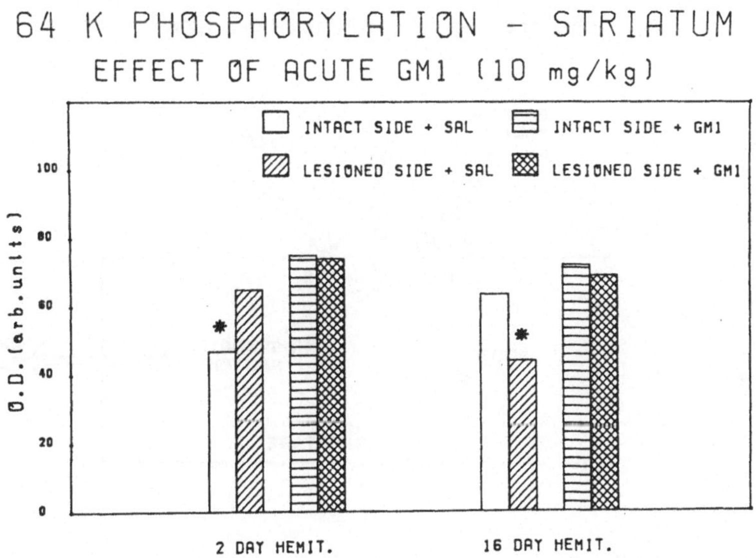

Figure 8. Acute effects of GM1 treatment on Ca^{++} induced 64K dalton protein phosphorylation in striatum. For experimental details, see text. Two time intervals after the lesion are shown (2 and 16 days). On the y-axis the optical density in arbitrary units are shown for the 64K dalton band. Means are given; s.e.m. was always below 15%. Statistical analysis according to Student's t-test was performed and the comparisons have always been made between intact and the corresponding lesioned side. The asterisk indicates a statistical significance $p < 0.025$.

hemitransection. Another important result of the present study is the fact that on the lesioned side the action of acute GM1 administration is to favor cyclic AMP and calcium induced protein phosphorylation processes, leading to a balance in the protein phosphorylation processes between the two striata (lesioned side and unlesioned side). Thus, acute and chronic GM1 treatment restores to a substantial degree cyclic AMP and CA^{2+} induced protein phosphorylation in the rat striatum of the lesioned side. The mechanisms underlying the restoration of protein phosphorylation are unknown, but one fact to be considered is the putative presence of trophic factors in the striatum of the lesioned side. GM1 may enhance the action of these trophic factors at their receptor sites in the membranes of the striatal nerve cells (see above). Of substantial interest is the recent demonstration that there exist ganglioside —Ca^{2+} dependent protein kinases in membrane fractions of cells (Nagai et al., 1986; Tsuji et al., 1985; Benfenati, Fuxe, Agnati, unpublished data). The most marked action of GM1 was its ability to restore calcium induced protein phosphorylation in the striatum of the lesioned side. Therefore, we speculate that there exist protein kinases in the nerve cell membranes which are controlled by calcium, gangliosides and also trophic factors (Fig. 9). The existence of this type of protein kinase could explain the present observations that acute and chronic ganglioside treatment after acute and chronic hemitransection restore Ca^{2+} induced protein phosphorylation in striatal membranes. A similar explanation can also be given for the ability of GM1 to restore cyclic AMP dependent

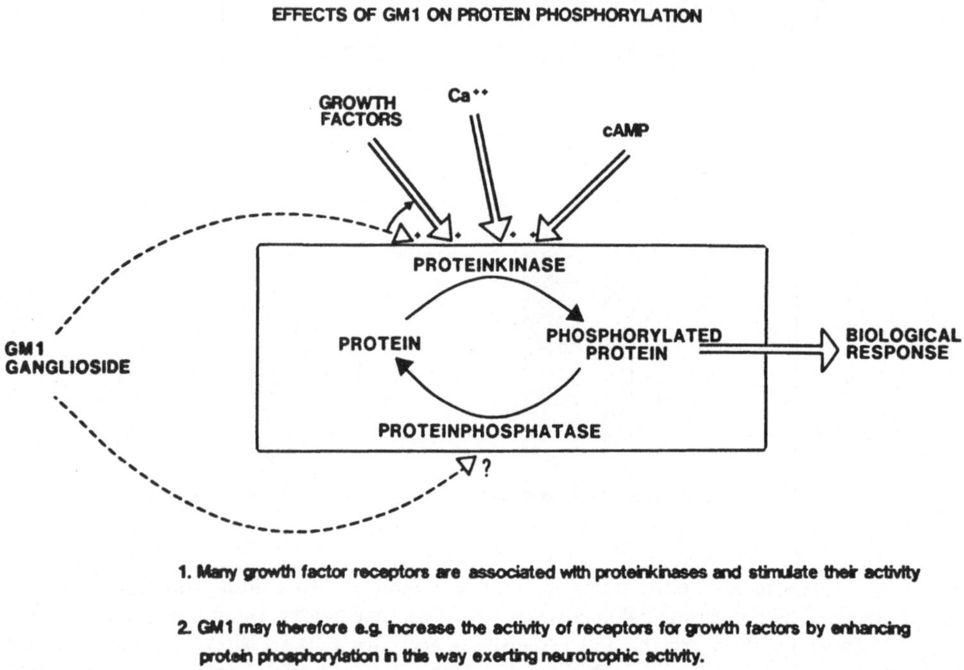
EFFECTS OF GM1 ON PROTEIN PHOSPHORYLATION
GROWTH FACTORS
Ca++
cAMP
PROTEINKINASE
PROTEIN
PHOSPHORYLATED PROTEIN
BIOLOGICAL RESPONSE
GM1 GANGLIOSIDE
PROTEINPHOSPHATASE
?
1. Many growth factor receptors are associated with proteinkinases and stimulate their activity
2. GM1 may therefore e.g. increase the activity of receptors for growth factors by enhancing protein phosphorylation in this way exerting neurotrophic activity.

Figure 9. Schematical illustration of an hypothesis on the mechanism of action of GM1 on protein phosphorylation. The present and previous results are compatible with the view that the GM1 ganglioside may enhance the activity of some membrane bound protein kinases dependent upon Ca^{++} ions and growth factors and the ganglioside GM1. Gangliosides may also in a similar way enhance the activation of cyclic AMP dependant protein kinases but to a lesser degree according to the present findings.

protein phosphorylation processes in striatum of the lesioned side.

Taken together the present results give evidence that acute and chronic GM1 treatment restores a number of striatal neurophysiological processes, since they are dependent upon cyclic AMP and calcium induced protein phosphorylation (Nestler and Greengard, 1984).

The facilitation of the activation of protein kinases may also represent the molecular mechanism underlying the ability of gangliosides to increase Na^+,K^+-ATPase activity in the membrane, which may be the major mechanism responsible for their ability to increase the survival of neurons following injury.

ACKNOWLEDGMENTS

This work has been supported by a grant from the Swedish Medical Council (04X-715), a grant from Knut and Alice Wallenberg Foundation and by a CNR grant. We are grateful for the excellent technical assistance of Lotta Frösell and Barbro Tinner and for the excellent secretarial assistance of Anne Edgren.

REFERENCES

Agnati LF, Andersson K, Wiesel F, Fuxe K (1979) A method to determine dopamine levels and turnover rate in discrete dopamine nerve terminal systems by quantitative use of dopamine fluorescence obtained by Falck-Hillarp methodology. J Neurosci Methods 1: 365-373.

Agnati LF, Fuxe K, Calza L, Benfenati F, Cavicchioli L, Toffano G, Goldstein M (1983a) Gangliosides increase the survival of lesioned nigral dopamine neurons and favour the recovery of dopaminergic synaptic function in striatum of rats by collateral sprouting. Acta Physiol Scand 119: 347-363.

Agnati LF, Benfenati F, Battistini N, Caviocchioli L, Fuxe K, Toffano G (1983b) Selective modulation of ^{3}H-spiperone labelled 5-HT receptors by subchronic treatment with the ganglioside GM1 in the rat. Acta Physiol Scand 117: 311-314.

Agnati LF, Fuxe K, Benfenati F, Battistini N, Zini I, Toffano G (1983c) Chronic ganglioside treatment counteracts the biochemical signs of dopamine receptor supersensitivity induced by chronic haloperidol treatment. Neurosci Lett 40: 293-297.

Agnati LF, Fuxe K, Toffano G, Calza L, Benfenati B, Zini I, Battistini N, Goldstein M, Zoli M (1984) Evidence for structural plasticity in the nigrostriatal dopamine neurons: Morphometrical and biochemical evidence for a trophic action of gangliosides on partially lesioned nigrostriatal dopamine neurons. In: Racagni G, Paoletti R, Kielholz P (eds): Clinical Neuropharmacology, Vol 7, Suppl 1. Raven Press, pp. 594-595.

Agnati LF, Fuxe K, Benfenati F, Zoli M, Owman C, Diemer NH, Kåhrström J, Toffano G, Cimino M (1985) Effects of Ganglioside GM1 treatment on striatal glucose metabolism, blood flow, and protein phosphorylation of the rat. Acta Physiol Scand 125: 43-53.

Agnati LF, Fuxe K, Davalli P, Corti A, Zini I, Merlo Pich E, Zoli M, Gavioli J (1986) Studies on the involvement of polyamines for the trophic actions of the ganglioside GM1 in mechanically and 6-hydroxydopamine lesioned rats. Evidence for a permissive role of putrescine, this volume.

Andersson K, Fuxe K, Agnati LF (1985) Determinations of catecholamine half-lives and turnover rates in discrete catecholamine nerve terminal systems of the hypothalamus, the preoptic region and the forebrain by quantitative histofluorimetry. Acta Physiol Scand 123: 411-426.

Consolazione A, Aldinio C, Agnati LF, Fuxe K, Savoini G, Toffano G (1985) Experimental conditions in which GM1 is or not able to facilitate functional recovery following injury of adult mammalian brain. In: ISN Satellite Symposium on Neuronal plasticity and gangliosides, Mantova, Italy, Poster Abstract P9.

Cuello AC, Stephens PH, Sofroniew MV, Pearson RCA, Tagari P, Powell TPS (1985) Effects of gangliosides on cholinergic neurones of the nucleus basalis following unilateral cortical lesions. In: ISN Satellite Symposium on Neuronal Plasticity and Gangliosides, Mantova, Italy, Abstract 38.

Facci L, Leon A, Toffano G, Sonnino S, Ghidoni R, Tettamanti G (1984) Promotion of neuritogenesis in mouse neuroblastoma cells by exogenous gangliosides. Relationship between the effect and the cell association of ganglioside GM1. J Neurochem 42: 299-305.

Fuxe K, Agnati LF (1984) Gangliosides as regulatory factors in central catecholamine neurons. In: Magistretti PJ, Morrison JH, Bloom FE (eds): Discussions in Neurosciences. Nervous system development and repair. FESN, Vol. 1, No. 2, pp. 84-89.

Ghidoni R, Venerando B, Fiorilli A, Tettamanti G (1985) Metabolism of exogeneously administered gangliosides and related glycolipids in the rat. In: ISN Satellite Symposium on Neuronal Plasticity and Gangliosides, Mantova, Italy, Abstract 11.

Gorio A, Di Gregorio F, Janigro D, Milan F, Vitadello M, Bianci R (1985) Gangliosides have a decay preventing activity on neuronal membrane functions. In: ISN Satellite Symposium on Neuronal Plasticity and Gangliosides, Mantova, Italy, Abstract 27.

Karpiak SE, Li YS, Aceto P, Mahadik SP (1986) Acute effects of gangliosides on CNS injury, this volume.

Laemmli UK (1970) Cleavage of structural proteins during the assembly of the head of bacterophage T_4. Nature 227: 680-685.

Leon A, Benvegnu D, Dai Toso R, Presti D, Facci L, Giorgi O, Toffano G (1984) Dorsal root ganglia and nerve growth factor: a model for understanding the mechanism of GM1 effects on neuronal repair. J. Neurosci Res 12: 277-287.

Nagai Y, Tsuji S, Nakajima J, Sasaki T (1986) Functional analysis of ganglioside-respondable human neuroblastoma cell lines, this volume.

Nestler Greengard P (1983) Protein phosphorylation in the brain. Nature 305: 583-588.

Sachs C, Champlain J, Malmfors T, Olson L (1970) The postnatal development of noradrenaline uptake in the adrenergic nerves of different tissues from the rat. Eur J Pharmacol 9: 67-79.

Toffano G, Savoini G, Moroni F, Lombardi G, Calza L, Agnati LF (1983) GM1 ganglioside stimulates the regeneration of dopaminergic neurons in the central nervous system. Brain Res 261: 163-166.

Toffano G, Savoini G, Aldinio C, Valenti G, Dal Toso R, Leon A, Calza L, Zini I, Agnati LF, Fuxe K (1984a) Effects of gangliosides on the functional recovery of damaged brain. In: Ledeen RW, Yu RK, Rapport MM, Suzuki K, Tettamanti G (eds): Ganglioside structure, function and biomedical potential. Plenum Press, New York, in press.

Toffano G, Agnati LF, Fuxe K, Aldinio C, Consolazione A, Valenti G, Savoini G (1984b) Effects of GM1 ganglioside treatment on the recovery of dopaminergic nigro-striatal neurons after different types of lesion. Acta Physiol Scand 122: 313-321.

Tsuji S, Nakajima J, Sasaki T, Nagai Y (1985) Bioactive gangliosides. IV. Ganglioside GQ1b/Ca^{2+}dependent protein kinase activity exists in the plasma membrane fraction of neuroblastoma cell line, GOTO. J Biochem 97: 969-972.

Varon S, Skaper SD, Katoh-Semba R (1986) Neuritic responses to GM1 ganglioside in several in vitro systems, this volume.

Gangliosides and neuronal plasticity
G. Tettamanti, R.W. Ledeen, K. Sandhoff,
Y. Nagai, G. Toffano (eds.)
Fidia Research Series, vol. 6
Liviana Press, Padova, © 1986

BEHAVIORAL AND NEUROCHEMICAL ALTERATIONS INDUCED BY EXOGENOUS GANGLIOSIDES IN BRAIN DAMAGED ANIMALS: PROBLEMS AND PERSPECTIVES

Gary L. Dunbar, William M. Butler, Barry Fass, Donald G. Stein

Brain Research Laboratory, Psychology Department, Clark University,
Worcester, MA 01610 USA

INTRODUCTION

Gangliosides are membrane glycosphingolipids and have a higher concentration within the nervous system than in any other tissue (Rapport, 1981). Although the function of gangliosides is not known, it has been proposed that they promote neural growth. Several lines of evidence support this proposal. For example, excessive accumulation of gangliosides during development results in the formation of meganeurites which possess dendritic-like growth cones (Purpura et al., 1978). Also, exogenous gangliosides enhance the survival of cells and neuritogenesis *in vitro* (e.g., Ferrari et al., 1983; Roisen et al., 1981).

Since gangliosides have been shown to possess neuritogenic properties, the possibility exists that they function as a neuronotropic agent. Indeed, some investigators have suggested that exogenous gangliosides promote structural repair after brain lesions *in vivo* (Sabel et al., 1984a; Toffano et al., 1983; Wojcik et al., 1982), which may have implications for recovery of function (Agnati et al., 1983; Karpiak, 1983). If gangliosides do enhance lesion-induced growth and behavioral recovery, they may offer some utility in the clinical management of patients with neurological disorders (Gorio et al., 1981; Massarotti, 1983; Pozza et al., 1981).

In our laboratory at Clark University, we have long been interested in trying to enhance behavioral recovery from traumatic injury to the brain. One area of our research has been to test the neuronotropic properties of exogenous gangliosides. Whereas other approaches, such as use of nerve growth factor or transplants of fetal brain tissue, are also effective in promoting behavioral recovery after brain damage

---

Abbreviations: LSN, lateral septal nucleus; FF, fimbria-fornix; i.m., intramuscular injection; SN, substantia nigra; i SN, ipsilateral substantia nigra; AC, midbrain angular complex; LGNd, lateral geniculate nucleus dorsalis; AGF2, inner ester of ganglioside GM1.

(Stein, 1981; Stein et al., 1985), they must be given intracerebrally (usually in or near the site of the lesion) and in temporal proximity to the injury (Freed et al., 1985; Stein, 1985). Exogenous gangliosides, by contrast, can be administered systemically and they result in elevated blood levels for up to a week after a single dose (Orlando et al., 1979; Tettamanti et al., 1981). Thus, gangliosides offer some practical advantages which make them an attractive tool for investigating behavioral recovery and lesion-induced neuroplasticity.

Because of their advantages, much of our research has been focused on whether ganglioside treatments can be used to reduce the severity of lesion-induced behavioral impairments and/or promote behavioral recovery. The purpose of the present chapter is to provide a report on the progress of our research on gangliosides. One line of research has addressed whether the stimulating effects of exogenous gangliosides on lesion-induced neuroplasticity might be mediated (at least in part) by effects on neural metabolic activity. Another line has been concerned with the potential behavioral significance of the facilitatory effects of ganglioside treatments.

GANGLIOSIDES AND NEURONAL METABOLISM

We recently have become interested in exploring what cellular events might underlie the effects of exogenous gangliosides on lesion-induced plasticity in the brain. As mentioned above, there is increasing evidence that ganglioside treatments promote the survival of injured neurons (Agnati et al., 1983; Toffano et al., 1984a) and enhance axonal sprouting (Sabel et al., 1984a; Toffano et al., 1983; Wojcik et al., 1982). Although the cellular mechanism of these effects is not clear at present, one possibility is that they involve alterations in neuronal metabolism. It has been well established that neurons exhibit dramatic alterations in protein synthetic and enzymatic activities after deafferentation or axotomy (Barron, 1983), and the pattern of such alterations may predict whether the affected neuron survives and/or regenerates its axon (Benowitz et al., 1981; Durham and Rubel, 1985; Harkonen and Kauffman, 1974; Skene and Willard, 1981; Steward and Rubel, 1985; Watson, 1965). Because of the apparent relationship between metabolism and a damaged neuron's ability to survive or regenerate, we decided to examine whether the effects of ganglioside treatments on neuroplasticity might be accompanied by detectable metabolic changes.

In an initial study (Stein and Fass, 1985; Fass and Stein, in press), we evaluated the differences in glucose-6-phosphate dehydrogenase (G-6-PDH; EC 1.1.1.49) between the lateral septal nucleus (LSN) of ganglioside-treated and untreated adult rats with a unilateral transection of the fimbria-fornix (FF). G-6-PDH is an oxidative enzyme involved in the metabolism of glucose via the hexose monophosphate pathway. The reason for studying this particular enzyme is that regenerating neurons exhibit reliable alterations in G-6-PDH activity at 2-4 days postlesion, followed by a return toward normal levels at 10-14 days which coincides with axonal regrowth (Ando et al., 1984; Harkonen and Kauffman, 1974). The recovery of G-6-PDH activity is thought to reflect the participation of this enzyme in the biosynthetic events important for regeneration (Barron, 1983; Harkonen and Kauffman, 1974). Thus, we hypothesized that if ganglioside treatments promote neuronal survival and/or axonal growth after a lesion, they also might induce detectable alterations in the activity of this metabolic enzyme.

The model system which we used in our study is the hippocamposeptal pathway. This system offers several important advantages for the purposes of our investigation. First, hippocampal afferents to the LSN are contained within a discrete fiber bundle; i.e., the FF (Meibach and Siegel, 1977; Swanson and Cowan, 1979). This pathway can be transected quite easily and selectively on one side only, and the resulting deafferentation is so much greater in the LSN ipsilateral to the lesion that the contralateral LSN can be used as an internal control (Field et al., 1980). Second, FF transections deafferent the LSN without axotomizing its intrinsic neurons. Efferents from the LSN either course ventrally to the medial septal nucleus or enter the stria medullaris, but they do not contribute to the FF (Meibach and Siegel, 1977; Swanson and Cowan, 1979). Thus, any changes in G-6-PDH within the LSN after FF transection could not readily be attributed to deafferentation. And third, unilateral FF transection has been shown to induce sprouting by surviving inputs to the LSN (Raisman, 1969). The time course of such sprouting has been studied at the electron microscopic level; it begins at about 6 days postlesion, proceeds maximally during the next 24 days, and then continues more slowly for the next 2 months (Field and Raisman, 1983). This time course indicates when to test for effects of ganglioside treatment on G-6-PDH activity. Consequently, we chose to use survival intervals of 2-4 days; i.e., prior to the onset of sprouting. The reason for selecting an early time period is that the effects of exogenous gangliosides on lesion-induced behavioral impairments and cerebral edema have been shown to occur within 4 days after surgery (Fass and Ramirez, 1984; Li et al., 1985; Karpiak and Mahadik, 1984). Such effects could be related to any potential alterations in metabolic activity as reflected by changes in G-6-PDH.

Our experiment was performed on 24 male albino rats, 300-450 gms at the time of surgery. Each animal sustained a unilateral transection of the FF (Grady et al., submitted). Twelve rats survived for 2 days, and the other 12 survived for 4 days. In each survival group, 6 of the animals were treated with GM1 ganglioside (Fidia Research Labs) and the other 6 were not treated. The regimen of GM1 treatments involved daily injections of 10 mg/kg i.m., beginning on the day before surgery and continuing through the day of sacrifice. The rats were killed with an overdose of anesthetic, decapitated, and their brains were quick-frozen in methyl butane chilled with dry ice. Thirty micron sections in the horizontal plane were cut in a cryostat, mounted onto subbed microscope slides, and then processed for G-6-PDH histochemistry (Troyer, 1980). The intensity of histochemical staining in the LSN at 3 standard dorso-ventral levels was measured with an optical densitometer (Nikon Magiscan), and a mean score was calculated. The optical density for the deafferented side then was expressed as a percent of the contralateral (control) value. Measurements also were obtained from the caudate nucleus just lateral to the site where readings were taken in the LSN, as an additional control.

As shown in Table 1, there was a dramatic reduction in the intensity of staining for G-6-PDH within the ipsilateral LSN of untreated rats at both time points. The size of the reduction was approximately 40%. By contrast, staining intensity in the caudate nucleus was only slightly altered, suggesting that the changes in the LSN were not merely artifacts or a nonspecific effect of the lesion. GM1 treatments did not affect

Table 1. *Effects og GMl Treatments on Histochemical Staining for G-6-PDH in the LSN and Caudate Nucleus after Unilateral FF Transection*

Group	LSN	Caudate Nucleus
2-day, no GM1	60.6 ± 3.7	104.2 ± 20.0**
2-day, GM1	56.8 ± 2.9	94.9 ± 6.1
4-day, no GM1	62.8 ± 3.5	90.8 ± 3.6**
4-day, GM1	77.0 ± 5.4*	95.1 ± 5.1

* Significantly different from the no GM1 group, Mann-Whitney U = 7, p = .05.
** N = 3; all other means based upon N = 6.
LSN, lateral septal nucleus; FF; fimbria-fornix.
Values are expressed as a percent of the intensity of staining in the contralateral homologue (mean ± S.E.M.)

G-6-PDH staining at 2 days after the transection. However, at 4 days, the reduction in staining was only 23% of the contralateral value. The difference between the treated and untreated cases at this time point is significantly different (Mann-Whitney U = 7, p = .05).

Our initial results indicate that FF transection induces a loss of G-6-PDH histochemical staining within the LSN at 2 and 4 days postlesion. This effect appears to be relatively specific; only slight changes in staining intensity were observed within the adjacent caudate nucleus. The loss of this enzyme within the LSN cannot readily be attributed to axotomy of intrinsic neurons, since their axons do not enter the FF (Meibach and Siegel, 1977; Swanson and Cowan, 1979). Thus, we propose that the present findings reflect an effect of deafferentation. It is not possible, however, to infer from these results whether the enzymatic changes are occurring pre- or postsynaptically. Although there is evidence that G-6-PDH normally is found in high concentrations within synaptosomes and not in mitochondrial fractions (Kauffman and Harkonen, 1977; Kimura et al., 1974), it remains to be determined where the lesion-induced changes are taking place. Nevertheless, we can speculate about the potential significance of the effects of ganglioside treatments on G-6-PDH.

One possibility is that the observed changes occur within surviving inputs to the LSN. In this case, the enhancement of enzyme activity in the treated animals may signify that these inputs are exhibiting an increase in their metabolic activity; i.e., the amount of enzyme within the presynaptic elements increases, but their number is the same in treated and untreated animals. Such an increase in metabolic activity may be necessary for these inputs to begin sprouting. Alternatively, the enhancement in G-6-PDH may reflect an increase in the number of surviving inputs; i.e., the amount of enzyme per presynaptic element is the same in treated and untreated animals, but the number of enzyme-containing elements changes. If this were the case, it would imply that GM1 treatments stimulate the onset of sprouting in the LSN at 4 days after unilateral FF transection, whereas ordinarily such sprouting does not begin until about 6 days postlesion (Field and Raisman, 1983). We recently have obtained biochemical evidence that exogenous gangliosides also might accelerate the onset of sprouting in the hippocampus of rats with lesions (Fass et al., 1985).

Another possibility is that the observed changes in G-6-PDH occur within the deafferented neurons of the LSN. Recent studies have demonstrated that deafferentation of the brainstem auditory system of chickens induces profound alterations in enzymatic and protein synthetic activities within the postsynaptic cells (Durham and Rubel, 1985). However, the magnitude of deafferentation in that system is much greater than in the LSN after unilateral FF transection; in the latter, only approximately 45% of the synapses are lost (Field and Raisman, 1983). Although this degree of synaptic removal may not be very severe compared to that in other model systems (Steward and Vinsant, 1983), it conceivably could be extensive enough to alter the activity of G-6-PDH by the observed amount (40%). If this were the case, the effect of exogenous gangliosides might be to restore metabolism in the affected neurons. By restoring the metabolic activity of deafferented cells, ganglioside treatments potentially could promote cell survival and/or affect the process of reinnervation by sprouting afferents. Indeed, to the extent that reinnervation is regulated locally by the deafferented targets (Fass and Steward, 1983), exogenous gangliosides might promote the rate or extent of reinnervation by means of their effects on the postsynaptic cell. We currently are undertaking experiments to test the above possibilities.

GANGLIOSIDES AND BEHAVIORAL RECOVERY

Our work, as well as that of other investigators (Agnati et al., 1983; Fass and Ramirez, 1984; Karpiak, 1983; Li et al., 1985; Toffano et al., 1983), has shown that gangliosides can reduce the severity of behavioral deficits following brain damage. In our first study with gangliosides, we demonstrated that daily injections of GM1 (30 mg/kg) could enhance the recovery or ability to solve spatial reversal learning that is lost after bilateral lesions of the caudate nucleus (Sabel et al., 1984b). In this study, rats treated with gangliosides took fewer days to reach criterion and made fewer escape errors in the two-choice discrimination learning task than did their saline-treated counterparts.

Next, we were able to show that gangliosides could also reduce deficits in spatial learning performance of rats with lesions of the dorsomedial frontal cortex (Dunbar et al., 1984). In this context, the GM1-treated animals did not show the significant impairments of their saline-treated counterparts on the number of perseverative errors and mean trials to criterion.

In a third behavioral experiment, we reported that GM1-treated rats with partial, unilateral transections of the nigrostriatal pathway showed less rotational asymmetry after injections of amphetamine or apomorphine than did saline-treated rats (Sabel et al., 1984a, 1985a). Of particular interest was our observation that, in comparison to saline-injected animals, the GM1-treated rats showed significant increases in retrograde labeling of the ipsilateral substantia nigra (iSN), the ipsilateral ventral tegmental area, and the contralateral SN after HRP injections into the denervated caudate. This intergroup difference occurred at 15 days postlesion, which also corresponded to the time we reported that amphetamine-induced rotational asymmetry was decreased in the GM1-treated rats.

Although our findings imply that the beneficial effects of ganglioside treatments are generalized, we have recently conducted two studies which provided evidence to the contrary. The first study indicated to us that certain limitations and problems of interpretation exist when assessing the effects of gangliosides on rotational behavior following unilateral transections of the nigrostriatal pathway. The second study suggested that gangliosides were unable to reduce deficits in brightness and pattern discrimination following occipital cortex lesions.

Nigrostriatal System

In an attempt to replicate our earlier findings and to test the behavioral effects of three different dosage levels of GM1 monosialoganglioside's internal ester (AGF2) (1,10, and 30 mg/kg), we employed the same surgical and testing procedures as previously reported (Sabel et al., 1985a). Using our previous measure of amphetamine-induced rotational behavior (which is a relative measure and considers both pre- and postoperative rotational activity), we found no significant differences among any of the groups of rats which sustained hemitransections. However, using only the number of postoperative ipsiversive rotations (i.e., an absolute measure), we discovered that the rats treated with 10 mg/kg of AGF2 consistently showed the greatest amount of rotational activity (Fig. 1).

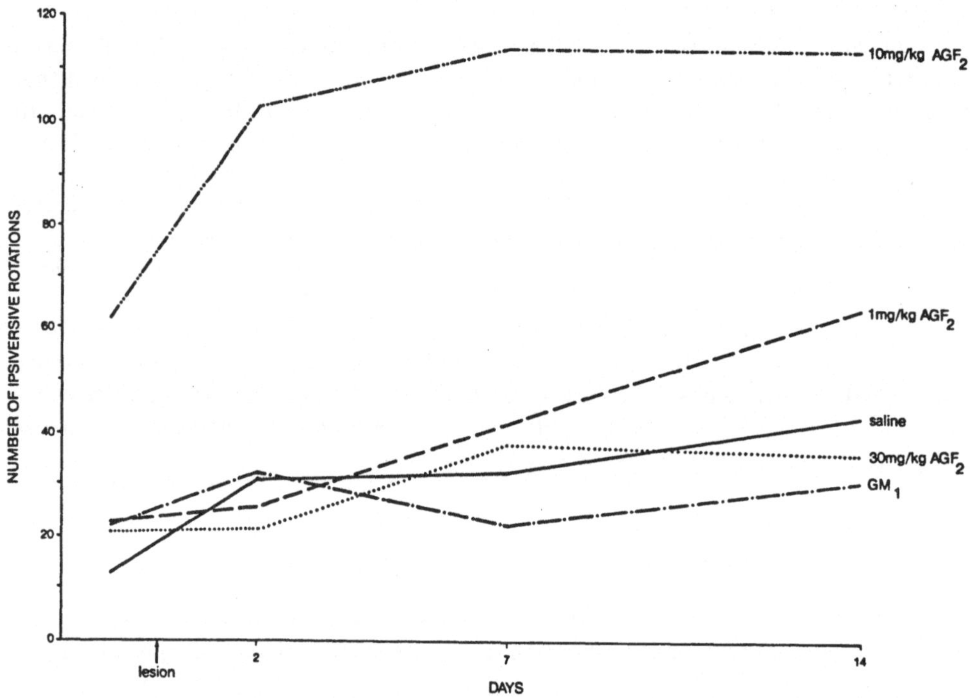

Figure 1. Hemitransected rats receiving injections of AGF2 (10 mg/kg) showed the greatest number of ipsiversive rotations on all three postoperative testing days. However, this group also showed a tendency toward more preoperative rotations. AGF2, internal ester of ganglioside GM1.

It is tempting to suggest that the AGF2 injections were disruptive, but inspection of the data (Fig. 1) reveal that this group of rats showed a tendency toward more preoperative rotations than the other treatment groups. Our interpretation of the data was further complicated by extreme within-groups variability after the transections. An analysis of individual cases showed that, in each group, some rats exhibited inconsistent directions in rotations across test days. For example, some rats rotated ipsiversive to the lesion on test day 2, contraversively on day 7, and then ipsiversively again on day 14.

We think that the lack of consistency we oberved is not due to the drug treatments we employed. Rather, the size of the transection may be critical, and it must be extensive if reliable, amphetamine-induced rotation is to occur. We proceeded to test this hypothesis by making larger hemitransections and evaluating the amphetamine-induced rotational activity in one group treated with GM1 (n = 9) and another treated with saline (n = 8). In this experiment, both groups showed a significant increase in ipsiversive rotations on all three days of postoperative testing (days 2,7, and 14; Fig. 2a), but there were no differences between the GM1- and saline-treated groups. However, when we used our original relative measure of rotational asymmetry, the GM-1 treated rats paradoxically showed significantly more rotational asymmetry on day 2 than did the saline-treated group (Fig. 2b).

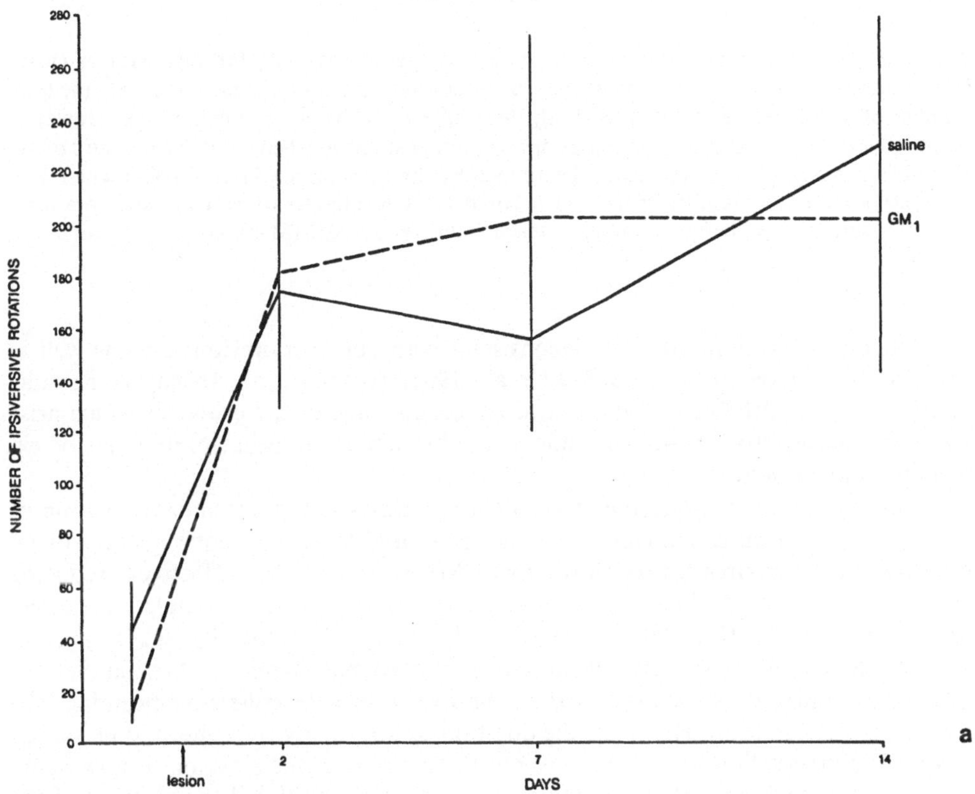

Figure 2a. Continued next page.

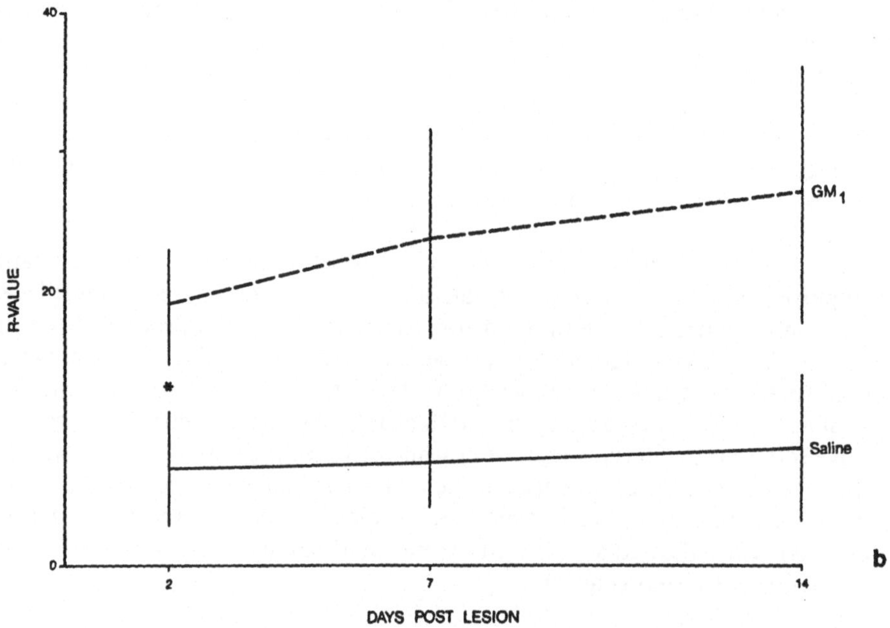

Figure 2. Whether or not treatment effect can be shown to exist for rats with extensive hemitransections may be a function of the type of dependent measure used. When only the total number of ipsiversive rotations is used (a), there appears to be no treatment effect. However, there is a significant increase in rotations at two days post-lesion when a relative measure of activity is applied (b). The R-value was determined by the formula $R = Ia/Ib - Ca/Cb$; where R = asymmetry score, I = number of ipsiversive rotations, C = number of contraversive rotations, a = after surgery, b = before surgery. Asterisk indicates $p < .05$, (Mann-Whitney U-test).

Obviously, these findings are inconsistent with our own previous data as well as those of others (Agnati et al., 1983; Li et al., 1985; Toffano et al., 1984a). Our results consequently required us to focus our attention on some of the problems which arise when the nigrostriatal rotational model is used to assess the behavioral effects of exogenous gangliosides.

First, it now seems clear that even slight variations in the size and/or location of the nigrostriatal transection can have either profound behavioral consequences or no effects at all. It has already been shown that GM1 treatments are ineffective in reducing the number of apomorphine-induced ipsiversive rotations when "complete" transections are made (Toffano et al., 1984b). However, when the damage is partial, ganglioside treatments are effective in reducing rotational asymmetry (Toffano et al., 1984b). Presumably, the spared fibers are needed to provide collateral sprouting into the denervated striatum. However, the question of how many fibers need to be spared to obtain reinnervation has not been resolved. Conversely, the question of how many fibers need to be destroyed in order to obtain consistent rotational asymmetry has not been resolved. We also need to know whether there are other critical pathways or structures which must be damaged (or avoided) when making the transection in order to reveal a ganglioside effect.

There is some evidence that other structures or pathways are involved in rotational behavior after hemitransections. For example, Starr and Summerhays (1985) found that unilateral lesions of the midbrain angular complex (AC) cause pronounced, apomorphine-induced, ipsiversive circling in rats. This finding suggests that incidental damage to pathways involving the AC could affect behavioral measures designed to reflect the activity of the nigrostriatal pathway. Since it is difficult to ascertain whether or not damage confined only to nigrostriatal fibers results in the same behavioral effects as transections that involve other structures and pathways, inferences concerning ganglioside-induced behavioral recovery need to be interpreted cautiously.

Second, the use of amphetamine (particularly with repeated administration) may make interpretation of lesion data difficult. Our concern arises from the recent finding that even a single injection of amphetamine (1 mg/kg, IP) can potentiate rotational behavior induced by a second injection for up to 12 weeks (Robinson, 1984). It has also been found that rats which had received continuous administration of amphetamine in high doses had lower striatal dopamine levels and damaged striatal dopamine-containing neurons (Ricaurte et al., 1984). Although the doses Ricaurte et al. employed were much higher than ours (16 mg/kg vs. 2 mg/kg), it is possible that amphetamines, even at low doses, could interact with recent brain lesions to accentuate the loss of nigrostriatal fibers, or retard the effects of gangliosides in promoting functional recovery.

Taken all together, these findings may help to explain why our GM1-treated rats showed no behavioral recovery in the rotational study where we used more extensive transections. In contrast, Li et al. (1985), who used essentially the same lesion parameters, were able to find significant reduction in the number of ipsiversive rotations 2 days after the lesion. The differences between the two studies could be due to the fact that in our experiment, rats were given amphetamine 1 day before surgery to obtain a baseline of preoperative rotational activity. Li et al. (1985) gave no injections until 2 days after surgery.

Even if no detectable amphetamine-induced neuronal damage occurred in our animals, the behavioral sensitization (reverse tolerance) from the previous injection that our rats received may have been able to offset any facilitating effect of the GM1 treatments. This drug potentiation effect would seem even more likely during later testing sessions when the animals have had 3-4 injections of the amphetamine.

A third issue of concern when using the nigrostriatal model involves selection of the most appropriate dependent variable for analysis. On the surface of it, this may seem like a straightforward problem; however, our recent experience has indicated that measures which compare preoperative with postoperative rotational activity (relative measures) reveal ganglioside effects which are not observed when only postoperative (absolute) measures are employed. In several of our studies using nigrostriatal transections, we found significant treatment effects on amphetamine-induced rotational activity when the relative measure was used, but no effect when the absolute score was employed. In another study, the reverse situation occurred; a significant ganglioside effect was obtained when we used an absolute measure, but not when our relative measure was used. It almost goes without saying that these problems make it very difficult to draw unambiguous conclusions about the effectiveness of treating nigrostriatal injury with gangliosides.

From our review of the literature, there seems to be little consensus as to what type of behavioral measures most appropriately assess ganglioside-induced recovery from brain injury. On the one hand, the major advantage of the "relative" rotational score is that it takes into account the finding that rats with lesions ipsilateral to their preoperative direction of rotation, show twice as much amphetamine-induced ipsiversive rotation as rats with lesions that are contralateral to their preferred direction of rotation (Glick et al., 1976). On the other hand, the drawbacks of the relative measure are that habituation and drug potentiation effects are carried over from preoperative testing and may confound postoperative results. Unless these factors can be ruled out, it becomes very difficult to draw definitive conclusions about how, or if, ganglioside treatments play a role in reduction of symptoms following nigrostriatal lesions.

Visual System Lesions

In order to test whether GM1 treatments would prove beneficial in promoting recovery after lesions of specific sensory systems, we designed an experiment in which rats sustained bilateral lesions of the occipital cortex. We chose this lesion paradigm to answer our question because it provided us with two measures of behavioral recovery: brightness and pattern discrimination. Previous research has demonstrated that the ability to perform a brightness discrimination is quickly recovered following bilateral occipital cortex ablation, while pattern discrimination takes much longer to reinstate (Horel et al., 1966; Meyer and Meyer, 1977). Having one behavior that is easily recovered after occipital lesions, and another which is more difficult to regain, provided the opportunity to evaluate whether GM1's behavioral effects are test-dependent. In addition to these well-known behavioral effects of the lesion, the extent of bilateral occipital lesions can be measured anatomically by examining neuronal degeneration in the lateral geniculate nucleus dorsalis (LGNd) (Hughes, 1976; Meyer and Meyer, 1977). We therefore assessed whether GM1 treatments affected degeneration in the LGNd after occipital lesions.

In this study, two groups of rats sustained bilateral removals of the occipital cortex excluding Krieg's area 20 (Hughes, 1976). A group of sham operates served as controls. Following surgery, one group of operates was injected daily with GM1 (30 mg/kg), while the other group was injected with saline. The treatments were administered for 14 days postlesion. On the 15th day, the rats began testing on a shock-motivated brightness discrimination task. This testing lasted 14 days, and was followed by 28 days of pattern discrimination. After testing was completed, the brains were processed in order to assess the lesions and the degree of degeneration in the LGNd.

The behavioral data revealed that the GM1-treated rats did not perform better than the saline-treated counterparts on either task. Indeed, on 14 of the 18 test days, the number of errors committed by the GM1 group in the brightness discrimination task was greater than that for the saline controls (Fig. 3a). In the pattern discrimination, the GM1-treated rats committed more errors on 17 of the 28 test days (Fig. 3b). Although these comparisons were not based upon a criterion for evaluating significant intergroup differences on individual test days, the finding that GM1 treatments did not result in behavioral improvements is inconsistent with previous reports that exogenous gangliosides facilitate recovery (Sabel et al., 1985b). Measurements of the lesions revealed no statistical differences between groups. In addition, severe degeneration was observed within the LGNd and there were no intergroup differences.

To our knowledge, these results are the first to indicate that exogenous gangliosides do not improve behavioral recovery after brain lesions. In trying to ac-

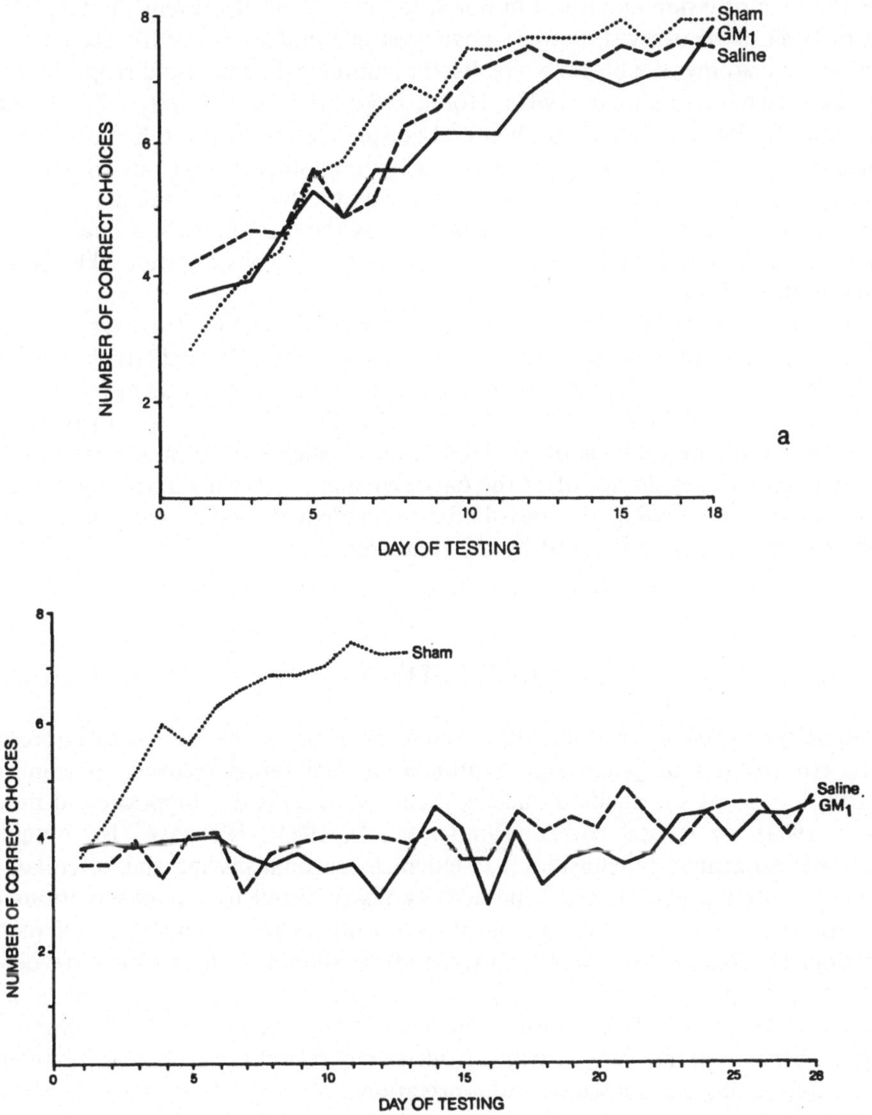

Figure 3. Ganglioside injections did not reduce behavioral deficits following occipital cortex lesions. GM1-treated rats made more errors on 14 of the 18 days of testing in a brightness discrimination task (a) and on 17 of the 28 test days on a pattern discrimination task (b).

count for these results, we noted the report by Toffano et al. (1984c) that in rats with unilateral transection of the nigrostriatal pathway, GM1's beneficial effect on recovery of striatal tyrosine hydroxylase activity is dependent upon a normal light/dark cycle. The recovery normally induced by GM1 treatments was absent in animals which were housed in darkness. Toffano et al. (1984c) proposed that "the mechanisms through which GM1 exerts its effect are regulated through the visual system, either directly or through hormone secretion via the circadian rhythm" (p. 405). If this hypothesis is accurate, the type of lesion employed in our study may in itself prevent the facilitating effects of GM1 on recovery. It is well known that bilateral lesions of the occipital cortex inhibit the learning of a black/white discrimination task, and also disrupt the usual dark preference of rats (Altman, 1962a; Horel, 1966; Meyer and Meyer, 1977). In addition, Altman (1926b) has found that bilateral occipital lesions disrupt the normal light-mediated diurnal activity rhythm of rats. This finding indicates that rats with such lesions are unable to maintain a normal diurnal rhythm due to an inability to use environmental visual cues, in much the same way as the dark-housed animals in Toffano's study were unable to use visual cues. Our rats conceivably were like those of Toffano et al. (1984c).

Based upon the above considerations, we propose than an intact visual system may be necessary for a brain-damaged animal to benefit from GM1's facilitatory effects on behavior. The animal must be given the opportunity to use the system in order for recovery to occur. This may be related to the regulation of the circadian rhythm per se. However, while the findings of Toffano et al. (1984c) may be important in interpreting our results, they do not offer the only explanation. We are currently pursuing other alternatives, including the possibility of inadequate ganglioside degradation and/or aberrant sprouting induced by the treatments.

PERSPECTIVES

Despite the problems encountered in these experiments, we have been impressed with the effectiveness of ganglioside treatments in facilitating recovery of cognitive tasks after lesions of the caudate nucleus (Sabel et al., 1984b), hippocampal inputs (Karpiak, 1983), or frontal cortex (Dunbar et al., 1984). We have also obtained evidence that exogenous gangliosides may enhance neuronal sparing and/or collateral sprouting by altering neural metabolic activity (as reflected by our observations on G-6-PDH). It is very likely that gangliosides do not provide benefits in all model preparations (Toffano et al., 1984b), so some effort should be directed toward determining which types (and/or loci) of brain injury are amenable to treatments with growth-promoting agents. In addition, some thought must be given to the development of specific, behavioral measures which can accurately reflect the effects of treatments without confounding the subsequent interpretations.

At the present time, gangliosides are the only systemically administered agents which seem to be capable of enhancing recovery from traumatic injuries to the central nervous system. Under these circumstances, it may be worthwhile for investigators concerned with the problem of functional recovery to develop and extend the parameters which would lead to more successful employment of this substance as a treatment for severe brain injury.

ACKNOWLEDGMENTS

The research described in this review was supported by a contract from Fidia Research Laboratories, Abano Terme, Italy. Dr. Gino Toffano graciously provided the samples of GM-1 and AGF2. We thank Dr. Robert O'Connell of the Worcester Foundation for Experimental Biology who permitted us the use of his optical densitometer. We also acknowledge Clark University for the student support it provided during the course of this research.

REFERENCES

Altman J (1962a) The effects of lesions in central nervous system visual structures on light aversion of rats. Am J Physiology 202: 1208-1210.

Altman J (1962b) Diurnal activity rythyms of rats with lesions of superior colliculus and visual cortex. Am J Physiology 202: 1205-1207.

Agnati LF, Fuxe K, Calza L, Benfenati F, Cavicchioli L, Toffano G, Goldstein M (1983) Gangliosides increase the survival of lesioned nigral dopamine neurons and favour the recovery of dopaminergic synaptic function in striatum of rats by collateral sprouting. Acta Physiol Scand 119: 347-364.

Ando M, Miwa M, Kato K, Nagata Y (1984) Effects of denervation and axotomy on nervous system-specific protein, ornithine decarboxylase, and other enzyme activities in the superior cervical sympathetic ganglion of the rat. J Neurochem 42: 94-100.

Barron KD (1983) Axon reaction and central nervous system regeneration. In: FJ Seil (ed): Nerve, Organ, and Tissue Regeneration: Research Perspectives. Academic Press, New York, pp. 3-36.

Benowitz LI, Shashoua VE, Yoon MG (1981) Specific changes in rapidly transported proteins during regeneration of the goldfish optic nerve. J Neurosci 1: 300-307.

Dunbar GL, Sabel BA, Firl AC, Stein DG (1984) GM1 gangliosides reduce spatial alternation deficits following bilateral lesions of the mediodorsal frontal cortex. Soc Neurosci Abst 10: 306-309.

Durham D and Rubel EW (1985) Afferent influences on brain stem auditory nuclei of the chicken: changes in succinate dehydrogenase activity following cochlea removal. J Comp Neurol 231: 446-456.

Fass B and Ramirez JJ (1984) Effects of ganglioside treatments on lesion-induced behavioral impairments and sprouting in the CNS. J Neurosci Res 12: 445-458.

Fass B, Ramirez JJ, Mahadik SP, Karpiak SE (1985) Effects of GM1 gangliosides on cholinergic enzymes and Na-K-ATPase in the hippocampus after fimbria-fornix transection in rats. Soc Neurosci Abst 11: 266-I.

Fass B and Steward O (1983) Increases in protein-precursor incorporation in the denervated neuropil of the dentate gyrus during reinnervation. Neurosci 9: 653-664.

Fass B and Stain DG (1986) Effects of fimbria-fornix transection and ganglioside treatments on histochemical staining for glucose-6-phosphatate dehydrogenase in the lateral septum. Synapse, in press.

Ferrari G, Fabris M, Gorio A (1983) Gangliosides enhance neurite outgrowth in PC12 cells. Devel Brain Res 8: 215-221.

Field PM, Coldham DE, Raisman G (1980) Synapse formation after injury in the adult rat brain: preferential reinnervation of denervated fimbrial sites by axons of the contralateral fimbria. Brain Res 189: 103-113.

Field PM and Raisman G (1983) Relative slowness of heterotypic synaptogenesis in the septal nuclei. Brain Res 272: 83-100.

Freed WJ, de Medinaceli L, Wyatt RJ (1985) Promoting functional plasticity in the damaged nervous system. Science 227: 1544-1552.

Glick SD, Jerussi TP, Fleisher LN (1976) Turning in circles: the neuropharmacology of rotation. Life Sci 18: 889-896.

Gorio A, Aporti F, Norido F, Canella R (1981) Ganglioside treatment in experimental diabetic neuropathy. In: Rapport MM, Gorio A (eds): Gangliosides in Neurological and Neuromuscular Function, Development, and Repair. Raven Press, New York, pp. 259-266.

Harkonen MHA and Kauffman FC (1974) Metabolic alterations in the axotomized superior cervical ganglion of the rat. II. The pentose phosphate pathway. Brain Res 65: 141-157.

Horel JA, Bettinger LA, Royce GJ, Meyer DR (1966) Role of neocortex in the learning and relearning of two visual habits by the rat. J Comp Physiol Psycol 61: 66-78.

Hughes HC (1976) Anatomical and neurobehavioral investigations concerning the thalamocortical organization of the rat's visual system. J Comp Neurol 175: 311-336.

Karpiak SE (1983) Ganglioside treatment improves recovery of alteration behavior after unilateral entorhinal cortex lesion. Exper Neurol 81: 330-339.

Karpiak SE and Mahadik SP(1984) Reduction of cerebral edema with GM1 ganglioside. J Neurosci Res 12: 485-492.

Kauffman FC and Harkonen MHA (1977) Metabolites and enzymes of the pentose phosphate pathway in isolated nerve endings. J Neurochem 28: 745-750.

Kimura H, Naito K, Nakagawa K, Kuriyama K (1974) Activation of hexose monophosphate pathway in brain by electrical stimulation in vitro. J Neurochem 23: 79-84.

Li YS, Rapport MM, Karpiak SE (1985) Acute effects of GM1 ganglioside on CNS injury: assessment of dosing schedule for optimal functional response. Soc Neurosci Abst 11: 175-18.

Massarotti M (1983) Ganglioside treatment of alcoholic neuropathies: experimental and clinical aspects. Pharmacol Biochem Behav 1: 51-54.

Meibach RC and Siegel A (1977) Efferent connections of the septal area in the rat: an analysis utilizing retrograde and anterograde transport methods. Brain Res 119: 1-20.

Meyer DR and Meyer PM (1977) Dynamics and bases of recoveries of functions after injuries to the cerebral cortex. Physiol Psychol 5: 133-165.

Orlando P, Cocciante G, Ippolito G, Massari P, Roberti S, Tettamanti G (1979) The fate of tritium labeled GM1 ganglioside injected in mice. Pharmacol Res Comm 11: 759-773.

Pozza G, Saibene V, Comi G, Canal N (1981) The effect of ganglioside administration in human diabetic peripheral neuropathy. In: Rapport MM, Gorio A (eds): Gangliosides in Neurological and Neuromuscular Function, Development, and Repair. Raven Press, New York, pp. 253-258.

Purpura DP, Pappas GD, Baker HJ (1978) Fine structure of meganeurites and secondary growth processes in feline GM1-gangliosidosis. Brain Res 143: 1-12.

Raisman G (1969) Neuronal plasticity in the septal nuclei of the adult rat. Brain Res 14: 25-48.

Rapport MM (1981) Introduction to the biochemistry of gangliosides. In: Rapport MM, Gorio A (eds): Gangliosides in Neurological and Neuromuscular Function, Development, and Repair. Raven Press, New York, pp. xv-xix.

Ricaurte GA, Seiden LS, Schuster CR (1984) Further evidence that amphetamines produce long-lasting dopamine neurochemical deficits by destroying dopamine nerve fibers. Brain Res 303: 359-364.

Robinson TE (1984) Behavioral sensitization: characterization of enduring changes in rotational behavior produced by intermittent injections of amphetamine in male and female rats. Psychopharmacol 84: 466-475.

Roisen FJ, Bartfield H, Nagele R, Yorke G (1981) Ganglioside stimulation of axonal sprouting in vitro. Science 214: 577-578.

Sabel BA, Dunbar GL, Stein DG (1984a) Gangliosides minimize behavioral deficits and enhance structural repair after brain injury. J Neurosci Res 12: 429-443.

Sabel BA, Slavin MD, Stein DG (1984b) GM1 ganglioside treatment facilitates behavioral recovery following bilateral brain damage. Science 225: 340-342.

Sabel BA, Dunbar GL, Butler WM, Stein DG (1985a) GM1 gangliosides stimulate neuronal reorganization and reduce rotational asymmetry after hemitransections of the nigro-striatal pathway. Exp Brain Res 60: 27-37.

Sabel BA, Dunbar GL, Fass B, Stein DG (1985b) Gangliosides, neuroplasticity, and behavioral recovery after brain damage. In: Will BE, Schmitt P, Dalrymple-Alford JC (eds): Brain Plasticity, Learning and Memory. Plenum Press, New York, pp. 481-493.

Skene JHP, Willard M (1981) Characteristics of growth-associated polypeptides in regenerating toad retinal ganglion cell axons. J Neurosci 1: 419-426.

Starr MS, Summerhayes M (1985) Dysfunction of the midbrain angular complex can accentuate circling behaviour in the rat. Exp Brain Res 58: 45-55.

Stein DG (1981) Functional recovery from brain damage following treatment with nerve growth factor. In: van Hof MW, Mohn G (eds): Functional Recovery from Brain Damage, Elsevier, Amsterdam, pp. 423-443.

Stein DG (1985) Fetal brain tissue transplant techniques: a cautionary note. Neurobiology of Aging 6: 157-160.

Stein DG and Fass B (1985) Time course of changes in histochemical staining for glucose-6-phosphate dehydrogenase (G-6-PDH) in the lateral septum after fimbria-fornix transections. Soc Neurosci Abst 11: 266-267.

Stein DG, Labbe R, Firl A Jr, Mufson EJ (1985) Behavioral recovery following implantation of fetal brain tissue into mature rats with bilateral, cortical lesions. In: Bjorklund A, Stenevi U (eds): Neural Grafting in the Mammalian CNS. Elsevier, Amsterdam, pp. 605-614.

Steward O and Vinsant SL (1983) The process of reinnervation in the dentate gyrus of the adult rat: a quantitative electron microscopic analysis of terminal proliferation and reactive synaptogenesis. J Comp Neurol 214: 370-386.

Swanson LW and Cowan WM (1979) The connections of the septal region in the rat. J Comp Neurol 186: 621-656.

Tettamanti G, Venerando B, Roberti S, Chigorno V, Sonnino S, Ghidoni R, Orlando P, Massari P (1981) The fate of exogenously administered brain gangliosides. In: Rapport MM, Gorio A (eds): Gangliosides in Neurological and Neuromuscular Function, Development, and Repair. Raven Press, New York, pp. 225-240.

Toffano G, Savoini G, Moroni F, Lombardi G, Calza L, Agnati LF (1983) GM1 ganglioside stimulates the regeneration of dopaminergic neurons in the central nervous system. Brain Res 261: 163-166.

Toffano G, Savoini GE, Moroni F, Lombardi G, Calza L, Agnati LF (1984a) Chronic GM1 ganglioside treatment reduces dopamine cell body degeneration in the substantia nigra after unilateral hemitransection in rat. Brain Res 296: 233-240.

Toffano G, Agnati LF, Fuxe K, Aldinio C, Consolazione A, Valenti G, Savoini G (1984b) Effect of GM1 ganglioside treatment on the recovery of dopaminergic nigro-striatal neurons after different types of lesions. Acta Physiol Scand 122: 313-322.

Toffano G, Savoini G, Aporti F, Calzolari S, Consolazione A, Maura G, Marchi M, Raiteri M, Agnati LF (1984c) The functional recovery of damaged brain: the effect of GM1 monosialoganglioside. J Neurosci Res 12: 397-408.

Troyer H (1980) Principles and Techniques of Histochemistry. Boston Little, Brown and Company.

Watson WE (1965) An autoradiographic study of the incorporation of nucleic acid precursors by neurons and glia during nerve regeneration. J Physiol 180: 741-753.

Wojcik M, Ulas J, Oderfeld-Nowak B (1982) The stimulating effect of ganglioside injections on the recovery of choline acetyltransferase and acetylcholinesterase activities in the hippocampus of the rat after septal lesions. Neurosci 7: 495-499.

Gangliosides and neuronal plasticity
G. Tettamanti, R.W. Ledeen, K. Sandhoff,
Y. Nagai, G. Toffano (eds.)
Fidia Research Series, vol. 6
Liviana Press, Padova, © 1986

STUDIES ON THE INVOLVEMENT OF POLYAMINES FOR THE TROPHIC ACTIONS OF THE GANGLIOSIDE GM1 IN MECHANICALLY AND 6-HYDROXYDOPAMINE LESIONED RATS. EVIDENCE FOR A PERMISSIVE ROLE OF PUTRESCINE

I. Zini, M. Zoli, L.F. Agnati, K. Fuxe[1], E. Merlo Pich, P. Davalli[2], A. Corti, G. Gavioli, G. Toffano[3]

Departments of Human Physiology and Biochemistry,
University of Modena, Modena, Italy;
[1] Department of Histology, Karolinska Institutet, Stockholm, Sweden;
[2] Department of Biochemistry, University of Parma, Parma, Italy and
[3] Fidia Research Laboratories, Abano Terme, Padova, Italy

INTRODUCTION

In 1983 we were able to demonstrate for the first time that chronic ganglioside GM1 treatment in mechanically lesioned rats could partly counteract the retrograde cell body degeneration taking place in the central nervous system (CNS) following lesion. The system analyzed was the nigrostriatal dopamine (DA) system (Agnati et al., 1983; 1984a,b; Toffano et al., 1983, 1984; Fuxe and Agnati, 1984). In these experiments with partially hemitransected rats it was also possible to demonstrate that chronic ganglioside GM1 treatment could increase the synaptic function of DA, since following a three week treatment the ipsilateral amphetamine induced rotational behaviour in these rats was partly counteracted (Agnati et al., 1983). In the present paper these behavioural experiments have been countinued by studying the action of chronic GM1 treatment on water intake in partially hemitransected rats and also on the ipsilateral bias of such animals using a sensorimotor test battery.

To characterize the biochemical mechanisms underlying the trophic action of GM1, we have studied the possible role of polyamines in the trophic activity of chronic ganglioside GM1 treatment. To this purpose putrescine levels and ornithine decarboxylase (ODC) activity have been measured in the substantia nigra and in the striatum

---

Abbreviations: ODC, ornithine decarboxylase; αDFMO, D,L-α-difluoromethyl ornithine; 60H-DA, 6-hydroxydopamine; DA, dopamine; i.p. intraperitoneal injection; NA, noradrenaline; CNS, central nervous system; PAP, peroxidase antiperoxidase.

following mechanical as well as 6-hydroxydopamine (60H-DA) induced lesions of the mesostriatal DA systems, and the ability of GM1 to modulate these two biochemical parameters has been tested in vivo and in vitro in intact and in lesioned animals (Agnati et al., 1985a,b,c). The role of polyamines has also been evaluated by the use of the irreversible ODC inhibitor D,L-α-difluoromethyl ornithine (α-DFMO).

BEHAVIOURAL EXPERIMENTS IN PARTIALLY HEMITRANSECTED MALE RATS. THE EFFECTS OF CHRONIC GANGLIOSIDE GM1 TREATMENT

In these experiments the effects of chronic ganglioside GM1 treatment (10 mg/kg, i.p., once daily) have been evaluated on daily water intake and on ipsilateral bias, using a sensorimotor test battery, in partially hemitransected male rats.

Materials and Methods

Male Sprague-Dawley rats (200 g body weight) were used and were divided into three groups. Two groups were partially hemitransected at the meso-diencephalic level (Köning and Klippel level A3200 μm) using a thin metal blade as previously described (Agnati et al., 1983). The operation was performed under ketamine anesthesia (80 mg/kg). Immediately following operation one group of rats was treated for three weeks with the ganglioside GM1 (10 mg/kg, i.p., once daily), while the other lesioned group of animals was treated with saline once daily. The third group was sham-operated and also treated with saline once daily. The rats were caged individually.

Water Intake

The experimental schedule in these experiments is summarized in Figure 1. Food and water consumption were recorded by weighing the animals, the water bottle and the standard laboratory chow pellets in the cage basket. In this way body weight, daily water and food consumption were measured for 24 days after the lesion. On day 6 after operation the rats were deprived of water in the presence of food and the subsequent consumption of food was measured. On day 7 the rats were tested for prandial drinking by removing all food from the cages and measuring the subsequent water consumption in the absence of food (Oltmans and Harvey, 1972). On day 8 food and water were again given *ad libitum* to the rats and the measurements of the 24 h consumption of water and food were continued. The same procedure was repeated on days 13, 14, 15 and 20, 21, 22 after operation.

Sensorimotor Test Battery

The sensorimotor test battery was modified from that described by Marshall and Teitelbaum (1974) and from that by Björklund et al. (1980), to test sensorimotor orientation and coordinated limb use on each side of the body. The test was performed on coded animals.

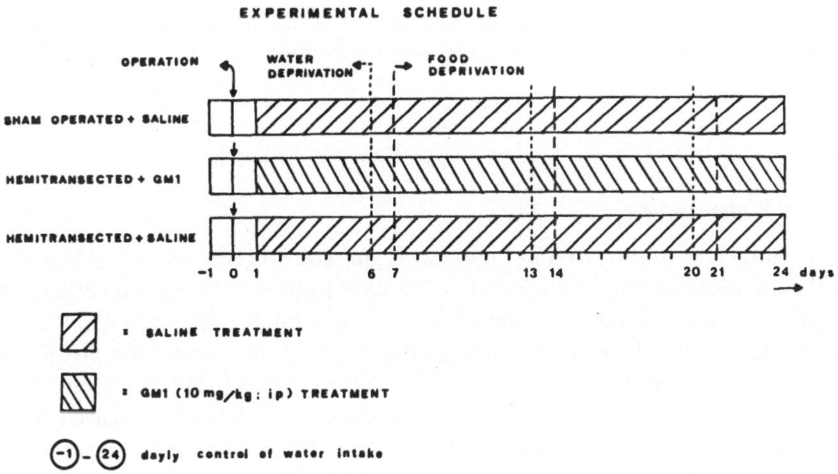

Figure 1. Experimental schedule for the studies on water intake in sham-operated animals treated with saline and in partially hemitransected rats treated with saline or the ganglioside GM1 (10 mg/kg, i.p., once daily).

The rats were placed on an unstable platform (40 × 40 cm) located approximately 70 cm from the bench surface in a way to minimize locomotion during testing. The general posture was scored according to the degree of postural asymmetry. The occurrence of spontaneous rotation on the platform before the beginning of the tests was also recorded. The degree of head orientation and biting of the stimulus probe was recorded on the two sides for each of the following stimuli:

(a). somaesthesia: a light pin prick was applied to 6 sites, rostrocaudally and dorsoventrally on the lateral surface of each side of the body.

(b). whisker touch: a wooden probe was gently brushed against the vibrissae, approaching the rat from the lower rear end of the animal in order to avoid the visual field.

(c). snout probe: the wood probe was lightly rubbed against the snout of the rat.

Limb reflexes were assessed in the following tests:

(a). forelimb suspension: the experimenter made the rat grasp a stainless steel bar with one forepaw and suspended the animal by this limb. A normal rat will rapidly grasp the bar with the free forepaw and try to pull itself up onto the bar. The latency to achieve successful pull-up was recorded with a failure criterion of 5 seconds.

(b). climbing grid: the rat was placed head downwards on a vertical wire grid and the latency to turn to a head up position was recorded.

(c). mouth probe: the rat was held vertically with head upwards and the wooden probe inserted into the side of the mouth. Normal rats reacted lightly or showed a liking to this stimulus. The grasping of the probe with ipsilateral forepaw or attempts to bite the probe were recorded.

The response to the above mentioned tests was rated on a 3-point scale: 0 (absent), 1 (weak) or 2 (strong). An index was obtained from individual test scores on each of the two sides of the brain, and the difference in the index between the two sides was calculated. The sensorimotor test battery was performed on post-operative days 6, 13 and 20.

Results

Studies on Water Intake

The results are summarized in Figure 2. It was shown that the partial hemitransection at the meso-diencephalic level acutely 1-2 days following the lesion caused a marked reduction of water intake compared with the sham-operated animals. This action induced by the partial hemitransection disappeared by day 3 postoperatively. It was shown that this early action was in part counteracted by the GM1 treatment. Thus, on day 2 the water intake in the GM1 treated group was significantly increased compared with the saline treated partially hemitransected animals. As seen in the figure, prandial drinking was reduced both on day 7 and on day 14 in GM1 and saline treated, partially hemitransected rats. However, on day 15 the water intake was significantly increased in the GM1 treated, partially hemitransected rats compared with the saline treated, partially hemitransected rats. Furthermore, on day 23 only the water intake in the saline

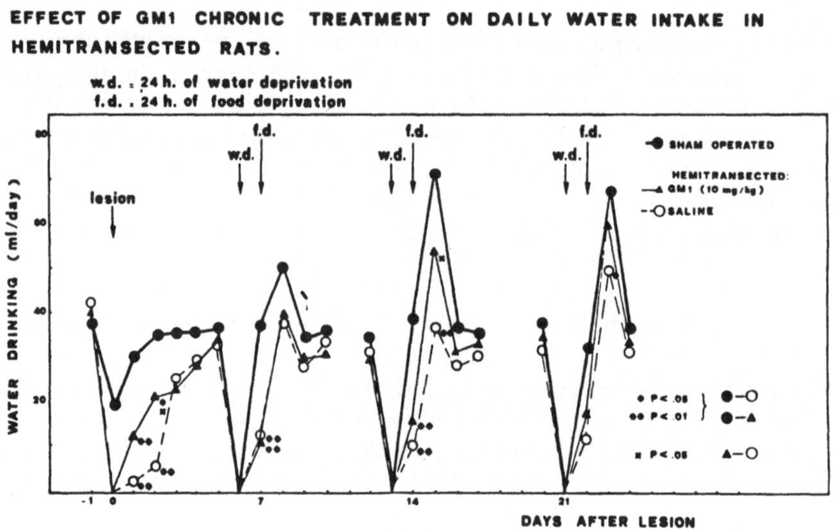

Figure 2. Effects of GM1 treatment on daily water intake in partially hemitransected rats. For experimental protocol, see text in Figure 1. The mean water intake each day is shown (n = 10). The significances shown have been obtained by the use of a multiple comparison test (Dunn's test). As stated in the text, ganglioside GM1 exerts an early action on the hypodipsia caused by the partial hemitransection, which is unrelated to regeneration. Subsequently the rats have been deprived of water on day 16 and 20 after the lesion, and subsequently of food on days 7, 14 and 21 after the lesion but with free access to water. From day 14 it can be seen that the GM1 treated partially hemitransected rats have a water intake value which is in between that observed in saline treated hemitransected rats and in sham-operated rats.

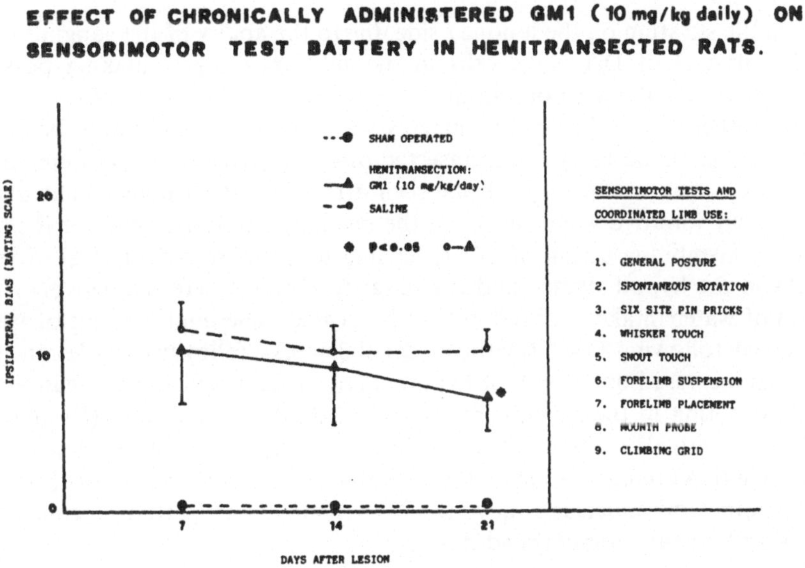

Figure 3. Three experimental groups similar to those described in the experiments shown in Figure 1 have been tested for ipsilateral bias and coordinated limb use (see panel on the right) 7, 14 and 21 days following the lesion. The statistical analysis has been carried out by means of Dunn's test (n = 10).

treated partially hemitransected rats was significantly reduced compared with that in sham-operated control animals.

Studies on Ipsilateral Bias

The results are summarized in Figure 3. Using the sensorimotor test and the test for coordinated limb use as indicated in the figure and described in the Materials and Methods section, it was found that the ipsilateral bias observed in partially hemitransected male rats was significantly reduced in operated animals which had been treated chronically with GM1 (10 mg/kg, i.p., once daily). However, the action of GM1 only became significant after 21 days of treatment.

Discussion

The present results obtained on ipsilateral bias in partially hemitransected rats give further evidence that chronic treatment with the ganglioside GM1 can increase striatal function on the lesioned side. Thus, the results further substantiate the view that chronic ganglioside GM1 treatment can produce functional recovery in striatum (Agnati et al., 1983). In view of the important role of the DA striatal systems in sensorimotor integration it seems likely that the reduction of ipsilateral bias is at least in part induced by the ability of the ganglioside GM1 treatment to increase DA synaptic function in the striatum on the lesioned side due to the ability of the ganglioside to increase the survival of DA nerve cells in the substantia nigra, making possible an enhancement of collateral sprouting in the striatum (Fuxe et al., 1986).

Also another type of behaviour, namely water intake, was found to be improved by GM1 treatment in partially hemitransected rats. Two types of action were observed: one seen after acute treatment and one seen after chronic treatment. The ability of GM1 to acutely improve water intake in the partially hemitransected rat is probably caused via a membrane action of GM1, leading to on increased survival of lesioned nerve cells in the hypothalamus and the basal forebrain where the networks for the regulation of water intake are predominantly located. The improvement of water intake observed following GM1 treatment for 2-3 weeks following the lesion may be related to an increased outgrowth of hypothalamic afferents or to an enhancement of collateral sprouting in the hypothalamus and in the basal forebrain (Kojima et al., 1985). In line with this observation, we have recently observed an increase in the noradrenaline (NA) reinnervation of the hypothalamus and especially of the paraventricular hypothalamic nucleus following GM1 treatment in partially hemitransected rats (Fuxe and Agnati, unpublished data).

THE INVOLVEMENT OF POLYAMINES IN THE TROPHIC ACTIONS OF GANGLIOSIDES ON THE CNS OF EXPERIMENTAL RATS

It has recently been shown by Agnati et al., (1985a) that polyamines in the CNS may play a permissive role with regard to the trophic actions of GM1 treatment in partially hemitransected animals. Thus, striatal putrescine levels were markedly increased 7 days after a partial hemitransection of the lesioned side, an increase which was not modulated by treatment with GM1. Furthermore, treatment with the irreversible ornithine decarboxylase (ODC) inhibitor α-DFMO (Fig. 4) was found to completely block the ability of GM1 treatment to increase the survival of DA cell bodies in the substantia negra upon a partial hemitransection at the meso-diencephalic level (Agnati et al., 1985a). These results are in good agreement with the well known view that polyamines play an important role for protein synthesis and for the survival of neurons

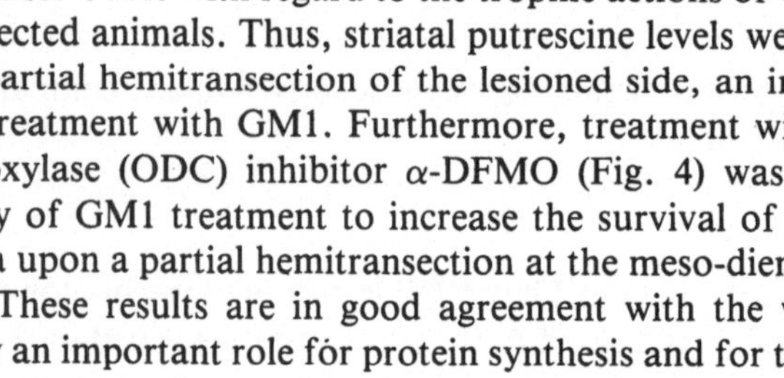

Figure 4. The chemical structure of D,L-α-difluoromethyl ornithine (α-DFMO) is shown. This inhibitor acts as a suicide inhibitor of ODC.

after injury (Gilad and Gilad 1983; Lewis et al., 1978; Shan, 1979; Tabor and Tabor, 1984).

In the present experiments further studies have been performed on the effects of GM1 treatment on striatal putrescine levels and on ODC activity in the substantia nigra and striatum of intact and partially hemitransected rats. A scheme illustrating the synthesis and metabolism of polyamines in shown in Figure 5. Spermidine and spermine levels were not measured in the present experiments. In a separate experiment the effects of the irreversible ODC inhibitor α-DFMO (Fig. 4) were used to evaluate the importance of polyamines for the trophic actions of the ganglioside GM1 exerted in the substantia nigra of partially hemitransected rats (Agnati et al., 1985a).

Materials and Methods

Normal male Sprague-Dawley rats were used (200 g body weight). Two groups were partially hemitransected at the meso-diencephalic level as described above. Immediately following the operation one group of lesioned rats was treated with the ganglioside GM1 (10 mg/kg, once daily), while the other group of lesioned rats was treated with saline. The third group was sham-operated and treated with saline (control rats).

Striatal putrescine content was determined as described by Seiler et al. (1978) (Agnati et al., 1985a,b). ODC activities in the substantia nigra and in the striatum were determined as described by Jänne and William-Ashman (1971) (Agnati et al., 1985c).

In the morphometrical experiments the area of the substantia nigra was determined by means of the IBAS Image Analyzer in transverse sections of the midbrain, stain-

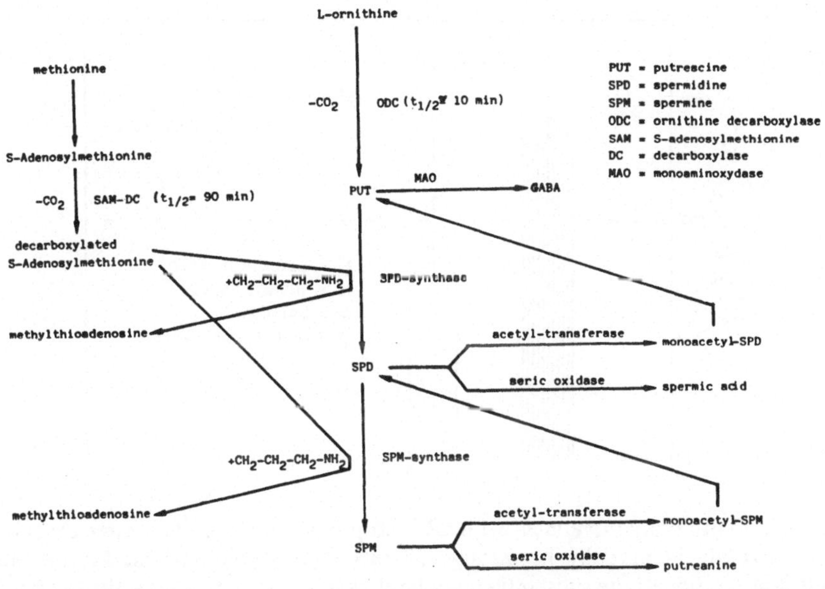

Figure 5. Biosynthetic pathway of polyamines: the rate limiting step is the conversion of L-ornithine to putrescine by means of ornithine decarboxylase (ODC).

ed for tyrosine hydroxylase (TH) immunoreactivity (Agnati et al., 1984b) using the PAP procedure of Sternberger (1979).

Results

Effects of GM1 Treatment on the Area of the Substantia Nigra in partially Hemitransected Animals with or without Concomitant Treatment with α-DFMO

The results are summarized in Figure 6. It is shown that the area of the substantia nigra on the lesioned side is significantly increased after GM1 treatment compared with the combined GM1 plus α-DFMO treatment. As seen, the area of the lesioned side is always related to the area of the intact side expressed as per cent of the ratio found in saline treated, partially hemitransected rats. It is also of substantial interest that the area found in the GM1 treated animals of lesioned side is clearly above the 100% value, while the value obtained in the group of animals treated both with GM1 and α-DFMO was below that of the 100% value. Thus, it seems possible that the inhibition of the ODC by means of α-DFMO enhances the degenerative processes in the substantia nigra that follow a partial hemitransection. Taken together the results underline the importance of polyamine synthesis in the substantia nigra for the trophic activity of gangliosides (Agnati et al., 1985a).

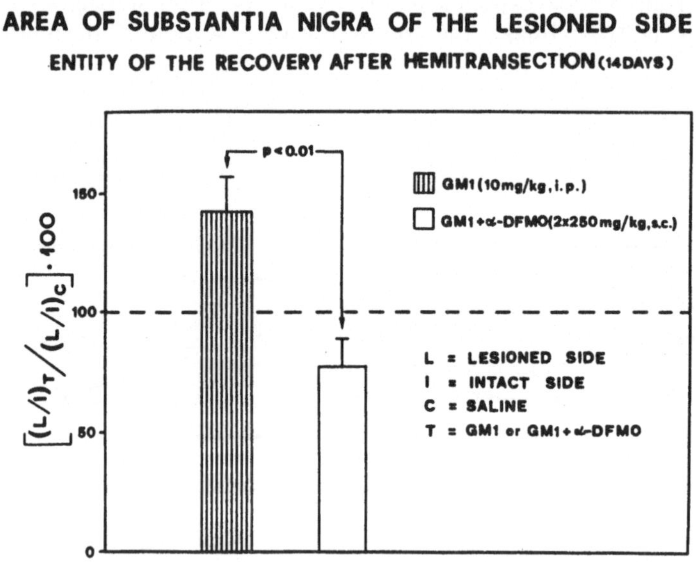

Figure 6. Failure of chronic treatment with GM1 to promote regeneration after chronic treatment with αDFMO in partially hemitransected rats. The 100% level marks the entity of retrograde degeneration at the substantia nigra level observed in partially hemitransected saline treated rats. The statistical analysis was performed by means of Student's t-test.
Sample size: n = 10.

Effects of GM1 Treatment on Striatal Putrescine Contents after Partial Hemitransection

The results are summarized in Figure 7. The results are expressed as percent of unoperated saline control animals, and the peak increase of striatal contents was observed 7 days following the partial hemitransection. The increases could still be observed 2 weeks and 3 weeks following the operation. However, no differences were observed between saline treated and GM1 treated operated animals. Furthermore, the putrescine contents were significantly increased on the intact side only at 21 day time interval in the GM1 treated animals.

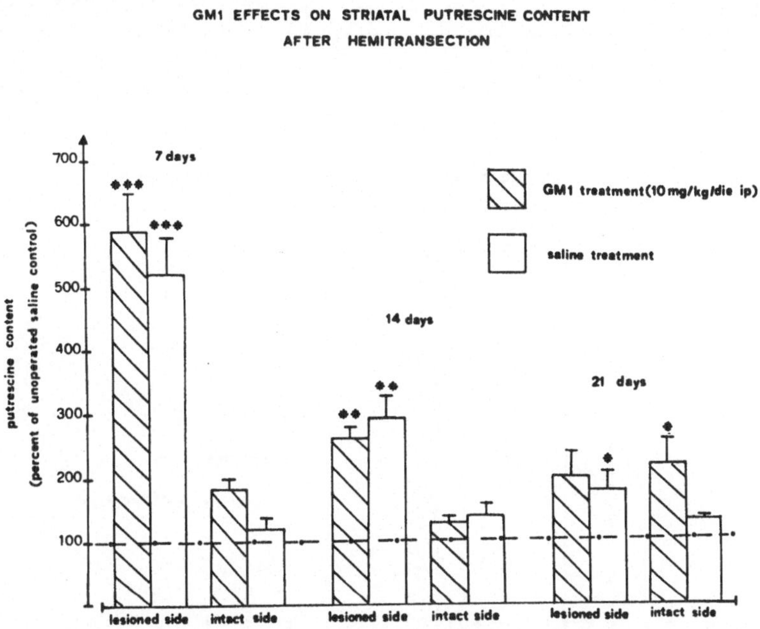

Figure 7. Increases in striatal putrescine content are observed after partial hemitransection (open bars) and partial hemitransection in combination with GM1 treatment (hatched bars) (10 mg/kg, i.p., once daily). The statistical analysis has been carried out by means of Wilcoxon test. All comparisons have been made with the respective control group (unoperated saline control group: putrescine levels in striatum = 5.41 ± 0.39 nmoles/g tissue; means ± s.e.m.).
Sample size: n = 15.

Effects of GM1 in Vitro on ODC Activity in Rat Brain

The results are shown in Figure 8. ODC activity *in vitro* was only affected by GM1 in a very high concentration of 10^{-3} M. This action was probably caused by a nonspecific effect in view of the high concentration of GM1 used.

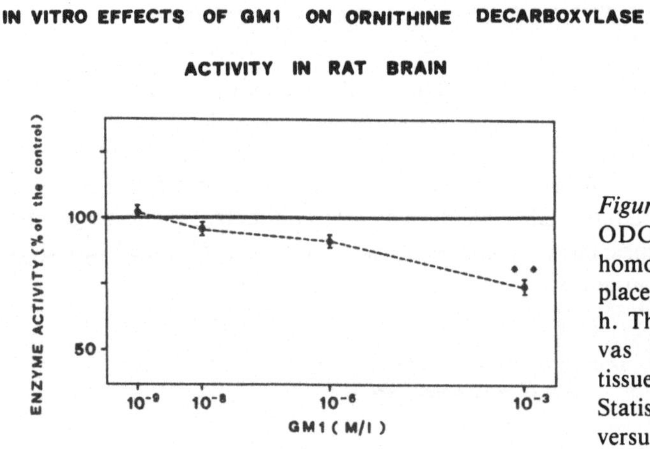

Figure 8. In vitro effects of GM1 on ODC activity in striatal homogenates. The incubation took place at 37°C for a time period of 1 h. The basal value of ODC activity vas 58±3 pmoles CO_2/h/g of tissue. Means ± s.e.m. are shown. Statistical analysis was treatment versus controls. ** = $p < 0.01$.

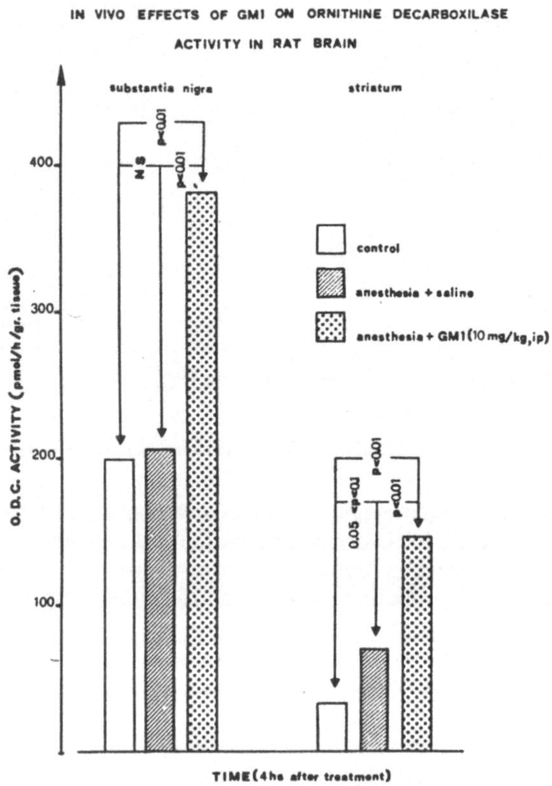

Figure 9. Acute GM1 treatment (4 h) is shown in vivo to enhance ODC activity in the substantia nigra and striatum of intact rats. The basal value of ODC activity in striatum was 26.7±0.65 pmoles/CO_2/h/g of tissue. The basal value of ODC activity in substantia nigra was 201±15 pmoles/CO_2/h/g of tissue. Statistical analysis was carried out by means of Dunn's test. (n = 15).

Effects of GM1 in Vivo on ODC Activity in Rat Brain of Intact Rats

The results are summarized in Figure 9. It was shown that following 4 h of treatment with GM1 (10 mg/kg, i.p.) a marked increase in ODC activity was observed in the substantia nigra and in the striatum. Furthermore, the anesthesia by itself produced a trend for an increase in ODC activity in the striatum.

Effects of a Partial Hemitransection on ODC Activity in the Substantia Nigra and Striatum of the Lesioned Side

The results are expressed as per cent of respective unoperated saline mean value and are summarized in Figure 10. The peak increase in ODC activity after partial hemitransection was observed already 4 h postoperatively in the substantia nigra followed by a gradual decline in ODC activity to reach the basal value 10 days after the operation. The peak increase of ODC activity in the striatum after a partial hemitransection was observed 1 day following the operation, followed again by a gradual decline in ODC activity to reach the basal value 10 days post operatively.

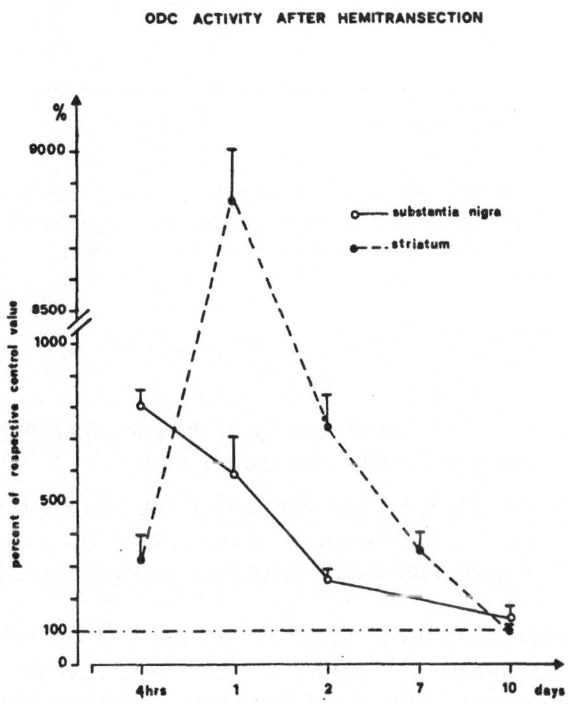

Figure 10. Effects of partial hemitransection on ODC activity on the lesioned side of substantia nigra and of the striatum at various time intervals following the lesion. Means ± s.e.m. are shown (n = 15).

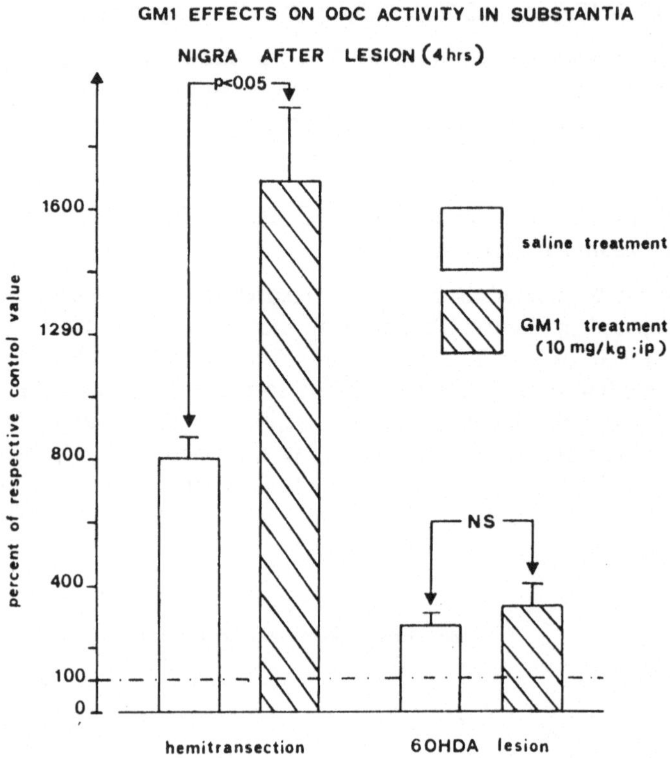

Figure 11. Effects of GM1 treatment on ODC activity in the lesioned side of substantia nigra in rats with mechanical (partial hemitransection) and neurotoxic (60H-DA) lesions of the mesostriatal DA system. Means ± s.e.m. are shown (n = 15). Mann Whitney U-test.

Effects of GM1 Treatment on ODC Activity in the Substantia Nigra and in the Striatum of mechanically and 6OH-DA Lesioned Rats

The 60H-DA induced lesion of the nigrostriatal DA neurons was performed by injection of 60H-DA (8 μg/4 μl) stereotaxically into the substantia nigra. GM1 was given i.p. in a dose of 10 mg/kg immediately after the operation and ODC activity in the substantia nigra and striatum was measured 4 h and 48 h after the operation in mechanically (partial hemitransection, see above) and 60H-DA lesioned male rats.

The results are summarized in Figure 11. Four hours after the operation, only in the partially hemitransected rats, could the ganglioside GM1 treatment produce a marked preferential increase in the substantia nigra of the lesioned side. On the other hand, in the 60H-DA lesioned rats GM1 treatment at the 4 h time interval had no ability to produce a preferential increase in ODC activity in the substantia nigra of the lesioned side.

Discussion

The results obtained in the present and previous experiments (Agnati et al., 1985a,b,c) have been summarized in Figure 12. In this summary we can see the relationship between the spontaneous and GM1 promoted regeneration of the mesostriatal DA neurons and the increases in striatal putrescine content and in nigral and striatal ODC activity in mechanically or neurotoxically lesioned animals with or without concomitant treatment with the ganglioside GM1. It is clear that the increase in ODC activity and the increase in putrescine levels represent a condition necessary but not sufficient for spontaneous regeneration and for GM1 promoted regeneration of the mesostriatal DA system. Thus, when ODC activation is prevented by treatment with α-DFMO or when only putrescine marked accumulation is prevented (60H-DA lesioned rats), no spontaneous or GM1 promoted regeneration of the mesostriatal DA system can be obtained as evaluated by measurement of TH immunoreactivity in the DA nerve cell bodies and nerve terminal systems of the mesostriatal DA neurons.

Thus, our present and previous experiments (Agnati et al., 1985a,b,c) give evidence that polyamines represent permissive factors in the mechanisms involved in

EXPERIMENTAL GROUPS	TH - IR		PERCENT CHANGES (Δ% of unlesioned saline control)		
			PUTRESCINE CONTENT (time of peak effect)	O.D.C. ACTIVITY (time of peak effect)	
	SUBSTANTIA NIGRA	STRIATUM	STRIATUM	SUBSTANTIA NIGRA	STRIATUM
HEMITRANSECTION	+ -	+ -	414% (7 days)	812% (4 hrs)	>5000% (24 hrs)
HEMITRANSECTION + GM1	+ +	+	480% (7 days)	1647% (4 hrs)	>3000% (24 hrs)
HEMITRANSECTION+α-DFMO	- -	-	25% (7 days)		
HEMITRANSECTION+α-DFMO+GM1	- -	-	40% (7 days)		
60H-DA	-	-	~0% (7 days)	200% (4 hrs)	58% (4 hrs)
60H-DA + GM1	-	-	~0% (7 days)	275% (4 hrs)	65% (4 hrs)

TH-IR evaluated at 7, 14, 28, 56 days after lesion.
Putrescine content evaluated at 7, 14, 21 days after lesion.
O.D.C. activity evaluated at 4, 24, 48 hrs and 7 days after lesion.

Figure 12. Summing up table obtained in studies on partially hemitransected rats with or without GM1 or α-DFMO treatment or combined treatment with both drugs as well as in 60H-DA lesioned rats with or without concomitant GM1 treatment. The results have been taken from the present paper and from other papers by our groups (Agnati et al., 1985a,b,c). In the table the status of the regenerative processes in the mesostriatal DA neurons has been evaluated by studies on tyrosine hydroxylase immunoreactivity (— — means reduced recovery; — means no recovery; — + means trend for recovery; + means observable recovery; + + means clearcut recovery). The percent changes observed in putrescine content and ODC activity represent the changes in percent compared to unlesioned saline control group.

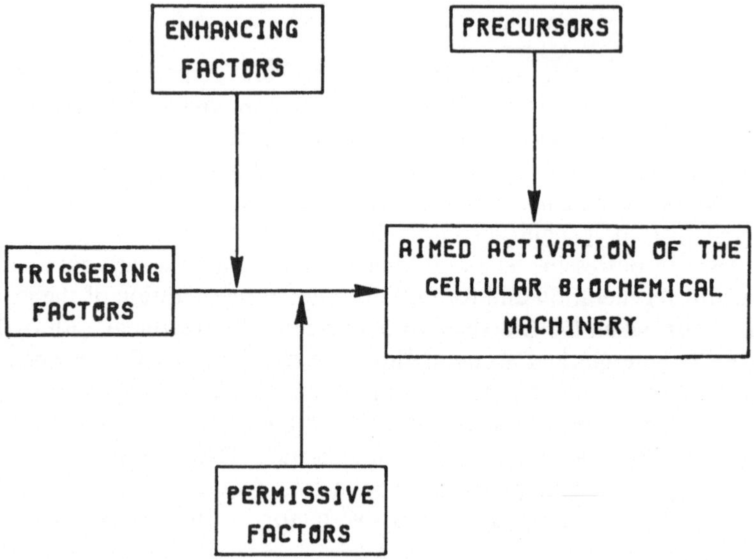

Figure 13. Hypothesis on the existence of different types of factors (triggering factors, enhancing factors, permissive factors) which are essential for the directed activation of neuronal biochemical machinery which leads to regenerative processes.

producing the regeneration of central neurons. In Figure 13 we have summarized a possible interplay of different types of factors involved in the regeneration of neurons. In order to obtain regeneration the directed activation of the cellular biochemical machinery of the nerve cell is needed. Such a direct activation may be induced by triggering factors (neurotrophic factors ?). However, in order for these triggering factors to act optimally they need enhancing factors (gangliosides ?), which may interact with the triggering factors, *e.g.*, at the level of the membrane bound protein kinases leading to increases of protein phosphorylation (Agnati et al., 1985d; Fuxe et al., 1986). Furthermore, the present and previous studies (Agnati et al., 1985a,b,c) clearly show that in order for the trophic factors and for the gangliosides to be effective they require the presence of permissive factors (putrescine ?). Only when all these factors are present in the lesioned area can a directed activation of the cellular biochemical machinery of the nerve cells take place so that macromolecules can be synthesized, which are essential for the survival of the nerve cells and also for their regenerative capacity.

ACKNOWLEDGMENTS

This work has been supported by a grant from the Swedish Medical Council (04X-715), a grant from Knut and Alice Wallenberg Foundation and by a CNR (Italy) grant. We are grateful for the excellent technical assistance of Lotta Frösell and Barbro Tinner and for the excellent secretarial assistance of Anne Edgren.

REFERENCES

Agnati LF, Fuxe K, Calza L, Benfenati F, Cavicchioli L, Toffano G, Goldstein (1983) Gangliosides increase the survival of lesioned nigral dopamine neurons and favour the recovery of dopaminergic synaptic function in striatum of rats by collateral sprouting. Acta Physiol Scand 119: 347-363.

Agnati LF, Fuxe K, Toffano G, Calza L, Benfenati F, Zini I, Battistini N, Goldstein M, Zoli M (1984a) Evidence for structural plasticity in the nigrostriatal dopamine neurons: Morphometrical and biochemical evidence for a trophic action of gangliosides on partially lesioned nigrostriatal dopamine neurons. In: Racagni G, Paoletti R, Kielholz P (eds): Clinical Neuropharmacology, Vol 7, Suppl, Raven Press, New York, pp. 594-595.

Agnati LF, Fuxe K, Calza L, Goldstein M, Toffano G, Giardino L; Zoli M (1984b) Computer assisted morphometry and microdensitometry of transmitter identified neurons with special reference to the mesostrial dopamine parthway. II. Further studies on the effects of the GM1 ganglioside on the degenerative and regenerative features of mesostriatal dopamine neurons. Acta Physiol Scand, Suppl 532: 37-44.

Agnati LF, Fuxe K, Zini I, Davalli P, Corti A, Calza L, Toffano G, Zoli M, Piccinini G, Goldstein M (1985a) Effects of lesions and ganglioside GM1 treatment on striatal polyamine levels and nigral DA neurons. A role of putrescine in the neurotropic activity of gangliosides. Acta Physiol Scand 124: 499-506.

Agnati LF, Fuxe K, Zoli M, Davalli P, Corti A, Zini I, Toffano G (1985b) Effects of neurotoxic and mechanical lesions of the mesostriatal dopamine pathway on striatal polyamine levels in the rat: Modulation by chronic ganglioside GM1 treatment. Neurosci Lett 61: 339-344.

Agnati LF, Fuxe K, Davalli P, Zini I, Corti A, Zoli M (1985c) Neurotoxic and mechanical lesions of the mesostriatal dopamine system of the male rat. Studies on striatal ornithine decarboxylase activity. Acta Physiol Scand 125: 173-175.

Agnati LF, Fuxe K, Benfenati F, Zoli M, Owman C, Diemer NH, Kåhrström J, Toffano, Cimino M (1985d) Effects of Ganglioside GM1 treatment on striatal glucose metabolism, blood flow, and protein phosphorylation of the rat. Acta Physiol Scand 125: 43-53.

Björklund A, Dunnett SB, Stenevi U, Lewis ME, Iversen SD (1980) Reinnervation of the denervated striatum by substantia nigra. Transplants: functional consequences as revealed by pharmacological and sensorimotor testing. Brain Res 199: 307-333.

Fuxe K and Agnati LF (1984) Gangliosides as regulatory factors in central catecholamine neurons. In: Magistretti PJ, Morrison JH, Bloom FE (eds): Discussions in Neurosciences, Nervous system development and repair. Foundation FESN, Vol. 1, No. 2, pp. 84-89.

Fuxe K, Agnati LF, Benfenati F, Zini I, Gavioli G, Toffano G (1986) New evidence for the morphofunctional recovery of striatal function by ganglioside GM1 treatment following a partial hemitransection of rats. Studies on dopamine neurons and protein phosphorylation, this volume.

Gilad GM and Gilad VH (1983) Polyamine biosynthesis is required for survival of sympathetic neurons after axonal injury. Brain Res 273: 191-194.

Kojima H, Gorio A, Janigro D, Jonsson G (1984) GM1 ganglioside enhances regrowth of noradrenaline nerve terminals in rat cerebral cortex lesioned by the neurotoxin 6-hydroxydopamine. Neuroscience Vol 13, No 4: 1011-1022.

Jänne J and Williams-Ashman HG (1971) On the purification of L-ornithine decarboxylase from rat prostrate and effects of thiol compounds on the enzyme. J Biol Chem 246: 1725-1732.

Lewis ME, Lakshmanan J, Nagaiah K, McDonnell PC, Guroff G (1978) Nerve growth factor increases activity of ornithine decarboxylase in rat brain. Proc Natl Acad Sci, USA Vol. 75, pp. 1021-1023.

Marshall JF and Teitelbaum P (1974) Further analysis of sensory inattention following lateral hypothalamic damage in rats. J Comp Physiol Psychol 86: 375-395.

Oltmans GA and Harvey JA (1972) LH syndrome and brain catecholamine levels after lesions of the nigrostriatal bundle. Physiol Behaviour 8: 69-78.

Seiler N, Knodgen B, Eisenbeiss F (1978) Determination of di- and polyamines by high performance liquid chromatographic separation of their 5-Dimethylaminonaphthalene-1-Sulphonyl derivatives. J Chromatogr 145: 29-39.

Shan GG (1979) Polyamines in the nervous system. Biochem Pharmacol 28: 1-6.

Sternberger LA (1979) Immunocytochemistry 2nd Ed, Wiley, New York.

Tabor CW and Tabor H (1984) Polyamines. Ann Rev Biochem 53: 749-90.

Toffano G, Savoini G, Moroni F, Lombardi G, Calza L, Agnati LF (1983) GM1 ganglioside stimulates the regeneration of dopaminergic neurons in the central nervous system. Brain Res 261: 163-166.

Toffano G, Agnati LF, Fuxe K, Aldinio C, Consolazione A, Valenti G, Savoini G (1984) Effects of GM1 ganglioside treatment on the recovery of dopaminergic nigro-striatal neurons after different types of lesion. Acta Physiol Scand 122: 313-321.

Gangliosides and neuronal plasticity
G. Tettamanti, R.W. Ledeen, K. Sandhoff,
Y. Nagai, G. Toffano (eds.)
Fidia Research Series, vol. 6
Liviana Press, Padova, © 1986

THE EFFECT OF GM1 ON CEREBRAL METABOLISM, MICROCIRCULATION AND HISTOLOGY IN FOCAL ISCHEMIA

Joel H. Greenberg, Martin Reivich, Rudolf Urbanics, Kortaro Tanaka, Eors Dora, Gino Toffano[1]

Cerebrovascular Research Center, Department of Neurology,
University of Pennsylvania, Philadelphia, PA 19104 and
[1]Fidia Research Laboratories, via Ponte della Fabbrica, 3a, Abano Terme,
Padova, Italy

INTRODUCTION

GM1 is one of the major gangliosides in the mammalian brain, and when it is exogenously administered is known to penetrate the blood-brain-barrier (Orlando et al., 1979) and to be actively incorporated into the neuronal membrane (Toffano et al., 1980). In fact, the interaction of the gangliosides with brain membranes is reportedly accompanied by various metabolic effects including increased adenylate cyclase activity (Partington and Daly, 1979), increased phosphodiesterase activity (Davis and Daly, 1980), enhanced dopamine release (Cumar et al., 1978) and modification of (Na+, K+) ATPase activity (Leon et al., 1981).

Cerebral ischemia induces a wide spectrum of metabolic derangements, ranging from severely depressed glucose utilization along with a depletion of high-energy phosphates to a persistent activation of anaerobic glycolysis which is closely associated with histological damage. Damage to plasma membranes is one of the major phenomena consistently characterizing irreversible ischemic cellular injury (Farber et al., 1981). Indeed, it is generally accepted that membrane failure triggers the processes causing irreversible structural damage during the ischemic event (Astrup, 1982). There is also growing evidence that irreversible cellular damage is manifested during the recovery period following ischemia (Siesjo, 1981). A postischemic decline in the activities of the (Na+, K+) ATPase and adenylate cyclase, which are critically dependent on the integrity of the membrane, notably on its lipid-layer composition, has been demonstrated (Schwartz et al., 1976).

Abbreviations: lCBF, local cerebral blood flow; ECoG, electrocortical activity; MCA, middle cerebral artery; MEG, middle ectosylvian gyrus; ^{14}C-2DG, ^{14}C-2-deoxy-glucose; lCMRgl, local cerebral metabolic rat for glucose; CVV, cerebrocortical vascular volume; MCAO, middle cerebral artery occlusion.

It is therefore reasonable to hypothesize that postischemic brain can be protected by intervention that either prevents or delays the development of membrane failure. In the present study, we have examined the effect of GM1 on cerebral glucose metabolism, local cerebral blood flow (lCBF), recovery of ECoG, NAD/NADH redox state, and histologic alterations at the end of a 4 hr recovery period following 2 hr of MCA occlusion in cats.

MATERIALS AND METHODS

These studies were undertaken in cats anesthetized with 40 mg/kg sodium pentobarbital. The animals were immobilized with gallamine triethiodide (5 mg/kg) which was repeated every hour. The trachea, the femoral arteries and veins and the left lingual artery were cannulated and the animals were placed on a ventilator. The heads of the animals were mounted in stereotaxic holders and the skin and muscles were removed from both sides of the skull. A 12-mm diameter hole was drilled in the left temporal bone above the middle ectosylvian gyrus (MEG) and the dura was opened. A cranial window was fixed into the burr hole with dental cement, as described previously (Dora, 1984). Electrical activity of the exposed brain cortex was measured with silver electodes built into the plastic ring of the cranial window (Dora, 1984) while EEG of the contralateral side was measured with copper screws fixed into the parietal bone.

The left middle cerebral artery (MCA) was exposed via a transorbital approach (O'Brien and Waltz, 1973). The MCA was occluded in 19 animals in close proximity to its branching from the internal carotid artery with a miniature Mayfield clip. After two hours of occlusion the clip was removed and the blood flow through the MCA was reinstated. In 5 of 19 animals with MCA occlusion GM1 (30mg/kg) was injected intravenously when the occlusion was released, i.e. after 120 min. of occlusion (GM/120 group). In 9 animals, GM1 was injected 30 minutes after the occlusion (GM1/30 group) while 5 animals were occluded but not treated (Untreated stroke group).

At 3 hr and 15 min after the release of the MCA occlusion, 250 μCi of ^{14}C-2-deoxyglucose (^{14}C-2DG, New England Nuclear) was injected intravenously as a bolus for the determination of the local cerebral metabolic rate for glucose (1CMRgl) (Sokoloff et al., 1977). Arterial blood samples were taken during the next 45 min, for the measurement of time course of the plasma ^{14}C-2-deoxyglucose activity and the glucose concentration.

After 4 hours of recirculation, the animals were sacrificed with an intravenous administration of potassium chloride solution, and the brains were quickly removed and cut into coronal blocks. Alternate blocks were immersion-fixed in formalin for 10 days and then processed for histological evaluation by light microscopy. The remainder of the blocks were frozen in Freon-22 and 20 μm thick sections were cut in a cryostat for autoradiography. The cut sections were dried and then placed on x-ray film together with calibrated 14-C standards.

Quantitative densitometric analysis of the autoradiograms were performed by using a computer-assisted microdensitometer system. For the calculation of lCMRgl, the operational equation of Sokoloff et al. (1977) was employed.

Histological changes were graded into four categories (Ginsberg et al., 1979):

histologically normal (Grade 0), slight ischemic changes with only a few scattered affected neurons (eg. shrunken cell bodies with triangular, darkly stained cytoplasm and a loss of discrete Nissl substance) (Grade 1), moderate changes with a typical microscopic field containing several affected neurons (Grade 2), a large portion of affected neurons which are often accompanied by edematous neuropil (Grade 3).

Cerebrocortical NADH fluorescence and reflectance (sum of the scattered and reflected light) were measured as described previously (Harbig et al., 1976; Dora, 1984). To avoid the changes in NADH fluorescence caused by the alterations in tissue blood content, a correction method based on artificial hemodilution was used (Harbig et al., 1976). For this purpose, 0.1-0.3 ml isosmotic and oxygenated dextran solution was injected into the lingual artery and the NADH concentration- dependent alterations in NADH fluorescence were calculated using the measured correction factor.

Reflected light was used to determine the changes in cerebrocortical vascular volume (CVV), mean transit time of cortical blood flow (tm), and blood flow (CBF) (Eke et al., 1979). In order to determine the reference value of tm, 0.1-0.3 ml isosmotic and oxygenated dextran solution was injected into the lingual artery. Mean transit time was calculated from the hemodilution-induced reflectance reactions by the area over height analysis (Meier and Zierler, 1954) with the reference value of tm considered as 100%. CBF changes were calculated by dividing the percentage values of CVV with the percentage values of tm (Eke et al., 1979).

Since there is a considerable variability in the severity of ischemia in the MCA occlusion model (Garcia, 1984), experimental animals were classified into 3 groups based on the severity of the depression in electrocortical activity (ECoG) of the ischemic hemisphere at 30 minutes after occlusion. When the ratio of the ECoG mean amplitude of the ischemic side over that of the non-ischemic side was less than 0.2, the animals were considered part of the severe group; when this ratio was between 0.2 and 0.7, the animals were classified as part of the moderate group. The GM1/120 group consisted of three moderate and two severe animals while the GM1/30 group consisted of three moderate and six severe. All groups were compared to the corresponding not treated MCA occluded animals (three moderate, two severe).

The pH, pCO2 and pO2 values of the arterial blood samples along with plasma glucose were determined during the control period, at the 60th min of MCA occlusion (MCAO), and at every hour during the recovery period.

RESULTS

The mean amplitude of ipsilateral ECoG as well as the ECoG ratio were depressed similarly during MCAO in all stroke groups (Fig. 1) Within 1 hour after the release of the MCAO, the mean amplitude of the ipsilateral ECoG and the ECoG ratio showed slight but similar increases in both the treated and untreated stroke groups. There was no significant difference between the untreated and treated (neither at 30 nor at 120 min.) groups. Similarly, the time of treatment didn't change the EEG activity and the GM1/30 and GM1/120 are lumped together in Figure 1.

Following the release of MCAO, a significant CBF increase occurred in the ganglioside-treated stroke groups, whereas CBF was slightly lower than its preischemic

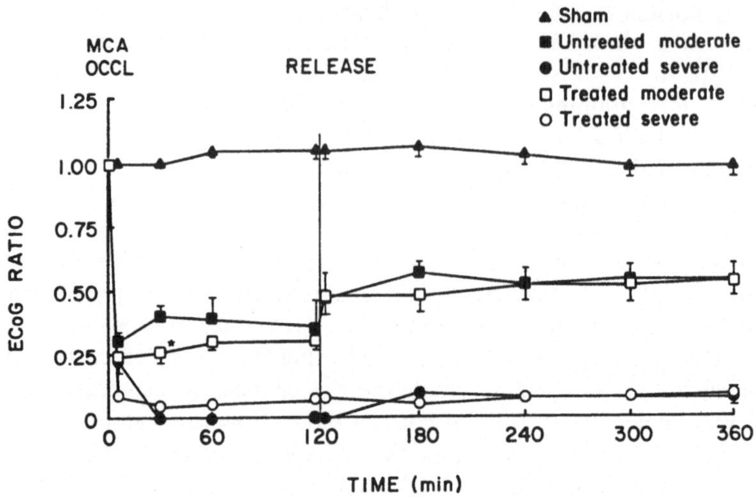

Figure 1. Changes in ECoG in untreated cats and in cats treated with GM1 during the occlusion and the recovery period. Differences between the treated and untreated stroked animals are indicated. Note the distinct separation between the moderate and severely stroked animals.

reference value in the untreated stroke groups. During the later phases of recovery, CBF remained approximately at the same high level in the ganglioside-treated stroke groups, while the untreated stroke groups slowly recovered to the preischemic reference level.

The NAD/NADH changes in the stroked groups were in opposite directions depending on the severity of the stroke occurring during the occlusion. The NAD/NADH redox state during MCAO became reduced in the moderate groups with NADH becoming spontaneously hyperoxidized in the severe animals. The GM1 treatment diminished the reduction in the moderate group and the hyperoxidation in the severe animals, but there was no significant difference between the corresponding treated und untreated groups. Note, however, that the untreated severe group was significantly different from the sham operated animals throughout the entire study, while the treated severe group was not significantly different from the sham after 240 min.

Data of either treated or untreated animals were divided into moderate and severe groups as described. The regions analysed in each animal for local cerebral glucose utilization were grouped into four territories: central MCA territory, peripheral MCA territory, non-MCA territory and cerebellar gray matter based on the blood supply of the MCA (Ginsberg et al., 1976).

The lCMRgl data of the treated and the untreated groups were expressed as a percentage of the corresponding sham data in each anatomical structure, and the average values of this percentage were calculated in each territory. For each region an analysis of variance was performed with two factors: treatment (untreated; GM1/30, GM1/120 treated) and severity (severe, moderate). The only region where either a significant treatment effect or interaction occurred was the left peripheral MCA territory in the moderate group (Fig. 2). For this region a significant treatment effect was found ($p < .025$) with the metabolism being lower in the GM1/30-treated group than in the controls.

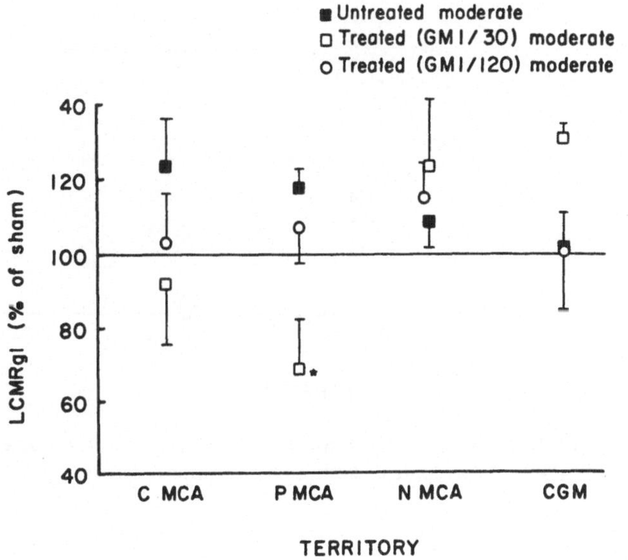

Figure 2. Local cerebral metabolic rate for glucose (lCMRgl) of the GM1-treated and untreated moderate animals in the occluded hemisphere.

The left hemisphere of both the treated and the untreated severe groups demonstrated a statistically significant depression in the central MCA territory as compared to either the sham value or to the contralateral hemisphere. There was no difference, however, between the treated and the untreated groups. In the moderate groups, (Fig. 2) the left hemisphere of the treated group had a near normal lCMRgl, while the corresponding area of the untreated group exhibited a statistically significant activation of lCMRgl compared to the control level. There was no statistical difference, however, in either hemisphere between GM1/30 and GM1/120 groups.

In the severe group, there was no statistically significant difference in the histological damage in the central MCA territory between the treated and the untreated groups, although the treated group showed a less homogeneous pattern (Fig. 3). In the untreated animals the histological damage was almost always grade 3, while in the treated group some of the tissie was only moderately damaged. A similar pattern was observed in the central and peripheral MCA territories of the moderately stroked animals.

DISCUSSION

The severity of ischemia (before the injection of ganglioside) was identical in the untreated and treated stroke groups. Although ganglioside treatment increased CBF

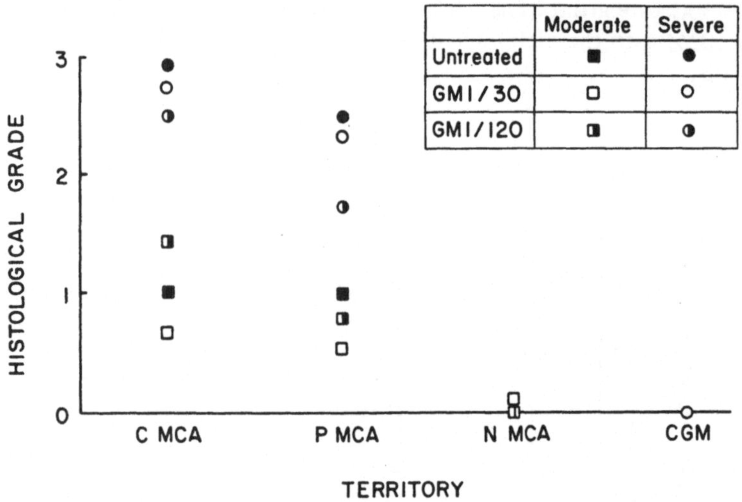

Figure 3. Histological findings in territories on the left, stroked side in the treated and untreated moderate and severe animals. Histological damage was graded into four categories (Grade 0 - 3) (see text).

significantly, the ECoG ratios of the treated and untreated stroke groups during the recovery period were not significantly different. This suggests that GM1 may not have the protective effect against ischemic neuronal damage that has been described for damaged peripheral nerves (Ceccarelli et al., 1976) or hippocampal lesions (Wojcik et al., 1982). It should be noted, however, that the recovery process may not have been followed for a sufficient period of time and that single dose treatment may not have the optimal effect.

The ganglioside treatment in our studies enhanced and extended the reactive hyperemia which, per se, could have induced secondary impairment of microcirculation, increased formation of edema, and, hence, further disintegration of the neurons (Yatsu, 1982; Garcia et al., 1983). The effect of GM1 on CBF is significant. Furthermore, GM1 does not decrease arterial blood pressure as do other cerebral vasodilators. The increase in CBF is due partly to the ability of GM1 to prevent the constriction of cerebrocortical vessels observed during the initial 120 min of recirculation in the untreated stroke group. Since GM1 did not alter the mean arterial blood pressure it can only be assumed that the decrease in mean transit time was primarily due to dilatation of large upstream arterial vessels (some major brances of the MCA or collateral vessels) that indirectly supply the cortical area monitored by the reflectometer.

Although the average lCMRgl values in the non-ischemic hemisphere of the treated group tended to be slightly higher than those of the non-treated group in many regions, there was no significant difference in lCMRgl suggesting that GM1 may not have an effect on the glucose metabolism of normal brain tissue. The possibility that the hemisphere contralateral to the MCA occlusion may differ metabolically from the normal brain (unpublished observations) makes it difficult to simply apply the above results to the normal brain.

In the ischemic hemisphere, there was a statistically significant difference in lCMRgl between the treated and the untreated groups in the peripheral MCA territory. The treated animals of the moderate group demonstrated a significant decrease of lMCRgl, whereas the average lCMRgl value of the corresponding untreated animals was higher than the sham values (Fig. 2). Histological findings in this region suggested that there was less damage to the treated (GM1/30) animals in the moderate group than that of the untreated animals (Fig. 3). Previous observations in our laboratory indicate that the brain tissue may have the capability of reducing its metabolic demand during ischemia so as to maintain a normal flow-metabolism couple as well as structural integrity (unpublished data). These data suggest that the depression of lCMRgl accompanied by less severe damage in the treated group may be due to a protective mechanism that has been triggered by GM1. In support of this hypothesis, gangliosides are known to inhibit some membrane-bound enzymes such as ($Na+$, $K+$) ATPase or $K+$-dependent nitrophenyl-phosphatase under certain in vitro conditions (Mirzoyan et al., 1971), and have also been reported to inhibit endogenous respiration of isolated brain mitochondria (Mirzoyan et al., 1979).

A previous histopathological study has shown that ischemic brain tissue may exhibit structural alterations as early as 15 min after MCA occlusion (Garcia et al., 1977), and that these changes mature and progress during the recirculation period (Ito et al., 1975; Kirino and Sano, 1984). It is reasonable, therefore, to assume that GM1 may exert some beneficial effects if it is administered during the early phase of ischemia before tissue damage has become irreversible. Since gangliosides are known to facilitate the recovery of the lesioned central nervous system over a period of 2 weeks or more (Wojcik et al., 1982; Toffano et al., 1983), the effect of GM1 treatment should also be examined with a chronic stroke model. Preliminary studies in a model in which GM1 is administered daily for one week following 2 hours of MCA occlusion, show a faster improvement in neurologic function than in untreated animals.

Although the ganglioside GM1 significantly increases CBF in the recovery period, it does not significantly alter the recovery kinetics of cortical NAD/NADH redox state. The degree of hyperoxidation and reduction did diminish, however, in the severe and moderate treated groups respectively. The treated severe group reached its preischemic level 240 min. after the MCA release and subsequently did not differ significantly from the sham operated animals, while the untreated severe group remained oxidized even 480 min. after the release of the MCA occlusion remaining significantly different from the sham during the entire observation period.

The secondary reduction of NADH that occurred in the late recovery phase in the moderate stroke group was not significantly diminished by the ganglioside treatment. This suggests that the secondary NAD reduction does not depend principally on the supply of oxygen since ganglioside treatment doubled CBF. Because the impairment of cortical microcirculation is very heterogenous, even at the microcirculatory level (Garcia et al., 1983; Garcia, 1984), the role of tissue hypoxia in the secondary NADH reduction cannot be excluded even though these measurements were made in a relatively large area, i.e. approximately 1 mm in diameter. We also must consider the possibility that the mitochondrial electron transport in dying cells is shut down and glycogenolysis and glycolysis are stimulated in some of the surviving cells. Cessation of mitochondrial electron transport and increased substrate supply can lead to NADH reduction (Chance and Williams, 1955; Dora, 1984).

In summary, these results show that the ganglioside GM1 injected as a single dose at 30 min. after MCA occlusion, or at the time of MCAO release, exerts no facilitatory effect on the recovery of the electrical activity of the affected brain cortex. Furthermore, it does not alter significantly the recovery kinetics of cortical NAD/NADH redox state in the recirculatory period following 120 min occlusion of the middle cerebral artery, although the NAD reduction in the moderate and the hyperoxidation in the severe group decreased. GM1 does significantly increase CBF in the ischemic area and produces a decrease of lCMRgl accompanied by less severe histologic damage in the periphery of the ischemic region, especially in the case of early treatment (30 min. after the MCA occlusion). The potentially protective effects must be examined further before any beneficial action of this substance during ischemia can be adduced.

REFERENCES

Astrup J (1982) Energy-requiring cell functions in the ischemic brain. J Neurosurg 56: 482-497.

Ceccarelli B, Aporti F, Finesso M (1976) Effect of brain gangliosides on functional recovery in experimental regeneration and reinnervation. Adv Exp Med Biol 12: 275-293.

Chance B, Williams RG (1955) Respiratory enzymes in oxidative phosphorylation. III. The steady state. J Biol Chem 217: 409-427.

Cumar FA, Maggio B, Caputto R (1978) Dopamine release from nerve endings induced by polysialogangliosides. Biochem Biophys Res Commun 84: 65-69.

Davis CW, Daly JW (1980) Activation of rat cerebral cortical 3',5'-cyclic nucleotide phosphodiesterase activity by gangliosides. Mol Pharmacol 17: 206-211.

Dora E (1984) A simple cranial window technique for optical monitoring of cerebrocortical microcirculation and NAD/NADH redox state. Effect of mitochondrial electron transport inhibitors and anoxic anoxia. J Neurochem 42: 101-108.

Eke A, Hutiray GY, Kovach AGB (1979) Induced hemodilution detected by reflectometry for measuring miroregional blood flow and blood volume in cat brain cortex. Am J Physiol 236: H759-H768.

Farber JL, Chien KR, Mittnacht S Jr (1981) The pathogenesis of irreversible cell injury in ischemia. Am J Pathol 102: 271-281.

Garcia JH (1984) Experimental ischemic stroke: A review. Stroke 15: 5-14.

Garcia JH, Kalimo H, Kamijyo Y, Trump BF (1977) Cellular events during partial cerebral ischemia. 1. Electron microscopy of feline cerebral cortex after middle-cerebral-artery occlusion. Virchows Arch B Cell Path 25: 191-206.

Garcia JH, Lowry SL, Briggs L, Mitchem HL, Morawetz L, Halsey JH, Conger KA (1983) Brain capillaries expand and rupture in areas of ischemia and reperfusion. In: Reivich M, Hurtig HI (eds): Cerebrovascular Diseases, Raven Press, New York, pp. 169-179.

Ginsberg MD, Reivich M, Frinak S, Harbig K (1976) Pyridine nucleotide redox state and blood flow of the cerebral artery occlusion in the cat. Stroke 7: 125-131.

Ginsberg MD, Graham DI, Welsh FA, Budd W (1979) Diffuse cerebral ischemia in the cat: III. Neuropathological sequelae of severe ischemia. Ann Neurol 5: 350-358.

Harbig K, Chance B, Kovach AGB, Reivich M (1976) In vivo measurement of pyridine nucleotide fluorescence from the cat brain cortex. J Appl Physiol 41: 480-488.

Ito V, Spate M, Walker JT Jr, Klatzo I (1975) Experimental cerebral ischemia in mongolian gerbils. I. Light microscopic observations. Acta Neuropath 32: 209-233.

Kirino T, Sano K (1984) Fine structural nature of delayed neuronal death following ischemia in the gerbil hippocampus. Acta Neuropathol 62: 209-218.

Leon A, Tettamanti G, Toffano G (1981) Changes in functional properties of neuron membranes by insertion of exogenous ganglioside. In: Rapport MM, Gorio A (eds): Gangliosides in Neurological and Neuromuscular Function, Development, and Repair, Raven Press, New York, pp. 45-54.

Meier P, Zierler KL (1954) On the theory of the indicator-dilution method for measurement of blood flow and blood volume. J Appl Physiol 6: 731-744.

Mirzoyan SA, Mkheyan EE, Sekoyan ES, Sotskii OP (1971) Effect of gangliosides on the cerebral circulation. Dok Biol Sci 210: 762-764.

Mirzoyan SA, Mkheyan EE, Sekoyan ES, Sotskii OP, Akopov SE (1979) Effect of gangliosides on Na^+, K^+-ATPase activity and conformation of microsomal membranes. Bull Exp Biol Med 86: 1607-1610.

O'Brien MD, Waltz AG (1983) Transorbital approach for occluding the middle cerebral artery without craniectomy. Stroke 4: 201-206.

Orlando P, Cocciante G, Ippolito G, Mossari P, Roberti S, Tettamanti G (1979) The fate of tritium labeled GM1 ganglioside injected in mice. Pharmacol Res Commun 11: 759-773.

Partington CR, Daly JW (1979) Effect of gangliosides on adenylate cyclase activity in rat cerebral cortical membranes. Mol Pharmacol 15: 484-491.

Schwartz JP, Mrsulja BB, Mrsulja BJ, Passonneau JV, Klatzo I (1976) Alterations of cyclic nucleotide-related enzymes and ATPase during unilateral ischemia and recirculation in gerbil cerebral cortex. J Neurochem 27: 101-107.

Siesjo PK (1981) Cell damage in the brain: A speculative synthesis. J Cereb Blood Flow Metabol 1: 155-185.

Sokoloff L, Reivich M, Kennedy C, et al. (1977) The (^{14}C) deoxyglucose method for the measurement of local cerebral glucose utilization: theory, procedure, and normal values in the conscious and anesthetized albino rat. J Neurochem 28: 897-916.

Toffano G, Benvegnu D, Bonetti AC, Facci L, Leon A, Orlando P, Ghidoni R, Tettamanti G (1980) Interactions of GM1 ganglioside with crude rat brain neuronal membranes. J Neurochem 35: 861-866.

Toffano G, Savoini G, Moroni F, Lombardi G, Galza L, Agnati LF (1983) GM1 ganglioside stimulates the regeneration of dopaminergic neurons in the central nervous system. Brain Res 261: 163-166.

Wojcik M, Ulas J, Oderfeld-Nowak B (1982) The stimulating effect of ganglioside injections on the recovery of choline acetyl transferase and acetylcholinesterase activities in the hippocampus of the rat after septal lesions. Neuroscience 7: 495-499.

Yatsu FM (1982) Acute medical therapy of strokes. Stroke 13: 524-526.

Gangliosides and neuronal plasticity
G. Tettamanti, R.W. Ledeen, K. Sandhoff,
Y. Nagai, G. Toffano (eds.)
Fidia Research Series, vol. 6
Liviana Press, Padova, © 1986

ACUTE EFFECTS OF GANGLIOSIDES ON CNS INJURY

Stephen E. Karpiak, Yu Shu Li, Paul Aceto, Sahebarao P. Mahadik[1]

Division of Neuroscience, New York State Phychiatric Institute,
Departments of Psychiatry, [1] Biochemistry and Molecular Biophysics,
College of Physicians & Surgeons, Columbia University,
722 W. 168th St., New York, NY 10032, USA

INTRODUCTION

Recently investigators have focused their research efforts on exogenous ganglioside effects on processes associated with CNS injury. These CNS studies were prompted by reports that ganglioside-treatments enhance peripheral nerve regeneration after damage (Gorio et al., 1981; 1983), and that when gangliosides are added to some cell or tissue cultures, increased neurite outgrowth occurs (Ferrari et al., 1983; Leon et al., 1982; Roisen et al., 1981).

There are now many reports that ganglioside-treated animals show facilitated recovery, as assessed by behavioral, biochemical and histological measures, following several types of CNS injuries. These studies, like the PNS and cell culture studies, postulated that the facilitated recoveries may be due to the stimulatory effect of gangliosides on neuronal sprouting/regeneration (Toffano et al., 1983; Sabel et al., 1984a). This remains a working hypothesis and is supported by several independent studies.

However, several of the CNS studies indicated that gangliosides may have a short-term (within 48 hours) effect on the CNS after injury, which cannot be explained by facilitated sprouting/regeneration. A series of studies in our laboratories support the observation that ganglioside injections have an acute effect on the CNS locus of injury. This acûte effect may itself account for the facilitated recovery, and/or may well be one of the reasons why subsequent sprouting is enhanced.

In the following chapter, we wish to review some of the reported short-term effects of gangliosides and to present our recent findings regarding the acute effects of ganglioside treatment on CNS injury.

Abbreviations: CNS, central nervous system; AChE, acetylcholine esterase; AGF2, inner ester of ganglioside GM1.

INDICATIONS FOR AN ACUTE GANGLIOSIDE EFFECT

Our studies showed that ganglioside injections may be affecting short-term events associated with CNS trauma. To study the effects of gangliosides on functional recovery, we assessed whether ganglioside injections could facilitate recovery of a learned alteration behavior after unilateral lesions of the entorhinal cortex in the rat. Following the lesion, rats lose their ability to perform this particular behavior, but recover it as afferents sprout into regions of the dentate gyrus that were denervated by the entorhinal lesions.

We found that 24 hrs after the entorhinal lesion control rats showed an 80% loss of the previously learned behavior, whereas ganglioside (total brain or GM1) treated rats showed only a 35% loss. This reduced behavioral deficit seen as early as 24 hrs after the lesion could not be explained by enhanced sprouting (Karpiak, 1983). Further, we found that 21% of the rats died following the suction ablation of the entorhinal cortex, whereas ganglioside-treated rats had only a 10% mortality rate (Karpiak, 1984).

Other studies on the effects of gangliosides on entorhinal lesions have also shown acute effects. Fass and Ramirez (1984) reported that at 72 hours after the lesion densitometric analyses of acetylcholinesterase (AChE) staining in the dentate gyrus showed that ganglioside-treated rats had a significant change in the levels of AChE staining as compared with controls. In another study, Fass and Ramirez (1984) performed bilateral entorhinal cortical lesion which resulted in hyperactive behavior. However, rats injected with gangliosides showed reduced levels of hyperactivity (50% of control levels) as soon as 48 hrs after the bilateral lesions.

Sabel et al. (1984b) reported that 48 hours after a unilateral partial hemitransection of the nigrostriatal pathway, ganglioside-treated rats showed a significant reduction in amphetamine induced asymmetric rotational behavior. Recently, we have confirmed this observation (Li et al., 1986).

None of the aforementioned acute effects can easily be explained by facilitated sprouting, particularly since there is no evidence in the adult mammalian CNS that lesion-induced sprouting and reinnervation occur within such a short time frame. Clearly, an alternative mechanism is required to account for these phenomena.

GANGLIOSIDE EFFECTS ON CEREBRAL EDEMA AND Na^+,K^+-ATPASE ACTIVITY

Since our data and that of others show that gangliosides are exerting a significant short-term effect, we began a series of studies to assess whether gangliosides might be reducing the extent of local damage in the CNS. We hypothesized that if gangliosides reduced the extent of damage (e.g. edema, cell loss, fiber degeneration) then the potential for recovery was increased.

Edema

In a first study (Karpiak and Mahadik, 1984) we examined the effects of GM1 ganglioside on CNS edema. We found that 48 hrs after a cortical lesion, rats treated with GM1 ganglioside showed a 33% reduction in edema at the site of injury. There were no effects on CNS water content outside the lesion area.

Na^+,K^+-ATPase

Subsequently we have examined the effects of GM1 ganglioside on levels of Na^+,K^+-ATPase and potassium. Na^+,K^+-ATPase and potassium are both markedly reduced in edemic tissue, indicating membrane failure and disruption of plasma membrane ionic balances. In addition to the functional significance of Na^+,K^+-ATPase, this enzyme is highly enriched in synaptic plasma membranes, and is therefore a good marker for synaptic membrane integrity. At 48 hrs after a cortical lesion, we found a 55.5% reduction in Na^+,K^+-ATPase activity and a 55% reduction in intracellular potassium. In contrast, GM1 ganglioside treated rats showed only a 31% reduction in Na^+,K^+-ATPase activity and a 28% reduction in potassium (Karpiak and Mahadik, 1984). In the same study, no changes were found in blood brain barrier permeability (determined with iodinated albumin) following ganglioside treatment.

TIMING OF INITIAL GANGLIOSIDE INJECTION CRITICALLY AFFECTS BEHAVIORAL OUTCOME

Since these short-term effects are of importance, it seems reasonable to assume that the "timing" of the first ganglioside injection in relation to the time of injury should be a critical parameter in affecting recovery. To test this assumption, we studied the reduction of amphetamine-induced asymmetric rotational behavior in rats induced by partial hemistransection of the nigrostriatal pathway. Sabel et al. (1984a,b) had previously reported that at 48 hrs after hemitransection, GM1 treated rats showed a markedly reduced number of ipsiversive rotations as compared to saline controls. The following study was done to determine if the time between the injury (hemitransection) and the first ganglioside injection affected the behavioral outcome, namely reduced asymmetric rotation.

Hemitransection

Male-Sprague Dawley Rats (200-250 gr) were anesthetized with ether and mounted in a Kopf No. 900 Stereotaxic with the incisor bar at —3.3mm. The surgical procedure was begun by placing a 4mm wide blade 1 mm posterior to bregma and 0.5mm lateral to midline. The blade was lowered 9mm from the calvarium at an angle of 22°, and then moved 0.5mm towards the midline and then 5mm away from the midline to achieve a partial hemitransection of the nigrostriatal pathway.

GM1 Ganglioside Dosing

Separate groups of rats were injected with 20mg/kg i.p. of GM1 ganglioside (Fidia Research Labs.) at varying time intervals (0,2,4,8 and 12 hours) after the surgical hemitransection. There was a saline control group for each group of rats tested at each time interval. All rats received a second injection at 24 hrs after surgery, and were tested for rotational behavior at 48 hrs after surgery.

Rotational Behavior

At 48 hrs after surgery rats were tested for amphetamine-induced rotational behavior. Rats were injected s.c. with 2mg/kg of amphetamine and rotations were monitored in automated rotometers for the 30-90 minute period after amphetamine injection. Scores for each animal were based upon the number of ipsiversive turns minus the number of contraversive turns. Typically, animals had few contralateral turns. Data for each time interval group were finally expressed as the percentage decrease in the number of turns when compared to corresponding saline controls.

Results

As compared to controls, ganglioside treated rats showed a significant reduction in the extent of asymmetric rotation (ANOVA $p<0.01$) (Table 1). The reduction in rotation was maximal (42%) when the first injection of ganglioside was given within 0-2 hrs after surgery (Turkey test: $p<0.01$). The effectiveness of GM1 ganglioside to reduce the rotation was weaker as the time between the lesion and the first injection was increased (greater than 2 hrs), with no statistical significance seen in rats injected 4,8 or 12 hrs after surgery, although all rats treated with GM1 showed a reduction in number of rotations compared to controls (Table 1).

Table 1. *Percentage Reduction[1] in Asymmetric Rotation with GMl Treatment: Critical Timing of First Post-Surgical Injection*

		Hours between surgery and first GMl injection			
	0-HRS	2-HRS	4-HRS	8-HRS	12-HRS
MEAN	38.6%	49.3%	22.0%	28.0%	27.3%
SEM	13.0	13.0	11.9	17.6	7.5

	Analysis by group	
	0 to 2 HRS	4 to 12 HRS
MEAN	42.4%	24.3%
SEM	9.4	7.5

[1] Data for each time interval group are expressed as the percentage reduction in mean scores compared to corresponding control levels.

STRIATAL Na^+,K^+-ATPASE ACTIVITY 48HRS AFTER NIGRO-STRIATAL TRANSECTION

Since we had previously seen changes in membrane Na^+,K^+-ATPase activity in the loci of injury (supra) we began to examine changes in this enzyme in the striata following hemitransection of the nigrostriatal pathway. Denervation of these nuclei after hemitransection leads to a loss of this enzyme activity.

Striatal Membrane Na^+,K^+-ATPase Analysis

Using the same surgical procedure as previously described, rats were hemitransected and injected with GM1 ganglioside (20mg/kg, i.p.) at 0.5 and 24 hrs postsurgery. At 48 hrs the brains were removed for ATPase analysis.

Na^+,K^+-ATPase was assayed in a crude total membrane preparation using our standard procedure modified from published reports (Atterwill et al., 1984; Averit et al., 1984). Animals were anesthetized (ether), perfused with PBS saline and their left and right striatum dissected out. A membrane fraction was prepared by homogenizing specimens in 5 ml of ice cold buffer (40 mM Tris-HCl pH 7.5; 10 mM $MgCl_2$; 10 mM EDTA), and centrifuging for 30 min at 100,000 xg. The resulting pellet was suspended in the homogenizing buffer. Five hundred μl of reaction mixture contained 40 mM Tris-HCl pH 7.5, 150 mM NaCl, 40 mM, KCl, 10 mM $MgCl_2$, 2 mM EDTA, 2 mM EGTA, 4 mM ATP, and 100-200 μg of protein. The inorganic phosphate released at 37°C for 30 min was determined by the method of Eibl and Lands (1969). Na^+,K^+-ATPase activity was then estimated as the difference between the activities in the absence and presence of 5 mM ouabain.

Results

By comparing the striatal Na^+,K^+-ATPase from the transected side with that from the untransected side, we found a 38% activity reduction in denervated striata from saline controls, whereas GM1 treated rats showed only a 9.9% decrease in activity ($p < 0.025$) (Tab. 2). The results indicate an acute effect of GM1 on this enzyme which may be indicative of either a protective or restorative effect on results of axonal and dendritic degeneration which occurs after the transection.

Table 2. *Striatal Na^+, K^+-ATPase Levels[1] 48 Hours after Nigro-Striatal Hemitransection: GMl Effects*

	Non-Transected Side	Transected Side	% Decrease
Saline			
Mean	5.38[2]	3.34	—38.0[3]
SEM	1.63	0.97	
GMl			
Mean	5.41	4.89	—9.9
SEM	1.67	1.58	

[1] Data represent separate experiments (Saline N = 6; GMl N = 4) each consisting of 6 rats from which striata were pooled for analyses. Percent decrease was calculated as the specific activity of the transected side compared to the untransected side.
[2] μmol/mg protein/hr
[3] GMl vs Saline: $p < 0.025$

REDUCTION IN LOSS OF Na^{+},K^{+}-ATPASE ACTIVITY IN ISCHEMIC BRAIN

Most of the studies concerning the effects of gangliosides on CNS injury have used animal models based upon "open-head" injuries (e.g. lesions, ablations, transections, and intracephalic mechanical/blunt injury). In order to more closely approximate the clinical condition, we are studying ischemia, assaying Na^{+},K^{+}-ATPase activity in animals treated with gangliosides.

Ischemia Model

The Mongolian Gerbil has been widely used as an experimental model in studies of cerebral ischemia since it has an incomplete circle of Willis, wherein there is little or no communicating vascular system between the hemispheres (Levine and Sohn, 1969). Consequently, following either a partial or complete ligation of the carotid artery (Levine and Pavan, 1966) ischemia is largely lateralized to one cerebral hemisphere in the majority of animals. This is advantageous, since the animal can be used as its own control, allowing comparisons between ischemic and "nonischemic" side of the brain.

Male gerbils (80-90 gr) were anesthetized (i.m.) with a mixture of Vetalar (87.5 mg/kg) and Rompun (7.5 mg/kg). An incision was made on the ventral surface of the neck, the salivary glands were moved laterally, and the right carotid sheath was exposed. Both the vagus and sympathetic nerve were separated from the right common carotid artery which was then ligated using No. 0 silk thread.

Na^{+},K^{+}-ATPase activity was analyzed as previously described. Our initial studies reported here focus on analysis of hippocampal ATPase activity. Other studies in progress are examining enzyme activities in other brain regions, together with neurological and behavioral tests.

GM1 and AGF2 Ganglioside Injections

Gerbils received injections (i.p.) 0.5 and 24 hrs after surgery. There were three groups: saline controls, GM1 ganglioside (20 mg/kg) and AGF2 ganglioside (20 mg/kg). AGF2 is the internal ester derivative of GM1. We are studying this ganglioside since it has been reported to have a longer half-life in serum than GM1 due to slow hydrolysis (Aldinio et al., 1984), and, as a less polar compound, larger quantities may cross the blood brain barrier.

Results

We found that 48 hrs following permanent ligation of the right carotid artery there was a 39.5% decrease in hippocampal Na^{+},K^{+}-ATPase activity on the ischemic side. However, in gerbils treated with GM1 there was a 20.3% decrease and in those treated with AGF2 there was a 10% decrease.

These results parallel our original observations that ganglioside injections reduce the loss of ATPase activity associated with CNS injury. It will be of interest to determine 1) how gangliosides are reaching the ischemic areas of the brain since the carotid artery has been ligated; 2) whether there are any morphological concomitants of the "spared" or "protected" levels of Na^+,K^+-ATPase and 3) whether this effect is reflected in neurological tests and behavioral assessments of learning performance.

DISCUSSION

We have reviewed and reported here a number of experimental findings which indicate that there is an acute effect of ganglioside injections on injury processes in the CNS. The functional data, namely the reduction in asymmetric rotation 48 hrs after a nigro-striatal lesion in ganglioside treated animals, show that the timing of the first GM1 ganglioside injection (0-2 hrs after surgery) significantly affects the behavioral outcome. Hence exogenous gangliosides need to be present during a very early phase of the injury process in order to achieve an optimal functional effect. Although injections later than 4 hrs do not produce a significant reduction in the rotational behavior, there is a consistent trend of reduction as compared to controls. Examination of less variable behaviors, or longer injection protocols, are likely to indicate more improved functional recovery, as has been reported in several other behavioral studies (Sabel et al., 1984; Karpiak et al., 1983; Fass and Ramirez, 1984; Ramirez et al., 1985).

Although there is no clear evidence to indicate what short-term process is being affected by the gangliosides, our analyses of edema, and particularly membrane associated Na^+,K^+-ATPase activity levels, may begin to provide some clues. The reduction in edema is paralleled by a reduction in the loss of Na^+,K^+-ATPase activity. The loss of this enzyme activity probably reflects a series of biochemical changes (e.g. lipid hydrolysis, phospholipase activation, levels and membrane action of arachidonic acid, ionic permeation) which occur as a result of injury and lead to membrane failure.

In three experiments (i.e. cortical lesions, denervated tissue, and ischemic tissue) the injury associated reductions of Na^+,K^+-ATPase activity are significantly lessened in ganglioside (GM1 and AGF2) treated animals. The maintenance of these higher levels of ATPase activity may well contribute to the facilitatory effect by gangliosides on short-term function and long-term recovery. It is not clear from these data as to whether gangliosides are affecting the enzyme itself or other membrane processes. Identification of the mechanism by which gangliosides are affecting these processes will allow us to determine if they are "protective" or "restorative". Preliminary data in our laboratories (Karpiak and Mahadik, 1985) indicate that Na^+,K^+-ATPase from GM1 treated rats shows a higher degree of activation at lower concentrations of K^+, than that in controls.

Stimulation of neuronal membrane Na^+,K^+-ATPase activity of gangliosides in vitro (Leon et al., 1981) and of striatal Na^+,K^+-ATPase in vivo (Mahadik et al., 1985) has been reported. More recently, using a model of hypoxia with hippocampal tissue slices, Janigro et al. (1984) reported that gangliosides reduced the hyperpolarization which occurs as a result of hypoxia. This may reflect an activation, or "protection" of the Na/K pump, resulting in maintenance of ionic balances across the plasma membrane. These authors speculated that the "early" effects of gangliosides may ultimately be supportive of subsequent neuronal regeneration, sprouting and neuritogenesis.

Gangliosides are exerting an acute effect on CNS injury. If this results in the reduction of "degenerative events" associated with that phase of injury, then we conclude that the opportunity for facilitated recovery is maximized. Therefore, the facilitated recovery may be due to *either* 1) a reduction in neuronal cell loss and axonal/dendritic degeneration, *or* 2) subsequent facilitated neuronal regeneration, *or both*.

REFERENCES

Aldinio C, Valenti G, Savoini G, Kirschner GE, Agnati LF, Toffano G (1984 Int J Dev Neurosci 2: 267-275.

Atterwill CK, Cunningham VJ, Balazs R (1984) J Neurochem 43: 8-18.

Averit N, Rigoulet M, Cohadon F (1984) J Neurochem 42: 275-277.

Eibl H and Lands W (1969) Anal Biochem 30: 51-57.

Fass B and Ramirez J (1984) J Neurosci Res 12: 445-458.

Ferrari G, Fabris M, Gorio A (1983) Dev Brain Res 8: 215-221.

Gorio A, Carmignoto G, Facci L, Finesso M (1980) Brain Res 197: 236-241.

Gorio A, Marini P, Zanoni R (1983) Neurosci 8: 417-429.

Janigro D, Di Gregorio F, Vyskocil F, Gorio A (1984) J Neurosci Res 12: 499-509.

Karpiak SE (1983) Exp Neurol 81: 330-339.

Karpiak SE (1984) in: Ledeen RW, Yu RK, Rapport MM, Suzuki S (eds): Ganglioside Structure, Function and Biomedical Potential. Plenum Press, New York, pp. 489-497.

Karpiak SE and Mahadik SP (1984) J Neurosci Res 12: 485-492.

Karpiak SE, Mahadik SP (1985) in: Dreyfus H, Massarelli R, Freysz L and Rebel G (eds): Cellular and Pathological Aspects of Glycoconjugate Metabolism. Inserm 126: 585-598.

Leon A, Toffano G, Sonnino S, Tettamanti G (1981) J Neurochem 37: 350-357.

Leon A, Facci L, Benvegnu D, Toffano G (1982) Dev Neurosci 5: 108-114.

Levine S and Pavan H (1966) Exp Neurol 16: 255-262.

Levine S and Sohn D (1969) Acta Pathol (Chicago) 87: 315-317.

Li Y, Mahadik SP, Rapport MM, Karpiak SE (1986) Brain Res 377: 292-297.

Mahadik SP, Korenovsky A, Karpiak SE (1985) Abst Am Soc Neurochem 16: 231.

Ramirez JJ, Karpiak SE, Kilfoil T, Henschel B, Grones W (1985) Soc Neurosci Abstr 11: 251.

Roisen FJ, Bartfeld H, Nagele R, Yorke G (1981) Science 214: 577-578.

Sabel BA, Dunbar GL, Stein DG (1984a) J Neurosci Res 12: 429-443.

Sabel B, Slavin MD, Stein DG (1984b) Science 225: 340-342.

Toffano G, savoini G, Moroni F, Lombardi G, Calza L, Agnati LF (1983) Brain Res 261: 163-166.

Gangliosides and neuronal plasticity
G. Tettamanti, R.W. Ledeen, K. Sandhoff,
Y. Nagai, G. Toffano (eds.)
Fidia Research Series, vol. 6
Liviana Press, Padova, © 1986

EFFECT OF GM1 ON THE ALTERATIONS INDUCED BY SELECTIVE NEUROTOXINS IN THE DEVELOPING CNS

G. Vantini, B. Figliomeni, R. Zanoni, A. Gorio[1], G. Jonsson[2], M. Fusco

Fidia Neurobiological Research Laboratories,
Via Ponte della Fabbrica, 3/A, 35031 Abano Terme, Italy;
[1]Scuola di Farmacologia, Istituto di Farmacologia e Farmacognosia,
Università di Milano, Italy, and
[2] Department of Histology, Karolinska Institutet, Stockholm, Sweden

The effect of the monosialoganglioside GM1 administration on the alterations induced in rat CNS by neurotoxins selective for transmitter-identified neurons has been studied by employing both neuro- and immunocytochemical techniques. 5,7-dihydroxytryptamine (5,7-HT), 6-hydroxydopamine (6-OH-DA) and capsaicin have been used to induce damage to serotonin (5-HT)-, noradrenaline (NA)- and substance P (SP)- containing neurons, respectively. In experiments employing 5,7-HT and 6-OH-DA it is found that the primary neurodegenerative actions of these neurotoxins on NA and 5-HT neurons are not modified by GM1 administration. In contrast under chronic conditions, GM1 diminishes the extent of the alterations induced by 5,7-HT and 6-OH-DA. Moreover, GM1 is able to reduce the capsaicin-induced decrease of SP nerve terminal density in the superficial layers of the dorsal horns of lumbar spinal sections. The present results are consistent with the view that GM1 has a protective and/or regrowth-stimulating effect on damaged central neurons.

INTRODUCTION

Gangliosides, a complex group of cell-surface sialoglycosphingolipids particularly abundant in neuronal tissues, are assumed to be involved in a variety of cell-surface events, such as regulation of cell-growth, synaptogenesis and neuronal regeneration (for review see Ledeen, 1983; 1984; Ando, 1983; Gorio, 1986). Research on the biologic functions of gangliosides in the CNS has been prompted in recent years by the discovery of the effects of ganglioside administration to neurons in culture and in vivo. These effects concern facilitation of neurite formation in vitro (Ferrari et al., 1983; Leon et al., 1982; Spoerri, 1983; Roisen et al., 1983) and acceleration of CNS neonatal development (Karpiak et al., 1984) as well as an enhancement of recovery following peripheral nerve damage (Ceccarelli et al., 1976; Gorio et al., 1980; 1983) and CNS injury (Karpiak, 1983; 1984; Toffano et al., 1983; 1984). Consistent with the view that gangliosides play a role in neuronal growth or regrowth processes is the finding that antibodies to gangliosides inhibit neurite outgrowth both in CNS and PNS explants (Spirman et al., 1982; Schwartz and Spirman, 1982; Sparrow et al, 1984) and cause behavioural dysfunction when administered intracerebrally during neonatal stage (Kasarskis et al., 1981).

Recently, several reports have shown that the monosialoganglioside GM1 promotes recovery of selective neurotoxin-induced damage of transmitter-identified neuronal cells in the rat CNS (Jonsson et al., 1984a; Kojima et al., 1984; Fusco et al., 1986; Hadjiconstantinou et al., 1986). The aim of the present article is to review some of our studies on the effect of GM1 on neurotoxin-induced lesions in the medulla pons and spinal cord of developing rats. Serotonin (5-HT)- and noradrenaline (NA)-containing nerve terminals have been lesioned by using 5,7-dihydroxytryptamine (5,7-HT) and 6-hydroxydopamine (6-OH-DA) respectively. Capsaicin has been employed to damage substance P (SP)-containing fibers in the superficial layers of the dorsal horns of the spinal cord.

MATERIALS AND METHODS

Newborn Sprague-Dawley rats of both sexes were injected subcutaneously with either 5,7-HT (50 mg/kg), 6-OH-DA (100 mg/kg) or capsaicin (50 mg/kg) within few hours after birth. Thereafter the rat pups were injected daily with the monosialoganglioside GM1 (nomenclature according to Svennerholm, 1980): the first dose was administered 2-3 hr after the neurotoxin injection. The rat pups of the 5,7-HT and 6-OH-DA groups received four injections of GM1 (30 mg/kg s.c.), while twelve injections of GM1 (20 mg/kg s.c.) were given in the capsaicin study. Controls were injected with equal amounts of appropriate solvent. Following neonatal administration of 5,7-HT, rats were sacrificed at 4, 28, 56 days of age. 6-OH-DA and capsaicin treated animals were sacrificed at 6 weeks and 2 months of age, respectively.

5-HT and NA Determination

Medulla pons and spinal cord (dissected into cervical, thoracic and lumbar segments) were rapidly removed, frozen on dry ice and stored at —80°C, in aluminum foil. NA and 5-HT content in the sampled tissue specimens were determined by HPLC with electrochemical detection using a RP C_{18} column (Keller et al., 1976; Mayer and Shoup, 1983, Felice et al., 1978; Jonsson et al., 1984a).

[^{3}H]-5-HT Binding Site Assay

Affinity (K_d) and density (B_{max}) of [^{3}H]-5-HT binding sites were evaluated in the thoracic and lumbar spinal cord of 5,7-HT lesioned rats. The assays were carried out following reported procedures (Peroutka and Snyder, 1979; Monroe and Smith, 1983) with minor modifications.

5-HT and SP Immunocytochemistry and Nerve Density Measurements

Following perfusion of the rats with 4% paraformaldehyde, spinal cords were dissected out, frozen and cut on a cryostat. The sections (10 μm) were mounted on glass slides and processed for immunocytochemical assays, according to the indirect immunofluorescence technique of Coons (1958). In brief, the sections were incubated for 20 hr at 4° C with monoclonal antibodies (Sera-LAB) to 5-HT and SP (Cuello et al., 1979; Consolazione et al., 1981), washed with phosphate buffered saline, for 20 min and subsequently incubated for 30 min with fluorescein isothiocynate-conjugated rabbit antibodies against rat immunoglobulins (MILES-YEDA). A Zeiss fluorescence microscope was used for examination and microphotography. In addition, spinal sections stained immunocytochemically for 5-HT and SP were analyzed by an interactive computer-assisted image analysis system (IBAS, Kontron/Zeiss) for quantitative determination of 5-HT and SP nerve terminal density. The calculated percent values represent a measure of the relative 5-HT or SP nerve density in the field under examination (Jonsson et al., 1984a).

RESULTS

6-OH-DA Induced Lesions

The most frequently used and best characterized neurotoxin for damaging catecholamine neurons is 6-OH-DA. Administration of 6-OH-DA in the neonatal stage is characterized by pronounced and permanent denervation of the most distal NA nerve terminal projections (e.g. in the cerebral cortex, hippocampus and spinal cord) as well as by hyperinnervation of the regions close to the NA cell-bodies such as in the medulla pons and cerebellum (Jonsson and Hallman, 1982a; Jonsson and Sachs, 1982). The effects of 6-OH-DA are very selective on the NA neurons, since the other monoamine systems are left unaltered.

Neonatal GM1 treatment has no apparent effect on the 6-OH-DA induced reduction of endogenous NA content in the spinal cord during the acute stage (data not shown). However, 6 weeks after induction of damage, there is a significant effect of GM1 on 6-OH-DA induced denervation of the spinal cord (Fig. 1). The NA levels are significantly higher in the spinal cord of GM1 treated animals as compared to the untreated animals. The effects of GM1 on 6-OH-DA induced denervation or hyperinnervation are also evident in other brain areas 1 month after damage (data not shown).

5,7-HT Induced Lesions

Neonatal administration of 5,7-HT produces profound alterations on the development of central 5-HT neurons (Jonsson and Hallman, 1982a). This neurotoxin causes a prominent and permanent degeneration of distant nerve terminal projections (e.g. in cerebral cotex, hippocampus, spinal cord), whereas it enhances the nerve density and determines hyperinnervation in brain regions close to the 5-HT cell-body groups (e.g. mesencephalon, medulla pons). For both 6-OH-DA and 5,7-HT it has been proposed

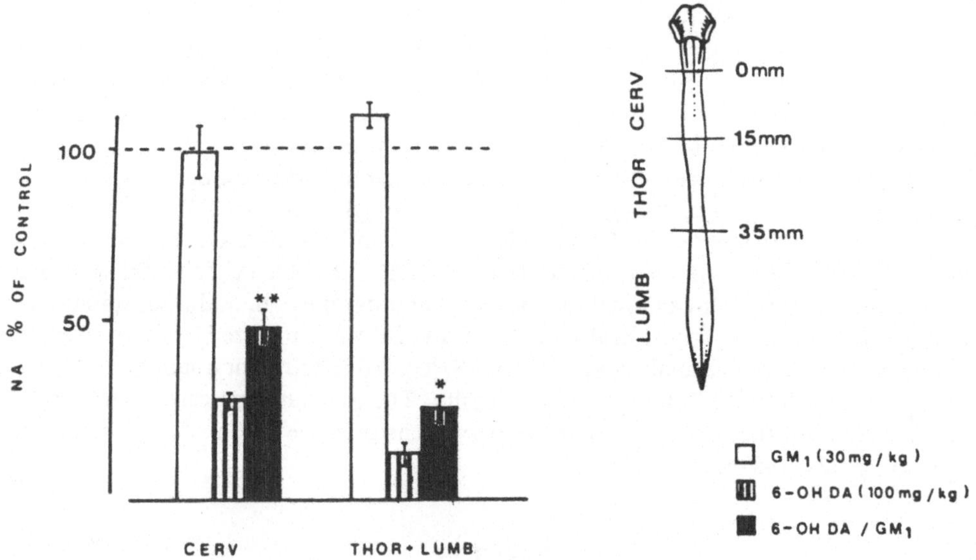

Figure 1. Effect of neonatal 6-OH-DA and/or GM1 treatment on the endogenous NA levels in different segments of the spinal cord of 6-week-old rats. Each bar represents the mean ± S.E.M. (n = 6) expressed as percentage of the controls. Cervical spinal cord control values: 427 ± 32 ng NA/g wet tissue. Thoracic + Lumbar spinal cord control values: 558 ± 44 ng NA/g wet tissue.. * $p < 0.05$ vs 6-OH-DA ** $p < 0.01$ vs 6-OH-DA.
(Duncan's new multiple range test).

that the permanent denervation of distant nerve terminal projections and the proliferative outgrowth of nerve terminals from the intact axons responsible for the hyperinnervation ("pruning effect") are intimately related processes. Such redistribution of NA and 5-HT nerve terminal fields could reflect a strict developmental program controlling the expression of the arborization size of these neurons (Jonsson and Hallman, 1982 b; Jonsson and Sachs, 1982; Schneider, 1981).

Neonatal administration of GM1 has no significant effect on the 5,7-HT-induced depletion of 5-HT in the thoracic and lumbar spinal cord of 4-day-old rats. At this stage the neurotoxin does not produce alterations in the medulla pons and cervical spinal cord. In 4 and 8-week-old animals treated at birth with 5,7-HT, GM1 treatment again has a counteracting effect on the 5,7-HT-induced alterations throughout the spinal cord. At these stages GM1 antagonizes all changes in 5-HT content, i.e. decrease in thoracic and lumbar segments and increase in medulla pons, due to neonatal treatment with 5,7-HT (Fig. 2). The partial restoration of 5-HT levels in the thoracic and lumbar spinal cord is paralleled by an increase of 5-HT nerve density in the same spinal cord regions (Fig. 3). A protective effect of GM1 against this type of lesion is also evident from the analysis of the spinal 5-HT binding sites. In thoracic and lumbar cord segments of 8-week-old rats treated with 5,7-HT the [^{3}H]5-HT binding site density is higher than in controls ("up-regulation"). Such an increase is partially reversed by

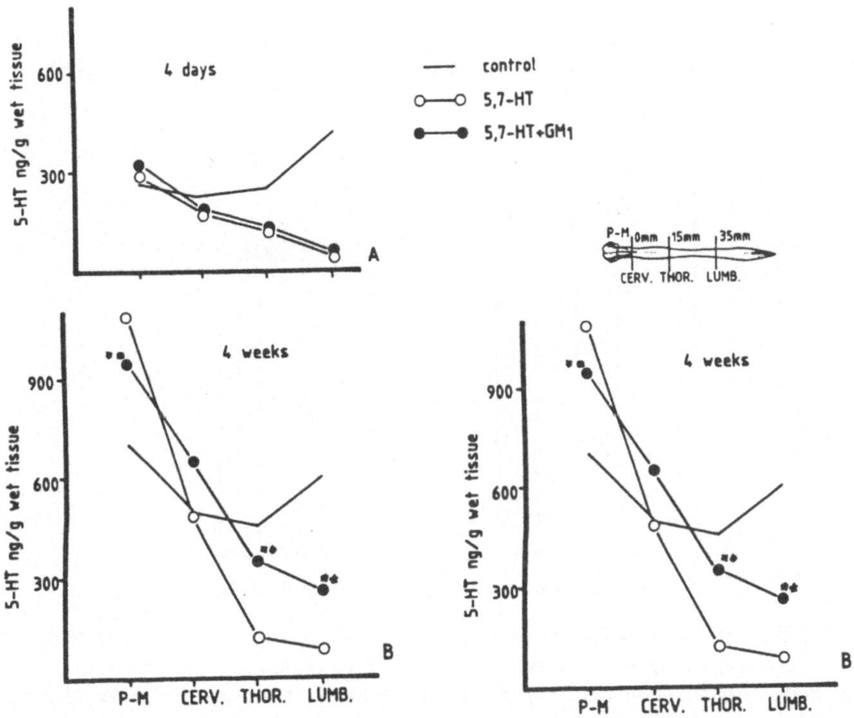

Figure 2. Endogenous 5-HT concentration in the medulla pons (P-M) cervical (CERV.), thoracic (THOR.) and lumbar (LUMB.) spinal cord of 4-day-, 4 and 8-week-old rats treated with 5,7-HT and 5,7-HT + GM1. Each point represents the mean of 6-9 determinations. S.E.M. of each mean did not exceed 12%.
* $p < 0.05$ vs 5,7-HT ** $p < 0.01$ vs 5,7-HT.

GM1 (Tables 1 and 2). We suggest that the 5-HT depletion and the low 5-HT nerve density induced by 5,7-HT are functionally linked to the "up-regulation" of the 5-HT binding sites. The normalizing effect of GM1 on the receptor density can be interpreted as a consequence of its ability to enhance the regrowth of 5-HT nerve terminals. It must be born in mind that GM1 does not interfere with the primary degenerative activity of the toxin (Jonsson et al., 1984a). Hence the effect of GM1 is probably due to enhancement of regeneration of 5-HT fibers.

Capsaicin Induced Injury

Neonatal administration of capsaicin induces a permanent and selective loss of SP from the superficial layers of the dorsal horns of the spinal cord, whereas it does not affect other regions of the CNS with a rich innervation of SP immunoreactive fibers (Cuello et al., 1981; Gamse et al., 1980; Nagy et al., 1980). In 2-month-old rats treated with capsaicin it was found that neonatal administration of GM1 has a conteracting effect on the capsaicin-induced decrease of SP nerve density in the superficial layers

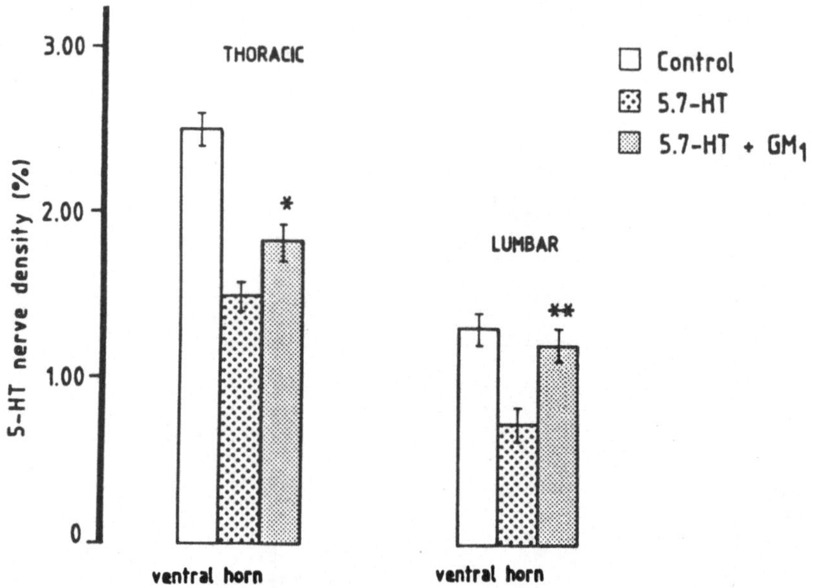

Figure 3. Effect of neonatal 5,7-HT and 5,7-HT + GM1 treatment on 5-HT nerve density expressed as percentage of specific fluorescence per unit area examined by an interactive computer-assisted image analysis system (IBAS). Each bar represents the mean ± S.E.M. of the results obtained from 4-5 spinal cords of 8-week-old rats. The nerve density relative to each animal was obtained by measuring specific fluorescence in the ventral horns of 4 spinal sections at lumbar and thoracic level.
* $p < 0.05$ vs 5,7-HT ** $p < 0.01$ vs 5,7-HT
(Duncan's new multiple range test)

Table 1. *Scatchard Plot Analysis of ^{3}[H]-5-HT Binding Sites in [illegible] Thoracic Spinal Cord of 8-Week-Old Rats. Each Value is the Mean ± S.E.M. fro[illegible] [illegible]pendent Experiments*

	B_{MAX} (pmol./mg. prot.)	K_D (nM)
Control	0.150 ± 0.012	10.4 ± 1.4
GM1	0.145 ± 0.015	10.5 ± 1.3
5,7-HT	0.174 ± 0.009	10.7 ± 1.4
5,7-HT + GM1	0.144 ± 0.006	9.7 ± 1.5

Table 2. *Scatchard Plot Analysis of ^{3}[H]-5-HT Binding Sites in the Lumbar Spinal Cord of 8-week-Old Rats. Each Value is the Mean ± S.E.M. from 4 Independent Experiments*

	B_{MAX} (pmol./mg. prot.)	K_D (nM)
Control	0.185 ± 0.027	7.6 ± 1.2
GM1	0.198 ± 0.009	6.0 ± 0.75
5,7-HT	*0.255 ± 0.018	7.4 ± 1.1
5,7-HT + GM1	0.218 ± 0.034	8.5 ± 2.3

* $p = 0.07$ vs Control

of the dorsal horns of lumbar sections (Fig. 4A). In the white matter there is a dense SP immunostaining in the region corresponding in part to the lateral spinal nucleus. According to Cuello et al. (1981) variations in this area from animal to animal preclude an accurate assessment of the effect of capsaicin. We have found that capsaicin causes a small non significant decrease of SP nerve density in the lateral spinal nucleus. GM1 has no significant counteracting effect on the capsaicin-induced decrease of SP nerve density in this area (Fig. 4B).

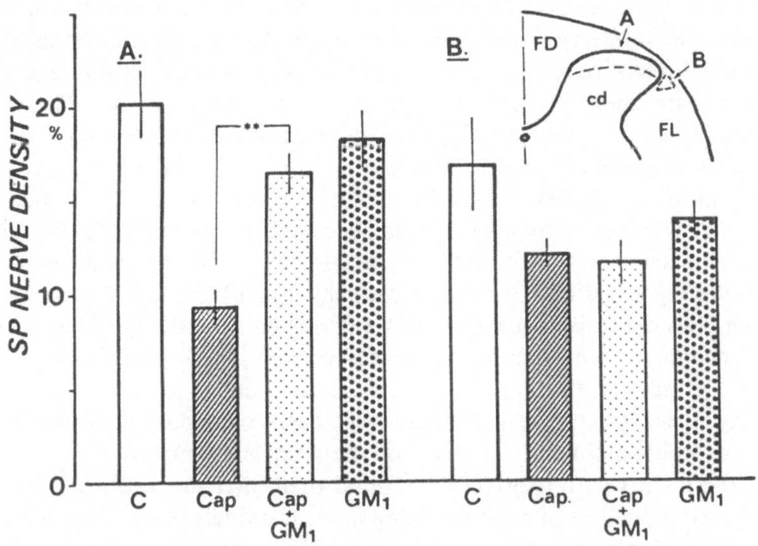

Figure 4. Effect of neonatal capsaicin and/or GM1 treatment on SP nerve density expressed as percentage of specific fluorescence per unit area analyzed by IBAS. The regions examined were the superficial layers of the dorsal horns (A) and the dorsal region of the lateral funiculus (B) at lumbar level. Each bar represents the mean ± S.E.M. of the data obtained from 5 rats. For each animal 4 lumbar sections were measured.
** $p < 0.01$ vs capsaicin (Cap)
(Duncan's new multiple range test)

DISCUSSION

The present results confirm previous data relative to the alterations induced in CNS by neonatal administration of 5,7-HT, 6-OH-DA (reviewed by Jonsson, 1983) and capsaicin (Jancso et al., 1977). 5,7-HT and 6-OH-DA produce their effects rapidly and already within two hours there is a detectable neuronal damage. The neonatal administration of these neurotoxins causes mixed-type alterations (denervations and hyperinnervations). However the absence of significant cell-body death in CNS provides possibilities for regrowth. It is likely that the hyperinnervation in regions close to cell-bodies is a consequence of the degeneration of the distally projecting axonal branches (Jonsson and Hallman, 1982a, b).

The results reported in the paper show that exogenous GM1 administration either prevents or enhances restoration of neurotoxin-induced 5-HT and NA denervation in the lumbar and thoracic spinal cord. All results obtained in the acute stage of the lesion show that the degree of 5-HT and NA terminal degeneration was not affected by GM1, thus suggesting that GM1 does not interfere with or modify the primary neurodegenerative action of these neurotoxins. The present findings confirm and extend previous data on the GM1 enhanced recovery in other brain regions obtained in experiments using a similar approach (Jonsson et al., 1984a, b). The GM1 induced decrease of 5-HT hyperinnervation in the medulla pons after 5,7-HT treatment is probably related to the recovery from damage of the distally projecting axonal branches. The restoration effects of GM1 after 5,7-HT and 6-OH-DA lesions are probably related to an enhancement of 5-HT and NA nerve terminal regrowth. The possibility that GM1 ganglioside can promote regeneration processes of different types of neurons has been already suggested by many authors (Toffano et al., 1983; Agnati et al., 1983; Oderfeld Nowak et al., 1984). In the lesions used in the present study the primary neuronal damage is followed by a retrograde degeneration, thus if GM1 diminishes the the extent of this degeneration it would indirectly cause also an enhancement of the process of nerve regeneration. Consistent with this hypothesis are the recent findings that GM1 increases the resistance to hypoxia and prevents the decay of the ATPase activity in hippocampal slice preparations (Janigro et al., 1984; Bianchi et al., 1986). Therefore, the protective effect of GM1 against the damage induced by 5,7-HT and 6-OH-DA may be explained in terms either of decreased secondary neuronal retrograde degeneration or enhanced neurite growth of the damaged neurons.

In the experiments with capsaicin we found that GM1 markedly antagonizes the capsaicin-induced reduction of SP-containing nerve terminals present in the substantia gelatinosa of the spinal cord. Otten et al. (1983) have shown that capsaicin destroys the SP-containing primary sensory neurons, probably by blocking the retrograde axonal transport of the nerve growth factor (NGF). Interestingly GM1 has been reported to potentiate the NGF-stimulated outgrowth of neurites from PC 12 cells in culture (Ferrari et al., 1983). Similarly GM1 enhances whereas antibodies to GM1 inhibit NGF-induced neurite outgrowth of embryonic chick dorsal root ganglionic cells in culture (Schwartz and Spirman, 1982; Doherty et al., 1985). Taken together these results suggest an interaction between GM1 and NGF. However, further studies are needed to establish whether the interaction between GM1 and NGF or other growth factors contributes significantly to the protective effects observed after exogenous GM1 administration.

REFERENCES

Agnati LF, Fuxe K, Calzà L, Benfenati F, Cavicchioli L, Toffano G, Goldstein M (1983) Gangliosides increase the survival of lesioned nigral dopamine neurons and favour the recovery of dopaminergic synaptic function in striatum of rats by collateral sprouting. Acta Physiol Scand 119: 347-363.

Ando S (1983) Gangliosides in the nervous system. Neurochem Int 5: 507-537.

Bianchi R, Janigro D, Milan F, Giudici G, Gorio A (1986) In vivo treatment with GM1 prevents the rapid decay of ATPase activities and mitochondrial damage in hippocampal slices. Brain Res 364: 400-404.

Ceccarelli B, Aporti F, Finesso M (1976) Effect of brain ganglioside on functional recovery in experimental regeneration and reinnervation. Advanc Exp Med Biol 21: 275-293.

Coons AH (1958) Fluorescent antibody methods. In: Canielli JF (ed): General cytochemical methods. 1, Academic Press, New York, pp. 399-422.

Consolazione A, Milstein C, Wright B, Cuello AC (1981) Immunocytochemical detection of serotonin with monoclonal antibodies. J Histochem Cytochem 12: 1425-1430.

Cuello AC, Galfre G, Milstein C (1979) Detection of substance P in the central nervous system by a monoclonal antibody. Proc Natl Acad Sci USA, 76: 3532-3536.

Cuello AC, Gamse R, Holzer P, Lembeck F (1981) Substance P immunoreactive neurons following neonatal administration of capsaicin. Naunyn-Schmiederberg's Arch Pharmacol 229: 219-224.

Doherty P, Dickson JG, Flanigan TP, Walsh FS (1985) Ganglioside GM1 does not initiate, but enhances neurite regeneration of nerve growth factor-dependent sensory neurones. J Neurochem 44: 1259-1265.

Felice LJ, Felice JD, Kissinger PT (1978) Determination of catecholamines in rat brain parts by reverse-phase ion-pair liquid chromatography. J Neurochem 31: 1461-1465.

Ferrari G, Fabris M, Gorio A (1983) Gangliosides enhance neurite outgrowth in PC12 cells. Develop Brain Res 8: 215-221.

Fusco M, Donà M, Tessari F, Hallman H, Jonsson G, Gorio A (1986) GM1 ganglioside counteracts selective neurotoxin-induced lesion of developing serotonin neurons in rat spinal cord. J Neurosci Res, in press.

Gamse R, Holzer P, Lembeck F (1980) Decrease of substance P in primary afferent neurons and impairment of neurogenic plasma extravasation by capsaicin. Br J Pharmacol 68: 207-213.

Gorio A, Carmignoto G, Facci L, Finesso M (1980) Motor sprouting induced by ganglioside treatment. Possible implications for gangliosides on neuronal growth. Brain Res 197: 236-241.

Gorio A, Zanoni R, Marini P (1983) Muscle reinnervation. III. Motoneuron sprouting capacity: enhancement by exogenous gangliosides. Neuroscience 8: 417-429.

Gorio A (1986) Ganglioside enhancement of neuronal differentiation, plasticity and repair. J Clin Neurobiol, in press.

Hadjiconstantinou M, Paxton RC, Neff NH (1986) Administration of GM1 ganglioside restores the dopamine content in striatum after chronic treatment with MPTP. Neuropharmacology, in press.

Jancsò G, Kiraly E, Jancsò Gàbor A (1977) Pharmacologically-induced selective degeneration of chemosensitive primary sensory neurons. Nature 270: 741-743.

Janigro D, Di Gregorio F, Vyskocil F, Gorio A (1984) Gangliosides'dual made of action: a working hypothesis. J Neurosci Res 12: 499-509.

Jonsson G, Hallman H (1982a) Response of central monoamine neurons following an early neurotoxic lesion. Bibl Anat 23: 76-92.

Jonsson G, Hallman H (1982b) Modulation of 6-hydroxydopamine induced alteration of the postnatal development of central noradrenaline neurons. Brain Res Bull 9: 635-640.

Jonsson G, Sachs C (1982) Changes in the development of central noradrenaline neurons after axonal lesions neonatally. Brain Res Bull 9: 641-650.

Jonsson G (1983) Chemical lesioning techniques: Monoamine neurotoxins. In: Björklund A, Hökfelt T (eds): Handbook of Chemical Neuroanatomy, Vol 1; Methods in Chemical Neuroanatomy. Elsevier, Amsterdam, pp. 463-507.

Jonsson G, Gorio A, Hallman H, Janigro D, Kojima H, Zanoni R (1984a) Effect of GM1 ganglioside on neonatally neurotoxin-induced degeneration of serotonin neurons in the rat brain. Develop Brain Res 16: 171-180.

Jonsson G, Gorio A, Hallman H, Janigro D, Kojima H, Luthman J, Zanoni R (1984b) Effects of GM1 ganglioside on developing and mature serotonin and noradrenaline neurons lesioned by selective neurotoxins. J Neurosci Res 12: 459-475.

Karpiak SE (1983) Ganglioside treatment improves recovery of alteration behavior after unilateral entorhinal cortex lesion. Exp Neurol 81: 330-339.

Karpiak SE (1984) Recovery of function after CNS damage enhanced by gangliosides. In: Ledeen RW, Yu R, Rapport MM, Suzuki K (eds): Ganglioside Structure, Function and Biomedical Potential. Plenum Press, New York, pp. 489-497.

Karpiak SE, Vilim F, Mahadik SP (1984) Gangliosides accelerate rat neonatal learning and levels of cortical acetylcholinesterases. Dev Neurosci 6: 127-135.

Kasarskis E, Karpiak S, Rapport MM, Yu R, Bass N (1981) Abnormal maturation of cerebral cortex and behavior in adult rats after neonatal administration of antibodies to GM1 ganglioside. Develop Brain Res 1: 25-35.

Keller R, Oke A, Mefford J, Adams RN (1976) Liquid chromatographic analysis of catecholamines routine assay for regional brain mapping. Life Sci 9: 995-1004.

Kojima H, Gorio A, Janigro D, Jonsson G (1984) GM1 ganglioside enhances regrowth of noradrenaline nerve terminals in rat cerebral cortex lesioned by the neurotoxin 6-hydroxydopamine. Neuroscience 13: 1011-1022.

Ledeen RW (1983) Gangliosides. In: Lajtha A (ed): Handbook of Neurochemistry, Vol. 3. Plenum, New York, pp. 41-90.

Ledeen RW (1984) Biology of gangliosides: neuritogenic and neuronotrophic properties. J Neurosci Res 12: 147-159.

Leon A, Facci L, Benvegnù D, Toffano G (1982) Morphological and biochemical effects of gangliosides in neuroblastoma cells. Dev Neurosci 5: 108-114.

Mayer GS, Shoup RE (1983) Simultaneous multiple electrode liquid chromatographic-electrochemical assay for catecholamines, indolamines and metabolites in brain tissue. J Cromatogr 255: 533-544.

Monroe PJ, Smith DJ (1983) Characterization of multiple [^{3}H]-5-Hydroxytryptamine binding sites in rat spinal cord tissue. J Neurochem 2: 349-355.

Nagy JI, Vincent SP, Staines WA, Fibiger HC, Reisine TD, Yamamura HI (1980) Neurotoxic action of capsaicin on spinal substance P neurons. Brain Res 186: 435-44.

Oderfeld-Nowak B, Skup M, Ulas J, Jezierska M, Gradkowska M, Zaremba M (1984) Effect of GM1 ganglioside treatment of postlesion responses of cholinergic enzymes in rat hippocampus after various partial deafferentiations. J Neurosci Res 12: 409-420.

Otten U, Lorez HP, Businger F (1983) Nerve growth factor antagonizes the neurotoxic action of capsaicin on primary sensory neurons. Nature 301: 515-517.

Peroutka SJ, Snyder SH (1979) Multiple serotonin receptors: differential binding of [^{3}H]-5-Hydroxytryptamine, [^{3}H]-lysergic acid diethylamide and [^{3}H]-spiroperidol. Mol Pharmacol 16: 687-699.

Roisen FJ, Bartfeld H, Nagel R, York G (1981) Ganglioside stimulation of axonal sprouting in vitro. Science 214: 577-578.

Schneider GE (1981) Early lesions and abnormal neuronal connections. TINS 4: 187-192.

Schwartz M, Spirman N (1982) Sprouting from chicken embryo dorsal root ganglia induced by nerve growth factor is specifically inhibited by affinity-purified antiganglioside antibodies. Proc Natl Acad Sci USA 79: 6080-6083.

Sparrow JR, McGuinnes C, Schwartz M, Grafstein B (1984) Antibodies to gangliosides inhibit goldfish optic nerve regeneration in vivo. J Neurosci Res 12: 233-243.

Spirman N, Sela BA, Schwartz M (1982) Antiganglioside antibodies inhibit neuritic outgrowth from regenerating goldfish retinal explants. J Neurochem 39: 874-877.

Spoerri PE (1983) Effects of gangliosides in the in vitro development of neuroblastoma cells: An ultrastructural study. Int J Dev Neurosci 1; 383-391.

Svennerholm L (1980) Gangliosides and synaptic transmission. In: Svennerholm L, Mandel P, Dreyfus H, Urban PF (eds): Structures and function of gangliosides. Plenum Press, New York, pp. 533-544.

Toffano G, Savoini G, Moroni F, Lombardi G, Calzà L, Agnati F (1983) GM1 ganglioside stimulates the regeneration of dopaminergic neurons in the central nervous system. Brain Res 261: 163-166.

Toffano G, Savoini G, Aporti F, Calzolari S, Consolazione A, Maura G, Marchi M, Raiteri M, Agnati LF (1984) The functional recovery of damaged brain: the effect of GM1 monosialoganglioside. J Neurosci Res 12: 397-408.

Ganglioside and neuronal plasticity
G. Tettamanti, R.W. Ledeen, K. Sandhoff,
Y. Nagai, G. Toffano (eds.)
Fidia Research Series, vol. 6
Liviana Press, Padova, © 1986

EARLY BIOCHEMICAL EFFECTS OF GM1 GANGLIOSIDE TREATMENT IN LESIONED BRAIN: DEPENDENCE ON DEGREE OF FIBER DEGENERATION

Barbara Oderfeld-Nowak, Małgorzata Skup, Małgorzata Grądkowska, Lech Kiedrowski

Laboratory of Neurochemistry, Department of Neurophysiology
Nencki Institute of Experimental Biology, Polish Academy of Sciences,
3, Pasteur Str. 02-093 Warsaw, Poland

INTRODUCTION

Several reports show that ganglioside treatment, and especially GM1 ganglioside, facilitates various forms of long-term recovery following lesions of the brain (Agnati et al., 1984; Kojima et al., 1984; Oderfeld-Nowak et al., 1984a; b; Toffano et al., 1983; Toffano et al., 1984a; b; Wójcik et al., 1982). Enhanced recovery has been hypothesized to be due mainly to increased sprouting processes. Recently, however, several laboratories have also reported beneficial effects of GM1 treatment upon the impaired functions in an acute phase of CNS injury, (Fass and Ramirez, 1984; Karpiak and Mahadik, 1984; Sabel et al., 1984), and this fact, considering the time period, can hardly be ascribed to facilitation of sprouting. Therefore, other phenomena are probably involved. In fact, several authors pointed out some early compensatory biochemical effects of GM1 treatment after brain lesions. An increase in choline uptake in the cerebral cortex after lesion in the nucleus basalis magnocellularis (Pedata et al., 1984), an increase in dopamine uptake in the striatum following partial hemitransection (Toffano et al., 1984b), an increase of the lowered activity of Na^+, K^+-ATPase associated with edema (Karpiak and Mahadik, 1984; Karpiak et al., 1986), restoration of striatal energy metabolism (Fuxe et al., 1986), were reported. Hypotheses concerning the effect of ganglioside on increasing of the impulse flow, and/or enhancing biosynthetic processes (Pedata et al., 1984; Toffano et al., 1984b), and effects on restabilization of membrane properties (Janigro et al., 1984; Karpiak and Mahadik, 1984; Karpiak et al., 1986) have been advanced. We are now reporting data which indicate yet another early

Abbreviations: CNS, central nervous system; i.m., intramuscular; AChE, acetylcholine esterase; ChAT, choline acetyltransferase; 5-HT, 5-hydroxytriptamine; GM1, monosialoganglioside; NGF, nerve growth factor.

biochemical facilitatory effect of GM1, namely, on post-lesion recovery of the hippocampal cholinergic and serotoninergic parameters.

MATERIALS AND METHODS

The model system used in the present investigations is a partially denervated rat hippocampus. Restricted, bilateral electrocoagulative lesions were made under nembutal anesthesia in male Wistar rats, 3 months old (the details of the lesion procedure are given elswhere — Skup et al., in preparation). The lesions were located in the supracallosal area and partly encroach upon the fornix superior (Figs 1,2), thus destroying part of the dorsal cholinergic (Lewis and Shute, 1978) and serotoninergic (Storm-Mathisen and Guldberg, 1974) projections to the hippocampus, while other parts of these inputs remain intact, as illustrated schematically in Figure 1. By producing such lesions, we were able to evoke a differential degree of degeneration of cut fibers along the longitudinal hippocampal axis, similar to that recently described by Gage et al. (1983b, c) and Dravid and Van Deusen (1984), who made aspirative and mechanical lesions, respectively, in the similar area.

Four groups of animals were run simultaneously: unoperated and operated - buffer injected, unoperated and operated - GM1 injected (30 mg/kg i.m. daily) (GM1 supplied by Fidia Research Lab., Italy). The animals were killed by decapitation on the

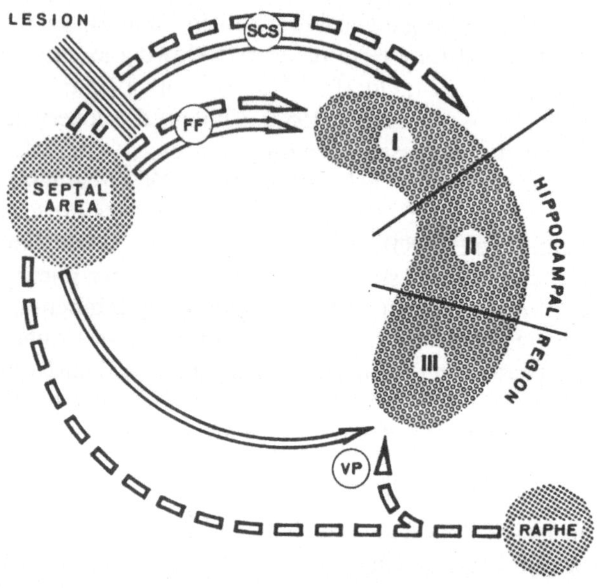

Figure 1. Schematic illustration of the cholinergic afferents ⇨ from septum, and serotoninergic afferents ▭ ⇨ from raphe nuclei reaching the hippocampal formation along two separate routes: a dorsal one: via the fimbria — fornix (FF) and supracallosal stria (SCS) and the ventral pathway (VP) via the piriform lobe. Lesion in supracallosal area is marked. Scheme of division of the hippocampus is shown (from dorsal to ventral end, consecutive slices I, II, III).

6th post-lesion day, chosen as the time of degeneration, when presumably sprouting processes do not yet take place (Oderfeld-Nowak et al., 1977). The lesion was verified in each brain and only those rats in the two operated groups were considered for biochemical estimations in which the lesions could be well matched. The hippocampi were divided into three equal parts along their longitudinal axis (Fig. 1, slices I, II, III); as markers for the cholinergic system AChE, ChAT and choline uptake were used, while 5-HT uptake served as marker for the serotoninergic system. Estimations were made by the following methods, with minor modifications: AChE - after Ellman et al

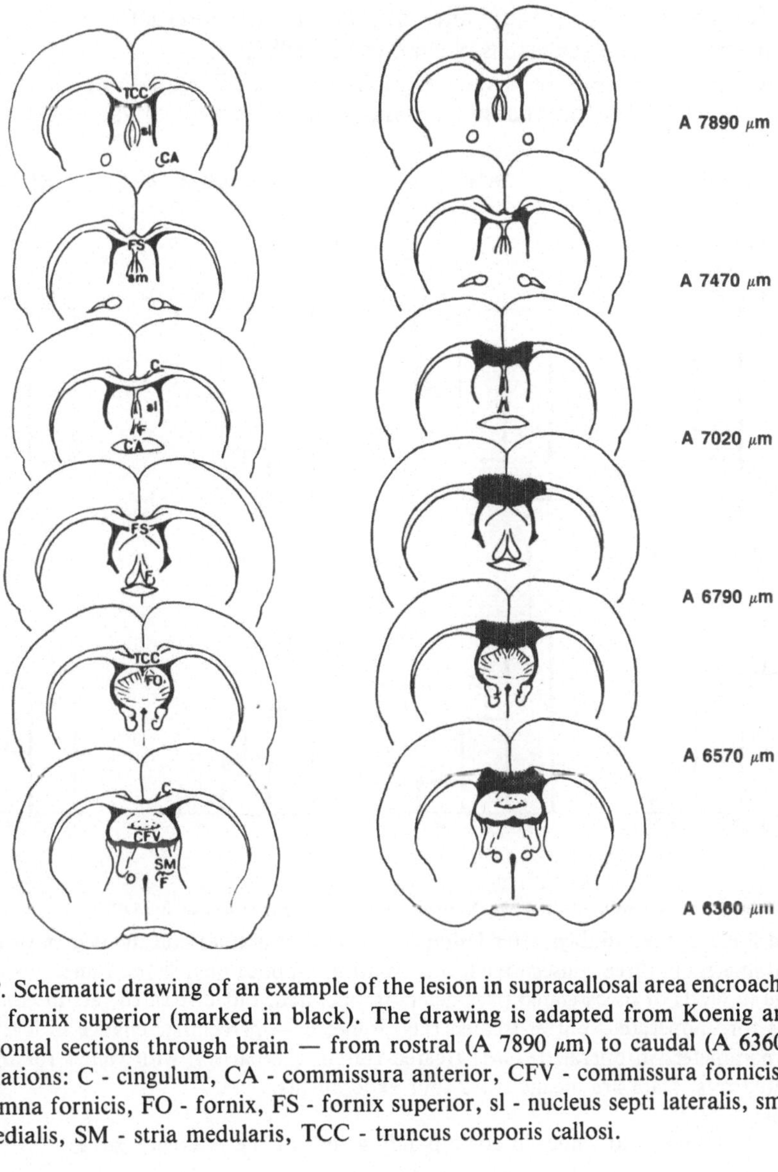

Figure 2. Schematic drawing of an example of the lesion in supracallosal area encroaching slightly upon fornix superior (marked in black). The drawing is adapted from Koenig and Klippel atlas: frontal sections through brain — from rostral (A 7890 μm) to caudal (A 6360 μm). Abbreviations: C - cingulum, CA - commissura anterior, CFV - commissura fornicis ventralis, F - columna fornicis, FO - fornix, FS - fornix superior, sl - nucleus septi lateralis, sm - nucleus septi medialis, SM - stria medularis, TCC - truncus corporis callosi.

(1961), ChAT - after Fonnum (1975), choline uptake — after Haga and Noda (1973), 5-HT uptake — after Ternaux et al. (1977) and protein — after Lowry et al. (1951). The details of the procedures are given elswhere (Skup et al., in preparation; Kiedrowski et al., in preparation).

RESULTS

The results obtained were partly included in reports presented at the Fifth European Society for Neurochemistry Meeting (Kiedrowski and Gradkowska, 1984; Oderfeld-Nowak et al., 1984a) and at the Tenth International Society for Neurochemistry Meeting (Oderfeld-Nowak et al., 1985).

In panel A of Figure 3 the decreases in AChE, ChAT, choline uptake and 5-HT uptake are shown as percent change in comparison with control values obtained for

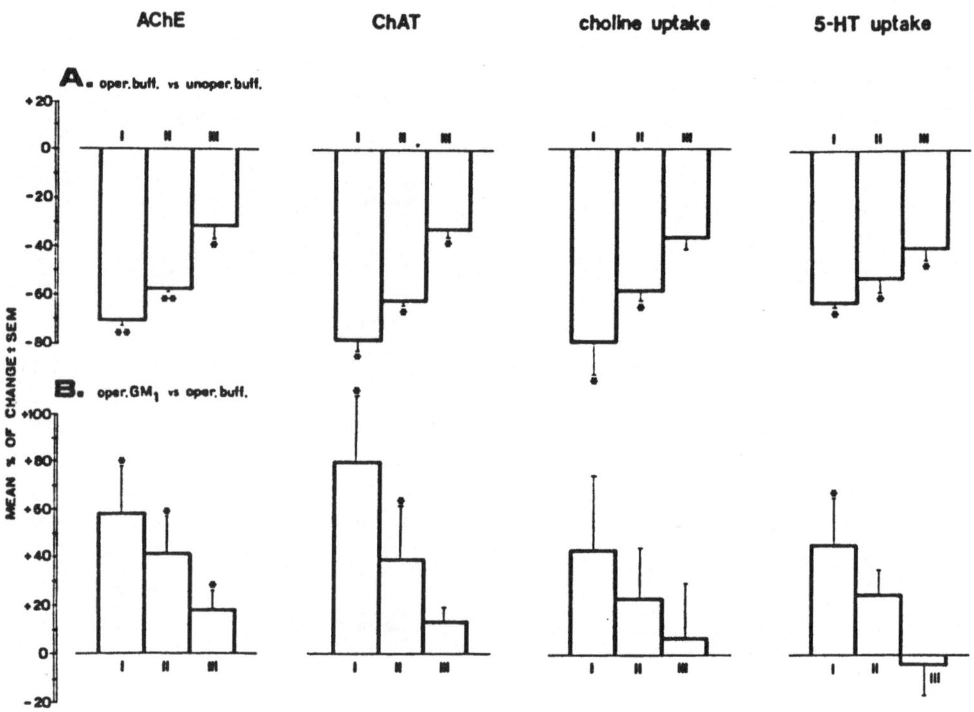

Figure 3. Relationship between GM1 influence on post-lesion changes in AChE, ChAT, choline uptake and 5-HT uptake (6 days after lesion in supracallosal area) and the degree of decrease in these parameters in three consecutive (dorso-ventral) hippocampal slices. Panel A — percentage change in levels of investigated parameters in operated, buffer-treated rats in comparison with values for unoperated, buffer-treated rats; Panel B — percentage change in levels of investigated parameters in operated, GM1-treated rats in comparison with values for operated, buffer-treated rats. Data are means from four experiments.

* $p < 0.05$

** $p < 0.001$ in respect to values in corresponding slice of the reference group unoperated, buffer-treated animals. It should be noted that while no effect of GM1 on any of the cholinergic parameters was observed in control animals, confirming our earlier data (Oderfeld-Nowak et al., 1984b), 5-HT uptake was slightly, but significantly, decreased (in longer treatment periods — up to 90 days — these changes became even more pronounced, Kiedrowski et al., in preparation).

The postoperative decrease in each parameter was the greatest in the dorsal part

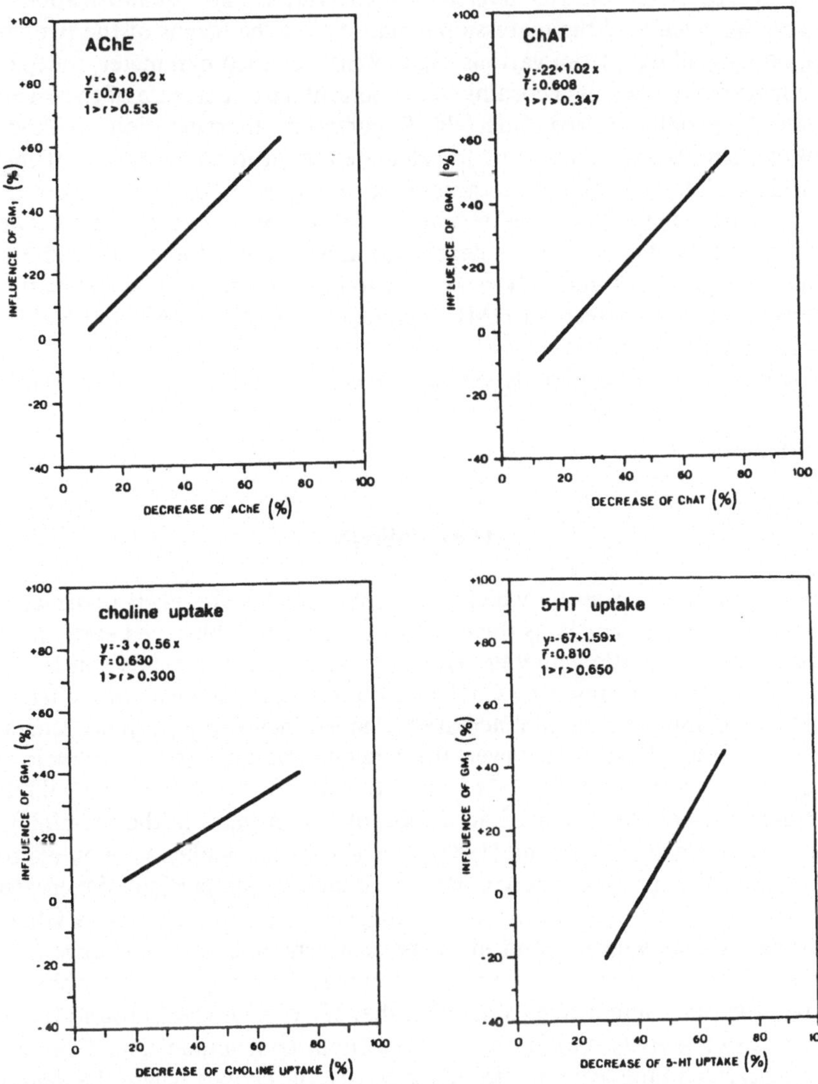

Figure 4. Correlation of GM1 influence on AChE, ChAT, choline uptake and 5-HT uptake in hippocampus (6 days after a lesion in supracallosal area) with the postlesion degree of decrease in these parameters. Data are means from eight experiments.
r̄ — mean correlation coefficient; confidence interval at 90% is given (for details see text).

(slice I). Changes in cholinergic and serotoninergic parameters reflected degenerative changes in the corresponding fibers.

Panel B of Figure 3 presents GM1 influence on the same parameters. Results are expressed as percent change in comparison with the levels found in operated, buffer-treated animals. Enhancement of all four parameters by GM1 was observed in all hippocampal slices, however, the most pronounced effect was seen in the dorsal part (slice I), i.e. where the degeneration of nerve fibers was most severe. This indicates that GM1 influence is dependent on the degree of denervation. For quantification of this dependence we calculated the regression equations for the means of the two variables (Fig. 4), pooling all data from various experiments for each parameter, independently of the hippocampal slice but according to the percentage of decrease of the investigated parameter. A specially devised (Oderfeld, Pogorzelski, in preparation) method, based on computer simulation, allowed us to calculate the mean correlation coefficients ($\bar{r}$) and confidence interval, taking into account also dispersion of the data, not included into the formula for the linear regression coefficient. Strong correlation between the GM1 effect and the degree of fiber degeneration was found for AChE and 5-HT uptake, moderate for ChAT and only weak for choline uptake. As indicated by the slopes of regression lines, the effects of GM1 in enhancing AChE, ChAT and 5-HT uptake levels were similar.

At longer post-lesion periods similar effects of GM1 progressively diminishing with time (up to 90 days) were observed (Skup et al., in preparation; Kiedrowski et al., in preparation).

DISCUSSION

The general mechanism by which exogenous gangliosides elicit pharmacological effects is still unknown, and so is the cause of the presently observed early increase of the uptakes and enzymatic activities. However, some considerations can be done.

The initial increase caused by GM1 in both uptakes may be a consequence of an increased rate of impulse flow and increased release of neurotransmitters (Pedata et al., 1984; Toffano et al., 1984b). However, the concomitant increase in enzymatic activity would speak in favour of an increased rate of synthesis, possibly with facilitation of axonal transport, and/or enhanced activation of the enzymes in the remaining nerve terminals. Interestingly, Gage et al. (1983a), have described a phenomenon of compensatory hyperactivity in noradrenergic nerve terminals in the partially denervated hippocampus, confined to the area of maximal denervation. The relation of GM1 effect to the degree of denervation found in the present experiments could suggest, in view of Gage et al. (1983a) observations, that GM1 may facilitate hyperactivity phenomena. To this point it seems interesting to mention that Hefti et al. (1983) found the relative increase in ChAT activity, caused by NGF (without concomitant AChE increase) in partially denervated hippocampus, to be the largest in the part where the denervation was the strongest.

The possibility also exists that GM1 exerts its protective role against secondary degeneration processes, causing a better survival of neuronal cells as postulated earlier (Agnati et al., 1984; Toffano et al., 1984a). Neither can facilitation of onset of sprouting in this early period be excluded, although other possibilities seem to be more likely. Whether GM1 affects only one or more of these processes remains to be determined. We are presently investigating some of these possibilities. Neither is it as yet possible to establish whether the long-term effects observed by us in other studies (Skup et al., in preparation) include and/or are the consequence of the presently noted short-term effects.

While it is not possible to distinguish among various possibilities of mechanisms responsible for early GM1 effects, the fact that they are correlated with the degree of degeneration of nerve fibers seems to support the hypotheses concerning the relation between GM1 action and that of neuronotrophic factors (Leon et al., 1985).

It is known that injury to the CNS elicits the release of neuronotrophic factors (Gage et al., 1984; Nieto-Sampedro et al., 1983; Schonfeld et al., 1984); their induction may vary with the type and degree of neuronal degeneration. It has been found (Schonfeld et al., 1984) that the highest level of the neuronotrophic factor in hippocampus appeared 6-7 days after its lesioning. A close correlation between the levels of this trauma-induced trophic activity and ChAT activity in the transplant was found, and a hypothesis that this released neuronotrophic factor may also stimulate the recovery within intrinsic cholinergic axons was advanced (Schonfeld et al., 1984). On the other hand, it has recently been reported that the facilitatory effects following unilateral hemitransection, depend on the titers of the neuronotrophic factor activity increase in the striatum and substantia nigra (Presti et al., 1985). Similar dependence between GM1 effect and the increasing neuronotrophic factor activity was observed in in vitro experiments (Leon et al., 1985).

In conclusion, the present data provide further support for short-term effects of GM1 on restoration of biochemical parameters in lesioned brain. The dependence of the observed effects on the degree of degeneration of nerve fibers suggests a relationship between GM1 effects and the action of endogenous neuronotrophic factors.

Finally, it may be supposed that the short-term biochemical compensatory effects of GM1 observed by us, together with other similar described effects, may provide a basis for functional improvement by GM1 after brain lesions.

ACKNOWLEDGMENTS

We are grateful to Professor J. Oderfeld and Dr A. Pogorzelski for preparing the computer program for our data and help with calculations, and to Fidia Research Laboratories for kindly providing GM1 ganglioside.

The excellent technical assistance of Mrs. Zaremba and Karczewska is greatly appreciated. The investigations were supported under Project 10.4.1.01.8 p.1 of the Polish Academy of Sciences.

REFERENCES

Agnati LF, Fuxe K, Calza L, Goldstein M, Toffano G, Giardino L, Zoli M (1984) Acta Phys Scand Suppl 532: 37-44.

Dravid AR and Van Deusen EB (1984) Brain Research 324: 119-128.
Ellman GL, Courtney KD, Anders M, Featherstone R (1961) Biochem Pharmac 7: 88-95.
Fass B and Ramirez JJ (1984) Neurosci Res 12: 445-458.
Fonnum F (1975) J Neurochem 24: 407-409.
Fuxe K, Agnati LF, Benfenati F, Zini J, Gavioli G, Toffano G (1986) This volume.
Gage FH, Björklund A, Stenevi U (1983a) Nature 303: 819-821.
Gage FH, Björklund A, Stenevi A, Dunnett SB (1983b) Brain Research 268: 39-47.
Gage FH, Björklund A, Stenevi U (1983c) Brain Research 268: 27-37.
Gage FH, Björklund A, Stenevi V (1984) Nature 308: 637-639.
Haga T and Noda H (1973) Bioch Bioph Acta 291: 564-573.
Hefti F, Dravid A, Hartikka J (1983) Brain Research 293: 305-311.
Janigro D. Di Gregorio F, Vyskocil F, Gorio A (1984) J Neurosci Res 12: 499-509.
Karpiak SE (1983) Exp Neurol 81: 330-339.
Karpiak SE and Mahadik SP (1984) J Neurosci Res 12: 485-492.
Karpiak SE and Mahadik SP (1986). This volume.
Kiedrowski L and Gradkowska M (1984) in: Abstracts Fifth Meeting of the European Society for Neurochemistry, Budapest, p. 161.
Kojima H, Gorio A, Janigro D, Jonsson G (1984) Neurosci 13: 1011-1022.
Leon A, Dal Toso R, Presti D, Benvegnu D, Ferrari G, Toffano G (1986). This volume.
Lewis PR and Shute CCD (1978) in: Iversen LL, Iversen SD and Snyder SH (eds): Handbook of Psychopharmacology, vol. 9. Plenum Press, New York, pp. 315-355.
Lowry OH, Rosebrough JJ, Farr AL, Randall RJ (1951) J Biol Chem 193: 265-275.
Nieto-Sampedro M, Manthorpe M, Barbin G, Varon S, Cotman C (1983) J Neurosci 3: 2219-2229.
Oderfeld-Nowak B, Potempska A, Oderfeld J (1977) Neurosci 2: 641-648.
Oderfeld-Nowak B, Skup M, Gradkowska M, Zaremba M (1984a) in: Vizi ES and Magyar K (eds): Regulation of transmitter function: basic and clinical aspects. Proceedings of the Fifth Meeting of the European Society for Neurochemistry, Akadèmiai Kiadò, Budapest, pp. 409-412.
Oderfeld-Nowak B, Skup M, Ulas J, Jezierska M, Gradkowska M, Zaremba M (1984b) J Neurosci Res 12: 409-420.
Oderfeld-Nowak B, Skup M, Gradkowska M, Pogorzelski A, Oderfeld J (1985) Abstracts of the Tenth Meeting of the International Society for Neurochemistry, J Neurochem 44 (Supplement), S36.
Pedata F, Giovannelli L, Pepeu G (1984) J Neurosci Res 12: 421-427.
Presti D, Facci L, Dal Toso R, Aldinio C, Leon A (1985) in: ISN Satellite Meeting, Neuronal Plasticity and Gangliosides, Mantova, Italy, Poster Abstract P25.
Sabel BA, Dunbar GL, Stein DG (1984) J Neurosci Res 12: 429-443.
Schonfeld AR, Heacock AM, Katzman R (1984) Brain Research 321: 377-380.
Storm-Mathisen J and Guldberg HC (1974) J Neuroschem 22: 7983-8003.
Ternaux JP, Héry F, Bourgoin S, Adrien J, Glowiński, Hamon M (1977) Brain Research 121: 311-326.
Toffano G, Savoini G, Moroni F, Lombardi G, Calza L, Agnati LF (1983) Brain Research 261: 163-166.
Toffano G, Savoini G, Moroni F, Lombardi G, Calza L, Agnati LF (1984a) Brain Research 296: 233-239.
Toffano G, Savoini G, Aporti F, Calzolari S, Consolazione A, Maura G, Marchi M, Raiteri M, Agnati LF (1984b) J Neurosci Res 12: 397-408.
Wójcik M, Ulas J, Oderfeld-Nowak B (1982) Neurosci 7, 2: 495-499.

Gangliosides and neuronal plasticity
G. Tettamanti, R.W. Ledeen, K. Sandhoff,
Y. Nagai, G. Toffano (eds.)
Fidia Research Series, vol. 6
Liviana Press, Padova, © 1986

EFFECT OF GMI GANGLIOSIDE AND OF ITS INNER ESTER DERIVATIVE IN A MODEL OF TRANSIENT CEREBRAL ISCHEMIA IN THE RAT

Jean Cahn, Marie-Gilberte Borzeix, Gino Toffano[1]

SIR international, Department of Experimental Therapy,
6, rue Blanche, 92120 Montrouge, France
[1]Neurobiological Research Laboratories, FIDIA Research Laboratories,
Via Ponte della Fabbrica, 3/A, 35031 Abano Terme, Padova, Italy

INTRODUCTION

Among the agents affecting neurite growth responses a particular role has been attributed to gangliosides. These are glycosphingolipids, sialic-acid containing membrane components (Ledeen, 1978) which are involved in development, differentiation and regeneration (Rapport and Gorio, 1981). GMl ganglioside intraperitoneally injected into rats after unilateral hemitransection produces a marked stimulation of collateral sprouting of dopaminergic axons in the striatum and maintains the number of dopaminergic cell bodies in the substantia nigra ipsilateral to the lesion (Toffano et al., 1983; 1984; Agnati et al., 1983).

More recently, Reivich (1984) observed that GMl ganglioside, intravenously injected into cats immediately after releasing the occlusion of the middle cerebral artery (MCA), may have a protective effect on the brain tissue against ischemic injury by reducing the tissue metabolism, so as to preserve the energy charge potential (ECP) and to prevent any structural damage.

All these results led us to hypothesize that GMl ganglioside could be helpful in conditions of transient cerebral oligemia in rat brain tissue (Borzeix, 1983; 1984). In addition, a comparison between GMl ganglioside and its inner ester derivative (Aldinio et al., 1984) was made. The therapeutic effect of these two compounds was investigated with regard to the edematous reaction of the brain, electrolytic imbalance, impaired cerebral function, the hypofusion, i.e. the reduction of the cerebral blood flow (CBF)

Abbreviations: KCA, middle cerebral artery; ECP, energy charge potential; CBF, cerebral blood flow; MABP, mean arterial blood pressure; ISI, initial slope index; ECoG, electrocorticographic; c.p., intraperitoneal injection, TH, tyrosine hydroxylase; DMP4, 6,7-dimethyl-5,6,7,8-tetrahydro pterine hydrochloride.

and the electro-corticographic disturbances developing over the acute phase — 3 days — following the release of ischemia, as well as toward the softening of the brain observed at the fourth week post-ischemia.

MATERIAL AND METHODS

The study was undertaken in 302 Sprague-Dawley male rats, weighing about 230 g at the beginning of the experiment.

Experimental Procedure

Both carotid arteries were dissected under light ether anesthesia. Twenty-four hours later, the two vessels were occluded over a 60-minute period. In parallel, MABP was slightly lowered (nearby 7.9 to 9.3 kPa) by a sub-cutaneous injection of sodium nitroprusside (1.1 mg per rat).

Cerebral Edema, Electrolytic Balance and Capability of Learning

These parameters were investigated over the first three days post-oligemia. The capability of learning was tested through a one-trial passive avoidance test (Burešova et al., 1964), the retention of the conditioning being controlled 24 and 48 hours after the punishment. The presence of edema in association with a vasospasm was observed macroscopically on removal of the brain; K^+ (flame photometry) and Ca^{++} (colorimetric method, o.cresolphtalein-complex) contents of the brain were measured in the dry brain tissue (Sarkar and Chauhan, 1967).

Cerebral Blood Flow (CBF) Measurement

Cerebral blood flow was measured at the third day post-oligemia. Since 1965, Lassen and Ingvar had developed a method based on the clearance of non-diffusible tracers (133 Xenon) injected by intra-carotid route. Detectors placed on the scalp recorded the gamma-emission of 133 Xenon. According to Høedt-Rasmussen et al. (1966), it was possible to calculate CBF from the desaturation curves using the Initial Slope Index (ISI) determination adapted to the rat by Nilsson and Siesjö (1976).

Electro-Corticographic (ECoG) Spectral Analysis

A quantitative ECoG analysis (Borzeix et al., 1979) was performed in terms of energy power over the four weeks following the release of ischemia. ECoG recordings were obtained from silver electrodes implanted through the skull in close contact with the dura-mater. Every day, three successive 10-minute sample recordings were split into 4 frequency bands (delta 0.5-3 Hz, theta 4-7.5 Hz, alpha 8-17 Hz, beta 18-32 Hz), then analyzed and averaged. After removal, the macroscopic examination of the brain allows one to identify the "injured" hemisphere. The evolution of the ECoG patterns was particularly followed up in this hemisphere.

Planimetric and Histologic Evaluation of the Tissue Damage

This study was undertaken at the end of the fourth week post-oligemia. Rats were sacrificed by intra-cardiac perfusion of paraformaldehyde in order to fix the brain in situ. A macroscopic examination was performed on 3 sections, i.e. the striatum, the thalamus, and both the anterior and posterior parts of the hippocampus. Furthermore, the surface of each cerebral hemisphere was measured by means of a digitizer table HIPADtm (Houston Instrument, Austin, TX) connected with an APPLE II Plus computer. The percentage of "deterged tissue" was measured with respect to the brain surface of normal rats. Moreover, an "hemispheric ratio" was calculated from both the left and the right hemispheres (the larger surface over the smaller one), since our experimental conditions show that the brain injury is essentially unilateral in the survivors. In "normal" animals, this ratio ranges from 1.0 to 1.10.

The Test Compounds

The post-oligemic rats were given GMl obtained and purified from bovine brain according to Tettamanti et al. (1973) or its inner ester derivative (Aldinio et al., 1984) by intraperitoneal route. Treatments began one hour post-oligemia at a rate of 2 injections per day, and lasted 3 to 28 days. Control animals were given the same volume of a phosphate buffer used to solubilize gangliosides.

RESULTS

Cerebral Edema, Electrolytic Balance and Learning

During the 60-minute oligemia, psychomotor reactions, such as piloerection (in all cases), severe hypotonia associated with ventral decubitus (in 93 percent of cases) and clono-tonic discharges (in 20 percent of rats) were observed. Following clipping off, the first event to be noticed about 2 hours after recirculation was the occurrence of clonic discharges in almost 20 percent of rats. Thereafter, a marked lateralization was observed, as well as a protrusion of the contralateral eyeball. Once the brain removed, it was possible to observe that the exophtalmus corresponded to the side of the most injured cerebral hemisphere exhibiting a marked edematous reaction and sometimes brain acmorrhagcs. At thc third day post-oligcmia, thc watcr contcnt of thc brain was increased, in parallel with a cellular potassium efflux and with a tremendous accumulation of calcium ions into the brain parenchyma (Borzeix, 1983, 1984). The ability of post-oligemic rats to acquire a passive avoidance learning was strongly impaired since only 21 percent of them were conditioned 24 hours after the punishment, and only 16 percent at the 48th hour, thus reflecting no long-term retention of the learning.

Under these conditions, GMl and its inner ester derivative protected the brain in the early post-ischemic period limiting the extent of the edematous reaction, the K^+ efflux and the accumulation of Ca^{++} ions into the brain. As a consequence, even the neurological disorders were reduced and the cerebral function was preserved, as shown by the improved ability of the post-oligemic treated rats to acquire and retain a learning test (Table 1).

Table 1. *Electrolyte Unbalance, Capability of Learning and Macroscopic Aspect of the Brain at the Third Day Post-Oligemia*

Series	Dose mg kg^{-1} IP (7×)	N	Electrolyte balance			Learning test			Cerebral edema
			H_2O	K^+	Ca^{++}	% conditioned rats after punishment			
			%	mM.kg^{-1} d.w	mM.kg^{-1} d.w	+ 24 h	+ 48 h	P	
Controls	—	10	78.4 ± 0.13	461 ± 3.8	5.5 ± 0.29	90	90	—	0
Oligemia	—	40	80.0 ± 0.19*	403 ± 7.7*	18.1 ± 1.61*	21	16	0.001*	64*
GM1	1	10	80.0 ± 0.33	408 ± 11.7	27.3 ± 5.99	50	10	NS	60
	2.5	10	80.0 ± 0.42	395 ± 18.9	22.8 ± 4.82	50	50	NS	70
	5	10	79.6 ± 0.48	435 ± 13.5**	11.8 ± 3.61**	70	30	NS	30
	10	10	78.6 ± 0.46**	427 ± 14.4	15.5 ± 4.38	47	37	NS	45
	30	10	79.2 ± 0.56	414 ± 15.3	15.6 ± 3.66	71	47	0.02**	65
GM1 inner ester	1	10	79.6 ± 0.36	419 ± 17.3	21.4 ± 5.59	60	60	0.01**	50
	2.5	10	79.2 ± 0.39**	417 ± 14.0	16.8 ± 4.77	78	67	0.001**	40
	5	10	78.3 ± 0.13**	458 ± 4.2**	5.5 ± *.40**	80	80	0.001**	10**
	10	10	79.3 ± 0.39	439 ± 11.6**	11.0 ± 3.60**	40	60	0.01**	20**
	30	10	78.5 ± 0.39**	428 ± 13.7	12.5 ± 3.29	50	45	0.02**	50

d.w. = dry weight $p \leqslant 0.05$ according to the continuity corrected Chi square test * versus controls ** versus oligemia

Delayed Hypoperfusion

According to the Initial Slope Index method, the CBF value assessed in young normal rats is about 90 $ml.min^{-1}.100\ g^{-1}$ of brain tissue (Nilsson and Siesjö, 1976). Three days after a 60-minute oligemic injury to the brain, the CBF was strongly impaired since it was lowered by almost 55 percent with respect to normal rats (Table 2). An effective treatment with GMl or its inner ester derivative resulted in marked improvement of CBF to the normal range (Table 2).

Table 2. *133 Xenon Clearance Method for CBF Determination in Post-Oligemic Rats*

Series	Dose $mg.kg^{-1}$ IP (7×)	N	ISI $ml.min^{-1}.\ 100\ g^{-1}$ Left hemisphere	Right emisphere
Controls	—	7	82.7 ± 7.35	97.8 ± 9.45
Oligemia	—	7	45.2 ± 6.71*	50.4 ± 8.69*
GM 1	5	7	76.5 ± 10.86**	88.3 ± 12.81**
GM 1 inner ester	5	7	74.3 ± 8.99**	86.5 ± 8.65**

$p \leqslant 0.05$ according to Student's T test   * versus controls   ** versus oligemia

ECoG Investigations

During the following hours and days a transient cerebral ischemia gave rise to marked disturbances of the EEG patterns which essentially consisted of an increase of delta slow waves, whereas the faster rhythms were depressed (Borzeix and Berbey, unpublished data).

Treatment with GMl (5 $mg.kg^{-1}$ i.p. twice a day) showed only a trend toward improvement of the EEG patterns, whereas the same dose of its inner ester derivative produced a significant improvement of the EEG patterns. Actually, the delta waves were significantly less enhanced in the treated subjects, than in the controls and the fast rhythms were also preserved, as reflected by the theta/delta ratio values (Tables 3 and 4).

Table 3. *Computerized Evaluation of the ECoG on the "Injured" Hemisphere in the Post-Oligemic Controls*

	N	delta (%)	theta (%)	alpha (%)	beta (%)	theta/delta ratio
T0	28	34.0 ± 1.05	28.2 ± 0.40	19.6 ± 0.39	19.3 ± 0.85	0.83 ± 0.033
+ 5 hours	8	53.8 ± 2.66	18.8 ± 1.82	13.2 ± 2.03	14.2 ± 1.02	0.36 ± 0.046
+ 24 hours	4	59.0 ± 3.49	18.6 ± 1.87	10.9 ± 1.67	11.6 ± 0.35	0.32 ± 0.047
+ 48 hours	3	55.1	19.2	12.8	12.9	0.41
+ 72 hours	3	44.2	23.7	16.6	15.6	0.59
+ 14 days	2	42.8	25.0	16.2	16.1	0.68
+ 28 days	2	36.8	27.0	17.4	18.9	0.74

Table 4. *Comparative Evolution of the ECoG Patterns*

	delta (%)			theta (%)			alpha (%)			beta (%)			theta/delta ratio		
	Controls	GM 1	GM1 inner ester	Controls	GM 1	GM1 inner ester	Controls	GM 1	GM1 inner ester	Controls	GM 1	GM1 inner ester	Controls	GM 1	GM1 inner ester
+5 hours	+7.8	+10.6 NS	+4.5 NS	—3.0	—4.0 NS	—1.9 NS	—2.5	—3.6 NS	—0.9 (*)	—2.3	—3.3 NS	—1.8 NS	—0.20	—1.13	—0.05 (*)
+24 hours	+8.3	+ 6.9 NS	+7.6 NS	—2.8	—3.0 NS	—3.4 NS	—1.6	—1.1 NS	—2.1 NS	—3.9	—2.8 NS	—2.2 NS	—0.21	—0.09 (*)	—0.11 (*)
+48 hours	+6.9	+10.5 NS	+3.2 NS	—2.3	—3.9 NS	—2.0 NS	—1.6	—2.1 NS	—1.5 NS	—3.6	—4.1 NS	—0.7 (*)	—0.17	—0.14	—0.05 (*)
+72 hours	+6.2	+ 7.1 NS	+1.6 (*)	—1.8	—2.8 NS	—1.4 NS	—0.7	—1.3 NS	—0.3 NS	—3.7	—3.0 NS	—0.6 (*)	—0.18	—0.11	—0.04 (*)
+14 days	+6.5	+ 6.2 NS	+1.5 (*)	—2.1	—2.2 NS	—0.6 (*)	—1.9	—1.8 NS	—0.4 (*)	—2.5	—2.2 NS	—1.4 NS	—0.19	—0.09 (*)	—0.02 (*)
+28 days	+5.3	+ 3.1 NS	+0.9 (*)	—1.5	—0.7 NS	—0.1 NS	—1.6	—0.6 NS	—0.9 (*)	—2.3	—1.9 NS	—1.8 NS	—0.16	—0.04	—0.003 (*)

Mean differences with regards to t0 (before oligemia)
* p value according to Student's "t" test versus the post-oligemic controls
NS: $p > 0.10$

Tissue Damage

At the end of the 4-week period, a cortical necrosis was observed in 40 percent of the controls, the surface of brain tissue deterged ranging from 56 percent in the hemisphere of the frontal region to 16 percent in that of the occipital cortex. Furthermore, the ratio of the hemispheric surfaces was increased in 70 percent of the controls, i.e. higher than 1.10 (see above). In the sub-cortical structures, the striatum and the thalamus were mainly damaged (from 40 to 45 percent of cases).

An effective treatment with GMl (5 mg.kg^{-1} i.p. twice a day) did not significantly protect the brain with respect to these parameters. Its inner ester derivative, however, (5 mg.kg^{-1} i.p. twice a day for 28 days) exerted a therapeutic effect since only 25 percent of the treated rats exhibited a loss of brain tissue. Moreover, the extent of the necrotic area was smaller than in the controls, as evidenced by a weaker imbalance of the "hemispheric ratio" which is impaired in only 35 percent of the rats (Table 5). Furthermore, at the sub-cortical level, only 30 percent of the rats presented a cell damage in both the striatum and the thalamic nuclei.

Table 5. *Evolution of the "Hemispheric Ratio"*

	N	Section 1	Section 2	Section 3
Controls	20	1.62±0.172	1.17±0.046	1.14±0.033
GM 1	21	1.65±0.180 NS	1.26±0.075 NS	1.22±0.073 NS
GM 1 inner ester	20	1.30±0.115 (*)	1.10±0.029 (*)	1.08±0.028 (*)

(*) $p \leq 0.05$ according either to Student's test or to Cochran's t test.

DISCUSSION

Previous reports have shown that treatment with exogenous gangliosides favoured the recovery of striatal and nigral dopaminergic parameters after a partial mechanical lesion of the ascending dopaminergic fibers (Toffano et al., 1983; 1984; Agnati et al., 1983). In particular, GMl ganglioside injected into rats with unilateral hemitransection increased the apparent V^{max} of tyrosine hydroxylase (TH) in the striatum of the lesioned side. The effect is dose-dependent and already apparent 14 days after the lesion (Toffano et al., 1983). Similarly, GMl inner ester derivative produces a significant recovery of the apparent V^{max} of TH for DMPH4 in the striatum of the lesioned side, the effect being also dose-dependent (Aldinio et al., 1984). The effect of GMl inner ester derivative was already apparent with 5 mg.kg^{-1}, and the recovery almost complete with 30 mg.kg^{-1}. In these cases, the treatment began two days after surgery and was given every third day.

Interestingly, when the effect of GMl was compared to that of its inner ester derivative at doses of 5 and 10 mg.kg^{-1}, the latter was more active in enhancing TH activity. This peculiarity is explained as possibly due to differences existing between the metabolism and/or the pharmacokinetics of the two compounds.

Our results, obtained under completely different experimental conditions, led us to a similar conclusion. GMl inner ester derivative was more effective than its parent compound GMl — given at the same dose — in protecting the brain against cell damage occurring during the early post-ischemic period. The beneficial effect of GMl internal ester derivative was dose-dependent following a bell-shaped profile. Its therapeutical dose-dependent effect increased at dosage of 1 to 10 $mg.kg^{-1}$ (twice a day) and then declined at higher dosage (30 $mg.kg^{-1}$).

Our results are also consistent with, and corroborate those provided by Reivich (1984). The latter showed that after middle cerebral artery (MCA) ligation in cats a single dose of GMl ganglioside (30 $mg.kg^{-1}$ intravenously) significantly decreased the cerebrovascular resistance, then increased the CBF and depresses the glucose consumption nearby the ischemic region. This resulted in an improved neurological and histological outcome in the treated animals.

CONCLUSION

Both GMl ganglioside and its inner ester derivative are helpful to the brain tissue during the post-ischemic period following a transient cerebral oligemia. Nevertheless, the therapeutic effect is generally much more pronounced with the inner ester derivative than with GMl. This effect involves the suppression of tissue edema, of the overload of Ca^{++} ions into the brain and of K^+ efflux from the cells; increased capability of learning and retention; improved cerebral blood flow; decreased electrocorticographic disturbances and a reduced extent of the cell damage.

ACKNOWLEDGMENTS

The Authors thank Fidia Research Laboratories for providing GMl and its inner ester derivative. They also thank J.P. Akimjak, B. Berbey, P. Charles, J.M. Dupont, D. Angignard and M. Filian for their technical assistance.

REFERENCES

Agnati LF, Fuxe K, Calzà L, Benfenati F, Cavicchioli L, Toffano G, Goldstein M (1983) Gangliosides increase the survival of lesioned nigral dopamine neurons and favour the recovery of dopaminergic synaptic function in striatum of rats by collateral sprouting. Acta Physiol Scand 119: 347-363.

Aldinio C, Valenti G, Savoini GE, Kirschner G, Agnati LF, Toffano G (1984) Monosialoganglioside internal ester stimulates the dopaminergic reinnervation of the striatum after unilateral hemitransection in rat. Int J Devl Neurosci 2(3): 267-275.

Borzeix MG, Labos M, Hartl C (1979) Comparative spectral EEG analysis during acute cerebral injuries of traumatic, ischemic or toxic origin in the unanesthetized rabbit. Acta Neurol Scad Suppl 72, 60: 392-393.

Borzeix MG (1983) Le syndrome post-oligémique chez le rat. Un modèle d'atteinte cérébrale chronique. Circulation et métabolisme du cerveau 1: 63-79.

Borzeix MG (1984) Le syndrome post-oligémique chez le rat et sa thérapeutique. Une étape préclinique dans l'exploration aigue puis chronique des conséquences d'un accident vasculaire allant jusqu'au rammollissement cérébral. In: Cahn J, Agnoli A, Cohadon F, Hoyer S, Lechner H (eds): Médicaments et les Maladies Cérébro-vasculaires (Drugs and Cerebrovascular Diseases). John Libbey Eurotext, pp. 27-35.

Burešova O, Bures J, Bohdanecky Z, Weiss T (1964) Effect of atropine on learning, extinction, retention and retrieval in rats. Psychopharmacologia 5: 255-263.

Høedt-Rasmussen K, Sveindottir E, Lassen NA (1966) Regional cerebral blood flow in man determined by intra-arterial injection of radio-active inert gas. Circ Res 18: 237-247.

Ledeen RW (1978) Ganglioside structure and distribution: are they localized at the nerve ending? J Supramol Struct 8: 1-17.

Nilsson B and Siesjö BK (1976) A method of Determining Blood Flow and Oxygen Consumption in the rat brain. Acta Physiol Scand 96: 72-82.

Rapport MM and Gorio A (1981) Gangliosides in Neurological and Neuromuscular Function, Raven Press, New York.

Reivich M (1984) The action of GM1 ganglioside on the hemodynamic and metabolic effects of ischemia. Clinical Neuropharmacology Vol. 6, Suppl. 1: p. 597.

Sarkar R and Chauhan UPS (1967) A new method for determining microquantities of calcium in biological materials. Anal Biochem 20: 155.

Tettamanti G, Bonali F, Marchesini S, Zambotti V (1973) A new procedure for the extraction, purification and fractionation of brain gangliosides. Biochim Biophys Acta 296: 160-170.

Toffano G, Savoini G, Moroni F, Lombardi MG, Calzà L, Agnati LF (1983) GM1 ganglioside stimulates the regeneration of dopaminergic neurons in the central nervous system. Brain Res 261: 163-166.

Toffano G, Savoini G, Moroni F, Lombardi G, Calzà L, Agnati LF (1984) Chronic GM1 ganglioside treatment reduces dopamine cell body degeneration in the substantia nigra after unilateral hemitransection in rat. Brain Res 296: 233-239.

Gangliosides and neuronal plasticity
G. Tettamanti, R.W. Ledeen, K. Sandhoff,
Y. Nagai, G. Toffano (eds.)
Fidia Research Series, vol. 6
Liviana Press, Padova, © 1986

EARLY RECOVERY OF STRIATAL DOPAMINE UPTAKE IN RATS WITH UNILATERAL NIGRO-STRIATAL LESION FOLLOWING TREATMENT WITH MONOSIALOGANGLIOSIDE

Maurizio Raiteri, Paola Versace, Mario Marchi

Istituto di Farmacologia e Farmacognosia, Università di Genova,
Viale Cembrano 4, 16148 Genova, Italy

INTRODUCTION

Evidence is accumulating that the central nervous system of the adult animal is capable of morphological, biochemical and functional recovery following injury (Cotman et al., 1981; Freed et al., 1985). Gangliosides, which are normally present in neuronal membranes, seem to play a role in the events related to neuronal plasticity. In fact, the administration of purified gangliosides was shown to enhance the recovery processes after lesioning of the central and peripheral nervous system (Rapport and Gorio, 1981; Toffano et al., 1983; Agnati et al., 1983; Sabel et al., 1984a; Jonsson et al., 1984). It has been recently reported that after unilateral hemitransection of the nigro-striatal pathway in the rat, long-term treatment with GM1 ganglioside stimulated the dopaminergic reinnervation in the striatum by enhancing the collateral sprouting of the nigro-striatal dopaminergic neurons (Toffano et al., 1983; 1984a; 1984b; Aldinio et al., 1984; Sabel et al., 1984b). The question arises whether the terminals of the nerve fibers regenerating under the action of gangliosides possess all the presynaptic characteristics of a "normal" striatal dopaminergic nerve terminal. In the present investigation we have studied the effect of repeated administration of monosialoganglioside inner ester on ^{3}H-DA uptake activity in striatal synaptosomes after unilateral hemitransection of the nigro-striatal pathway.

MATERIALS AND METHODS

Animals, Surgery and Treatment

Adult male Sprague Dawley rats (180 - 220 g) were used. Animals were anesthetized with ketamine (Ketalar, Parke Davis; 100 mg/kg, i.p.) and placed in stereotaxic

Abbreviations: DMI, desmethylimipramine; DA, dopamine; DOPA, dihydroxyphenylanine.

head holder. The unilateral lesion of the nigro-striatal fibers was performed as previously described (Toffano et al., 1983). Briefly, a 4 mm-wide stainless steel blade was inserted next to the midline (0.5 mm lateral to the midline and 1 mm posterior to the bregma) and lowered 9 mm with an angle of 68°. After the lesion the animals were caged in groups with normal light-dark cycles. The animals were treated for different periods of time with the inner ester derivative of GM1 ganglioside (daily dose 30 mg/kg; i.p.), with L-DOPA (daily dose 2 × 100 mg/kg, p.os.) plus carbidopa (daily dose 2 × 40 mg/kg; p.os.), or with saline, starting the 2nd day after surgery.

In one set of experiments some animals were kept in darkness starting immediately after the lesion and treated with the GM1 monosialoganglioside inner ester as described above.

Preparation of Synaptosomes and DA Uptake Determination

The animals were sacrificed 24 h after the last injection. The brain was removed; the corpus striatum, both of the lesioned and of the unlesioned side, was rapidly dissected out and homogenized (1 : 40 [w/v]) in 0.32 M sucrose buffered at pH 7.5 using a glass-teflon homogenizer. A crude synaptosomal fraction was prepared essentially according to Gray and Whittaker (1962). The uptake of ^{3}H-DA was assayed as follows: the synaptosomes (500 μl; 100-200 μg protein) were incubated 10 min at 37°C in a rotary waterbath, in an oxygenated standard medium having the following composition (mM): NaCl, 125; KCl, 3; $CaCl_2$, 1.2; $MgSO_4$, 1.1; Na_2HPO_4, 5; glucose, 10 and Tris-HCl buffer, 10; pH 7.5. Different concentrations (from 0.01 to 0.32 μM) of ^{3}H-DA (Amersham Radiochemical Centre; spec. activity, 46 Ci/mmol) were then added and incubation was continued for 1 min. At the end of the incubation time aliquots (3 × 100 μl) of the synaptosomal suspension were transferred onto Whatman Glass microfiber filters (GF/B). The synaptosomes were then washed twice with two 5 ml portions of medium. Each filter was removed, placed in a scintillation vial with 0.5 ml of 2% sodium dodecyl sulfate and counted for radioactivity in a Packard liquid scintillation counter. Blank values were obtained by maintaining the samples at 0° C. In some experiments desmethylimipramine (DMI), at the concentration of 1 μM, was present in the synaptosomal suspension from the beginning of the preincubation, to prevent uptake of ^{3}H-DA into noradrenergic nerve terminals.

The uptake of ^{3}H-DA was measured on days 5, 8, 15 and 32 after surgery (i.e. after 3, 6, 13, 30 days of treatment) and the apparent V_{max} and K_m values were calculated. The GM1 monosialoganglioside inner ester was kindly supplied by Fidia Research Laboratories, Abano Terme, Italy. Its purity, assessed by different methods, was over 99% (Aldinio et al., 1984).

RESULTS

When the uptake of ^{3}H-DA was measured in synaptosomes prepared from control saline-treated rats, 5 and 8 days after hemitransection of the nigro-striatal pathway, it was found that the apparent V_{max} in the striatum of the lesioned side was reduced to 10% of that in the striatum of the contralateral unlesioned side (Fig. 1). The

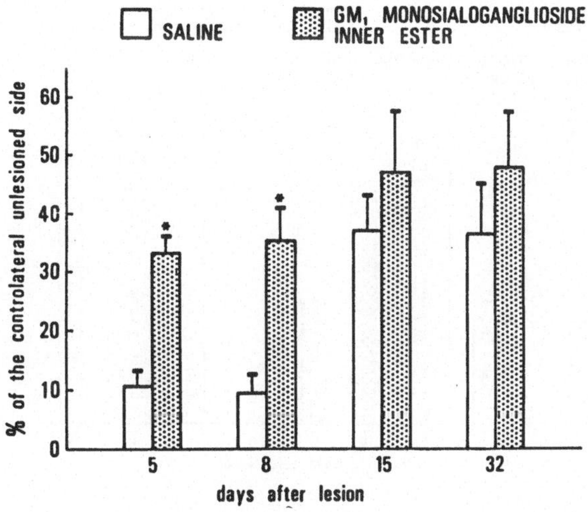

Figure 1. Effect of GM1 monosialoganglioside inner ester treatment on ^{3}H-DA uptake in striatal synaptosomes after lesion of the nigro-striatal pathway. The data represent the V_{max} mean values (± S.E.M.) of ^{3}H-DA uptake in the striatum of the lesioned side expressed as percent of the respective contralateral unlesioned side. All means are from 6 to 8 determinations run in triplicate. The V_{max} of ^{3}H-DA uptake determined on day 5 after surgery in the unlesioned side of the saline-treated group (194 ± 37 pmol/min/mg prot) was not statistically different from that of the ganglioside treated one (163 ± 32 pmol/min/mg prot) and showed no significant changes with time (8, 15 and 32 days after the lesion).
* $P < 0.005$ vs saline (Student's t-test).

apparent V_{max} of ^{3}H-DA uptake recovered spontaneously to about 40% of the contralateral side (day 15 after surgery) and did not show any further increase either on day 32 (Fig. 1) or on day 90 (Fig. 3). The apparent K_m values (0.11 ± 0.02 and 0.14 ± 0.02 μM in the lesioned and unlesioned side, respectively, on day 5 after surgery) showed no significant changes with time.

In the animals treated with GM1 monosialoganglioside inner ester, the apparent V_{max} of ^{3}H-DA uptake showed a marked recovery in the lesioned side already on day 5 after lesion (33% of the contralateral side), with no changes in the apparent K_m values. No significant differences existed on day 15 and 32 after surgery between the ^{3}H-DA uptake in the rats receiving saline and that in the ganglioside-treated animals (Fig. 1).

Figure 2 shows that, in the presence of DMI, the values of ^{3}H-DA uptake, determined in striatal synaptosomes 8 and 15 days after the lesion, did not differ significantly from those obtained in the absence of the noradrenaline uptake inhibitor. Figure 2 also shows that treatment with L-DOPA plus carbidopa did not modify the uptake of ^{3}H-DA after hemitransection. The effect of GM1 monosialoganglioside inner ester observed 8 days after surgery (i.e. 6 days of treatment) was not altered when L-DOPA + carbidopa were co-administered.

Figure 3 illustrates the results obtained using animals kept in the normal light-dark cycle compared with those obtained employing animals always maintained in darkness.

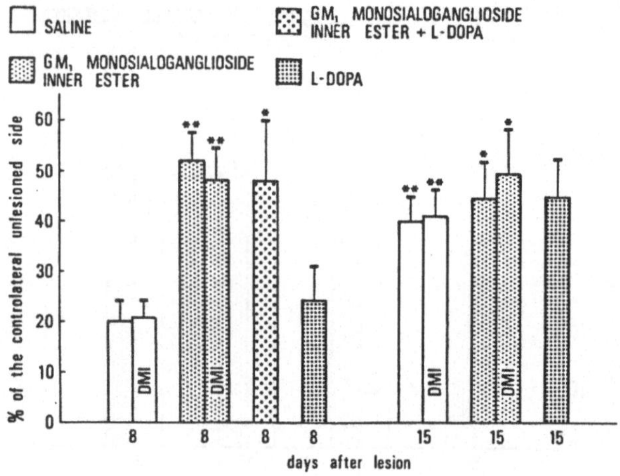

Figure 2. Effect of treatment with GM1 monosialoganglioside inner ester and L-DOPA on ^{3}H-DA uptake in striatal synaptosomes after lesion of the nigro-striatal pathway. The data represent mean values of ^{3}H-DA uptake in the striatum of the lesioned side expressed as percent of the respective contralateral unlesioned side. All means are from 6 to 8 determinations run in triplicate. * $P < 0.01$ ** $P < 0.005$ vs saline treated animals on day 8 after lesion (Student's t-test).

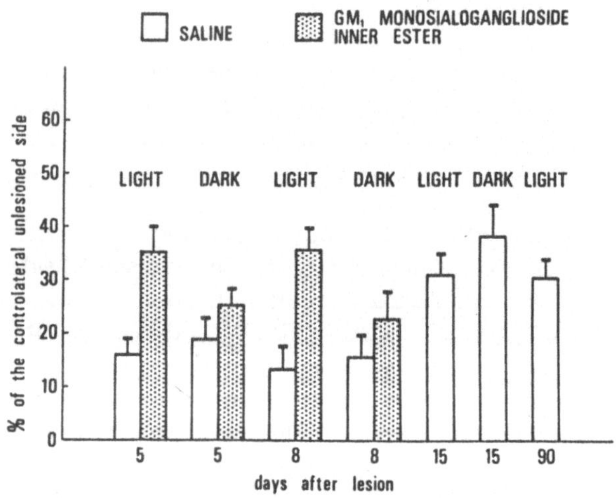

Figure 3. Effect of treatment with GM1 monosialoganglioside inner ester on ^{3}H-DA uptake in striatal synaptosomes from lesioned animals kept in light or in darkness. The data represent mean values of ^{3}H-DA uptake in the striatum of the lesioned side expressed as percent of the respective contralateral unlesioned side.

No significant differences existed on day 5 and 8 after surgery between the striatal ^{3}H-DA uptake on the lesioned side of the saline treated animals kept in a normal light/dark cycle with those kept in darkness. The ^{3}H-DA uptake recovered spontaneously to 30 - 40% of the contralateral side on day 15 after surgery also in the animals maintained always in darkness. The marked recovery in ^{3}H-DA uptake shown on days 5 and 8 after lesioning in the animals treated with GM1 monosialoganglioside inner ester was not evident when the animals were kept for 5 and 8 days in darkness.

DISCUSSION

The surgical hemitransection of the nigro-striatal pathway in the rat brain has been recently used as a model to investigate the regenerative capacity of central dopaminergic neurons and its possible stimulation by various factors. In particular, Toffano et al. (1983) and Aldinio et al. (1984) studied tyrosine hydroxylase in the rat corpus striatum of animals which had been lesioned and then treated chronically with GM1 monosialoganglioside (Toffano et al., 1983, 1984) or its inner ester (Aldinio et al., 1984). These authors found that after 13 days of ganglioside treatment (but not after 3 or 6 days) both the apparent V_{max} of tyrosine hydroxylase activity and the density of tyrosine hydroxylase terminals determined by immunofluorescence in the lesioned striatum were higher than after treatment with saline. No spontaneous recovery of tyrosine hydroxylase activity was apparently observed up to 32 (Toffano et al., 1983) or 56 (Aldinio et al., 1984) days after the lesion. The authors suggested that gangliosides facilitate the collateral sprouting of the surviving nigro-striatal dopaminergic neurons.

The first result of the present work showed that a partial spontaneous recovery of ^{3}H-DA uptake occurred in the lesioned striatum 15 days after the hemitransection. Secondly, 3 days of ganglioside treatment caused a recovery of ^{3}H-DA uptake which was quantitatively similar to that occurring spontaneously 15 days after surgery but was already present 5 days after hemitransection.

Under similar experimental conditions, Sabel et al. (1984b) reported that clear anatomical signs of ganglioside-induced reinnervation into the denervated striatum were detectable only about 2 weeks after the lesion. Therefore, the early increase of ^{3}H-DA uptake observed in our study does not seem to reflect a proliferation of dopaminergic striatal projections, but rather an increase of function of spared terminal fibers. Although a protecting action of gangliosides on the degenerating neurons can not be excluded, this seems to be less likely in view of the above-mentioned behavior of tyrosine hydroxylase.

Considering that transmitters may have, in some conditions, trophic functions (for instance, serotonin during early development; Lauder, 1983) we tested the possible effect of an increased DA synthesis and release by treating the lesioned rats with L-DOPA plus carbidopa. The results of Figure 2 tend to exclude that such a treatment could have stimulatory effect on regeneration or improve the effect of GM1 ganglioside.

The effect of subacute treatment with GM1 monosialoganglioside inner ester on the recovery of ^{3}H-DA uptake at 5 and 8 days after the lesion was drastically reduced when animals were kept in darkness. However, the spontaneous recovery of ^{3}H-DA uptake in the saline treated animals kept in darkness at 15 days after the lesion was unaffected. It therefore appears that some mechanisms through which the GM1 monosialoganglioside inner ester exerts its effect are connected with the normal light/dark cycle. It was previously observed that lesioned animals kept in darkness for 30 days had lower striatal TH activity compared to those maintained in a normal light/dark cycle; moreover, the increase in TH activity caused by chronic GM1 monosialoganglioside treatment disappeared when the lesioned rats were kept in darkness (Toffano et al., 1984b).

The effect of ganglioside found in the present work was present only in the lesioned striatum and therefore was not due to a generalized stimulatory action on cathecholamine uptake which, on the unlesioned side, was unaffected. It is also interesting to note that the chronic treatment did not produce further recovery of the ^{3}H-DA uptake on day 15 or 30 after lesion. This suggests that the effect of gangliosides is only dependent on the early events occurring at the level of the injured tissue. Accordingly, the edematous reaction in experimental stroke was almost completely suppressed by the monosialoganglioside inner ester 3 days after cerebral oligemia (Borzeix, 1984).

It is possible that the incorporation of gangliosides in the membrane of the neuronal terminal causes the early events which are supportive of the functional recovery of DA uptake. Whether this early effect may trigger the later sprouting of spared DA terminals remains to be established.

ACKNOWLEDGMENTS

This work was supported by grants from the Italian National Research Council and the Italian Ministry of Education. Mrs Maura Agate is thanked for her excellent secretarial services.

REFERENCES

Agnati LF, Fuxe K, Calza L, Benfenati F, Cavicchioli L, Toffano G, Goldstein M (1983) Gangliosides increase the survival of lesioned nigra dopamine neurons and favour the recovery of dopaminergic synaptic function in striatum of rats by collateral sprouting. Acta Physiol Scand 119: 347-363.

Aldinio C, Valenti G, Savoini G, Kirschner G, Agnati LF, Toffano G (1984) Monosialoganglioside internal ester stimulates the dopaminergic reinnervation of the striatum after unilateral hemitransection in rats. Int J Dev Neurosci 2: 267-275.

Borzeix MG (1984) Drugs effect in an experimental stroke resulting from a transient cerebral oligemia. In: The need for treatment for stroke, Alzheimer's disease and abnormal brain aging. MEM-G.M.D. vol. 1, n. 3; pp. 44.

Cotman CW, Neito-Sampedro N, Harris EW (1981) Synapse replacement in the nervous system of adult vertebrate. Physiol Rev 71: 684-784.

Freed WJ, de Medinaceli L, Wyatt RJ (1985) Promoting functional plasticity in the damaged nervous system. Science 227: 1544-1552.

Gray EG, Whittaker VP (1962) The isolation of nerve endings from brain: an electron microscope study of cell fragments derived by homogenization and centrifugation. J Anat 96: 79-87.

Jonsson G, Gorio A, Hallman H, Janigro D, Kojima H, Zanoni R (1984) Effect of GM1 ganglioside on neonatally neurotoxin induced degeneration of serotonin neurons in the rat brain. Dev Brain Res 16: 171.

Lauder JM (1983) Hormonal and humoral influences on brain development. Psychoneuroendocrinol 8: 121.

Rapport MM, Gorio A (1981) Gangliosides in neurological and neuromuscular function, development and repair. Raven Press, New York.

Sabel BA, Slavin MD, Stein DG (1984a) GM1 ganglioside treatment facilitates behavioral recovery from bilateral brain damage. Science 225: 340-342.

Sabel BA, Dunbar GL, Stein DG (1984b) Gangliosides minimize behavioral deficits and enhance structural repair after brain injury. J Neurosci Res 12: 429-443.

Toffano G, Savoini G, Moroni F, Lombardi G, Calza L, Agnati FL (1983) GM1 ganglioside stimulates the regeneration of dopaminergic neurons in the central nervous system. Brain Res 261: 163-166.

Toffano G, Savoini G, Moroni F, Lombardi G, Calza L, Agnati LF (1984a) Chronic GM1 ganglioside treatment reduces dopamine cell body degeneration in the substantia nigra after unilateral hemitransection in rat. Brain Res 296: 233-239.

Toffano G, Savoini G, Aporti F, Calzolari S, Consolazione A, Maura G, Marchi M, Raiteri M, Agnati LF (1984b) The functional recovery of damaged brain: the effect of GM1 monosialoganglioside. J Neurosci Res 12: 397-408.

Gangliosides and neuronal plasticity
G. Tettamanti, R.W. Ledeen, K. Sandhoff,
Y. Nagai, G. Toffano (eds.)
Fidia Research Series, vol. 6
Liviana Press, Padova, © 1986

MOTORNEURON SPROUTING AND SPINAL PLASTICITY IN AMYOTROPHIC LATERAL SCLEROSIS: THE "WINDOW OF OPPORTUNITY" FOR A GANGLIOSIDE TREATMENT

P. Pinelli,[1] C. Pasetti, L. Mazzini, F. Pisano, A. Villani

[1]Neurological Clinic, University of Milan, Milan, Italy;
Medical Center of Rehabilitation, Dept. of Neurology, Veruno (Novara), Italy

INTRODUCTION

Amyotrophic Lateral Sclerosis (ALS) is a disease of unknown pathogenesis affecting the first and the second motorneuron with a progressive course which leads to the death of the patient within a very few years (Rowland, 1982). Abnormal ganglioside composition has been found in spinal cord by Dawson et al., (1986) who have hypothesized that in ALS abnormal gangliosides build up in motorneurons and lead to their eventual degeneration. On the other hand, a biochemical abnormality occurring early in the disease is reflected in the impairment of thiamine phosphorylation, detected at the level of cerebral spinal fluid (Poloni et al. 1982; 1983); this finding seems to parallel a lowered energetic metabolism of the motorneurons in the earliest phase of the pathogenetic process.

A large amount of experimental research carried out both in vitro and in vivo has shown that gangliosides administered under appropriate conditions (the "window of opportunity" according to Varon et al., 1986) can enhance collateral sprouting of lower motorneurons (Gorio et al., 1981) and extrapyramidal neurons (Fuxe et al., 1986); moreover they can increase energy metabolism, blood flow and protein phosphorylation in lesioned motor centers (Dunbar et al., 1986).

Hence we should expect that administration of equivalent doses of exogenous gangliosides in ALS patients should exert the same positive trophic and energetic effects on spinal neurons, if a "window of opportunity" could be available. To test these events a methodology is required which allows reliable follow-up measurements not only of the total motor performance of the patient but also of the occurrence of neuronal

Abbreviations: ALS, amyotrophic lateral sclerosis; EMG, electromyography; BBG, bovina brain gangliosides; m.u., motor units.

sprouting in his paretic muscles (Borenstein and Desmedt, 1973; Stalberg and Ekstedt, 1973; Swash and Schwarts, 1983). Since the transmission of impulses between the corticospinal tract and the alpha motorneurons represents a crucial process for motor units recruitment, interneuron and propriospinal neuron excitability should also be evaluated. A previous trial with gangliosides in ALS (Pinelli et al., 1985) failed to show an enhancement of sprouting; on the other hand an improvement of the voluntary recruitment of motor units and a decrease in the amount of fasciculations were found even if for a limited period of a few months.

In the present paper we analyse more detailed methods of electrophysiological investigation of the motor function and we re-evaluate the strategy of treatment taking into account the principle of "window of opportunity".

METHODOLOGY

Electromyography

A double channel MS 92a Medelec electromyograph has been used. Surface skin electrodes (Disa Type 13L26) have been applied over triceps brachii and quadriceps femoris. Voluntary maximal activity has been obtained by asking the patient to do the extension movement in presence of visual and auditive feedbacks: these were provided on the EMG screen and loud-speaker, on a galvanometer and on a light display connected to the transducer for strength measurement. The strength and the rectified and filtered EMG were measured of the best performance carried out among at least three trials. Adrian monopolar concentric electrodes were inserted into the muscles for evaluating the parameters of single motor unit (M.U.) action potentials during slight effort; their frequency of discharge was also measured.

A muscle was defined as affected by "very severe paresis" when only less than six motor units could be recruited at maximal effort; in these conditions a count of the motor units was possible together with measurement of their frequency of discharge and endurance of recruitment. T reflexes were recorded by means of surface electrodes at rest and during Jendrassik manoeuvre (or while clenching the contralateral fist). M responses, F and H responses were recorded by surface electrodes during electrical stimulation of the femoral and the radial nerves with 0.2 or 0.5 msec impulses at threshold and maximal intensity.

Subjects

The most detailed findings reported in this paper concerned 5 normal subjects, 2 out of 7 ALS patients treated with placebo and 2 out of the 26 ALS patients treated with 100 mg/day BBG (bovine brain gangliosides). The age of normals and patients ranged from 42 to 68 years; among the normal subjects 4 were males; among the patients of the placebo group one was male; both of the BBG group were males. The comparison of strenght, collateral sprouting, motor units recruitment and occurrence of fasciculations was effected in all controls, placebo group and BBG group.

RESULTS

Quantitative Evaluation of Lower Motor Neuron Sprouting

The basic calculation of changes in motor unit action potentials related to collateral sprouting was carried out according to the principles adopted by Buchthal (1957). Among the motor unit action potentials recorded from 20 points of the muscle three types were separately counted; then their percentage of occurrence among the total number of motor unit action potentials was taken into account as an index of sprouting. The three types were identified according to the following parameters: (a) *First type*: motor unit action potentials with a duration longer than 25% of the mean normal value for each muscle at corresponding age: it was called *D index*. (b) *Second type*: motor unit action potentials with normal mean duration value, but with amplitude >3 mV: it was called *A index*. (c) *Third type*: motor unit action potentials with linked potential, that is a potential following the first one at an interval equal to or longer than the duration of the first potential; the potential was considered abnormal when the linked potential showed an amplitude more than 25% of the main phase, or when more than one linked potential occurred even of low amplitude: it was called *L index*.

Normal Controls

In normal subjects, in triceps and quadriceps muscles no L abnormal motor unit potential was found; the percentage for D index in 5 subjects was 0-10%; the percentage of A index was 0-15%.

Muscles with Moderate Paresis

In muscles with moderate paresis (quadriceps f.) of the placebo group a temporary increase in strength at maximal effort was found during the course of the disease when the decrease in maximal strength reached the value of 50%. The percentages are reported for the three indices of sprouting in the left quadriceps femori of one patient (P.A.) (Fig. 1).

In muscles with moderate paresis (quadriceps f.) of BBG group the increase in strength at maximal effort was found just after the end of the first period of BBG administration. The changes in strength and the three indices of sprouting in the left quadriceps f. of one patient (B.A.) are reported in Figure 2.

Muscles with very Severe Paresis

The muscles with severe paresis of patients in the placebo group showed a progressive decrease in the number of M.U. action potentials recruited at maximal voluntary effort; the last motor unit action potentials left did not show signs of collateral sprouting.

In the muscle Triceps brachii of the patient P.A. (Fig. 3) the strength remained constant between the sixth and ninth month while the percentage of potentials with linked potentials showed a small increase in the examination of the sixth month.

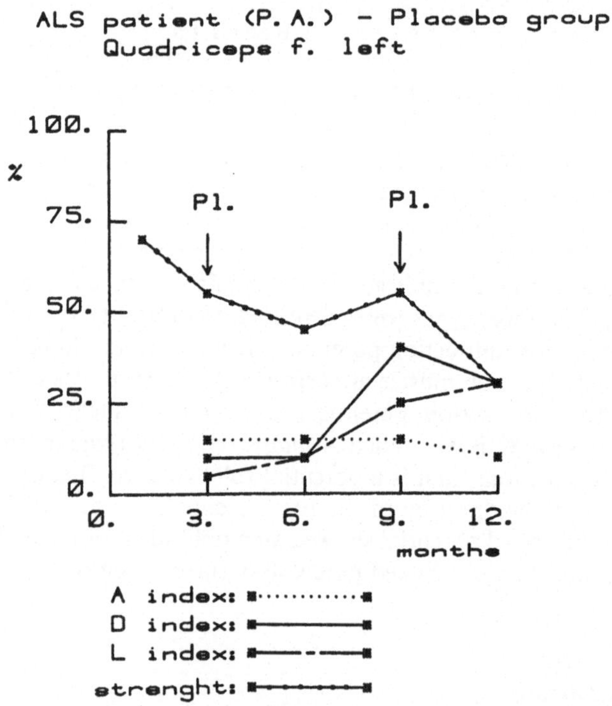

Figure 1.

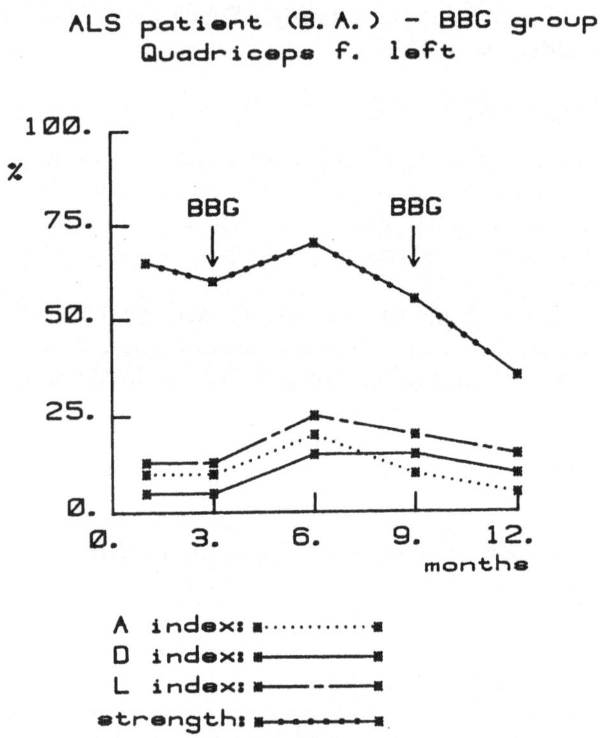

Figure 2.

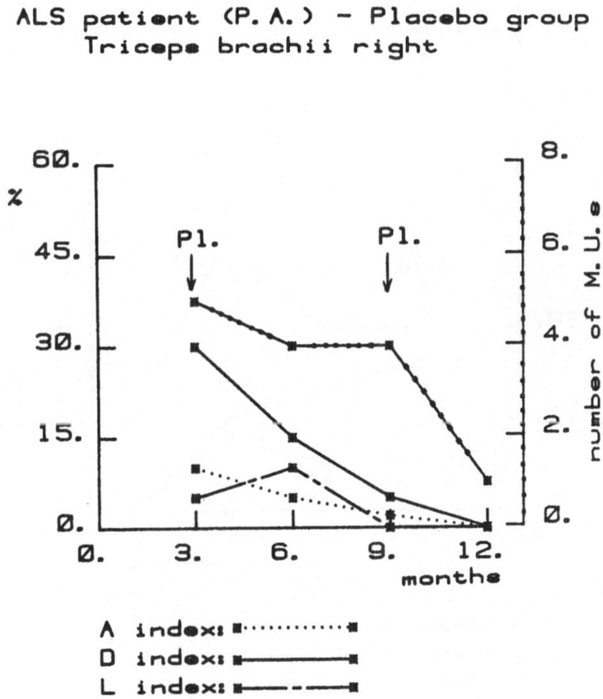

Figure 3.

In muscles with very severe paresis of patients treated with BBG an increase in the number of M.U.s recruited at maximal effort was observed three months after the end of first BBG treatment, just at the beginning of the second BBG treatment. No significant increase in percentage of M.U. potentials with sprouting was found at the same period.

In triceps brachii of the case B.A. (Fig. 4) the improvement found in the number of recruited M.U.s lasted less than three months. A small increase in percentage of linked potentials was found associated with the improvement in recruitment: it seemed to represent the last small sprouting of the surviving motor neurons. The newly recruited motor units corresponded to previously blocked motor neurons and their action potentials showed normal mean values.

T Responses

A significant correlation between the ratio T/M and the increase in voluntary strength was found in 5 ALS patients treated with BBG 100. In three patients T/M was 0.7 and in the other 2 patients 0.5.

M Responses

An increase in M response was found in both the placebo and BBG groups in muscles showing an increase in percentage of sprouting indices. This was the case in

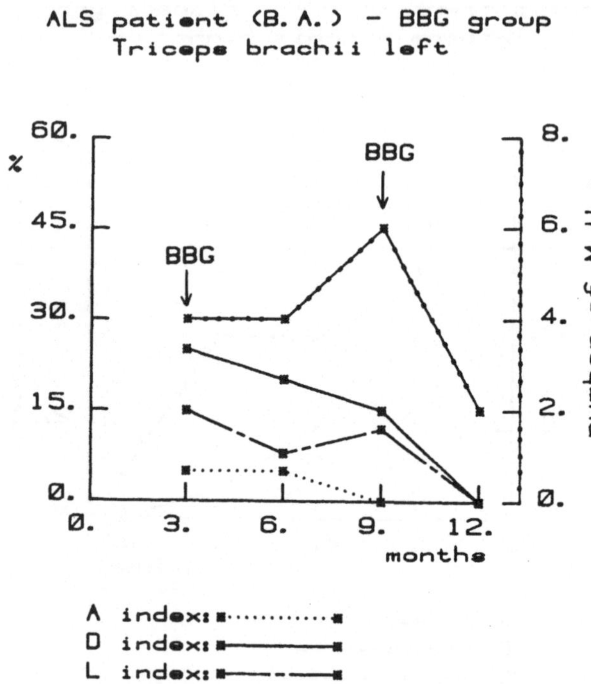

Figure 4.

muscles with moderate paresis while it failed completely in muscles with the most severe paresis.

Endurance

In cases with very severe paresis and signs of impairment of upper motor neurons the time of endurance of the voluntary effort was sometimes limited to a short time (3-10 seconds) with a maximal frequency of discharge of 15/sec. In the BBG group the lower time endurance of M.U. recruitment was 35 sec with a frequency reaching a maximum of 17/sec.

Fasciculations

The amount of different M.U. action potentials spontaneously discharging (fasciculations) in the muscles of patients treated with BBG was decreased (less than 25%) during the three months of interval after the BBG treatment. No significant change in percentage of fasciculations was found in the placebo group.

DISCUSSION

The statistical evaluation of the motor unit action potentials with changes related to increased density and area of innervation did not show enhanced collateral sprouting of motor units in ALS patients under BBG 100 treatment with respect to the placebo group. On the other hand, an unexpected positive finding was represented by an increase in number of motor units voluntarily recruited; an analysis of the value of the other parameters (M, T, T/M) did not show any sign of increased excitability of alpha motor neurons but could rather suggest a restoration of the transmission of impulses between endings of corticospinal neurons and spinal neurons as the cause of the improvement. However this effect could be shown to persist for only a few weeks after the treatment.

With the aim of changing the "window of opportunity" in such a way that the administered BBG could exert their effect on sprouting and possibly prolong the improvement in motor unit recruitment, our study can support the following considerations: the negative results of the treatment with the high dosage of BBG for long periods (3 + 3 months) recommended by Bradley et al. (1984) seems to make unlikely the hypothesis that a ganglioside metabolic impairment represents the cause of ALS; the phenocopies of Federico and Dawson may deserve some interest only with regard to the final neuropathological lesions, while the physiopathogenetic process of ALS appears to be quite peculiar.

We might deduce that the pathogenetic agent killing the motorneurons in ALS is able to counteract and overcome the effects that the exogenous gangliosides can exert in experimental conditions on plastic changes.

The discovery by Gurney (1985) of the occurrence in ALS patients of antibodies against muscle molecules normally promoting motorneuron sprouting could explain why in the last phase of the disease the compensatory collateral sprouting fails and BBG are not able to enhance it. Moreover the impairment of thiamine phosphorylation shown in cerebral spinal fluid in the earliest phase of the disease (Poloni et al., 1983) seems to indicate that biochemical alterations of the motorneurons in ALS represent the primary pathological event preceding the autoimmunological attack. These two abnormalities can be responsible for the inefficay in ALS of the trophic and energetic properties of the BBG.

In order to readapt the "window of opportunity" in ALS we should normalize these two conditions: a) the antibodies should be eliminated by means of intensive immunosuppression and clearing treatments; b) the energetic machinery supporting the basal metabolism of the motor neurons should be repaired: the latter seems to represent the most critical point in the planning for a rational treatment of ALS. The key for opening this window has not yet been envisaged.

ACKNOWLEDGMENTS

This research was supported by a Grant of CNR, Italy (n CT 84.00794.04.115.03927). We are grateful to Mr. Danilo Pianca for his technical assistance and manuscript preparation.

REFERENCES

Borenstein S, Desmedt JE (1973) Electromyographical signs of collateral reinnervation. In: Desmedt JE (ed): New developments in electromyography and clinical neurophysiology. Brussels, vol. 1, pp. 130-140.

Bradley WG, Hedlund W, Cooper C, Desousa GJ, Gabbai A, Mora JS, Munsat TL, Scheife R (1984) A double- blind controlled trial of bovine brain gangliosides in ALS. Neurology 34: 1079-1082.

Buchthal F (1957) An introduction to electromyography. Scandinavian University Books, Copenhagen.

Dawson G, Hancock LW, Horwitz AL, Stefansson L, Antel J (1986) Possible association of degenerative motor neuron disease (ALS) with abnormal ganglioside metabolism. This volume.

Dunbar GL, Butler W, Fass B, Stein DG (1985) Gangliosides enhance behavioral recovery and neuronal repair in brain damaged rats. This volume.

Fuxe K, Agnati LF, Benfenati F, Zoli M, Diemer NH, Owman C, Toffano G, Ruggeri M, Jansson AM, Goldstein M (1966) Effects of ganglioside GM1 treatment on lesioned nigral dopamine neurons and on striatal energy metabolism, striatal blood flow and cyclic AMP and Ca^{++} induced protein phosphorylation. This volume.

Gorio A, Carmignoto G, Ferrari G (1981) Axon sprouting stimulated by gangliosides: a new model for elongation and sprouting. In: Rapport MM, Gorio A (eds): Gangliosides in neurological and neuromuscular function, development and repair. Raven Press, New York, pp. 177-195.

Gurney M (1985) Initial steps toward a molecular biology of ALS. In: Kato (ed): Proceedings of ALS International Congress: Therapeutic Psychological and Research Aspects of ALS. Varese, Italy, March 27-31.

Pinelli P, Mazzini L, Villani A (1985) A follow-up electromyographic investigations of ALS patients treated with high dosage gangliosides. In: Kato (ed): Proceedings of ALS International Congress: Therapeutic Psychological and Research Aspect of ALS. Varese, Italy, March 27-31.

Poloni M, Rocchelli B, Patrini C, Rindi G (1982) Thiamin monophosphate in patients with ALS. Arch Neurol 39: 507.

Poloni M, Patrini C, Pinelli P (1983) Cerebrospinal fluid, plasma and erythrocytes thiamine content in ALS. A.L.S. Workshop, Munich, October 14-15.

Rowland LP (1982) Diverse forms of motor neuron disease. In: Rowland LP (ed): Advances in Neurology. Human Motor Neuron Disease, pp. 1-13.

Stalberg E, Ekstedt J (1973) Single fiber EMG and microphysiology of the motor unit in normal and diseased human muscle. In: Desmedt JE (ed): New developments in electromyography and clinical neurophysiology. Brussels, vol. 1, pp. 113-129.

Swash M, Schwarts MS (1983) Staging motor neuron disease: single fiber EMG studies of assymetry, progression and compensatory reinnervation. In: Clifford Rose F (ed): Research progress in motor neuron disease. Pittman.

Varon S, Skaper SD, Katoh-Semba R (1986) Neuritic responses to GM1 ganglioside in several in vitro systems. This volume.

Gangliosides and neuronal plasticity
G. Tettamanti, R.W. Ledeen, K. Sandhoff,
Y. Nagai, G. Toffano (eds.)
Fidia Research Series, vol. 6
Liviana Press, Padova, © 1986

SUBACUTE PHASE OF STROKE TREATED WITH GANGLIOSIDE GM1

S. Bassi, M.G. Albizzati, M. Sbacchi, L. Frattola[1], M. Massarotti[2]

Neurological Clinic, Bassini Hospital, Cinisello Balsamo (Milano) Italy;
[1] Neurological Clinic, University of Milan, S. Gerardo dei Tintori Hospital, Monza;
[2] Fidia Research Laboratories, Abano Terme (Padova) Italy

INTRODUCTION

Various effects of treatment with GM1 ganglioside were shown:

a) enhancement of neuronal sprouting and neuritogenesis (potentiating the effects of trophic factors) (Cotman et al., 1981; Gorio et al., 1980; Wojcik et al., 1982; Toffano et al., 1983; Ferrari et al., 1983);
b) functional recovery of dopaminergic and cholinergic activity after lesion (Toffano et al., 1983);
c) reduction of cerebral edema (Karpiak, 1983);
d) prevention of progressive neural damage following trauma by means of complex intracellular mechanisms (activation of membrane bound enzymes) (Seifert, 1981).

These proofs obtained on experimentally lesioned animals are the experimental bases for a possible therapeutical use of GM1 in Central Nervous System Diseases. Particularly, starting from the data showing that this drug is able to stimulate the potential capacity for structural reorganization and functional recovery of brain after damage, we carried out a study in the post-acute phase of stroke. Moreover, recent reports on GM1 effect in reducing cerebral edema and on its implication in the activation of membrane linked enzymes, induce to study this drug also in the acute phase of stroke. We are not yet able to present data of a study in progress on the acute phase of stroke we are now carrying out; instead we report here the results of a study carried out in the subacute phase of cerebrovascular diseases.

Abbreviations: CT-scan, Computerized assial Tomography; EEG, electroencephalography; i.m., intramuscular injection.

PATIENTS AND METHODS

Thirty-eight patients with acute stroke of ischemic or hemorrhagic nature entered the study. The following conditions had to be satisfied for a patient to enter the trial:

1) time between the onset of neurologic symptoms and hospitalization not exceeding 48 hours;
2) first cerebrovascular event (all types of strokes);
3) presence of a single focal lesion (verified by CT-scan);
4) no evidence of acute myocardial infarction nor of malignant hypertension;
5) no previous treatment with anti-coagulant or anti-platelets drugs.

On admission all patients underwent detailed neurological and physical examination, complete blood and urine analyses, EEG and CT-Scan. An anti-edema therapy with corticosteroids was stopped at the tenth day of hospitalization.

Fifteen days after stroke, the patients entered the study, and were randomly allocated to the two treated groups. The first group was treated with GM1 (20 mg i.m. twice a day for 6 weeks). In this group, four patients did not complete the trial: one did not adhere to the treatment schedule, the other three worsened and underwent corticosteroid therapy. The second group received only placebo injected i.m. as a saline solution twice a day. In this group fifteen subjects completed the trial, three died and one relapsed. The clinical features of the two groups were adequately homogeneous. Neurological deficits were scored from 0 (death) to 100 (normal), as indicated by Mathew et al. (1972) as modified by Frithz and Werner (1975), at the beginning, at 2-week intervals, and at the end of the study. Brain CT-scan was performed before treatment and at the end of the six weeks. Routine emato-chemical tests were performed every week, EEG was recorded at the beginning, at the third week and at the end of treatment.

Neither objective nor subjective adverse drug reactions were seen in the patients subjected to the treatment.

RESULTS AND DISCUSSION

Both the GM1 treated and the placebo groups showed an improvement in the neurologic score. However, comparison of the average score (± S.D.) of patients of the two groups showed statistical significance ($P < 0.001$) at regression analysis (Table 1). Particularly, GM1 induced a more pronounced improvement than did placebo. Moreover, the GM1 treated group showed a persistent effect of the drug after four weeks, whereas the placebo group did not.

Table 1. *Clinical Course in Cerebral Stroke: GM1 Treatment Versus Placebo*

	Mean scores ± S.D.		Number of patients	
Patient group	Before treatment	After treatment	Score improvement ≤ 14	≥ 19
Placebo	73.9 ± 13.6	81.9 ± 12.7	15	0
GM1	74.1 ± 15.4	87.3 ± 9.6	9	6

More in detail, the maximum score improvement in the placebo group was 14 points (observed in only one subject), whereas in the GM1 group six patients showed an increase greater than 19 points (3 S.D. above the mean improvement in the placebo group): this result is very important from a clinical point of view, since it comes from conscious patients with a wide range of neurological deficits.

The morphological evolution of the lesion, assessed by CT-scan, was not different in the two groups. The results of EEG study showed a trend similar to that of the clinical picture. In fact, both groups showed a reduction in EEG abnormality.

The results of this study, although preliminary, seem to be particularly relevant: considering the examined pathology they are worth to be accurately examined and extended to a greater number of patients. On the other hand, the efficacy of GM1 treatment on the subacute phase of stroke, together with the experimental evidence reported above (Karpiak, 1983; Jonsson et al., 1986), prompted us to carry on a clinical study in the acute phase of stroke.

REFERENCES

Cotman CW, Nierto-Sampedro N, Harris EW (1981) Synapse replacement in the Nervous System of adult vertebrate. Physiol Rev 71 (3): 684-784.

Ferrari G, Fabris M, Gorio A (1983) Gangliosides enhance neurite outgrowth in PC12 cells. Dev Brain Res 8:215-221.

Frithz G and Werner J (1976) Studies on cerebrovascular strokes. Acta Med Scand 199:133-140.

Gorio A, Carmignoto G, Facci L, Finesso M (1980) Motor nerves sprouting induced by ganglioside treatment. Possible implications for gangliosides on neuronal growth. Brain Res 197: 236-241.

Jonsson G, Gorio A, Hallman H, Janigro D, Kojima H, Zanoni R (1986) Effect of GM1 ganglioside on neonatally neurotoxin induced degeneration of serotonin neurons in the rat brain. Dev Brain Res, in press.

Karpiak SE (1983) Ganglioside treatment improves recovery of alternation behaviour following unilateral enthorinal cortex lesion. Exp Neurol 81: 330-339.

Mathew NT, Meyer YS, Rivera VU, Hartmann A (1972) Double blind evaluation of glycerol therapy in acute cerebral infarction. Lancet II, 1327-1329.

Seifert W (1981) Gangliosides in nerve cell cultures. In: Rapport MM, Gorio A (eds): Gangliosides in Neurological and Neuromuscular Function. Development and Repair. Raven Press, New York, pp. 99-117.

Toffano G, Savoini GE, Moroni F, Lombardi GA, Calza L, Agnati L (1983) GM1 gangliosides stimulate the regeneration of dopaminergic neurons in the central nervous system. Brain Res 261: 163-166.

Wojcik M, Ulas J, Oderfeld-Nowak B (1982) The stimulating effect of ganglioside injections on the recovery of choline acetyltransferase and acetylcholinesterase activities in the rat after septal lesions. Neuroscience 7(2): 495-499.

Gangliosides and neuronal plasticity
G. Tettamanti, R.W. Ledeen, K. Sandhoff,
Y. Nagai, G. Toffano (eds.)
Fidia Research Series, vol. 6
Liviana Press, Padova, © 1986

GANGLIOSIDE THERAPY OF PERIPHERAL NEUROPATHIES: A REVIEW OF CLINICAL LITERATURE

Marino Massarotti

Clinical Research Dept., Fidia Research Laboratories, Abano Terme, Italy

INTRODUCTION

The present report aims at presenting a synthetic review of the most important clinical trials which use gangliosides as therapeutic agents in peripheral neuropathies of various etiology. The complex of biochemical and pharmacological research studies on gangliosides indicates that these molecules, natural components of cellular membranes (in particular, of neuronal membranes) are involved in neuronal development, differentiation and regeneration processes. The ganglioside preparation has been shown to possess a reinnervation-stimulating activity due to enhanced nerve sprouting, an essential feature of muscular reinnervation processes and of restoration of synaptic contacts. Electrophysiological and functional evidence of early recovery from nerve damage, due to parenteral ganglioside treatment, has been obtained in several animal models, including motor nerve function after nerve crush, sensory nerve function after transection, cochler impairment by noise, diabetic neuropathy in mutant diabetic mice and toxic damage to peripheral nerves (Ceccarelli et al., 1976; Gorio et al., 1980; 1983; 1984; Leon et al., 1981).

EARLY CLINICAL STUDIES

General Criteria for Evaluation

The clinical application of the ganglioside mixture named Cronassial (contains the 4 major gangliosides: GM1-GD1a-GD1b-GT1b) has been proposed for the first time in Italy about 10 years ago. However, an objective difficulty does exist to consider all the accumulated material and also to evaluate the results in detail. Therefore, some global observations on 32 studies, described under the aspecific name of "early clinical studies" and published in a period ranging from 1976 to 1982 are shown (see Tables 3 and 8).

Some studies had a very poor study design and must be considered as preliminary studies of phase 1 and 2. Others can be considered as studies of more advanced phase with a more sophisticated study design and a higher number of patients. The list includes all the studies published in the above mentioned period, revising every single trial and giving a personal judgement independently of the conclusions of the Authors.

Table 1 shows the coding adopted, as far as both technical aspects and "quality" of the results are concerned.

Table 2 shows the total number of patients studied, the total number of patients treated with gangliosides, the number of patients studied under the profile of neurophysiological examinations and both clinical subjective and objective symptoms*. Most patients have been studied with both neurophysiological and clinical

Table 1

Coding of the technical aspects	Coding related to "quality" of the results
++++ Randomized, Double blind	*** Good: consistent improvement of all studied parameters
+++ Single blind	** Fair: improvement in some studied parameters
++ Open controlled	
+ Open	* Equivocal: major difficulties to try some conclusions on the efficacy

Table 2

Total number of studied patients		1608
Total number of treated patients with gangliosides		1035
Acute peripheral neuropathies (11 studies)	Patients studied as to neurophysiological examinations	215
	Patients studied as to subjective and objective symptoms	493
Chronic peripheral neuropathies (21 studies)	Patients studied as to neurophysiological examinations	347
	Patients studied as to subjective and objective symptoms	399

*1) Neurophysiological examinations:
Motor and sensitive conduction velocity
Latency
Motor and sensory action potential
2) Subjective symptoms:
Pain, paresthesia, cramps
3) Objective symptoms:
Motor deficit, sensory deficit (including hypovibration and hyporeflexion).

examinations. The first observation to be made is that the number of patients studied and treated in the phase of ganglioside early studies is very high. This is due to the fact that at the first approach to the therapy with gangliosides very little was known about the dosage, the duration of therapy, and the predictive power of different clinical parameters. Secondly, since peripheral neuropathies can be caused by a very large number of factors, it has been necessary to evaluate the possible therapeutical activity of gangliosides in different etiological conditions. As it is shown in the table, the studies are divided into two main groups: acute neuropathies and chronic neuropathies. This is of course only a schematic division, based on the different characteristics of etiological factors, clinical behaviour, natural history, etc. The biological attitude to respond to the therapy is also different in the two groups, and this justifies the expectation to different results. On the other hand, animal experiments suggest pharmacological activity of gangliosides both in acute and chronic situations.

Acute Peripheral Neuropathies

Table 3 shows studies related to acute peripheral neuropathies and subdivided per years of publication and different etiological factors. The total number of studied patients is 795. The total number of treated patients is 501.

Table 3. *Acute Peripheral Neuropathies. Early Clinical Studies with Gangliosides*

	Traumatic	Post-Surgery	Infective	Cochlear-Vestibular	Bell's Palsy
1976	1 Marangolo and Ventura				
1977					
1978	2 Saraceni and Alicicco	3 Viva et al.	4 Leoni and Bosco		5 Negrin and Fardin
1979					
1980	6 Scoppio and Veneziano 7 Gai et al.	8 Bevilacqua 9 Di Gesù et al.			
1981				10 Marino et al.	
1982		11 De Blas Orlando et al.			
Total number of studied patients:		795			
Total number of treated patients:		501			

Table 4 takes into account, from left to right, the study number, the technical coding, the number of patients treated with gangliosides in the correspondent study, the dose per day and the duration of treatment, the total administered dose of gangliosides and finally the quality of the results, according to the coding system shown before, as to subjective and objective symptoms. The first observation is that some studies have fulfilled the criteria, so that they can be included in the category of "good". The second one is that the results classified as "good" are achieved with the highest total administered dose. In other words, it seems that, under a thresold of 1200 mg, the results fall to "fair" or "equivocal".

Table 4. *Acute Peripheral Neuropathies. Subjective and Objective Symptoms*

Total administered dose vs. results

Study number	Technical coding	No. patients treated with gangliosides	Treatment dose/die × duration	Total administered dose	Results coding
7	+ + + +	22	40 mg × 50 days	2000	***
2	+ +	40	20-30 mg × 60 days	1200-1800	***
7	+ + + +	16	20 mg × 40 days	800	**
5	+ +	24	10 mg × 30-90 days	300-900	**
1	+ +	180	10-20 mg × 60 days	600-1200	**
9	+ +	45	20 mg × 15 days	300	**
11	+ +	25	20 mg × 30 days	600	**
10	+ +	15	20 mg × 40 days	800	**
4	+	27	20-40 mg × 10 days	200-400	**
6	+	43	20 mg × 20 days	400	*
3	+	56	20 mg × 20 days	400	*
Total		493			

Table 5 shows the accuracy of the two previous observations, also considering the neurophysiological parameters.

Table 6 on symptoms shows that "good" results are obtained only using the highest dose for the longest period of time. Both factors seem to be important for the final results. The difference is not so clear between "fair" and "equivocal" groups. What appears for the symptoms seems to be true even for the neurophysiological parameters, putting together "fair" and "equivocal" results (Table 7).

Chronic Peripheral Neuropathies

Chronic peripheral neuropathies are subdivided into different etiological groups, with a total number of 534 treated patients (Table 8).

Table 9, showing the results on neurophysiological parameters, clearly confirms the trend described in the previous tables, indicating that "good" results are achieved only using a total dose higher than 1200 mg.

Table 5. *Acute Peripheral Neuropathies. Neurophysiological Examinations*

Total administered dose vs. results

Study number	Technical coding	No. patients treated with gangliosides	Treatment dose/die × duration	Total administered dose	Results coding
7	+ + + +	22	40 mg × 50 days	2000	***
2	+ +	40	20-30 mg × 60 days	1200-1800	***
7	+ + + +	16	20 mg × 40 days	800	**
5	+ +	24	10 mg × 30-90 days	300-900	**
8	+ +	30	10 mg × 30 days	300	**
10	+ +	15	20 mg × 40 days	800	**
11	+ +	25	20 mg × 30 days	600	**
6	+	43	20 mg × 20 days	400	*
Total		215			

Table 6. *Acute Peripheral Neuropathies. Subjective and Objective Symptoms*

Dose/die vs. results
Duration vs. results

	Dose mg/die		Duration/days	
Good	40 20-30	mean: 32.5	50 60	mean: 55
Fair	10 20-40 20 10-20 20 20 20	mean: 19.2	30-90 10 40 60 15 30 40	mean: 36.4
Equivocal	20 20	mean: 20	20 20	mean: 20

Table 7. *Acute Peripheral Neuropathies. Neurophysiological Examinations*

Dose/die vs. results
Duration vs. results

	Dose mg/die		Duration/days	
Good	40 20-30	mean: 32.5	50 60	mean: 55
Fair	10 10 20 20 20	mean: 16	30-90 30 40 40 30	mean: 40
Equivocal	20	20	20	20

Table 8. *Chronic Peripheral Neuropathies. Early Clinical Studies with Gangliosides*

	Alcoholic	Diabetic	Toxic	Uremic	Entrapment
1976	1 Marzot and D'Agostini 2 Mazzoni				3 Ferromilone et al.
1978		4 Sandrini et al.	5 Azzoni 6 Dantona et al.	7 Catizone et al.	
1979	8 Sandrini et al.				9 Mingione et al.
1980	10 Mamoli et al.	11 Ossez Gonzales et al.			12 Cubells et al.
1981	13 De Mattos et al.	14 De Mattos et al. 15 Pozza et al.	16 Cotroneo et al.		17 Trontelli et al.
1982	18 Bassi et al.	19 Montenero et al. 20 Bassi et al.			
1983				21 Lindner et al.	
Total number of studied patients:		813			
Total number of treated patients:		534			

This is true for all studies classified as "good" but one, number 15, which obtained "good" results with a total dose of 800 mg. This study was performed on patients with diabetic neuropathy. The thresold is very clear also considering the results on subjective and objective symptoms (Table 10). Separating dose from duration of treatment, with respect to the quality of the results obtained measuring neurophysiological parameters, noteworthy is that the dose is the parameter which apparently correlates with the results. The latter are independent from the duration of therapy, which does not show important variations in the three categories (Table 11).

This is true also considering objective and subjective symptoms. In fact the dose per day proved to be crucial to obtain good results (Table 12).

Table 9. *Chronic Peripheral Neuropathies. Neurophysiological Examinations*

Total administered dose vs. results

Study number	Technical coding	No. patients treated with gangliosides	Treatment dose/die × duration	Total administered dose	Results coding
18	+ + + +	15	50 mg × 40 days	2000	***
20	+ + + +	15	50 mg × 40 days	2000	***
21	+ + + +	10	20 mg × 60 days	1200	***
15	+ +	20	20 mg × 40 days	800	***
16	+ +	15	50 mg × 30 days	1500	***
10	+ + +	17	20 mg × 28 days	560	**
1	+ +	40	10 mg × 32 days	320	**
12	+ +	71	20 mg × 25 days	500	**
9	+ +	60	10 mg × 90 days	900	**
13	+	3	20 mg × 28 days	560	**
4	+	33	20 mg × 30 days	600	**
7	+ +	10	20 mg × 30 days	600	*
17	+ +	23	20 mg × 30 days	600	*
14	+	7	20 mg × 28 days	560	*
8	+	4	20 mg × 30 days	600	*
11	+	4	10 mg × 90 days	900	*
Total		347			

Table 10. *Chronic Peripheral Neuropathies. Subjective and Objective Symptoms*

Study number	Technical coding	No. patients treated with gangliosides	Treatment dose/die × duration	Total administered dose	Results coding
18	+ + + +	15	50 mg × 40 days	2000	***
20	+ + + +	15	50 mg × 40 days	2000	***
21	+ + + +	10	20 mg × 60 days	1200	***
19	+ +	105	40 mg × 30 days	1200	***
16	+ +	15	50 mg × 30 days	1500	***
10	+ + +	17	20 mg × 28 days	560	**
7	+ +	10	20 mg × 30 days	600	**
5	+ +	7	20 mg × 28 days	560	**
2	+ +	12	10 mg × 36 days	360	**
9	+ +	60	10 mg × 90 days	900	**
13	+	3	20 mg × 28 days	560	**
1	+ +	40	10 mg × 32 days	320	*
3	+ +	12	10 mg × 30 days	300	*
17	+ +	23	20 mg × 30 days	600	*
14	+	7	20 mg × 28 days	560	*
8	+	4	20 mg × 30 days	600	*
11	+	4	10 mg × 90 days	900	*
6	+	40	10 mg × 30 days	300	*
Total		399			

Table 11. *Chronic Peripheral Neuropathies. Neurophysiological Examinations*

Dose/die vs. results
Duration vs. results

	Dose mg/die		Duration/days	
Good	50 50 20 50 20	mean: 38	40 40 40 30 60	mean: 42
Fair	20 20 10 20 10 20	mean: 16.6	28 30 32 25 90 28	mean: 38.8
Equivocal	20 20 10 20 20	mean: 18	28 30 90 30 30	mean: 41.6

Table 12. *Chronic Peripheral Neuropathies. Subjective and Objective Symptoms*

Dose/die vs. results
Duration vs. results

	Dose mg/die		Duration/days	
Good	50 50 40 20 50	mean: 42	40 40 30 60 30	mean: 40
Fair	20 20 20 10 10 20	mean: 16.6	28 30 28 36 90 28	mean: 40
Equivocal	20 20 10 10 10 10 20	mean: 14.2	28 30 90 30 32 30 30	mean: 38.5

Comparative Remarks

Table 13 summarizes the data into the two main groups. In the trials involving chronic peripheral neuropathies the results are apparently better, from a quantitative point of view, than those obtained in patients with acute forms. Particularly among 6 trials judged as "good", 3 are dedicated to diabetic neuropathies.

In the scientific research studies conclusions are always a starting point to further steps toward knowledge. Consequently the above-described conclusions were utilized to design the current phase of clinical research with gangliosides.

Table 13

Acute neuropathies	Chronic neuropathies
11 trials (501 patients)	21 trials (534 patients)
Good results: 2 trials	Good results: 6 trials
No. 2: Entrapment	No. 15: Diabetic
No. 7: Post-surgery	No. 19: Diabetic
	No. 20: Diabetic
	No. 16: Toxic
	No. 18: Alcoholic
	No. 21: Uremic

RECENT CLINICAL STUDIES

The more recently concluded clinical trials have been designed in order to go deeply into the clinical field of diabetic peripheral neuropathies. The material includes:

a) the trial sponsored by the Italian Group for Diabetic Peripheral Neuropathies (multicentre, randomized, cross-over, double-blind, controlled vs placebo; 140 patients) (Crepaldi et al., 1983);
b) the trial sponsored by W.H.O. (multicentre, randomized, double-blind, controlled vs placebo; 162 patients) (in press);
c) 3 randomized, double-blind, controlled vs placebo trials (2 in the U S A, 1 in the U K) (Abraham et al., 1984; Horowitz, 1984; Naarden et al., 1984).

The patients globally studied were 373; among them 256 were treated with gangliosides.

Even if some points deserve further clarification and discussion, the results are quite exciting (Tables 14 and 15). In fact, considering the outstanding difficulties from a neurophysiological and clinical point of view, in studying peripheral neuropathies, it seems almost clear that most parameters of chronic nerve damage, (including sensitive and motor systems, subjective and objective symptoms) are positively influenced in double-blind controlled conditions by the parenteral administration of gangliosides (Table 16).

Table 14

Diabetic P.N.	Electrophysiology	Subjective symptoms	Objective signs
Italian Mult. Study (1983) 20 mg/6 weeks			
Protocol I (pts. 97)	Median S.C.V. = $p<0.02$ Sural S.C.V. = $p<0.02$ Ulnar D.L. = $p<0.01$ Peroneal M.C.V. = $p<0.03$		
Protocol II (pts. 43)	Median S.C.V. = $p<0.06$ Sural S.C.V. = $p<0.05$ Ulnar M.C.V. = $p<0.002$	Paresthesia = $p<0.02$	
WHO Mult. Study (1985) 40 mg/6 weeks			
Florence (pts. 39)	Sural S.C.V. = $p<0.01$	Pain = $p<0.05$	Vibration = $p<0.05$
Lisbon (pts. 44)	Sural S.C.V. = $p<0.02$		
Dakar (pts. 16)			
Shanghai (pts. 63)			Vibration = $p<0.05$

M.C.V. = Motor Conduction Velocity
S.C.V. = Sensory Conduction Velocity
D.L. = Distal Latency
M.A.P. = Motor Action Potential

Table 15

Diabetic P.N.	Electrophysiology	Subjective symptoms	Objective signs
A. Naarden et al. (1984) (20 pts.) 40 mg/3 months	Median S.C.V. (right side only) = $p<0.005$	Improvement = N.S.	Improvement = N.S.
	latency = $p<0.005$		
	Median M.C.V. = $p<0.05$		
	latency = $p<0.05$		
	Peroneal M.C.V. = $p<0.07$		
	latency = N.S.		
Horowitz (1984) (25 pts.) 40 mg/6 months	Improvement at Sural, Peroneal Median S.C.V. = N.S.		Refl. and Perception of Sensory Stimuli = $p<0.04$
Abraham et al. (1984) (26 pts.) 20 mg/6 weeks	M.A.P. Peroneal = $p<0.05$	Improvement = N.S.	

Table 16

Median	S.C.V.	$p<0.02$	(Italian M. Study - 140 pts.)
	S.C.V.	$p<0.005$	(Naarden et al. - 20 pts.)
	M.C.V.	$p<0.05$	(Naarden et al. - 20 pts.)
	Latency	$p<0.005$	(Naarden et al. - 20 pts.)
Ulnar	M.C.V.	$p<0.002$	(Italian M. Study - 140 pts.)
	Latency	$p<0.005$	(Naarden et al. - 20 pts.)
Sural	S.C.V.	$p<0.02$	(Italian M. Study - 140 pts.)
	S.C.V.	$p<0.01$	(W.H.O. M. Study - Florence - 39 pts.)
	S.C.V.	$p<0.02$	(W.H.O. M. Study - Lisbon - 44 pts.)
Peroneal	M.C.V.	$p<0.03$	(Italian M. Study - 140 pts.)
	M.C.V.	$p<0.06$	(Naarden et al. - 20 pts.)
	M.A.P.	$p<0.05$	(Abraham et al. - 26 pts.)
Subjective Symptoms	Paresthesia	$p<0.02$	(Italian M. Study - 43 pts.)
	Pain	$p<0.05$	(W.H.O. M. Study - Florence - 39 pts.)
And			
Objective Signs	Vibration	$p<0.05$	(W.H.O. M. Study - Shanghai - 63 pts.)
	Vibration	$p<0.05$	(W.H.O. M. Study - Florence - 39 pts.)
	Reflexes and Sensory Perception	$p<0.04$	(Horowitz - 25 pts.)

CONCLUSIONS

Early Clinical Trial

1. Eight out of 32 trials (1608 studied patients, 1035 treated patients), have demonstrated a "good" therapeutical efficacy.
2. The results "good" are achieved with the highest global ganglioside quantity administered (dose/day x duration).
3. In acute neuropathies both parameters (dose/day and duration) play an important role in the quality of results; in chronic neuropathies the parameter dose/day seems to be more important than the duration of treatment.
4. The best quantitative results are obtained in chronic neuropathies. In particular diabetic neuropathies show the highest number of "good" evidence in comparison with other chronic neuropathies.

Recent Clinical Trial

1. Considering 2 multicentre and 3 single trials on diabetic peripheral neuropathies (373 studied patients; 256 treated), the results show that neurophysiological

parameters (both sensory and motor), subjective symptoms and objective signs are affected by the therapy.

2. The fact that the results are not exactly overlapping in the different centers may be explained by different clinical features of the patients; different types (dosage and duration) of treatment; low number of patients included in the two groups with consequent low power of the study.

Before concluding it is important to note that, reviewing 1291 patients treated with gangliosides in a period of about 10 years (1035 in early studies plus 256 in recent studies on Diabetic Peripheral Neuropathies), no evidence of local and general side-effects related to the drug was observed. Few patients complained a minimal and transient pain in the side of the injection at the beginning of the treatment. However, none of them stopped the drug because of this complaint (present in some cases also in placebo-treated subjects).

REFERENCES

Abraham RR et al. (1984) A double blind placebo controlled trial of mixed gangliosides in diabetic peripheral and autonomic neuropathy. In: Ledeen R W, Yu K, Rapport M, Suzuki K (eds): Ganglioside structure, function and Biomedical Potential. Plenum Press, New York, pp. 607-624.

Azzoni P (1978) L'impiego dei gangliosidi nella prevenzione della neurotossicità da Vincristina. Il Politecnico. Sez Medica 85, (4): 255-262.

Bassi S, Albizzati MG, Calloni E, Frattola L (1982) Electromyographic study of diabetic and alcoholic polyneuropathic patients treated with gangliosides. Muscle and Nerve 5: 351-356.

Bevilacqua F (1980) L'uso dei gangliosidi dopo stapedectomia nelle otosclerosi ad alto rischio. Il Valsalva 7, (3): 206-213.

Catizone L, Merlini L, Fusaroli M (1978) La neuropatia periferica dell'uremico in emodialisi periodica: risposta clinica ed elettrofisiologica al trattamento con gangliosidi di corteccia cerebrale. Clin Terap 85, (4): 395-409.

Ceccarelli B, Aporti F, Finesso M (1976) Effects of brain gangliosides on functional recovery in experimental regeneration and reinnervation. Adv Exp Med Biol 71, Porcellati G (ed), Plenum Press 275-293.

Cotroneo L, Genta PA, Gilioli R (1981) Studio clinico-elettromiografico sull'effetto dei gangliosidi cerebrali nelle neuropatie periferiche professionali: dati preliminari. 44° Congresso Nazionale della Società Italiana di Medicina del Lavoro e di Igiene Industriale. Padova, ottobre 1981.

Crepaldi G et al. (1983) Ganglioside treatment in diabetic peripheral neuropathy: a multicenter trial. Acta Diab Latina, vol. XX, n. 3, pp. 265-276.

Cubells JM, De Blas A, Hernando C, Rodriguez del Barrio E (1980) Los gangliósidos de corteza cerebral bovina en el tratamiento de las lesiones radiculares. Med Clin (Barcelona) 75, (4), 156-160.

Dantona A, Labianca R, Tabiandon D (1978) L'uso dei gangliosidi nel trattamento e profilassi delle neuropatie periferiche da farmaci antiblastici. Ricerca Scientifica ed Educazione Permanente Suppl 9, 155-158.

De Blas Orlando A (1983) Los gangliósidos en el postoperatorio de hernia discal. Rev Clin Espan 168, (3), 193-198.

De Mattos JP, Sepulveda FCA, Villaça LF (1981) Emprego dos gangliosidios do córtex cerebral nas neuropatias periféricas. Revista Seara Médica Neurocirúrgica 10, (2), 1-11.

Di Gesù G, Di Carlo G, Fiasconaro G, Palazzolo M, Vetri G, Li Voti G (1980) Il trattamento con gangliosidi delle lesioni ricorrenziali nella chirurgia tiroidea. Atti dell'Accademia delle Scienze Mediche di Palermo 14, 1-7.

Ferromilone F, Nordera GP, Franciosi A (1976) Sperimentazione clinica controllata con un preparato a base di gangliosidi. Gazzetta Medica Italiana 135, (10), 559-564.

Gai AM, Bellucci Sessa M, Angeli S (1980) Valutazioni prognostiche in funzione di nuovi protocolli farmacologici in alcune lesioni del sistema nervoso periferico. Europ Medicophys 16, (3), 221-231.

Gorio A, Carmignoto G, Facci L, Finesso M (1980) Motor nerve sprouting induced by ganglioside treatment. Possible implications for gangliosides on neuronal growth. Brain Res 197, 236-241.

Gorio A, Marini P and Zanoni R (1983) Muscle reinnervation — III. Motoneuron sprouting capacity, enhancement by exogenous gangliosides. Neuroscience 8, (3), 417-429.

Gorio A et al. (1984) Ganglioside treatment of genetic and alloxan-induced diabetic neuropathy. In: Ledeen RW, Yu K, Rapport M, Suzuki K (eds): Ganglioside structure, function and Biomedical Potential. Plenum Press, New York, pp. 549-564.

Horowitz S H (1984) Ganglioside (Cronassial) Therapy in diabetic neuropathy. In: Ledeen RW, Yu K, Rapport M, Suzuki K (eds): Ganglioside structure, function and Biomedical Potential. Plenum Press, New York, pp. 593-600.

Leon A, Facci L, Toffano G, Sonnino S, Tettamanti G (1981) Activation of (Na^+, K^+) ATPasi by nanomolar concentrations of G_{M1}. J Neurochemistry 37, 350-357.

Leoni A, Bosco G (1978) I gangliosidi nella terapia dell'Herpes Zoster. Il Policlinico - Sez Medica 85, 428-439.

Linder G and Terenziani S (1985) Treatment of uremic neuropathy with gangliosides a controlled trial vs placebo. Clinical Trials Journal, in press.

Mamoli B, Brunner G, Mader R, Schanda H (1980) Effects of cerebral gangliosides in the alcoholic neuropathies. Europ Neurol 19, 320-326.

Marangolo M, Ventura F (1976) Osservazioni cliniche sull'uso dei gangliosidi nelle affezioni dei nervi periferici. Acta Neurol 31, (6), 759-772.

Marzot G, D'Agostini N (1972) Relazione sulla sperimentazione clinica nel preparato GL/5 per uso i.m. dei Laboratori Fidia di Abano Terme. Report of Arcispedale S. Anna Ferrara - Italy.

Mingione A, Monteleone M, Paruzzi G, Soragni O, Cristiani G, Moretti C, Mega W, Scanabissi F (1979) Research in the use of cerebral gangliosides in the neurolysis of the upper-limb. Electromyogr Clin Neurophysiol 19, 353-359.

Montenero P (1982) Cronassial - Studio controllato su 132 soggetti diabetici con neuropatia periferica sensitiva. Comunicazione dell'Istituto di Scienze dell'Alimentazione e Dietetica di Roma, 13 febbraio.

Naarden A et al (1984) Treatment of painful diabetic polyneuropathy with mixed gangliosides. In: Ledeen RW, Yu K, Rapport M and Suzuki K (eds): Ganglioside structure, function and Biomedical Potential. Plenum Press, New York, pp. 581-592.

Negrin M, Fardin P (1978) Influenza dei gangliosidi di corteccia cerebrale sull'evoluzione clinico EMGrafica della paralisi facciale a frigore. Min Med 69, (48), 3277-3282.

Osset Gonzalez-Rico M, Lopez Vazquez JF, Fernandez Vila B, Cid Feijoo A, Vila Pastor B (1980) Ensayo de los gangliosidos en el tratamiento de las neuropatias perifericas. Traumatologia, Cirurgia y Rehabilitacion 10, (3), 193-205.

Pozza G, Saibene V, Comi G, Canal N (1981) The effect of gangliosides administration in human diabetic peripheral neuropathies. In: Rapport M and Gorio A (eds): Gangliosides in neurological and neuromuscular function, development and repair. Raven Press, New York, pp. 253-258.

Sandrini G, Arrigo A, Mola M, Micieli G, Nappi G, Savoldi F (1979) Alcoholic neuropathy: electromyographic findings during treatment with cerebral gangliosides. Communication in: International Congress of Neurotoxicology. Varese, Italy 27-30 Septemper.

Saraceni V, Alicicco E (1978) Effetto dei gangliosidi somministrati per ionoforesi nella patologia del sistema nervoso periferico. La Clinica Terapeutica 85, 517-525.

Scoppio M, Veneziano L (1980) Ruolo dei gangliosidi in alcune affezioni del S.N.P. Il Policlinico - Sez Medica 87, (4).

Viva E, Gazzotti A, Monza S (1978) Contributo della terapia con gangliosidi in chirurgia maxillo facciale. Min Stomat 27, (3), 177-184.

SUBJECT INDEX

Acetylcholine
— hippocampal afferents, 428
— release in aging, 111
Activator proteins, 172
AGF2, 370
ALS (Amyotrophic Lateral Scelrosis), 113, 453
— fasciculations in, 454
— ganglioside abnormalities in, 121
— ganglioside treatment, 453
— motor unit action potential, 455
Astroglial cells, 272
— morphology and GMl, 274
— proliferation and gangliosides, 276

Behaviour
— recovery by GMl, 365, 369, 409
— rotational asymmetry, 369, 408
— sensorimotor test, 382
— spatial reversal learning, 369
— water intake, 382

Ca^{++}-dependent kinase, 95
— ganglioside-stimulated, 97
Ca^{++}-gangliosides interactions, 125, 128, 130, 138, 240
— monolayer technique, 127, 130
— surface potential, 127
— surface requirements, 127, 130, 133
Calcium, 125
— after cerebral ischemia, 437
— deposit in synaptic cleft, 128, 133
— electronmicroscopical demonstration, 126
— extracellular, 125, 128, 130, 133
— ganglioside interaction, 127, 128, 130, 240
Carrier proteins
— for gangliosides, 199
Caudate nucleus, 369
Cell proliferation
— regulation by gangliosides, 201
Ceramide, 185
Cholera toxin, 139, 143
C-kinase (Ca^{++}/phospholipid-dependent kinase), 95, 99
CNTF (ciliary neuronotrophic factor), 218
— in ciliary ganglia, 217
— interaction with GMl, 218
Culture systems
— astroglial cells, 271
— ganglionic explant, 216, 289, 317
— neuroblastoma, 231, 285, 309
— primary neurons in monolayers, 218, 245, 295, 310
— rat pheochromocytoma, 221, 287
— sensory neurons, 336
— spinal cord-dorsal root ganglion, 327

EGF (Epidermal Growth Factor)
— mitogenesis, 205
— receptor phosphorylation, 205
— receptors, 201

FGF (Fibroblast Growth Factor)
— growth stimulation, 202
— receptors, 201
Fimbria fornix, 366
Fragmentation pattern
— GDla, 4
— GDlb, 4
— GM1, 4, 6, 7
— GM3, 4

Gangliosides
— abnormalities in metabolism, 117
— age-related change, 105, 107
— analysis, 185
— and neuronotrophic interaction, 283
— and protein phosphorylation, 283
— bidirectional axonal flow, 161
— biosynthesis in Golgi apparatus, 191, 197
— Ca^{++} interactions, 127, 240
— degradation, 172, 190
— distribution, 67
— distribution in adult human brain, 71, 73
— distribution in developing human brain, 68, 74
— effect on synapse morphology, 313
— hydrogenation of, 19-21
— in PNS, 168

— in SC-DRG explants, 330
— in cell proliferation, 200, 213
— in motoneurons, 163
— in peripheral neuropathies, 465
— in primary cultures, 316
— in synaptic transmission, 137
— insertion of, 173
— intracellular flow, 178, 181
— metabolism, 171, 179, 188, 299
— mode of action, 237
— molecular weight, 8, 11
— monoclonal antibody to, 78, 83
— NMR, 62
— ontogenesis, 258, 260, 300
— pathological role in ALS, 121, 459
— preparation, 23
— preparation of deacylated GM1, 32
— preparation of radiolabelled, 40
— preparation of species with threo configuration, 23
— storage disease, 113, 117
— structure-function relationship, 257
— transfer by "carrier proteins", 199
Gangliotriaosylceramide, 209
— and transferrin receptor, 210
GD1 α, 57
— chemical composition, 59, 63
— hydrolysis, 61
— in rat ascites hepatoma cells, 57
— isolation, 58
— methylation, 60
GD3
— developmental expression, 85, 89
— in mitotically active cells, 83, 91
— monoclonal antibodies, 83
Glucose-6-phosphate dehydrogenase, 366
Glycosphingolipids, 77
— antibodies to, 78
— carbohydrate binding sites, 77
— cellular localization, 81
— labelling in sensory neurons, 162
— subcellular localization, 81
— transport in sensory axons, 164
GM1
— amide preparation, 31
— and CNTF, 218
— and NGF, 216, 321
— and behaviour, 365, 369, 382, 409
— and long term potentiation (LTP), 147
— and neuritc response, 215, 335
— and neuronotrophic agents, 250
— and neurotoxin lesions, 415
— and population spike, 147
— antibodies to, 148, 291
— association to cell membranes, 251
— effect in hippocampal Ach parameters, 427
— effect on Na^+, K^+-ATPase, 409, 411
— effect on ODC, 389,391
— effect on edema, 408, 413, 437
— effect on lesioned dopaminergic neurons, 348, 393, 445
— effect on polyamines, 389
— effect on regeneration, 419
— effect on striatal DA uptake, 447
— effect on striatal protein phosphorylation, 358
— in PC12 cells, 222, 224
— in cerebral ischemia, 397, 412, 435, 461
— in development, 246, 248
— in hippocampus synapses, 140, 147, 149
— in primary CNS neurons, 219, 296
— inner ester, 25, 370, 437, 447
— interaction with EGF receptors, 206
— interaction with PDGF receptors, 203
— metabolism, 183
— minimum energy conformation, 53
— pattern of fragmentation, 4, 6, 7
— preparation of GM1 acetyl, 34
— preparation of deAc-GM1, 32
— preparation of deAc-deAcyl-GM1, 32
— preparation of defined acyl chains species, 35
— preparation of radiolabelled, 42, 184
— preparation with fluorescent acyl chain, 37
— preparation with paramagnetic acyl chain, 37
— uptake in liver an brain, 186
GM2, 190
— gangliosidosis, 174, 178
— metabolism in fibroblasts, 173
GM3, 190
— effect on EGF receptors, 206
— effect on mitogenesis, 206
— inhibition of FGF-dependent growth, 202
— interaction with PDGF receptors, 203
GQ1b, 231
— and NGF, 232
Growth factor, 201

Hexosaminidase, 113
— deficiency, 114, 117, 120
— enzymology, 114
— SDS-PAGE, 116
Hydrogenation of gangliosides, 19-21

Kinases, 95
— Ca^{++}-dependent, 95, 97
— Ca^{++}/phospholipid-dependent (C-kinase), 95, 99
— calmodulin-dependent, 95, 101

Lactone (of gangliosides), 3, 25
— FAB-MS fragmentation pattern, 29

— NMR spectrum, 30
— ring formation, 3,8
— synthesis of GM1 inner ester, 25
Lateral septal nucleus, 366

Mass spectrometry, 1, 3
— chemical ionization, 3
— direct chemical ionization, 6
— fast atom bombardment, 7
— field desorption, 5
Membranes
— composition changes of synaptosomes, 106
— GM1 association, 251
— microviscosity of synaptosomes, 108
— plasma, 185
Metabolism
— abnormalities in gangliosides, 117
— gangliosides, 171, 179, 188
— neuronal, 366
Mitogenesis, 213
— EGF dependent, 205
— inhibition by gangliosides, 213
Myelin basic proteins, 98
— phosphorylation, 99

N-acetylgalactosamine, 57
Na^+, K^+-ATPase
— and GM1, 409, 411
Neuraminidase, 138, 139, 145
Neurofilament, 336
— monoclonal antibody to, 336
Neuritogenesis, 258, 309, 335
Neuronotrophic agents, 215, 245
— and GM1, 250
— CNTF, 218
— NGF, 21, 221, 284, 337
NGF (Nerve Growth Factor), 216, 221, 284
— in DRG, 337
— in PC12 cells, 221, 224
— in ganglionic explants, 216
— interaction with GM1, 216, 321, 338
Nigrostriatal system, 370
— hemitransection, 382, 408
— lesion by 6OH-DA, 392

Oligosaccharides
— conformational analysis, 47
— NMR spectroscopy, 51
Ornithine decarboxylase (ODC), 285
— effect of GM1 on, 389
— inhibition by α DFMO, 386

Paraproteinemia, 153
PDGF (platelet-derived growth factor)
— -dependent phosphorylation, 203
— receptors, 201, 203
Peripheral neuropathies, 465
— acute, 467
— chronic, 468
— treatment with gangliosides, 465
Permethylated gangliosides, 4
— FAB spectrum, 12, 13
— pattern of fragmentation, 4, 6, 7
Phosphorylation
— and gangliosides, 240
— EGF receptor, 205
— PDGF, 203
Polyamines, 386
— effect of GM1 on, 389
Polyradiculoneuropathy, 153
Polysialogangliosides
— expression of associated antigen, 91
— in cholinergic synapses, 149
— in GABA-ergic synapses, 149

Ruthenium red, 140, 142

Signal transduction, 17
Sodium periodate, 139, 140
Sphingomyelin, 185
— biosynthesis, 195
Sphingosine, 186
Sprouting, 368
Substantia nigra, 369
Supracallosal area
— lesions, 429
Synapses, 315
— formation induced by gangliosides, 316
Synaptosomes, 105, 106, 108
— DA uptake, 446

Tetanus toxin, 140, 143, 260
Transferrin, 209
— internalization, 211
— effect on cell proliferation, 209
— receptor, 210
Transport
— axonal, 161
— intracellular, 178
Tunicamycin, 262

Visual system
— lesions, 374

Finito di stampare
nel mese di dicembre del 1986
dalla tipolitografia «La Grafica & Stampa Editrice s.r.l.» di Vicenza
per conto della Liviana Editrice s.p.a. di Padova